Pollution:
Causes, Effects and Control

Fourth Edition

Edited by

Roy M. Harrison
The University of Birmingham, UK

RS•C

ROYAL SOCIETY OF CHEMISTRY

ISBN 0-85404-621-6

A catalogue record for this book is available from the British Library.

First published 1983

Second Edition 1990
Reprinted 1992, 1993, 1995

Third Edition 1996
Reprinted 2000

Fourth Edition © 2001 The Royal Society of Chemistry

Published by The Royal Society of Chemistry,
Thomas Graham House, Science Park, Milton Road, Cambridge CB4 0WF, UK
Registered Charity Number 207890

For further information see our web site at www.rsc.org

Typeset by Paston PrePress Ltd, Beccles, Suffolk
Printed and bound by Bookcraft Ltd, UK

Preface

The subject of pollution remains high in the public consciousness and has been a significant factor on the political agenda of both developed and developing countries for a number of years. The subject is now seen as a priority area for research and for technological developments. It is therefore a fast-moving field and one where books require updating on a rather frequent basis. The First Edition of this book was published in 1983 and arose from the collation of course notes from a Residential School held at Lancaster University in 1982, supplemented with additional chapters to give a fuller overview of the field. Subsequent editions have expanded the coverage so as to provide a fairly full overview of the field of chemical and radioactive pollution. The level of treatment remains much the same, being essentially introductory, although covering some more advanced aspects. The very high sales achieved by the book suggests that this has been very popular and *Pollution* is used both as a teaching text and a reference book by practitioners requiring broad knowledge of the field.

In a fast-moving field it is necessary to scrutinize contents carefully and to ensure thorough updating. The Fourth Edition of *Pollution* has one wholly new chapter on *Clean Technologies and Industrial Ecology* reflecting the growing importance of pollution prevention as opposed to end-of-pipe controls. Whilst authorship of the majority of the other chapters remains in the same hands, a large proportion of these chapters have been thoroughly revised to reflect new developments in the field. Additionally, some glaring omissions have been addressed, such as the inclusion of microbiological contamination as well as chemical pollution of water in the chapter on *Drinking Water Quality and Health*. The subject of sewage sludge treatment is dealt with at greater length in the chapter on *Sewage and Sewage Sludge Treatment* reflecting the pressures generated by the cessation of disposal at sea. Similarly, the subject of local air quality management now features strongly in the chapter on *Air Pollution, Sources, Concentrations and Measurements* following the implementation of the UK National Air Quality Strategy since its launch in 1997.

Once again, the chapter authors have been selected on the basis of their established reputation in the field and their ability to write with clarity of presentation. I am delighted that a high proportion of those who wrote for the Third Edition have updated their contributions for the Fourth Edition. A number of those in fast-moving areas have completely re-written their contributions.

v

Comparing the Fourth with the First Edition of this book, I am struck by the explosion in knowledge in this vital area. Environmental pollution is now a very major area of research, consultancy and technological development, and I hope that this book goes some way towards providing an authoritative knowledge base for those working within the field.

Roy M. Harrison
Birmingham

Contents

Contributors

B. J. Alloway, *Department of Soil Science, University of Reading, Whiteknights, PO Box 233, Reading RG6 2DW, UK*

D. Arthur, *Environmental Resources Management, Eaton House, Wallbrook Court, North Hinksey Lane, Oxford OX2 0QS, UK*

J. Ayres, *Department of Respiratory Medicine, Birmingham Heartland Hospital, Bordesley Green East, Birmingham B9 5SS, UK*

R. Chester, *Oceanography Laboratories, Department of Earth Sciences, University of Liverpool, Bedford Street North, Liverpool L69 3BX, UK*

R. Clift, *Centre for Environmental Strategy, University of Surrey, Guildford, Surrey GU2 5XH, UK*

B. Crathorne, *Water Research Centre, Henley Road, Medmenham, Marlow, Bucks. SL7 2AD, UK*

D. Edge, *Anglian Water Services, Anglian House, Ambury Road, Huntingdon PE18 6NZ, UK*

G. Eduljee, *Environmental Resources Management, Eaton House, Wallbrook Court, North Hinksey Lane, Oxford OX2 0QS, UK*

J. Fawell, *Warren Associates, 8 Prince Maurice Court, Hambleton Avenue, Devizes, Wiltshire SN10 2RT, UK*

S. France, *Water Research Centre, Henley Road, Medmenham, Marlow, Bucks. SL7 2AD, UK*

S. Harrad, *Division of Environmental Health and Risk Management, University of Birmingham, Edgbaston, Birmingham B15 2TT, UK*

P. T. C. Harrison, *MRC Institute for Environmental Health, University of Leicester, 94 Regent Road, Leicester LE1 7DD, UK*

R. M. Harrison, *Division of Environmental Health and Risk Management, University of Birmingham, Edgbaston, Birmingham B15 2TT, UK*

C. N. Hewitt, *Division of Environmental Sciences, Institute of Environmental and Biological Sciences, University of Lancaster, Bailrigg, Lancaster LA1 4YQ, UK*

C. Holman, *Brook Cottage, Camp Lane, Elberton, Bristol BS35 4AQ, UK*

A. James, *Department of Civil Engineering, University of Newcastle upon Tyne, Newcastle upon Tyne NE1 7RU, UK*

C. John, *Ty Newydd, Neenton, Bridgnorth, Shropshire WV16 6RJ, UK*

J. Lester, *School of Environment, Royal School of Mines Building, Prince Consort Road, South Kensington, London SW7 2BP, UK*

P. W. Lucas, *Department of Biological Sciences, Institute of Environmental and Biological Sciences, University of Lancaster, Bailrigg, Lancaster LA1 4YQ, UK*

A. R. MacKenzie, *Division of Environmental Sciences, Institute of Environmental and Biological Sciences, University of Lancaster, Bailrigg, Lancaster LA1 4YQ, UK*

R. Macrory, *Faculty of Laws, University College London, Bentham House, Endsleigh Gardens, London WC1H 0EG, UK*

T. A. Mansfield, *Department of Biological Sciences, Institute of Environmental and Biological Sciences, University of Lancaster, Bailrigg, Lancaster LA1 4YQ, UK*

C. F. Mason, *Department of Biology, University of Essex, Wivenhoe Park, Colchester, Essex CO4 3SQ, UK*

M. R. Preston, *Oceanography Laboratories, Department of Earth Sciences, University of Liverpool, Bedford Street North, Liverpool L69 3BX, UK*

Y. Rees, *Water Research Centre, Henley Road, Medmenham, Marlow, Bucks. SL7 2AD, UK*

D. Slater, *OXERA Environmental Ltd, Blue Boar Court, Alfred Street, Oxford OX1 4EH, UK*

G. Stanfield, *WRC-NSF Limited, Henley Road, Medmenham, Marlow, Bucks. SL7 2AD, UK*

S. Walters, *Department of Public Health and Epidemiology, University of Birmingham, Edgbaston, Birmingham B15 2TT, UK*

M. Williams, *AEQ – 4th Floor D14, Department of Environment, Transport and the Regions, Ashdown House, 123 Victoria Street, London SW1E 6DE, UK*

CHAPTER 1

Chemical Pollution of the Aquatic Environment by Priority Pollutants and its Control

B. CRATHORNE, Y. J. REES AND S. FRANCE

1.1 INTRODUCTION

It is difficult to imagine a modern society without the benefits of chemicals and the chemical industry. Pharmaceuticals, petrochemicals, agrochemicals, industrial and consumer chemicals all contribute to our modern lifestyles. However, with the rise of chemical manufacture and use has come increasing public awareness and concern regarding the presence of chemicals in the environment.

There is an important distinction between the presence of chemicals in the environment (contamination) and pollution. Although these terms tend to be used in similar ways in everyday speech and journalism, in scientific areas there is a broad consensus that the term 'contamination' should be used where a chemical is present in a given sample with no evidence of harm and 'pollution' used in cases where the presence of the chemical is causing harm. Pollutants therefore are chemicals causing environmental harm.

The effects of water pollution can be summarized as:

- aesthetic: visual nuisance caused, *e.g.* litter, discoloration and smells
- temperature: usually heat
- deoxygenation: lack of oxygen in the water
- toxicity: acute or chronic toxicity causing damage to aquatic or human life
- sublethal toxicity: such as endocrine disruption or changes in biodiversity
- acidity/alkalinity: disturbance of the pH regime
- eutrophication: nutrients giving rise to excessive growths of some organisms

Any chemical can become a pollutant in water causing one or more of these effects if it is present at a high enough concentration. For example, serious pollution incidents result from spills of sugar and milk, substances which

1

contain a high organic content. In fact, the majority of pollution incidents in the UK continue to be due to gross organic pollution. In 1998 sewage accounted for 24% of the 17 863 substantiated pollution incidents in England and Wales with oil being the single most frequent cause, accounting for 30%.[1] Such organic pollution is caused by effluents containing biodegradable organic chemicals which generally act as pollutants, not because they contain chemicals at concentrations that are toxic, but rather the reverse. They contain chemicals that provide food for microorganisms which multiply rapidly as a result of the increased food input. The microorganisms, in the process of growing and oxidizing the organic chemical foodstuff, use up the dissolved oxygen rapidly which in some cases leads to the death of higher organisms, like fish.

Despite the fact that any chemical can be a pollutant, certain chemicals have been identified in regulation or by international agreement as being 'priority chemicals for control'. Such chemicals have generally been selected based on the following criteria:

- the chemicals are frequently found by monitoring programmes
- they are toxic at low concentrations
- they bioaccumulate
- they are persistent
- they are carcinogens

While a few years ago it would have been possible to quote the List I substances in the Dangerous Substances Directive (see Section 1.3.1) as being *the* priority pollutants, it is no longer possible to identify a single list. Different chemicals are priority pollutants in different contexts.

For many of these priority pollutants, the elimination of pollution is not considered sufficient by many and, applying the precautionary principle, targets of 'no contamination' have been set. This is demonstrated by a recent agreement between the North Sea countries, through the Oslo and Paris Commission, to reduce discharges, emissions and losses of specific hazardous substances continuously, with the ultimate aim of reducing concentrations in the marine environment to near background values for naturally occurring substances and close to zero for man-made synthetic substances.

1.2 POLLUTION CONTROL PHILOSOPHY

Preventing pollution of the environment by priority chemicals is very complex as chemicals are released and can gain entry to the environment at any stage in their life cycle (Figure 1.1), from development and testing, through manufacture, storage and distribution through to use and finally disposal.

Releases to the environment can broadly be categorized into point source releases, *i.e.* specific inputs from an industrial site or sewage works, *etc.*, and diffuse or non-point source releases. Controls have tended in the past to concentrate on tackling the largest point sources and introducing strict requirements on discharges to water and sewer. A great deal has been achieved but,

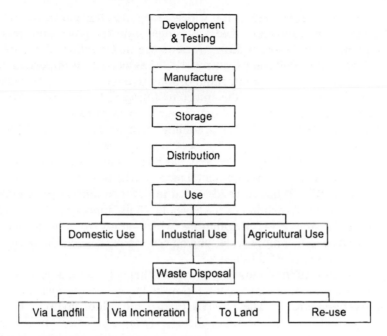

Figure 1.1.   *Life-cycle of chemicals in products*

whilst the source and single media based controls have been effective, we are left with other problems. In the last decade or so the emphasis on priority pollutant control has shifted away from control of point sources towards control of diffuse sources. The main reason for this is the success achieved by control of point source pollution. A combination of factors has caused this, principally:

- new technology enabling priority pollutant use to be avoided, *e.g.* new pesticide discovery and new production processes
- improved pollution control technology

Given the success in this area, the relative contribution of diffuse sources has risen and attention has come to focus on these areas to facilitate further improvements. By their nature, diffuse sources require different types of regulatory control from those for point sources.

Experiences over the last few decades have also highlighted a need to look more holistically at environmental controls and specifically to:

- take a more integrated view of the environment.
- consider the impacts of mixtures on the environment.
- look beyond national borders.
- move away from an end-of-pipe approach.
- introduce a wider range of control measures.

Taking a more integrated view. It is widely recognized that control of pollution is not a single environmental media or single industry issue, and integrated pollution control and cross industry reviews are now routine. The implementation of Integrated Pollution Control (IPC) has played an important role in introducing a more holistic control philosophy to environmental management. IPC applies to releases from the 'most polluting industrial processes'. The system tightened pollution control requirements by introducing the concepts of Best Available Techniques Not Entailing Excessive Costs (BATNEEC). Releases of the most polluting substances must be prevented or, where this is not possible, minimised and rendered harmless. For processes resulting in releases to more than one environmental medium the Best Practicable Environmental Option (BPEO) has to be identified in order to minimize the impact of the process on the environment as a whole. This approach will be further advanced under the IPPC Regulations which apply to a wider range of processes and require a more wide ranging criteria in the consideration of best available techniques (see Section 1.3.3).

A good example of the complex interactions that can occur, at least in theory, is the report of the potential accumulation of HCFC breakdown products in seasonal wetland areas.[2] The HCFCs (hydrochlorofluorohydrocarbons) are expected to break down in the atmosphere fairly rapidly. This is the reason they are preferred over the CFCs (chlorofluorohydrocarbons). However, trifluoroacetic acid, which some HCFCs are expected to produce, is very stable and will wash out from the atmosphere in rain. In areas like seasonal wetlands which have high evapotranspiration rates the trifluoroacetic acid may concentrate to levels which may damage plants. At present, this is based on theory and modelling but it does illustrate the complexity of environmental processes.

Considering the impacts of mixtures. An issue currently being considered is how best to deal with mixtures of chemicals. Chemicals that enter the environment usually do so as wastes or by-products, and as such are mixed with other chemicals. They may interact when they are mixed in an additive, synergistic or antagonistic way. They may produce breakdown products, by-products or react to form new substances in the waste stream or in the environment. Whilst controls on individual chemicals can be effective, they cannot take account of these real environmental situations.

Direct Toxicity Assessment (the use of ecotoxicity tests to assess complex chemical samples) is being considered as a control tool for chemical mixtures for both process and emissions control. In addition, in order to provide a more integrated view of the state of the environment, ecological monitoring is essential. Given that the ultimate aim of chemical control is protection of the environment it is only by measuring improvements in ecological quality that the ultimate effectiveness of chemical control measures and other control measures can be determined. This focus on ecological quality is the basis for the new approach to water management in the recently adopted Water Framework Directive (see Section 1.3.4).

Looking beyond national borders. Recognition that pollution does not recognize national borders, particularly for priority pollutants which tend to persist in the environment for long periods and can therefore be transported long distances, has led to growing international action on priority pollutant control. There are many initiatives to protect the aquatic environment from priority pollutants particularly for the protection of marine waters (see Section 1.3).

Moving away from end-of-pipe controls. Government, regulatory agencies and industrial initiatives have recognized a hierarchy of approaches to priority pollutant control:

- replace: use another, more environmentally friendly chemical
- reduce: use as little of the priority pollutants as possible
- manage: use in a carefully managed way to minimize accidental or adventitious loss and waste

This is reflected in the requirements under IPC which require that emissions of the most harmful substances are 'prevented or if that is not possible minimized and rendered harmless'.

Introducing a wider range of control measures. Traditionally the instruments for controlling chemicals have been regulations setting limits for discharges to water or banning a chemical from specific or all uses (see Sections 1.3 and 1.4). However, regulation has been recognized as being a rather blunt instrument with which to achieve continuous improvement, particularly for diffuse sources. Specifically in relation to priority pollutant control it has been difficult to find ways of encouraging replacements of priority chemicals by other, more benign substitutes. It has also been recognized that if the regulation can work with the 'grain' of the market, *i.e.* the normal market forces – supply and demand, consumer choice *etc.* – it will be more effective. Governments are aware of a need to increase the tools for controlling chemicals by introducing voluntary schemes, information systems and economic incentives to encourage more environmentally friendly practices (see Section 1.5).

The important points to emphasize here are that the regulation and control of priority pollutants:

- is not just a case of tightening up on discharge consents as diffuse sources can be important and can dominate in some cases
- is an international issue
- requires a portfolio of complementary activities based on a hierarchy of replace–reduce–manage

1.3 REGULATION OF DIRECT DISCHARGES

In the UK, all direct discharges to water and indirect industrial discharges to sewer are controlled *via* permitting systems. The legislation establishing these systems is described in more detail in Chapter 20, but briefly the main Acts are:

- The Environmental Protection Act 1990 – for discharges from the most polluting industrial processes.
- The Water Resources Act 1991 – for all discharges direct to controlled waters except those covered under IPC.
- The Water Industry Act 1991 – for discharges to sewer.

The priority pollutants controlled under these permitting systems and the standards which apply are heavily influenced by international initiatives, in particular several key EU Directives and agreements made at the North Sea Conferences and by the Oslo and Paris Commission (OSPAR).

The permitting systems adopt the two general approaches for controlling direct discharges to water; the use of Environmental Quality Objectives/ Environmental Quality Standards (EQO/EQS) and Uniform Emission Standards (UES). This difference lies at the heart of a long running debate in Europe with the UK favouring the EQO/EQS approach and most other countries favouring UESs.

The EQO/EQS approach seeks to define the use that is to be made of a given water body, examples would be 'use for drinking water abstraction' or 'use for the support of salmonid fish populations', which defines an Environmental Quality Objective (EQO). In order to secure the objective in the presence of dangerous chemicals, Environmental Quality Standards (EQS) are needed. These are concentrations of the chemicals concerned below which there is expected to be no impact on the EQO. Emission limits for effluents can then be established by taking account of dilution capacity within the receiving waters on the basis that the EQS limits must not be exceeded outside the immediate impact zone (also called the mixing zone).

The UES, or limit value, approach sets limits for the concentration of dangerous substance in the effluent, without taking specific account of the available dilution capacity or the presence of other inputs to the same water body. The UES limits are usually expressed as effluent concentrations (monthly flow-weighted averages) with additional limits on daily values – the daily values being usually a factor of two to four higher. Values are also expressed as a total amount of substances per unit of production or use, *e.g.* 40 g CCl_4 per tonne of production.

At a technical level there has been a growing recognition that the most sensible approach to pollution control is a fusion of the two approaches as adopted in the Integrated Pollution Control system (see Section 1.3.3).

1.3.1   Dangerous Substances Directive

The Dangerous Substances Directive (76/464/EEC) was adopted in 1976 to provide a framework for eliminating or reducing pollution of inland waters by particularly dangerous substances. In this Directive chemicals are either placed on List I, which has come to be known as the 'Black' List, or on List II, the 'Grey' List. Different control procedures are applied to chemicals on these lists. Those on List I have limit values and EQSs agreed at Community level. These

Table 1.1 *Categories of substances on List I*

List I contains certain individual substances which belong to the following families and groups of substances, selected mainly on the basis of their toxicity, persistence and bioaccumulation, with the exception of those which are biologically harmless or which are rapidly converted into substances which are biologically harmless:

1 organohalogen compounds and substances which may form such compounds in the aquatic environment.
2 organophosphorus compounds.
3 organotin compounds.
4 substances in respect of which it has been proved that they possess carcinogenic properties in or *via* the aquatic environment.
5 mercury and its compounds.
6 cadmium and its compounds.
7 persistent mineral oils and hydrocarbons of petroleum origin.
8 persistent synthetic substances which may float, remain in suspension or sink and which may interfere with any use of the waters.

appear in daughter Directives, *e.g.* Directive of 26 September 1993 on limit values and quality objectives for cadmium discharges (93/513/EEC). (Confusingly, in Directives the Commission uses 'quality objective' to mean the same as EQS in the discussion above!) The List I chemical categories are given in Table 1.1 and specific chemicals agreed for control as List I chemicals are provided in Table 1.2. In 1982 the Commission also published a list of 129 potential List I chemicals selected by the Commission on the basis of production volume and estimates of toxicity, persistence and bioaccumulation.

List II chemicals are to be controlled by using the EQO approach using quality standards set nationally. Member States are also required by the Directive to establish programmes to reduce pollution by these substances. The families of chemicals identified as List II are given in Table 1.3 and those for which National Quality Standards have been set are given in Table 1.4.

1.3.2 The North Sea Conferences and OSPAR

The North Sea Conferences were set up by governments of countries bordering the North Sea as a forum for agreeing policies aimed at reducing pollution from dangerous substances. At the second conference, in London in 1987, participating countries agreed to reduce inputs of dangerous substances to rivers and estuaries by 50% over the period 1985–1995. In response, the Department of the Environment and the Welsh Office issued a consultation paper entitled 'Inputs of Dangerous Substances to Water: Proposals for a Unified System of Control' in July 1988. This paper set out the Government's proposal for tightening controls over the input of dangerous substances to water. Dangerous substances were defined as those which represented the greatest threat to the aquatic environment due to their persistence, toxicity, their ability to bioaccumulate

Table 1.2   *Substances on List I and other priority lists*

CAS No.	Substances	DS List I	UK prescribed substances	IPPC	OSPAR	WFD
	DDT and metabolites (DDD, DDE)	✔	✔		✔	
309-00-2	Aldrin	✔	✔		✔	
60-57-1	Dieldrin	✔	✔			
72-20-8	Endrin	✔	✔			
465-73-6	Isodrin	✔				
58-89-9	Hexachlorocyclohexane	✔[1]	✔		✔[2]	✔
	Polychlorinated biphenyls				✔	
1912-24-9	Atrazine		✔		✔	✔
959-98-8	Endosulfan		✔		✔	✔
122-34-9	Simazine		✔		✔	✔
1582-09-8	Trifluralin		✔		✔	
470-90-6	Chlorfenvinphos				✔	✔
25154-52-3	Nonylphenol				✔	✔
1806-26-4	Octylphenol					✔
	Azinphos-methyl		✔			
140-66-9	1,1,3,3-tetramethyl-4-butylphenol (4-*tert*-octylphenol)					✔
117-81-7	Di(2-ethylhexyl) phthalate (DEHP)					✔
	Butylbenzyl phthalate			✔		
	Diethylhexyl phthalate			✔		
	Di-*n*-octyl phthalate			✔		
107-06-2	1,2-dichloroethane (EDC)	✔	✔		✔	✔
15972-60-8	Alachlor				✔	✔
71-43-2	Benzene				✔	✔
	Brominated diphenyl ether					✔
32534-81-9	Diphenyl ether, penta-bromo derivative					✔
40088-47-9	2,2',4,4'-tetrabromodiphenyl ether					✔
1163-19-5	Bis(pentabromophenyl)-ether					✔
85535-84-8	Chloroalkanes C10–13					✔
2921-88-2	Chlorpyrifos				✔	✔
75-09-2	Dichloromethane				✔	✔
330-54-1	Diuron				✔	✔
	Dichlorvos		✔			
	Malathion		✔			
76-44-8	Heptachlor				✔	✔
118-74-1	Hexachlorobenzene (HCB)	✔	✔			✔
87-68-3	Hexachlorobutadiene (HCBD)	✔	✔		✔	✔
34123-59-6	Isoproturon				✔	✔
	Monochloronitrobenzenes					✔

[1] Gamma isomer only.
[2] Not beta isomer.

(continued)

Table 1.2 *Continued*

CAS No.	Substances	DS List I	UK prescribed substances	IPPC	OSPAR	WFD
98-95-3	Nitrobenzene				✔	✔
	PAHs					✔
193-39-5	Indeno[1,2,3-*cd*]pyrene					✔
56-55-3	Benzo[*a*]anthracene					✔
191-24-2	Benzo[*g,h,i*]perylene					✔
205-99-2	Benzo[*b*]fluoranthene					✔
207-08-9	Benzo[*k*]fluoranthene					✔
50-32-8	Benzo[*a*]pyrene					✔
218-01-9	Crysene					✔
91-20-3	Naphthalene					✔
120-12-7	Anthracene					✔
206-44-0	Fluoranthene					✔
86-73-7	Fluorene					✔
83-32-9	Acenaphthene					✔
85-01-8	Phenanthrene					✔
87-86-5	Pentachlorophenol	✔	✔		✔	✔
	Trichlorobenzene (TCB)	✔	✔		✔	✔
67-66-3	Trichloromethane (chloroform)	✔			✔	✔
7440-38-2	Arsenic			✔	✔	✔
7440-43-9	Cadmium	✔	✔		✔	✔
7440-50-8	Copper				✔	✔
7439-92-1	Lead				✔	✔
7440-02-0	Nickel				✔	✔
7439-97-6	Mercury	✔	✔			
	Metals and their compounds			✔		
140-66-9	1,1,3,3-Tetramethyl-4-butylphenol				✔	
87-61-6	1,2,3-Trichlorobenzene				✔	
	1,2,4,5-Tetrachlorobenzene				✔	
120-82-1	1,2,4-Trichlorobenzene				✔	
	1,3,5-Trichlorobenzene				✔	
541-73-1	1,3-Dichlorobenzene				✔	
	1,4-Pentadien-3-ol, 3-methyl-1-(2,6,6-trimethyl-1-cyclohexen-1-yl)-				✔	
121-14-2	2,4-Dinitrotoluene				✔	
121-73-3	3-Chloronitrobenzene				✔	
	4,4'-Dianilinodiphenyl (benzidine)				✔	
	4-*tert*-Butyltoluene				✔	
	5-isocyanato-1-(isocyanatomethyl)-1,3,3-trimethyl-					
208-96-8	Acenaphthene (PAH)				✔	✔
120-12-7	Anthracene (PAH)				✔	

(continued)

Table 1.2 *Continued*

CAS No.	Substances	DS List I	UK prescribed substances	IPPC	OSPAR	WFD
	Benzene,1,1'-(2,2,2-trichloro-ethylidene)bis[4-chloro-				✔	
	Benzene, 1,1'-oxybis-, pentabromo deriv.				✔	
	Benzene, 2,4-dichloro-1-(4-nitrophenoxy)-				✔	
	Benzene, hexachloro-				✔	
	Benzene, pentabromomethyl-				✔	
	Benzenethiol, pentachloro-				✔	
	Benzo[a]anthracene (PAH)				✔	
	Benzo[a]pyrene (PAH)				✔	
	Benzo[b]fluoranthene (PAH)				✔	
	Benzo[g,h,i']perylene (PAH)				✔	
	Benzoic acid, 5-(2,4-dichlorophenoxy)-2-nitro-, methyl ester				✔	
	Benzo[k]fluoranthene (PAH)				✔	
	Biocides and plant health products			✔		
	Bis(pentabromophenyl) ether				✔	
	Carbon tetrachloride	✔				
	Chlorinated paraffins, short-chain				✔	
	Chlorobenzene				✔	
	Cyanides			✔		
	Cyclohexane				✔	
	Diphenyl ether				✔	
	Diphenylmethane				✔	
	Fenitrothion		✔		✔	
	Fluoranthene (PAH)				✔	
	Hexachloroethane				✔	
	Hexachloronaphthalene				✔	
	Indeno[1,2,3-cd]pyrene (PAH)				✔	
	Kelthane (Dicofol)				✔	
	Methoxychlor			✔		
	Naphthalene (PAH)				✔	
	Organohalogen compounds			✔		
	Organophosphorus compounds			✔		
	Organotin compounds			✔		
	Perchloroethylene (PER)	✔				
	Persistent hydrocarbons and persistent and bioaccumulable organic toxic substances			✔		
	Phenol, 2,4,6-tris(1,1-dimethylethyl)-				✔	

(continued)

Table 1.2 *Continued*

CAS No.	Substances	DS List I	UK prescribed substances	IPPC	OSPAR	WFD
	Phenol, 4,4'-(1-methyl-ethylidene)bis[2,6-dibromo-				✔	
	Phoxim				✔	
129-00-0	Pyrene				✔	✔
	Substances with carcinogenic and mutagenic properties			✔		
	TCDD, PCDD, PCDF				✔	
	Tetraethyl lead				✔	
	Tributyltin compounds		✔			
	Triphenyltin compounds		✔			
	Toxaphene				✔	
	Trichloroethylene (TRI)	✔				
	Trichlorophenol (all isomers)				✔	
	Tricresylphosphate				✔	

Table 1.3 *Categories of substances on List II*

List II contains

- substances belonging to the families and groups of substances in List I for which the limit values referred to in Article 6 of the Directive have not been determined.
- certain individual substances and categories of substances belonging to the families and groups of substances listed below and which have a deleterious effect on the aquatic environment, which can, however, be confined to a given area and which depend on the characteristics and location of the water into which they are discharged.

Families and groups of substances referred to

1 The following metalloids and metals and their compounds:

(1) Zinc	(6) Selenium	(11) Tin	(16) Vanadium
(2) Copper	(7) Arsenic	(12) Barium	(17) Cobalt
(3) Nickel	(8) Antimony	(13) Beryllium	(18) Thallium
(4) Chromium	(9) Molybdenum	(14) Boron	(19) Tellurium
(5) Lead	(10) Titanium	(15) Uranium	(20) Silver

2 Biocides and their derivatives not appearing in List I.
3 Substances which have a deleterious effect on the taste and/or smell of the products for human consumption derived from the aquatic environment, and compounds liable to give rise to such substances in water.
4 Toxic or persistent organic compounds of silicon and substances which may give rise to such compounds in water, excluding those which are biologically harmless or are rapidly converted in water into harmless substances.
5 Inorganic compounds of phosphorus and elemental phosphorus.
6 Non-persistent mineral oils and hydrocarbons of petroleum origin.
7 Cyanides, fluorides.
8 Substances which have an adverse effect on the oxygen balance, particularly ammonia, nitrites.

Table 1.4   *List II chemicals for which UK National Standards have been set*

Lead[3]	4-Chloro-3-methylphenol[5]
Chromium[3]	2-Chlorophenol[5]
Zinc[3]	1,2-Dichlorophenol[5]
Copper[3]	2,4-D (ester)[5]
Nickel[3]	2,4-D (non-ester)[5]
Arsenic[4]	1,1,1-Trichloroethane[5]
Boron[3]	1,1,2-Trichloroethane[5]
Iron*[3]	Bentazone[5]
pH*[3]	Benzene[5]
Vanadium[3]	Biphenyl[5]
Tributyltin compounds[4]	Chloronitrotoluenes[5]
Triphenyltin compounds[4]	Demeton[5]
Cyfluthrin[3]	Dimethoate[5]
Sulcofuron[3]	Linuron[5]
Flucofuron[3]	Mecoprop[5]
Permethrin[3]	Naphthalene[5]
Polychlorochloromethylsulfonamidodiphenylether mothproofers[3]	Omethoate[5]
	Toluene[5]
Atrazine and simazine[4]	Triazaphos[5]
Azinphos-methyl[4]	Xylene[5]
Dichlorvos[4]	
Endosulphan[4]	
Malathion[4]	
Trifluralin[4]	

* Neither of these falls within the scope of the substances listed in Table 1.3 but DoE Circular 7/89 requires that they be treated in the same way as List II substances.

and their likely presence in the aquatic environment. The aim of the proposals in the consultation paper were stated as:

(i)   To reduce inputs to the aquatic environment of those substances which represent the greatest potential hazards.

(ii) To improve the scientific basis of the system for identifying the most dangerous substances.

(iii) To develop a more integrated approach to pollution control.

(iv) In controlling dangerous substances in water, to take full account not only of 'point source' discharges from production plants, but also of 'diffuse sources' of entry into the aquatic environment.

A further important objective was to propose a unified system for controlling the discharge of dangerous substances and reconcile the approach favoured in the UK, the EQO/EQS system, and the approach favoured by other states in the EC, that of uniform emission standards (UES), the more stringent of the two being applied as a control measure in any given case. With these considerations in mind, the Government set out proposals for a unified approach to controlling the discharge of dangerous substances. The essential points of the proposals were as follows:

(i) Identification of a limited range of the most dangerous substances selected according to clear scientific criteria, whose discharge to water should be minimized as far as possible – the 'Red List'.

(ii) The setting of strict environmental quality standards for all Red List substances.

(iii) The introduction of a system of 'scheduling' of industrial processes discharging significant amounts of Red List substances, and the progressive application of technology-based emission standards, based on the concept of Best Available Techniques Not Entailing Excessive Costs (BATNEEC).

(iv) Measures designed, where possible, to reduce inputs of Red List substances from diffuse sources.

The control regime for the Red List chemicals is more stringent than that for Black List substances and marks a change in philosophy by becoming more 'precautionary' and requiring 'BATNEEC' (see Section 1.3.3) for all Red List substances. The proposals also imply that because there is always some uncertainty about the long-term environmental effects of Red List chemicals we should aim to reduce their environmental concentrations as much as is possible – regardless of the evidence that they are causing environmental damage.

Other participating countries produced their own national priority lists of hazardous substances different to those on the UK Red List. At the third conference in the Hague in 1990, the individual priority lists were amalgamated and redefined to produce a common list of 32 substances for which the reduction target of 50% applied. A further target was set at the fourth conference in Esjberg, Denmark, in 1995 based on the objective of ensuring a sustainable, sound and healthy North Sea ecosystem. It was agreed to reduce continuously discharges of hazardous substances thereby moving towards the ultimate aim of reducing concentrations in the environment of natural hazardous substances to near background levels and of synthetic substances to zero within 25 years.

The UK is a party to the Convention for the Protection of the Marine Environment of the North East Atlantic – the OSPAR Convention – which, in 1998, agreed a strategy with regard to hazardous substances. Its objective is to "prevent pollution of the maritime area by continuously reducing discharges, emissions and losses of hazardous substances (that is substances which are toxic, persistent and liable to bio-accumulate or which give rise to an equivalent level of concern) with the ultimate aim of achieving concentrations in the marine environment near background values for naturally occurring substances and close to zero for man-made synthetic substances."[6]

The Convention will implement the strategy progressively "by making every endeavour to move towards the target of the cessation of discharges, emissions and losses of hazardous substances by the year 2020."

Selection and prioritization mechanisms and a risk assessment methodology for the marine environment have been developed and used to select hazardous substances for priority action (see Table 1.2).

1.3.3 Integrated Pollution Control (see also Chapter 21)

In the Environmental Protection Act (1990), the Government introduced a new system, Integrated Pollution Control (IPC). The system was designed to implement the commitment made earlier to use the precautionary approach for control of the most dangerous substances and also moved to a more integrated system by considering all three media (air, water and land) together. IPC applies to emissions from the 'most polluting processes' (defined in Regulations) and is designed to tighten requirements on them by introducing a number of measures:

- BATNEEC (Best Available Techniques Not Entailing Excessive Costs) must be applied to emissions of 'the most polluting substances' (also defined in Regulations[7] and, for water, virtually identical to the Red List to prevent or, if that is not possible, minimize emissions. Other emissions will be minimized and rendered harmless by the use of BATNEEC (*i.e.* they must meet the EQS and tighter standards should be imposed where technology allows). The use of the term 'techniques' emphasizes that thought must be given to all aspects of the process including the nature of raw materials) the technology of the process, treatment of wastes and training of operators.
- BPEO (Best Practicable Environmental Option) will have to be identified if emissions are to more than one medium in order to minimize the effect of emissions to the environment as a whole.

In September 1996 the Council of Ministers adopted a Directive on Integrated Pollution Prevention and Control (IPPC) (96/61/EEC) which requires the implementation of systems similar to that operated in the UK across the whole of Europe. The substances subject to the most stringent controls are listed in an Annex to the Directive and are similar to those on Lists I and II of the Dangerous Substances Directive (Table 1.2).

1.3.4 Water Framework Directive

The Water Framework Directive (2000/60/EC)[8] will fundamentally change EU water policy. It will apply to all waters and will introduce a Europe-wide system of river basin management. The Directive introduces a combined approach for the control of point and diffuse sources and will establish a priority list of pollutants for control under the Directive. An amended proposal for a Decision establishing this list was issued in January 2001. The main objective of the Directive is the achievement of good status in all waters of the Community. Good status will be defined in terms of ecological, chemical and physical quality and should be achieved by the implementation of various measures laid down in a river basin management plan. To achieve "good physico-chemical status" the concentration of specific pollutants should not exceed the EQS laid down for that substance. These will be established at an EU level and priority hazardous substances for action have been defined (see Table 1.2). Discharges of these substances should be eliminated within 20 years of the Directive's entry into force. The WFD will, once adopted, encompass the requirements of the DSD and the Directive will be repealed from the date of entry into force of the Framework Directive, with the exception of Article 6.

1.4 REGULATION OF DIFFUSE SOURCES

Pollution from diffuse sources is normally less immediately apparent and more difficult to control than that from point sources. However, for many pollutants diffuse sources are of equal or greater importance, for example pesticides and nitrates from agriculture, urban run-off and deposition from the atmosphere all make significant contributions. The differing nature of the sources means that several approaches are required to control diffuse inputs to water.

1.4.1 Product Controls

A major contribution to controlling diffuse sources is the use of 'product controls'. This type of approach shifts the emphasis from 'end-of-pipe', to focus on the manufacture and use of the chemicals at a stage before they become wastes. The general aims of these controls are either to make the product more environmentally acceptable or to restrict or prohibit the use of certain substances in product formulations.

1.4.1.1 New Substances. Under an EC Directive that came into effect in 1979 (revised in 1992: 92/32/EEC), a new chemical can only be placed on the market if the manufacturer or importer submits a notification to a competent authority (the competent authority is a body or bodies nominated by Member States). The notification includes the results of tests to evaluate possible harmful effects on humans and the environment (only in place since 1992), and may include an assessment of the risks that may arise. The authority passes the results to the European Commission, which relays them to the authorities in other Member States. If no objections are raised within 60 days of the notification, the manufacturer has assured access to the whole EU market. The Directive thus allows notifiers to make their own assessment before marketing, although this is audited by Member States.

1.4.1.2 Existing Substances. The Existing Substances Regulation (793/93/EEC) came into effect in 1993 and introduced a scheme for assessing 'existing' chemicals, *i.e.* those 20 000 or so chemicals already on the market. Manufacturers or importers who make or supply more than certain quantities of existing substances are required to send existing data on the chemicals so that an evaluation of the risk they pose both to the environment and to human health can be undertaken. These data are used to draw up lists of priority substances to be examined in more detail by Member States and, if necessary, proposals are made for risk reduction. So far over one hundred chemicals have been listed for priority attention (three priority lists have been agreed).

1.4.1.3 Dangerous Substances and Preparations. Recognizing the need to reduce diffuse inputs of dangerous substances as well as point sources, various Member States introduced measures to totally ban or restrict the use of some substances. For example, the UK restricted the use of benzene in toys and, in France, restrictions were introduced for the use of polychlorinated biphenyls (PCBs). Controls at a European level were introduced to prevent distortion in trade through a framework Directive (76/769/EEC) to ban or restrict the marketing and use of certain dangerous substances and preparations. Since the original Directive, restrictions on other substances continue to be introduced through daughter Directives (see Table 1.5). The Directives provide a mechanism to implement a precautionary approach in a quick and consistent manner across the whole of the European Union.

Table 1.5 *Dangerous substances subject to restrictions on marketing and use (Directive 76/769/EEC and amendments)*

Substance	Directive
PCBs	76/769/EEC, 85/467/EEC, 89/677/EEC
PCTs	76/769/EEC, 82/828/EEC, 85/467/EEC, 89/677/EEC
Chloro-1-ethylene (monomer vinyl chloride)	76/769/EEC
Liquids in the following categories according to Directive 67/548/EEC, highly toxic, toxic, harmful, corrosive, explosive, extremely flammable, highly flammable, flammable and any liquid with a flashpoint < 55 °C	79/663/EEC, 89/677/EEC
Tris(2,3-dibromopropyl) phosphate	79/663/EEC
Benzene	82/806/EEC, 89/677/EEC
Tris-(aziridinyl)-phosphinoxide	83/264/EEC
PCBs	83/264/EEC
Soap bark powder and its derivatives containing saponines	83/264/EEC
Powder of the roots of *Helleborus viridis* and *Helleborus niger*	83/264/EEC
Powder of the roots of *Veratrum album* and *Veratrum nigrum*	83/264/EEC
Benzidine and/or its derivatives	83/264/EEC
o-Nitrobenzaldehyde	83/264/EEC
Wood powder	83/264/EEC
Ammonium sulfide or ammonium hydrogen sulfide	83/264/EEC
Ammonium polysulfide	83/264/EEC
Methyl bromoacetate	83/264/EEC
Ethyl bromoacetate	83/264/EEC
Propyl bromoacetate	83/264/EEC
Butyl bromoacetate	83/264/EEC
Asbestos	83/478/EEC, 85/610/EEC, 91/659/EEC, 1999/77/EC

(continued)

Table 1.5 *Continued*

Substance	Directive
2-Naphthylamine	89/677/EEC
Benzidine and its salts	89/677/EEC
4-Nitrophenyl	89/677/EEC
4-Aminobiphenyl and its salts	89/677/EEC
Neutral anhydrous lead carbonate	89/677/EEC
Lead hydrocarbonate	89/677/EEC
Lead sulfates	89/677/EEC
Mercury	89/677/EEC
Arsenic	89/677/EEC
Organostannic compounds	89/677/EEC
Di-/J-oxo-di-*n*-butylstanniohydroxyborane	89/677/EEC
Pentachlorophenol	91/173/EEC, 1999/51/EC
Cadmium	91/338/EEC, 1999/51/EC
Monomethyltetrachlorodiphenylmethane	91/339/EEC
Monomethyldichlorodiphenylmethane	91/339/EEC
Monomethyldibromodiphenylmethane	91/339/EEC
Nickel	94/27/EC
Flammable solvents in aerosols	94/48/EC
Chlorinated solvents	94/60/EC
Wood preservatives	94/60/EC
Carcinogens	94/60/EC, 97/10/EC, 97/56/EC, 1999/43/EC
Mutagens	94/60/EC, 97/10/EC, 97/56/EC, 1999/43/EC
Teratogens	94/60/EC, 97/10/EC, 97/56/EC, 1999/43/EC
Chlorinated solvents	96/55/EC
Hexachloroethane in the non-ferrous metals industry	97/16/EC
Petroleum or coal-tar derived carcinogens, mutagens and teratogens	97/56/EC
Coloured lamp oils	97/64/EC
Tin	1999/51/EC

1.4.1.4 Pesticides. The way in which pesticides are released to the environment is largely through their use. This is a diffuse source which is controlled through product registration. Recognition of the need to protect water is a key element in the process and approvals can be reviewed or revoked at any time. The Commission is currently introducing a system to harmonize approval procedures throughout the EU. The Directive concerning the placing of plant protection products on the market (91/414/EEC) contains a 'positive' list of active ingredients that may be used in the formulation of plant protection products. Inclusion on the list is dependent upon the active ingredient satisfying a number of conditions, including the assessment of its impact on human health and the environment.

Another Directive (79/117/EEC) restricts the marketing and use of certain pesticides and lists the substances that may not be present in pesticide formulations. The list currently bans several mercury and persistent organo-chlorine compounds, for example DDT, aldrin, endrin, dieldrin and chlordane, as well as other compounds such as nitrogen and ethylene oxide. Pesticides that are marketed must conform with various classification, packaging and labelling requirements (defined in Directive 78/631/EEC). Amongst other things, these specify application methods, timing and rates, and disposal methods to ensure that pollution of the environment is avoided. In the UK, those applying pesticides must meet certain training requirements and guidance is given by the Ministry of Agriculture, Fisheries and Food in a Code of Practice for the use and storage of pesticides.[9]

1.4.1.5   Biocides.   Biocides are a diverse range of chemical additives, commonly used to control growth of microorganisms in many industrial processes. Increasing concerns about possible effects on non-target organisms, including man, led the Council of Ministers and European Parliament to adopt a Directive to introduce a Community-wide scheme for the Authorization of Biocides (98/8/EC), very similar to that already in operation for pesticides. The Directive seeks to harmonize existing national regulations.

1.4.1.6   Detergents.   In Western Europe in the 1960s, one of the most visible examples of water pollution was foaming in rivers due to the use of 'hard' detergents. These are poorly biodegradable, both naturally and in sewage treatment works. Foaming is not only an aesthetic problem but also impairs photosynthesis and oxygenation in rivers as well as reducing the operating efficiency of sewage treatment plants.

   In an attempt to alleviate the problem, the Community developed the Detergent Directive (73/404/EEC) to restrict the sale of detergents with a poor biodegradability. The Directive banned the marketing of detergents based upon surface active agents (surfactants) with an average biodegradability less than 90%. The Directive applies to all types of surfactant – anionic, nonionic, cationic and amphoteric. More detailed 'daughter' directives stipulate the test methods that should be used to assess whether anionic and nonionic surfactants comply with the requirements.

1.4.2   Controlling Land Use

Most land uses or activities, whether industrial, urban or agricultural, have an impact on water quality. Whilst it is clearly impracticable to prohibit all such activities, it is also important to protect and enhance the environment. Increasingly, controls are being introduced to ensure that the environmental impact of land uses are taken into account both for new developments and those already in existence. The general aims of land use controls are to reduce the risk of pollution from priority activities such as the storage of large amounts of hazardous substances by regulating the way in which the activity is carried out (for example, requiring certain storage measures, or controlling fertilizer use). Such controls can be applied on a national basis, as with planning controls in

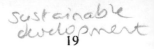

the UK, or may be used to protect particularly sensitive areas or catchments, for example in water pollution zones or nitrate sensitive areas.

1.4.2.1 Planning. Planning permission is required for any development of land in the UK under the Town and Country Planning Act 1990 and the Planning and Compensation Act 1991. Through this system, Local Authorities are able to control the location of "building, engineering, mining or other activities or other operations in, on, over or under land, or the making of any material change in the use of any buildings or other land". Planning controls provide a primary opportunity for implementing the precautionary principle. They are the key to balancing the need for development whilst protecting the environment thereby achieving 'Sustainable Development' which the UK Government is determined to make the basis of its environmental policies for the future.

Increasingly, planners are required to take into account the environmental impact of proposed developments prior to providing planning permission. A number of procedures and guidance documents have been introduced. Most notably the DoE has issued a planning policy guidance note on planning and pollution control[10] which is intended to provide guidance on industrial and waste disposal related developments. The National Rivers Authority (NRA), now subsumed into the Environment Agency, also published guidance for Local Planning Authorities on methods for protecting the water environment.[11] In addition, the Environment Agency must be consulted for developments which may affect water quality and for certain projects, developers are required to carry out an environmental impact assessment as a part of an application for planning permission.

A new system enabling Local Authorities to control the storage of hazardous substances was introduced in the Planning (Hazardous Substances) Act 1990. The subsequent Regulations[12] list 71 substances, selected on the basis of their explosive or inflammable properties, for which consent will be needed if they are to be stored on a site in quantities above that prescribed in an Annex to the Regulations.

1.4.2.2 Controls on Industrial Sites. The Control of Industrial Major Accident Hazards (CIMAH) Regulations in the UK (1984),[13] amended in 1990[14] and 1994,[15] are concerned with reducing the hazard posed to man and the environment from accidents at industrial plants which involve 'one or more dangerous substances'. The Regulations were introduced as a result of the Seveso Directive (82/501/EC), adopted by the Community after the accident at Seveso in Italy. This Directive was replaced by the COMAH Directive (96/82/EC), which entered into force on 3 February 1997. This has a wider scope than the Seveso Directive and aims to:

- simplify the application system and to encourage more consistent implementation
- require the development of land use policies around the COMAH sites

- provide improved freedom of access to information and encourage greater public participation
- ensure an increased emphasis on safety management.

It was implemented in the UK by the Control of Major Accident Hazards Regulations 1999 (SI 1999/743).

1.4.2.3   Water Protection Zones.   For certain areas requiring extra protection, the Secretary of State for the Environment may prohibit or restrict activities carried out by designating water protection zones (WPZs). This approach has proved to be effective at, for example, reducing the pollution of groundwater from diffuse sources in other European countries. In the UK, several voluntary water protection zones have been established and their effectiveness is currently being assessed. The first statutory WPZ around the river Dee was designated in June 1999.[16] The Order prohibits, without the consent of the Environment Agency, the keeping or use within the catchment area of 'controlled substances' above defined threshold amounts. The 'controlled substances' are listed in Table 1.6.

Table 1.6   *Controlled substances according to the Dee Water Protection Zone Designation Order*[16]

'Controlled substance' means any substance which is

1 a dangerous substance;
2 a fuel, lubricant or industrial spirit or solvent which is a liquid under normal conditions or which is kept as a liquid within a site;
3 a medicinal product;
4 food which is a liquid under normal conditions;
5 feeding stuff which is a liquid under normal conditions;
6 an inorganic fertilizer;
7 a cosmetic product;
8 a substance identified by its manufacturer as being toxic, harmful, corrosive or irritant, but does not include:

- controlled waste that is kept, treated or disposed of under a Waste Management Licence;
- radioactive waste;
- any fuel, whether kept within a site and used exclusively for the production of heat or power;
- any substance contained in an exempt pipe-line;
- any substance at a site for a period of 24 hours or less;
- any substance which is a gas or vapour under normal conditions.

The minimum quantities subject to control are:

- in the case of food and feeding stuffs other than defined dangerous substances, an amount in excess of 500 litres;
- in other cases, an amount equal to or in excess of, 50 litres when the substance is present in a single container but otherwise 200 litres.

The Environment Agency is able to grant protection zone consents, either unconditionally or subject to conditions, likely to be based on BATNEEC, or to refuse consent.

1.4.2.4 Nitrate Sensitive Areas and Vulnerable Zones. Section 94 of the Water Resources Act 1991 gives the Secretary of State powers to establish nitrate sensitive areas (NSAs). Farmers in these areas are compensated if they adopt 'environmentally friendly' farming practices that result in a decrease in nitrate application to land. The initial scheme, in which ten areas were designated, has been extended to a further 22 areas. A further 69 areas,[17,18] termed Nitrate Vulnerable Zones, where farming practices must be modified to reduce the inputs of nitrate, have been designated to ensure compliance with the Nitrates Directive (91/692/EEC). Unlike NSAs, landowners in these areas are bound to conform to specified agricultural practices and do not receive compensation payments.

1.4.2.5 Groundwater Protection. To control groundwater contamination outside of areas designated as WPZs or NSAs, the Environment Agency published a revised Groundwater Policy[19] as guidance for use by planning and waste regulatory authorities in 1998. Two of the central principles of the policy are the classification of groundwaters according to their vulnerability and the definition of three-tier 'Source Protection Zones' which will form the basis for controls on activities posing a potential threat to groundwater.

1.5 ALTERNATIVE CONTROL PROCEDURES

Many of the newer regulatory initiatives can be considered as market-based instruments. They are roughly aggregated together into the subsections titled Regulations, Information and Initiatives.

1.5.1 Voluntary Schemes

The activities dealt with in this section are both voluntary schemes which have been introduced by regulation and schemes initiated by government departments or industry aiming to maximize the benefits to both industry and the environment of new technologies and approaches. The initiatives are concerned with environmental protection in general, but often have priority chemicals as one of the focus areas.

The introduction of an EC Ecolabelling scheme was agreed by the Council of Ministers in 1991. It was designed to enable consumers to select goods and services based on their environmental impact, the assumption being that given a choice consumers will select environmentally friendly goods which will encourage manufacturers in this direction. The UK Regulations came into force in November 1992. They created an Ecolabelling Board which could award an Ecolabel to a product, which thereby alerts consumers to the more environmentally friendly alternatives. However, the United Kingdom Ecolabelling Board (Abolition) Regulations entered into force on 19 April 1999, revoking the 1992

Regulations and replacing the Board with Secretary of State as the Competent Body. The Ecolabel criteria are still being established, but priority pollutants are considered as a significant component, for example the German Federal Environmental Agency is advocating a negative list of substances for detergents. More specifically 'only traces' of mercury are permitted in light bulbs awarded an Ecolabel.

The Eco-Management and Audit Scheme (EMAS) has a similar objective to Ecolabelling. EMAS was set up by the EC under Regulation EC/1836/93 "allowing voluntary participation by companies to a Community Eco-Management and Audit Scheme". Like the Ecolabelling scheme, EMAS is voluntary and is also providing a means by which consumers can choose to buy from, or deal with, industries that have better environmental performance. EMAS applies to sites, not companies, and in order to comply with the Scheme the following environmental components must be verified by an independent audit of each site: an environmental policy, programme, management system, review, audit procedure and statement. A proposal for the revision of the EMAS scheme was put forward in 1998 and is likely to be adopted in the near future. It will extend the scope of the current regulation to all organizations with significant environmental effects.

A report in 1991[20] identified that industry is, in general, unaware of the benefits that could result from a systematic approach to reduction of emissions and the introduction of cleaner technology. A number of case study projects have now been undertaken which show the dual benefits of cost reduction in industry and pollution load reduction for the environment. In some cases the savings are quite staggering, for example in the Aire and Calder study 51 opportunities to reduce waste were identified, resulting in estimated savings of over £400 000. Over 68% had payback periods of less than twelve months and over 89% of less than two years. To encourage technology transfer DTI and DoE launched the Environmental Technology Best Practice Programme in 1994. The initial objectives relate to waste minimization and cleaner technology, both of which can be expected to impact on priority pollutants in water.

In response to growing concerns about the impact of chemicals in the environment, the UK Chemical Industries Association (CIA) made a commitment to develop a programme of 'responsible care'. The Association published new guidelines in May 1999, setting out commitments which it asked chief executives to accept as a condition of membership. In addition, CIA has developed a new initiative, 'confidence in chemicals', which addresses risk assessment, product stewardship and long range research.

The Royal Society of Chemistry has developed a Green Chemistry Network to promote concepts such as waste minimization, solvent selection, intensive processing and alternative synthetic routes from sustainable resources. The challenge for chemists is to develop products, processes and services in a sustainable manner to improve quality of life, the natural environment and industry competitiveness. To achieve this, the Green Chemistry Network promotes awareness and helps with education, training and the practice of green chemistry in industry, universities and schools.

1.5.2 Information

The Environment Agency has a number of systems aimed at improving public access to information on releases to air, land and water. The most significant of these is the Pollution Inventory which was launched in May 1999 as a replacement for the Chemical Release Inventory from processes regulated by the Agency under Integrated Pollution Control (IPC). Data on releases of over 150 substances from over 2000 sites reported to the Agency are available. Information is also provided on substance properties, sources, uses and effects to help the user understand the implications of the emissions. In 2001 the PI will be extended to include landfill, sewage treatment works and nuclear installations and all processes regulated under the Pollution Prevention and Control Act, 1999, implementing the reporting requirements under the European Council Directive 96/61/EC on Integrated Pollution Prevention and Control (the IPPC Directive).

The Agency also maintains publicly available registers for all authorizations and for monitoring data relating to emissions, including waste regulation and discharges to water.

Individual companies have also initiated schemes to provide annual reports on their environmental performance, alongside those relating to their financial performance. In some cases these reports show aggregate releases of priority substances and comparisons with previous years' releases and future targets.

All these activities help to provide pressure from public, shareholders and pressure groups to reduce discharges of dangerous substances at both a local and national level.

1.5.3 Economic Instruments

In the UK there are few current examples of economic instruments being applied to environmental regulation of priority substances in water. The case of differential tax on unleaded and leaded petrol is a good example applied to priority pollutants in the atmosphere. The Environment Agency in England and Wales and the Scottish Environmental Protection Agency are only allowed to use charges for discharge consents to cover their administration costs. So the scope for using consent charges as incentives is limited. However, the analytical costs associated with priority substances in effluents can be significant and the differential between charges for effluents containing these substances and effluents not containing them can be seen as a financial incentive, albeit possibly not substantial enough to cause a company to modify its process.

1.6 CASE STUDIES

Certain chemicals or groups of chemicals have been chosen in this section to illustrate the wide range of problems which arise in considering the effects of chemicals on the environment. The examples illustrate the varying sources of pollutants, their fate and analysis and factors which can influence what type of control procedures have been or may need to be adopted.

1.6.1 Polybrominated Diphenyl Ethers (PBDEs)

The general chemical formula of brominated diphenyl ether is

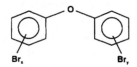

A large number of congeners exist (209), but the main interest is in the production of penta-, octa- and deca- varieties, used commercially as flame retardants. In this respect they are very effective and European consumption of PBDEs has increased considerably over the last ten years. For example, it has been estimated that production and imported quantities of PBDE amounted to around 8500 tonnes in 1986, rising to around 11 000 in 1991.[21] As the use of one of the major alternative fire retardants (polybrominated biphenyls, PBBs) has been reduced, the use of PBDEs has increased and worldwide production was estimated at 40 000 tonnes in 1994.[22]

In general terms PBDEs are not considered to be highly toxic although there are few data available; they exhibit low acute oral toxicity to rats, and there is no evidence that they give rise to mutations. Recent data from Sweden has, however, indicated that PBDEs can be detected in many fish species and their predators, such as the osprey and grey seal.[22] There is some evidence of carcinogenicity in rats, but they are not thought to present a carcinogenic risk to humans.[21]

There are various potential sources of release and exposure to PBDEs, although release from products is considered to be low. Production sites appear to be a source of environmental exposure with release possible in the form of dust or particles. No studies are available on the fate of PBDE-containing products in landfill and it is possible that leaching in the long-term may occur. Few data are available on the fate of PBDEs in the environment although some studies have been carried out on the commercially produced isomers. Data suggest that their solubility in water is low, that the higher brominated congeners are persistent and that they are likely to bind to particles and sediment (log K_{ow} values ranging from around 4.3 to 10).[21] The presence of PBDE congeners in the environment has been reported, with highest values found in some sediments near to production sites[21] (up to 1 g kg^{-1} has been reported although more typical levels appear to be below 100 μg kg^{-1} away from production sites). Concentrations in aqueous samples are much lower and the higher brominated congeners are frequently undetectable. PBDEs have, however, been regularly detected at low concentrations in the fatty tissue of a wide range of aquatic life, animals and humans. The analysis of PBDEs in environmental samples is relatively straightforward and several methods based on solvent extraction, clean-up and separation/detection using GLC with electron capture detection or GC–MS have been published. The main concern behind the proposals to restrict the use of PBDEs is, in fact, the possibility that they may contain, or form, brominated dioxins and furans during combustion. Numerous studies have been conducted on this and the results vary according to the products tested, temperatures applied and other conditions such as the presence of oxygen or catalyst (see also Chapter 17).

The way in which PBDEs are controlled as environmental contaminants presents regulators (and ultimately society itself) with a difficult dilemma. Concern for fire safety is a high priority and each year many people die directly from burns or indirectly through inhalation of fumes. There is no question that the use of fire retardants can, and does, save lives. Many fire regulations actually specify a certain level of fire resistance which currently could not be met for many materials without the use of a fire retardant.

As mentioned previously, PBBs are now being phased out of use due to concern over their persistence in the environment and their toxicity to a range of animals and possibly man. Questions have now been raised over the use of PBDEs, which leaves the materials/products industry short of obvious alternative products. Some alternatives are available, *e.g.* tetrabromobisphenol A, compounds based on phosphate or phosphate and chlorine and compounds of zirconium and aluminium. However, these cannot replace PBDEs for all applications so there remains the possibility that restrictions on the use of PBDEs would lead to an *increase* in the number of fires and hence human injury/death as a result. While it is recognized that brominated dioxins and furans may be emitted when treated textiles and fabrics are burned, the available evidence suggests that the quantities produced are low in comparison with total dioxin emission. There has been considerable discussion about the relative toxicity of dioxins and furans but, in general, governments tend to take a precautionary approach, reducing release and exposure wherever possible.

The need for, and nature of, any controls over PBDEs therefore require a careful balance of the risks posed by (a) the environmental hazards of PBDEs (particularly as a source of dioxins) and (b) the hazards presented by using less efficient flame retarding substances. In simple terms the lives saved by using PBDE fire retardants has to be weighed against the effects from exposure to PBDEs and/or dioxins. We currently do not have either the data (*e.g.* on toxicity or the hazards presented from alternative fire retardants) or indeed the techniques available to carry out such a risk assessment. Perhaps the most likely form of control will be voluntary agreements amongst manufacturers in introducing a range of risk reduction measures, *e.g.* during manufacture of the chemical and its products, particularly in minimizing the production of unwanted by-products, and controls over disposal. In fact, the Organization for Economic Cooperation and Development (OECD) has accepted (June 1995) a voluntary agreement of this nature put forward by the main producers.

1.6.2 Oestrogenic Chemicals (see also Chapter 19)

Chemicals which possess oestrogenic activity (*i.e.* are able to mimic the action of, or inhibit, hormones such as oestrogen) form an extremely wide and diverse group. Chemicals in this category include chlorinated pesticides such as DDT, detergents such as alkylphenol ethoxylates and their breakdown products, plasticizers such as phthalates, and dioxins. These chemicals, in fact, display very weak oestrogenic activity compared to the hormones themselves; however, there is growing concern in the medical/scientific community that this activity is giving rise to adverse effects on the reproductive health of humans and wildlife.[23-26]

There is evidence of a decrease in the quality of human sperm over the last 40 years and there is good evidence for an increase in rates of testicular cancer and other male reproductive abnormalities.[23,24] A wide range of evidence is also available linking exposure from chemicals with oestrogenic activity to effects on the reproduction of wildlife. This ranges from studies on alligators in a lake in Florida exposed to the organochlorine insecticide dicofol, studies on gulls in California having high tissue levels of DDT, to the effects of tributyltin on a range of molluscs.[23] In the UK, recent research has shown that male fish exposed to sewage effluents can undergo changes in their biochemistry and produce an egg yolk protein, vitellogenin, normally found only in fertile females.[26]

A wide range of industrial chemicals have been implicated as giving rise to, or at least potentially contributing to, these effects. Many of these chemicals are already banned or restricted (*e.g.* DDT, tributyltin, PCBs) whilst others (*e.g.* nonylphenol, various phthalates) are the subject of risk assessments *via* the EC Existing Chemicals Regulations. Particular attention has been paid recently to nonylphenol ethoxylates, largely because they are in widespread use as industrial cleaning agents or detergents. Annual UK consumption was estimated to be at about 18 000 tonnes in 1993.[27] Nonylphenol ethoxylates are the most common products from a group of similar chemicals (alkylphenol ethoxylates) manufactured from alkylphenols and ethylene oxide. They are invariably produced as a complex mixture of chemicals with a varying number of ethylene oxide units. This has led to difficulties with their analysis in environmental samples but methods are now available based on either HPLC or GLC coupled with various detectors, including mass spectrometry. The ethoxylates degrade to the parent phenol during waste water treatment and in the environment, and both chemicals can be detected in the environment. For example they have been detected in both sewage effluents and receiving waters at relatively high concentrations (approx. 40–400 μg l^{-1});[28] at present it is not clear whether such concentrations are within the oestrogenically active range. More research is required to provide evidence of a link between observed activity and measured water concentrations.

It is becoming increasingly likely that if the observed changes in human and animal reproduction are established to be caused primarily by environmental contamination, they are not due to a single chemical or exposure to a single source but to a range of chemicals and exposure routes. This clearly gives rise to problems for the control of chemicals having oestrogenic activity. What is certainly needed (and this is the subject of much research worldwide) is a 'test' for oestrogenic activity to assess the potential for damage to reproductive systems in both humans and animals. However, before a meaningful test can be developed and implemented more research needs to be carried out on the link between reproductive health and exposure to environmental oestrogens. In fact the recent 5th North Sea Conference (June 1995) called for internationally co-ordinated research on this subject.

The question arises as to what controls, if any, should be placed on chemicals with oestrogenic activity in the meantime. Environmental pressure groups argue that we cannot afford to wait for research and that chemicals shown to have oestrogenic activity should be phased out or banned now. This is already happening to a certain extent and most industrialized countries have, for example, endorsed an international agreement under PARCOM to phase out alkylphenol ethoxylates (APEs) in domestic and industrial cleaning uses by the year 2000. The Soap and Detergent Industries Association phased out the use of APEs in domestic cleaning products and plans to do the same for industrial cleaning products by the end of 2000.[29] However, the European chemical industries organic surfactants group (CESIO) claimed in 1998 that the cost of phasing out APEs would be close to £800 million with most of the costs falling on SMEs.

There will almost certainly be increasing pressure on both governments and industry to phase out suspect chemicals, and more widespread product and other controls may need to be implemented in the future as and when chemicals with oestrogenic activity are identified.

1.6.3 Pesticides

Pesticides are designed to control living organisms, thus they can be expected to have an environmental impact and many examples of effects can be cited. However, most of these were due to accidents or deliberate misuse and there is little evidence to suggest that when used properly pesticides cause any adverse effects in the aquatic environment. Recent studies in the UK and other countries have demonstrated, however, that low concentrations of a wide range of pesticides can be detected in the aquatic environment.

The EC Directive Relating to Water intended for Human Consumption (80/778/EEC) stipulated a maximum admissible concentration (MAC) of 0.1 μg l^{-1} for individual pesticides in drinking water, regardless of toxicity. Since some pesticides are not efficiently removed by conventional drinking water treatment, 0.1 μg l^{-1} is now one level of concern for pesticides in the aquatic environment in general. A further level of concern is that needed to protect the aquatic environment, although for most pesticides 0.1 μg l^{-1} tends to be the lower value.

Much more information is now available on pesticide concentrations in water than was the case a few years ago. Water companies regularly monitor drinking water for a range of pesticides (based on those used within the catchment concerned) and many companies also carry out routine monitoring of their raw water sources. The Environment Agency has a statutory duty to monitor pesticide concentrations in water, sediment and biota and is also required to undertake non-statutory monitoring of pesticides, tailored to local problems. It is estimated that the cost of this programme is in excess of £4 million/year.[30] In addition to regular monitoring, the Environment Agency has adopted a modelling system for the prediction of pesticide pollution in the environment (POPPIE). The model predicts ground and surface water quality concentrations and assesses leaching/run-off potential for new pesticides and is used to better focus monitoring programmes and identify potential problem pesticides.

Under the Control of Pollution Act (COPA) 1974, Water Authorities, as pollution control authorities, had powers to set up areas in order to protect water resources and these powers could be applied to pesticide usage. Under the provisions of the Water Act 1989, these powers were transferred to the National Rivers Authority and now to the Environment Agency. A variety of actions are possible under these powers; for example pesticide users could be advised, wherever possible, to reduce pesticide usage in specified, sensitive areas. This action would probably have little effect on the agricultural use of pesticides but might reduce non-agricultural use, for example by Local Authorities, British Rail, golf clubs, *etc*. More formal restrictions on pesticide use could be achieved by setting up water protection zones (WPZs), for example to protect water sources used for drinking water supplies. However, it might be difficult to identify such zones and the resource costs for policing and monitoring pesticide use in a WPZ would be high. Control of pesticide inputs to water bodies requires a co-ordinated approach. This has to involve liaison and co-operation between regulators and also educating pesticide users to adopt 'best practice' to minimize pesticide pollution of water.

It is impossible to assess, at the moment, how long it would take for any pollution control measures to reduce the concentration of pesticides currently in the aquatic environment, particularly in groundwaters. In the future, the registration requirements for new pesticides should prevent widespread contamination of the environment.

1.6.4   Mercury (see also Chapter 19)

Mercury is a metal, liquid at normal temperatures and pressures, which forms salts in two ionic states: mercury(I) and mercury(II). Mercuric salts [mercury(II)] are the more prevalent of the two. Mercury also forms a range of relatively stable organometallic compounds.

Natural degassing of the earth's crust releases considerable quantities of mercury into the environment and this is probably the major source. Estimates of the load vary between 25 000–125 000 tonnes,[31] although a more recent estimate was much lower, in the range 2700–6000 tonnes. In the same study, anthropogenic release was estimated at 630–2000 tonnes. World production of mercury, from mining and smelting operations, was estimated at around 10 000 tonnes in 1973, with industries such as chloralkali, electrical and paint production the largest users. The use of mercury is now declining although it was widespread as a cathode in the production of chlorine and caustic soda, in electrical appliances, control instruments and dental amalgams. Current uses of mercury are predominantly in the production of chlorine and in the extraction of gold from ore, a use which has been estimated to release 50–70 tonnes per annum to the environment.[32] Mercury is still also used in dental amalgams – estimated at around 10–20 tonnes annually.[33] Organomercury compounds have been used as fungicides, antiseptics, preservatives, electrodes and reagents.

Previous industrial use has led to the release of significant amounts to the environment, *e.g.* from the chloralkali industry, wood pulping and burning fossil fuels. Elemental mercury is virtually insoluble in water, while the solubility of mercury compounds varies widely. It is known that inorganic mercury can be methylated in the environment by bacterial action under aerobic and anaerobic conditions.[31] Methyl mercury is released from the bacteria, rapidly becomes bound to proteins in aquatic biota and enters the food chain.

Some sediments close to areas of production are heavily contaminated and pollution of water from old mines is still a problem. General environmental levels thus vary considerably depending on the location. Atmospheric levels of mercury range from 2–10 ng m^{-3}, while the concentration in rainwater can vary from 5–100 ng l^{-1}. A large number of data are available on mercury levels in the aquatic environment; background levels of around 1–15 ng l^{-1} are considered representative, but levels as high as 5 μg l^{-1} have been reported in contaminated areas. Fewer data are available on concentrations of methyl mercury and other organomercury compounds. The analysis of inorganic mercury is straightforward either by flameless atomic absorption spectrometry or atomic fluorescence spectrometry. Alkyl mercury compounds are normally determined by GLC.[31]

The metal is toxic at low concentrations to a wide range of organisms, including humans. The organic form of mercury can be particularly toxic, and the methyl and ethyl forms have been the cause of several major epidemics of poisoning in humans resulting from the ingestion of contaminated food, *e.g.* fish. Two major epidemics in Japan were caused by the release of methyl and other mercury compounds from an industrial site followed by accumulation of the chemicals in edible fish. These incidents gave rise to the effects of the poisoning becoming known as Minimata disease, following the name of the bay into which the discharges of mercury were released. The use of organomercury fungicides as seed dressing caused the death of many seed-eating birds in Europe, with many raptors also dying from eating the corpses.[34]

Mercury is heavily regulated, being covered by a wide range of legislation. In the EC it is designated as being of List I status under the Dangerous Substances Directive, with corresponding limit values set at a Community level, and the marketing and use of mercury and some of its compounds are covered under the 8th Amendment to the Marketing and Use Directive. In the UK, mercury and its compounds are on the Red List and classified as prescribed substances and are therefore subject to strict control with respect to industrial processes and discharge to sewer, *etc.* Nevertheless contamination of surface waters and other environmental compartments by mercury is still considered to be a problem, with attention being drawn recently to the accumulation of mercury in fish in acid upland waters. This highlights the great difficulty in controlling a chemical whose input to the environment is now largely from diffuse sources. By its very nature, as a relatively volatile metal, it can be released into the environment from many of man's activities not related directly to mercury use, *e.g.* burning fossil fuels, steel, cement and phosphate production and metal smelting. Atmospheric transport of elemental mercury (and dimethyl mercury) is thought to be a significant process and can give rise to environmental contamination far from its original source.

Future controls on mercury will therefore have to concentrate on the many relatively minor uses/sources of this chemical. For example, a considerable amount of mercury resides and is used in dental amalgams, control instruments (barometers, *etc.*) and laboratory apparatus (thermometers, *etc.*). An inventory of mercury-containing equipment followed by controlled withdrawal and disposal, and the use of mercury traps are measures not being adopted. In Germany and the Netherlands the installation of such traps by dentists is now required by law and under the Paris Commission (PARCOM) have been recommended for widespread introduction.

1.7 CONCLUSIONS

The difference between a 'priority' pollutant and a 'non-priority' pollutant, in terms of environmental impact, is not distinct. The distinction is also not without problems in relation to overall environmental improvement, because attention tends to be drawn to priority pollutants even when non-priority substances are causing the most damage. As indicated at the start of this chapter, any chemical can become a pollutant and a move to a risk-based approach to the assessment and control of chemicals seems an inevitable and desirable development. The relative risks between the beneficial effects of 'use' of a chemical and the potential for pollution are becoming less clear as the more obvious priority pollutants are brought under control. This is well illustrated by the brominated diphenyl ethers (Section 1.6.1) where the clear, undoubted benefits of fire retardancy have to be balanced against poorly characterized concerns about long-term toxicity. At the same time, our concerns about the potential effects of environmental contaminants are focusing on the longer term and, in particular, trying at an early stage to identify more subtle impacts, for example that represented by oestrogens (Section 1.6.2). Such effects are often difficult to detect but, nevertheless, are possibly more sinister than fish deaths or foaming rivers which provided the driving force for priority pollutant control a few decades ago.

1.8 REFERENCES

1. 'Water Pollution Incidents in England and Wales 1998', Environment Agency, 1999.
2. T. L. Tromp, M. K. W. Ko, J. M. Rodriquez and N. D. Sze, *Nature*, 1995, **376**, 327.
3. 'Water and the Environment', DoE Circular 7/89, HMSO, London.

4. 'The Surface Waters (Dangerous Substances) (Classification) Regulations 1997', S.I. No. 1997/2560, HMSO, London.
5. 'The Surface Waters (Dangerous Substances) (Classification) Regulations 1998', S.I. No. 1998/389, HMSO, London.
6. OSPAR Strategy with Regard to Hazardous Substances (1998-16).
7. 'The Environmental Protection (Prescribed Processes and Substances) Regulations', S.I. No. 472, HMSO, London, 1991.
8. 'Directive Establishing a Framework for Community Action in the Field of Water Policy', *Official J.*, L327, 22 December 2000.
9. 'Code of Practice for the Safe Use of Pesticides on Farms and Holdings', MAFF, London, 1990.
10. 'Planning Policy Guidance: Planning and Pollution Control', PPG23, HMSO, London, 1994.
11. 'Guidance Notes for Local Planning Authorities on the Methods of Protecting the Water Environment through Development Plans', NRA, 1994.
12. 'The Planning (Hazardous Substances) Regulations', S.I. No. 656, HMSO, London, 1992.
13. 'The Control of Industrial Major Accident Hazards Regulations', S.I. No. 1902, HMSO, London, 1984.
14. 'The Control of Industrial Major Accident Hazards (Amendment) Regulations', S.I. No. 2325, HMSO, London, 1990.
15. 'The Control of Industrial Major Accident Hazards (Amendment) Regulations', S.I. No. 118, HMSO, London, 1994.
16. 'The Water Protection Zone (River Dee Catchment) Designation Order 1999', S.I. No. 1999/915, HMSO, London.
17. 'The Protection of Water Against Agricultural Nitrate Pollution Regulations 1996', S.I. No. 1996/888, HMSO, London.
18. 'The Protection of Water Against Agricultural Nitrate Pollution (Scotland) Regulations 1996', S.I. No. 1996/1564, HMSO, London.
19. 'Policy and Practice for the Protection of Groundwater', Environment Agency, 1998.
20. Centre For Exploitation of Science and Technology, 1991.
21. 'Environmental Health Criteria, 162', International Programme on Chemical Safety, WHO, Geneva, 1994.
22. KEMI, The Swedish National Chemicals Inspectorate, Solna, Sweden, 1994.
23. 'Chemically Induced Alterations in Sexual and Functional Development: The Wildlife/Human Connection', eds. T. Colborn and C. Clement, Princeton Scientific Publishing, New Jersey, 1992.
24. M. J. Wilkinson, I. Milne, R. Mascarenhas and J. Fawell, 3/94, WRc, 1994.
25. Institute of Environmental Health, University of Leicester, 1995.
26. S. Jobling and J. P. Sampler, *Aquat. Toxicol.*, 1993, **27**, 361.
27. ENDS Report, 1993, No. 222, 9.
28. S. S. Talmage, 'Environmental and Human Safety of Major Surfactants', Lewis Publishers, Michigan, USA, 1994.
29. ENDS (1998) 'Phase-out for APEs Draws Closer', The ENDS Report No. 280, May 1998.
30. 'Pesticides in the Aquatic Environment', Environment Agency, January 1999.
31. 'Environmental Health Criteria 86', International Programme on Chemical Safety, WHO, Geneva, 1993.
32. W. C. Pfeiffer and L. Drude de Lacorda, *Environ. Health Lett.*, 1988, **9**, 4, 325.
33. M. Hutton and C. Symon, *Sci. Total Environ.*, 1986, **57**, 129.
34. U. Sellstrom, B. Jansson, A. Kierkegaard, C. de Witt, T. Odsjo and M. Olsson, *Chemosphere*, 1993, **26**, 1703.

CHAPTER 2

Chemistry and Pollution of the Marine Environment

M. R. PRESTON and R. CHESTER

2.1 INTRODUCTION

The World Ocean acts as a giant 'chemical system' operating within the global biogeochemical cycles that control both the movement and fate of material on the planet. Within these cycles, the importance of the World Ocean is heightened because it acts as a large-scale 'dumping ground' for material which originates in other geospheres.

Material reaches the oceans following a series of source ↔ sink pathways which, according to Chester,[1] may be considered to consist of a number of individual, but inter-related, stages. These are: Stage 1 (source), the initial release of material into the environment; (ii) Stage 2 (transport), the introduction of material to the ocean reservoir; (iii) Stage 3 (internal reactivity), the biogeochemical processes which operate within the ocean reservoir; and Stage 4 (sink), the removal of material from the ocean reservoir.

Much of the material involved in this overall source ↔ sink pathway has a natural origin, but in addition material generated in association with man's activities (anthropogenic origin) can have an important effect on oceanic cycles. However, it is necessary to distinguish between two different types of anthropogenic material.

Type I – naturally occurring materials, for which the 'anthropogenic effect' arises only when man's activities result in its excessive release into the environment (*e.g.* by mining, smelting, waste incineration, sewage disposal *etc.*).

Type II – non-naturally occurring material which has been created in the laboratory or industrial plant, and which is subsequently released into the environment.

Both types of anthropogenic material can impinge on marine biogeochemical cycles, and once they have been released into the environment they follow the

32

source ↔ sink pathway identified above. It follows, therefore, that a detailed understanding of the impact of anthropogenic material on marine systems can only be obtained from (i) a knowledge of the natural processes which occur following the release of material into the environment, and (ii) an awareness of the types of environmental stresses that are superimposed on the natural processes by the presence of the anthropogenic material itself.

The present chapter is intended to provide an overview of the most important features of the interactions between potential pollutants and the marine system, so that the relationships between natural processes and pollutant-driven environmental changes can be evaluated. In the treatment adopted, some general features of the source ↔ sink flux pathways in the World Ocean are identified. Following this, pollutants are divided into a number of categories, each of which is treated individually.

2.2 GENERAL FEATURES OF THE OCEANIC ENVIRONMENT

The World Ocean, including coastal and marginal seas, covers almost three-quarters of the Earth's surface ($361\,110 \times 10^3$ km^2), and has an average depth of ~ 3800 m.

Four principal parameters must be considered when attempts are made to describe the distribution of material in the oceans. These are: (i) source terms, by which material is delivered to the ocean reservoir; (ii) circulation patterns, which govern the transport of material within the reservoir; (iii) biogeochemical processes, which govern the reactivity of material in the reservoir; and (iv) sink terms, by which the material is removed from the reservoir. Some of the complexities of the system as they apply to coastal regions are shown in Figure 2.1.

2.2.1 Source Terms

The ocean reservoir is continuously subjected to series of material fluxes which are delivered *via* a number of transport pathways. With respect to the input of pollutant material river run-off, atmospheric deposition and a number of anthropogenic pathways are the dominant natural input mechanisms.

Points of impact

(i) River run-off delivers both particulate and dissolved material to the surface ocean at the land/sea margins with discharge usually being *via* estuaries; *i.e.* regions where fresh and saline waters mix. Estuaries are regions of intense physical, chemical and biological activity and the river flux undergoes considerable modification as it passes through them.

(ii) Atmospheric deposition delivers both particulate material ('dry' deposition) and a combination of dissolved and particulate material ('wet' deposition, or precipitation scavenging) to the whole ocean surface; *i.e.* atmospheric inputs are not confined to the land/sea margins, and so need not pass through the estuarine environment. However, the strength of

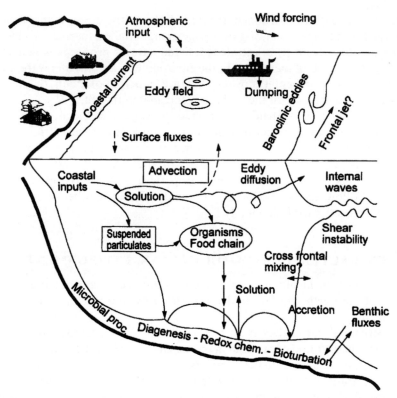

Figure 2.1 *Some of the processes determining pollutant behaviour in the coastal zone*[2]

the atmospheric signal is strongest in coastal regions closest to the continental material sources.

(iii) Anthropogenic pathways by which pollutants are delivered to the oceans include: (1) dumping (*e.g.* for the disposal of sewage sludge, radioactive waste, dredge spoil, military hardware, off-shore structures), (2) deliberate coastal discharges (*e.g.* coastal pipeline discharges from power stations, industrial plants, oil refineries, radioactive reprocessing plants, and (3) off-shore operational discharges from tankers or other ships.

2.2.2 Circulation Patterns

Once material is delivered to the oceans it is subjected to transport *via* the marine circulation systems. On a global-scale circulation in the surface ocean is mainly wind-driven, whereas in the deep-ocean it is gravity-driven. However, the strengths of both the river run-off and the atmospheric deposition signals bringing pollutant material to the oceans are strongest in coastal and marginal seas, and here the water circulation patterns are constrained by local influences.

2.2.3 Sea Water Reactivity: Biogeochemical Processes

Sea water is not simply a static reservoir in which the material supplied to it has accumulated over geological time. Thus, rather than being thought of as an 'accumulator', sea water should be regarded as a 'reactor' from which material is continually being removed on time scales (residence times) which vary considerably from one element to another.

The driving force behind the removal of elements from sea water to their marine sink, mainly sediments, is particulate ↔ dissolved reactivity. Vertical water column profiles of elements can provide data on the type of particulate ↔ dissolved reactivity that affects them in the oceans, and on this basis the elements in sea water can be divided into three principal types.

(i)   Conservative-type unreactive major elements. These exhibit largely invariant concentration profiles down the water column, and have relatively large oceanic residence times (usually $> \sim 10^6$ y). Conservative-type elements are mainly the sea salt-forming elements, such as sodium, potassium and chlorine.

(ii) Scavenging-type reactive trace elements. These are usually trace elements which are involved in passive particle scavenging throughout the water column and are not recycled, which results in a surface enrichment–subsurface depletion down-column concentration profile. Scavenging-type trace elements have oceanic residence times which can be as low as a few hundred years, and include Al, Mn and Pb.

(iii) Nutrient-type reactive trace elements. These are involved in active biological removal mechanisms. They exhibit a surface depletion–subsurface enrichment, and as a result of their involvement with biota in the major oceanic biological cycles they are involved in a major recycling stage similar to that exhibited by nutrients. Nutrient-type elements have residence times in the order of thousands of years.

Pollutants delivered to the oceans enter the circulation-driven transport and biogeochemically-driven removal processes, and are finally deposited in the sediment sink. Thus, pollution can affect the water, the biota and the sediment compartments of the ocean reservoir, and tends to have its greatest impact in coastal and marginal seas.

A wide range of pollutant, and potentially pollutant, substances can affect the marine environment and, thus, the individual pollutants described in this chapter have been selected to cover examples of a variety of substances having different effects on the marine environment. The pollutants selected in this way are oil, sewage/nutrients, persistent organic compounds, trace metals and artificial radio-nuclides. Each type of pollutant is treated individually below.

2.3   SOURCES, MOVEMENTS AND BEHAVIOUR OF INDIVIDUAL POLLUTANTS OR CLASSES OF POLLUTANT

2.3.1   Oil

The world demand for oil is enormous, at a level of $> 3200 \times 10^6$ t a^{-1}, which represents around 39% of the world's commercial energy demands.[3] Production of this oil is concentrated in the Middle East, which accounts for 28.6% of world production and 66% of proven reserves. Much of this oil is moved from its point of production to its purchasers in tankers which are frequently of very considerable size. Over one third of the present tanker fleet (between 400 and 460 vessels) has a capacity $> 250\,000$ t and over 60% has a capacity $> 100\,000$ t. These tankers are too large to follow the older trade routes (*e.g.* through the Suez Canal) and so, in the case of tankers heading towards western Europe, follow the African coast from the Middle East, around the Cape of Good Hope up through the Atlantic. In most cases the tankers follow a minimum depth contour (*e.g.* 200 m) which frequently takes them relatively close to coastlines. It is therefore nor surprising that most tanker accidents result in widespread coastal pollution which subsequently receives a great deal of public attention.

2.3.1.1   The Composition of Crude Oil.   Crude oil is a very complex mixture of many different chemicals. Consequently the effects of an oil spill on the marine environment depend on the exact nature and quantity of the oil spilled, as well as such other factors as the prevailing weather conditions and the ecological characteristics of the affected region.[4] An indication of the physical-chemical properties of the major components is shown in Table 2.1.

2.3.1.2   Fluxes of Oil to the Marine Environment.   The fact that there are many direct and indirect ways in which oil can reach the marine environment means that it is very difficult to construct reliable estimates of the total amounts of oil reaching the seas each year. Recent estimates range from between 1.7 and 8.8×10^6 t a^{-1}, with a best estimate of $\sim 3.25 \times 10^6$ t a^{-1}. However, recent improvements in tanker operations have reportedly reduced contributions from

Table 2.1   *Typical physical-chemical properties for hydrocarbon groups*

Hydrocarbon	Molecular weight (approx.)	Aqueous Solubility (g m^{-3})	Vapour pressure (Pa)	Density (kg m^{-3})	Oil–water partition coeff.
Lower alkanes (C$_3$–C$_7$)	72	40	70 000	800	20 000
Higher alkanes (> C$_8$)	120	0.8	2000	800	1 000 000
Benzenes	100	200	1500	800	4000
Naphthenes	160	20	5	800	40 000
Higher polycyclics	200	0.1	0.003	800	8 000 000
Residues	–	0	0	800	∞

Adapted from Doerffer.[5]

Table 2.2 *Sources of oil to the marine environment*[7]

Source	Estimate (m t a^{-1})	Total (m t a^{-1})
Transportation		
Tanker operations	0.7	
Tanker accidents	0.42	
Bilge and fuel oil	0.3	
Dry docking	0.03	
Non-tanker accidents	0.02	
	Sub-total	1.47
[*Revised 1991 total*][6]		[0.568]
Fixed installations		
Coastal refineries	0.1	
Offshore production	0.05	
Marine terminals	0.02	
	Sub-total	0.17
Other sources		
Municipal wastes	0.70	
Industrial waste	0.20	
Urban run-off	0.12	
River run-off	0.04	
Atmospheric fallout of oil-derived compounds	0.30	
Ocean dumping	0.02	
	Sub-total	1.38
Natural inputs	0.25	0.25
Total		3.27
		[2.368]
*Natural sources/fluxes of organic carbon**		
Primary production by marine phytoplankton	36 000	
Atmospheric fallout of naturally derived species	220	

* From Chester.[1]

this source from around 1.47×10^6 t a^{-1} in 1981 to 0.0568×10^6 t a^{-1} in 1989.[6] The contributions from different sources are given in Table 2.2.

It is important to note that there are considerable natural sources of both crude oil and other hydrocarbons to the marine environment. Natural oil seeps are known to have existed for thousands of years, particularly in regions such as the Persian Gulf or Gulf of Mexico. This has had the result that a variety of marine organisms (particularly bacteria) have evolved which can feed on and degrade crude oil.[8] Oil is therefore a Type I pollutant, unlike the chlorinated pesticides or polychlorinated biphenyls (PCBs) which are both totally synthetic and recent in origin and where degradation in the environment is not greatly assisted by pre-adapted organisms (Type II).

2.3.1.3 The Behaviour and Fate of Spilled Oil. Once crude oil is released onto the sea surface a number of different processes immediately begin to act on it

Table 2.3 The effects of weathering processes on an oil slick

Process	Rate	Effects	Controlling factors
Evaporation	Rapid	Major influence in first 24–48 h removes volatiles (e.g. < C_4 n-alkane) and reduces acute toxicity	Oil composition, wind speed and temperature, water temperature and roughness
Dissolution	Fairly rapid	Removes more polar components from slick producing high sub-slick concentrations	Oil composition, wind speed/water roughness, water temperature
Dispersion	Variable though more rapid in early stages	Formation of oil in water emulsions. Dispersion enhances degradation rates	Oil composition, wind speed/water roughness
Emulsification	More rapid in early stages	Formation of water in oil emulsions ('chocolate mousse'). Oil in this state is untreatable by chemical techniques	Oil composition, wind speed/water roughness
Photochemical oxidation	Dependent on light	Oxidation produces aliphatic/aromatic acids, alcohols, ethers, dialkyl peroxides which are more soluble (and possibly more toxic) than parent molecules	Oil composition, light intensity
Adsorption	Dependent on suspended particle loading	Removal of oil from surface slick to underlying sediment, potential impact on benthos	Dependent on suspended particle loading
Biodegradation	Initially fairly rapid then more slowly	Rapid removal of n-alkanes followed by other susceptible species	Oil composition, local bacterial/fungal populations, degree of oil dispersion, dissolved oxygen content and availability of nutrients

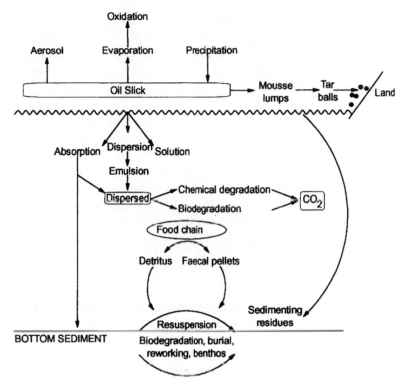

Figure 2.2 *The fate of a marine oil slick*

which influence its composition and environmental toxicity. These include evaporation, dissolution and advection, dispersion, photochemical oxidation, emulsification, adsorption onto suspended particulate material, biodegradation and sedimentation (Figure 2.2) each of which has different effects on the oil (Table 2.3). In addition, the action of surface waves and currents drive the oil slick away from its point of release.

2.3.1.4   The Environmental Impact of Marine Oil Spills.   The most obvious problems of oil pollution are those associated with the aftermath of major events. These have included tanker accidents such as the Amoco Cadiz wreck off the Brittany coast in 1978 and, more recently, the Exxon Valdez wreck in Prince William Sound, Alaska 1989; the IXTOC I blowout in the Bay of Campeche in 1979, and the major releases in the Persian Gulf in the 1980s and early 1991 associated with the eight year Iran–Iraq war and the subsequent war in Kuwait. These high profile events clearly demonstrate the major features of oil pollution damage. The general effects of oil spills on marine organisms are shown in Table 2.4.

The damage to coastlines is very much a function of the coastal ecosystem as well as the prevailing weather conditions. Various indices of coastal variability have been formulated with high energy coastlines such as exposed rocky

Table 2.4   *The types of damage caused to marine organisms by oil spills*

Affected organisms	Nature of damage
Plankton	Minor local damage, possible growth inhibition of phytoplankton by shading effect of slick
Seaweeds	Major damage to slick-affected inter-tidal species. Recovery rapid but removal of grazers may cause excessive growth in future years
Invertebrates	Large scale mortality in littoral communities through acute toxicity and smothering. Recovery of populations may take years
Fish	Normally only minor casualties but pollution of spawning grounds/ migration routes can cause greater damage
Seabirds	Diving birds badly affected by oil. Death through drowning, hypothermia or toxic effects of ingested oil
Marine mammals	Rarely affected but coastal populations (*e.g.* seal colonies) are vulnerable

headlands, or eroding wave-cut platforms, being amongst the least vulnerable; with sheltered tidal flats, salt marshes, mangrove stands and coral reefs being most sensitive.[8,9]

Experience has shown that in some circumstances attempts to clean-up oil spills have caused more damage than if the oil had been left entirely alone; this was particularly the case in the Torrey Canyon incident.[10] Various guidelines have therefore been produced which indicate how to select the most appropriate treatment strategy from the various options available. Particularly clear examples of these include two field guides published by CONCAWE[9,11] on inland and coastal oil spill control and clean-up techniques.

2.3.1.5   Control and Clean-up Techniques.   Prevention.   Nearly all marine pollution incidents involving oil pollution are avoidable. Routine oil discharges from tank washing, and accidents involving tankers, have historically been amongst the most important sources of oil to the marine environment and have led to the development of a number of strategies designed to reduce the levels or risks of pollution. For the most part, pollution prevention has received the most attention with techniques such as the Load on Top (LOT), Crude Oil Washing (COW) and Segregated or Clean Ballast Tanks (SBT or CBT) being introduced. These techniques are now widely used, and when combined with the oil reception facilities required under international conventions, such as Annex 1 of the International Convention for the Prevention of Pollution from Ships,[12] have dramatically reduced routine oil discharges to the oceans. More recent requirements for double hull construction to be incorporated into all new tankers have been contentious both on grounds of cost and effectiveness, and it remains to be seen how great the benefits from this strategy will be. In principle the double hulls should allow for better protection against spillage in the event of either grounding or collision.

Table 2.5 *A summary of oil spill treatment techniques*

Treatment process	Advantages/disadvantages
Chemical	Modern treatments effective on appropriate oil types. Dispersion enhances degradation. Only applicable to fresh oil, requires good logistical support, may enhance local toxicity whilst protecting coastlines
Physical (containment)	Effective for small spills in enclosed or calm waters. Poor efficiency in rough weather
Physical (recovery)	Effective for small spills in enclosed or calm waters. Poor efficiency in rough weather. Significant disposal problems with recovered oil
Adsorption–sinking	Removes oil to sediment. Trades short term protection for long term benthic contamination
Burning	May work on some fresh oils but not generally used. Leaves tarry residues

2.3.1.6   Oil Spill Treatment Technologies.   Oil spill treatment technologies can be divided into a number of main types: chemical (including chemical enhancement of microbial processes), physical containment, and recovery systems, adsorption and burning (Table 2.5). Useful reviews of this subject have been provided by ITOPF[13] and Doerffer.[5]

2.3.2   Sewage

Domestic sewage, with or without the presence of industrial wastes, probably represents the commonest and most widespread contaminant of inshore and nearshore waters (see Chapter 5 for details of the nature of sewage and sewage treatment techniques). Sewage poses aesthetic and health risks to human populations and also acts as a vector whereby a considerable variety of other contaminants reach the marine environment. Sewage may reach the seas in a number of different forms; ranging from untreated raw sewage discharges, through various degrees of treated discharge, to the dumping of associated sewage sludge at marine sites. The potential problems associated with sewage discharge are significant. Over 10^6 tonnes of dry sewage sludge are produced in the UK alone each year,[14] and there is no doubt that some disposal strategies are working at or close to their practical limits. Until recently, a significant amount of both untreated and treated sewage and sewage sludge reached coastal waters; for example, over 5×10^6 wet t a^{-1} (equivalent to approximately 5×10^5 dry t a^{-1}).[15]

2.3.2.1   Problems Associated with BOD.   Whether a high BOD (biochemical oxygen demand) waste discharge to a natural water causes a serious environmental problem depends almost entirely on the characteristics of the receiving system. In essence, if the input of BOD is greater than the ability of the receiving water to supply new oxygen then there will be major problems of oxygen depletion and, in extreme cases, total anoxia. If the BOD input and the new oxygen supply are similar in magnitude then some oxygen depletion may be seen, and only if the renewal of oxygen is much greater than the supply of BOD will the discharge be innocuous (at least in this respect). Overall, therefore, the major problems arise from the discharge of untreated sewage to rivers, estuaries or enclosed coastal waters

One of the UK rivers most affected by high BOD discharges is the River Mersey[16] which has a legacy not only of high, direct BOD discharges to the estuarine system but also to the catchment area. This has led to significant oxygen depletion in the upper estuary, particularly on spring tides where the higher tidal energy causes sediment with a high BOD to be resuspended in the water column thus stripping dissolved oxygen from the overlying water (see Figure 2.3). Major effluent treatment schemes introduced to the region over the last thirty years have reduced the BOD coming into the estuary from about $42 \, mg \, l^{-1}$ to about $4 \, mg \, l^{-1}$. In addition, within the estuary itself new treatment schemes have been constructed with a view to intercepting the large number of direct, raw sewage outfalls constructed in Victorian times, and directing the effluent to a treatment plant. This treatment has been very successful, with regular fishing competitions now being held on a river that in the 1970s was largely devoid of life.

The effects of severe oxygen depletion on organisms within the estuary are considerable. Many benthic animals, estuarine fish species, and those migratory

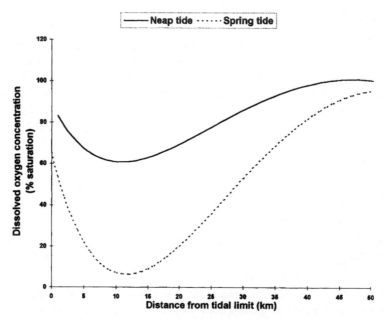

Figure 2.3 *Dissolved oxygen in the River Mersey Estuary (typical profiles before the introduction of modern sewage treatment systems)*

species which pass from land to sea or *vice versa* as part of their natural life cycle, are effectively barred from transit through anoxic waters. Life in underlying sediments is also considerably restricted. Only when oxygen levels can be maintained at a consistently high level over all tidal states can the full, natural range of organisms be sustained.

Problems of oxygen depletion associated with sewage discharge are quite common, and may extend beyond the river and estuarine environment into coastal waters. Oxygen depletion has been reported in areas of the New York Bight,[17] the German Bight, southern and eastern Kattegat and eastern Skagerrak as well as several Norwegian fjords.[15] Whilst the hydrographic conditions prevailing in these regions may sometimes lead to low oxygen levels as a result of natural processes, there can be little doubt that organic matter from sewage, agriculture, industry and sundry diffuse sources has also had a significant impact.

2.3.2.2 Sewage and Nutrients. Sewage effluents contain large quantities of micronutrient elements such as nitrogen and phosphorus. The greater the degree of treatment, the greater the mineralization of the nutrients so that, for example, a treated waste will contain primarily nitrate and an untreated one ammonia and organic nitrogen species. Nutrient inputs to marine systems have been a matter for considerable concern because of the potential eutrophication effects that they may cause. A simple view of eutrophication is that additional nutrients lead to the formation of excess biomass which, in turn, leads to an increase in BOD with subsequent oxygen depletion effects as the biomass decays.

It is not only an increase in nutrient concentrations that can cause deleterious effects. Anthropogenically induced alterations in nutrient ratios can also cause changes in prevailing phytoplankton species because of their differing physiological properties and requirements for nutrients. As a general rule marine phytoplankton assimilate nitrogen and phosphorus in a ratio of about 16:1 and under normal conditions in marine systems nitrogen is the growth limiting element. However, efforts to reduce nutrient inputs over recent years have primarily focused on phosphorus, with the result that nitrogen to phosphorus ratios have tended to increase. As a consequence, phosphorus has then become the growth limiting element in some regions (*e.g.* parts of the North Sea), so that phosphorus supplies are effectively exhausted during the spring/early summer phytoplankton bloom leaving a considerable excess of nitrogen unconsumed.[15] This excess nitrogen may be advected to other regions where it raises the total nitrogen levels, alters the prevailing N:P ratio, and has been implicated not only in changing dominant plankton species but also in stimulating toxin production in some species.

Examples of the problems associated with excessive nutrient enrichment have been seen in a number of regions in recent years. For example, a major bloom of the phytoplankton *Phaeocystis* in the upper Adriatic in the summer of 1990 produced large quantities of a scum-like organic material which washed up on beaches. There it had the appearance of sewage waste, had a very unpleasant smell as it decomposed, attracted flies and, not surprisingly, had a devastating effect on the local tourist industry. In 1988, similar problems with excessive plankton blooms caused major damage to coastal mariculture units on the Swedish and Norwegian coasts.[15]

2.3.2.3   Marine Disposal of Sewage Sludge.   Disposal of sewage sludge in an environmentally acceptable way has increasingly been recognized as a problem in recent years. In major societies the quantities involved are considerable and practical, and/or social, pressures have resulted in the marine disposal option being used by some countries, notably the UK which annually dumped over 5 million tonnes of sewage sludge in the North Sea alone (27% of current production). The UK ceased all sewage sludge dumping at the end of 1998 (a decision taken at the 1990 North Sea Conference) although public resistance to the construction of alternative, incinerator based options has caused disquiet in some areas.

Sewage sludge dumping takes place at designated sites, which are defined areas rather than a single point. The sludge particles gradually settle through the water column, where they may act as a food source to mid-water feeding fish species, finally settling on the sediment surface. The problems arising from this practice are essentially two-fold. First, the sewage acts as a vector for the transfer of contaminants to the biota (notably trace metals and persistent organic chemicals). Second, the high organic input to the sediments cause changes in both species diversity (decreases) and biomass (increases) in the benthic community.

The extent of the changes induced depend to a considerable extent on the hydrodynamics of the receiving site. There is essentially a choice between a 'dispersive' site, where strong current systems disperse the sludge widely at low concentration, and an 'accumulating' site, which is hydrodynamically quiet and where there will be higher sludge concentrations over a limited area. With the exception of the Garroch Head site which received sludge from the Glasgow region, all UK sites are relatively dispersive in character.

2.3.2.4   Marine Sewage Disposal and Public Health.   Sewage debris on beaches and in coastal waters is clearly an aesthetic problem, but it also poses a health risk. There is not only a direct risk of infection by sewage derived pathogenic organisms, but there are also considerable risks associated with the consumption of contaminated and improperly prepared sea food. Some examples of such infections are given in Table 2.6.

It should be noted that the lifetimes of most terrestrially derived bacteria in sea water are relatively short; typically of the order of 12–24 hours. This limited lifetime is due to the natural antibiotic properties of sea water which derive from a combination of the high salt concentration, low concentrations of inorganic and organic chemicals with antibiotic properties and exposure to natural UV radiation in surface waters. However, not all bacteria or viruses necessarily die. Some bacteria may, for example, enter a dormant phase (non-platable bacteria) which does not show up in routine test procedures, but which can become active again if ingested by bathers.

Table 2.6　*Sewage derived pathogenic organisms and their effects on humans*[18]

Aetiological agent	Mode of transmission to humans*	Diseases/symptoms
Salmonella typhi	Fish or shellfish	Typhoid
S. paratyphi	Fish or shellfish	Paratyphoid
S. typhimurium	Fish or shellfish	Salmonellosis; gastroenteritis
S. enteritidis	Fish or shellfish	Salmonellosis; gastroenteritis
Vibrio parahaemolticus	Fish or shellfish	Diarrhoea, abdominal pain
Clostridium botulinum	Fish or shellfish	Botulism (high fatality rate)
Staphylococcus aureus	Fish or shellfish	Staphylococcal intoxication, nausea, vomiting abdominal pain, prostration
Clostridium perfringens	Fish or shellfish	Diarrhoea, abdominal pain
Erysipelothrix insidiosa	Skin lesions	Erysipeloid – severe wound inflammation
Hepatitis virus	Shellfish	Infectious hepatitis
Heterophyes heterophyes	Fish or shellfish	Heterophyiasis: abdominal pain, mucous diarrhoea (eggs may be carried to brain, heart etc.)
Paragonismus westermani (P. ringeri)	Crabs, crayfish or contaminated water	Flukes in lungs and other organs
Anisakis matina	Marine fish (notably herring)	Anisakiiasis: eosinophilic enteritis
Angiostrongylus cantonensis	Shrimp or crabs	Eosinophilic meningitis

* Note that in most cases contamination of the organism is not transmitted to humans unless the food has been inadequately stored or cooked.

2.3.2.5　The European Bathing Water and Seafood Directives.　　Within the European Union sea water quality standards as they affect human health are set by the European Bathing Water Directive (EEC/76/160). This directive requires certain microbial quality standards to be met for beaches designated as bathing beaches. The mandatory limit for faecal coliforms under this directive is 2000 per 100 ml.

The Seafood Directive (EEC/91/492), which became operative in January 1993, sets limits for the number of faecal coliforms, *Salmonella*, and toxins such as those which cause Paralytic Shellfish Poisoning (PSP) or Diarrhetic Shellfish Poisoning (DSP). These toxins are produced by dinoflagellates, such as *Alexandrium* and *Protogonyaulax*, and are accumulated in shellfish, consumption of which is capable of causing human deaths. The relationship between blooms of these dinoflagellates and pollution is not clear because such events do not take place exclusively in waters recognized as being polluted; however, such a link cannot be entirely discounted.[15]

2.3.3　Persistent Organic Compounds (see also Chapter 17)

The European Commission listing of potential environmental pollutants is dominated by organic chemicals (see Chapter 1). The criteria for inclusion of a chemical in that, and other priority pollutant lists, are that they have relatively high environmental toxicity (to humans and other organisms), they are persistent and liable to undergo significant biomagnification and they are produced in sufficient quantities to represent a potential threat.

Of the huge number of chemicals that might be considered to be potential pollutants only comparatively few have been studied in any detail in marine systems and many of these are halogenated species.

2.3.3.1 Halogenated Compounds. Chlorinated pesticides (see also Chapter 17). Dominant amongst the older formulations were the organochlorine pesticides such as DDT, dieldrin, aldrin and endrin (the 'drins'), lindane (γ-hexachlorocyclohexane), hexachlorobenzene (HCB) and toxaphene. Despite widespread bans on many persistent organochlorine pesticides in the industrialized world, usage on a global scale is still important because the cheapness and effectiveness of the chemicals makes them attractive options for developing countries.

Other persistent organic chemicals which have proved to be of some concern in marine systems include polychlorinated dibenzo-*p*-dioxins (PCDD) and polychlorinated dibenzofurans (PCDF products of PCB combustion; some 2378 PCDD/F isomers exist); including 2,4-D (2,4-dichlorophenoxyacetic acid), 2,4,5-T (2,4,5-trichlorophenoxyacetic acid), and MCPA (2-methyl-4,6,-dichlorophenoxyacetic acid).[19] 2,4-D and 2,4,5-T became notorious through their use as defoliants in the Vietnam War (Agent Orange) and because of the contamination of the product with the dioxin 2,3,7,8-TCDD (2,3,7,8-tetrachlorodibenzo-*p*-dioxin). The 'dioxins' and the PCDF have become widespread contaminants of marine systems, although their links with any deleterious effects in the oceans remain slight.

PCBs (see also Chapter 17). PCBs are widely distributed amongst all marine systems. Their toxicity to marine organisms varies considerably, but the co-planar PCBs exhibit the greatest mammalian toxicity. However, such acute toxicity is unknown in marine organisms which are more likely to suffer sub-lethal effects from the biomagnification of PCBs through the food web (see Table 2.7).

The Origins of the Environmental Hazards. The environmental problems associated with the chlorinated pesticides and PCBs derive from similarities in their physical, chemical and biological properties. These may be summarized as persistence, widespread distribution amongst most environmental compartments, have a propensity to undergo biomagnification and high toxicity (including non-lethal effects; note, however, that DDT has a low mammalian toxicity). In other words, nearly all of the properties that make a pollutant a high risk.

The organochlorines are relatively water insoluble and involatile. They also have high octanol–water partition coefficients (*e.g.* log K_{ow} = 4.46–8.18 for different PCB congeners) and are therefore lipophilic. The low apparent

Table 2.7 *The distribution of PCBs between different environmental compartments*[20]

Environment	PCB load (t)	Percentage of PCB load
Terrestrial and coastal		
Air	500	0.13
River and lake water	3500	0.94
Sea water	2400	0.64
Soil	2400	0.64
Sediment	130 000	34.73
Biota	4300	1.15
Total (A)	143 100	38.23
Open ocean		
Air	790	0.21
Sea water	230 000	61.45
Sediment	100	0.03
Biota	270	0.07
Total (B)	231 000	61.77
Total load (A + B)	374 000	100

volatility is, however, misleading. Under certain conditions (notably high temperatures and humidity) evaporative losses of applied pesticides can be high, and this accounts for their widespread occurrence through subsequent transport as vapour or condensates on atmospheric aerosol particles (Table 2.8).

The lipophilic nature of the organochlorines leads not only to their rapid uptake and storage in fatty tissues, but also to their slow elimination because the fat reserves are only called upon at stages in the life cycle when energy demands are high or food supplies low. It is this feature which leads to their biomagnification. The extent of biomagnification is large. Concentrations of organochlorines in sea water are typically of the order of $pg\,l^{-1}$ to low $ng\,l^{-1}$, rising to 10s of $ng\,g^{-1}$ in marine invertebrates, low $mg\,g^{-1}$ values in mussels and up to 10–100s of $mg\,l^{-1}$ in fatty tissue of top predators such as seals, pelicans and terrestrial hawk species.

Table 2.8 *Estimated fluxes of PCBs to the ocean surface*[21]

	Flux ($ng\,m^{-2}\,a^{-1}$)		
	Arochlor 1242	Arochlor 1254	Total
Particles			
Dry	16	41	57
Wet	250	650	900
Gas phase	2209	1709.5	3919
Total	2475	2400.5	4876
Total flux to oceans ($\times 10^6\,g\,a^{-1}$)	8.6	8.3	16.9

The Effects of Marine Organochlorine Pollution.   At the planktonic level primary production rates are reduced at DDT or PCB concentrations above ~ 1 mg l^{-1}. This is a low concentration but still very much higher than those likely to be found in sea water. Amongst marine invertebrates and fish 96 h LC$_{50}$ values generally fall within the range 1–100 mg l^{-1}. However, bivalve molluscs are very resistant with 96 h LC$_{50}$ values $> 10\,000$ mg l^{-1}.

Amongst the top predators the main deleterious effects have been eggshell thinning in birds (DDE) and interference with the reproductive and immune system in mammals (PCBs). A considerable number of top predator birds showed major declines in abundance during the period when DDT usage was at its greatest and since the banning of these chemicals in the industrialized world these populations have mostly demonstrated a major recovery. Where populations have not recovered other factors, such as decline in suitable habitats, are probably responsible.

A number of seal populations, most notably in the Dutch Wadden Sea and the Baltic Sea, have exhibited reproductive abnormalities attributed to PCBs which have had a significant impact on populations. The main symptoms of PCB poisoning are changes in the uterus, implantation or abortion/premature pupping. There have also been suggestions that PCBs may cause carcinogenic, teratogenic and immunological effects. An outbreak of phocine distemper virus in the common seal populations in much of the North Sea in 1988 has sometimes been attributed to depression of immune systems by PCBs, but the evidential basis for this is weak and such epidemics have also been reported in seal populations in other, less contaminated, regions.

2.3.3.2 Other Persistent Organic Compounds.   Organochlorines are by no means the only organic compounds which are of concern as marine contaminants. There are many other chemicals which are sufficiently common, toxic and persistent to represent potential threats. These include, for example other pesticides, polycyclic aromatic compounds (PAH), plasticizers (*e.g.* phthalate esters), detergent residues, organic solvents *etc.* It is not possible within this chapter to review the behaviour and effects of all of these compounds. However, one aspect of organic pollution which has become particularly important recently is the relationship between the reported number of abnormalities in male sex development in wildlife and humans coinciding with the introduction of so-called 'oestrogenic' or 'endocrine disrupting' chemicals.[22–24] Compounds apparently responsible for this effect include several members of the DDT family of compounds, chlordecone (Kepone), various sterols and possibly nonyl phenol. Of these, *p,p'*-DDE is the most potent. Research in this area is only just beginning, but the consequences of widespread disruption of fertility by organic contaminants is clearly a very serious one. See *e.g.* IEH,[25] Hester and Harrison[26] and Vos *et al.*[27] for recent reviews of this subject, and also Chapters 1, 17 and 19 of this volume.

2.3.4 Trace Metals

A large number of trace metals are transported to the oceans from natural sources. However, these natural sources are supplemented by releases from anthropogenic processes which, for some metals, can exceed natural inputs.

Trace metals are found in the water, biota and sediment compartments of the marine system, but potentially the most hazardous environmental effects to human health arise when they enter the food chain. The relationship between the total concentration of a trace metal in the environment and its ability to cause toxic effects in organisms is complex, and two important constraints must be considered; *i.e.* the speciation of the metals, and the condition of the organisms.

(i) *Metal speciation.* All organisms have a requirement for certain trace metals which must be present in their diet (or growth medium) to sustain healthy development. Such 'essential' metals include iron, copper, vanadium, cobalt and zinc. Other, non-essential, metals exert neither beneficial nor deleterious effects if present at sufficiently low concentrations, but become increasingly harmful as the concentrations increase. However, it is the speciation of a trace metal, *i.e.* the way in which it is partitioned between host associations in the water and particulate phase, and not its total concentration which constrains its effect on the environment. For example, not all forms of a trace metal are 'bioavailable', and numerous incidents have demonstrated that it is the organic forms of metals which tend to be of the most damaging to marine organisms.

(ii) *Condition of the organisms.* A number of factors influence the effect that potentially toxic substances have on organisms. These include the following. (a) The stage of development, *e.g.* egg, larva, adult. (b) Size and sex. (c) History, *e.g.* previous exposure to toxic substance. (d) Location, which is very relevant for intertidal organisms. (e) Diet. (f) General environmental conditions, such as temperature, salinity, pH, Eh, light intensity, dissolved oxygen, that might affect metal/organism interactions.

The hazardous effects of trace metals on the oceanic biogeochemical system are illustrated below with respect to mercury, lead and tin; all of which have been shown to be capable of generating stress in the marine environment.

2.3.4.1 Mercury (see also Chapters 1 and 19). Mercury is on the 'Black List' of chemicals, and probably represents the best known example of trace metal pollution in the marine environment through its role as the causative agent of Minimata disease. This was an extreme example, however, and mercury contamination of coastal waters affected by industrial effluents is not an uncommon problem. For a full discussion of mercury in the environment the reader is referred to Nriagu.[28]

Mercury has a very wide range of industrial uses and the global production of mercury for industrial purposes is ~ 6000 t a^{-1}, which has increased threefold since 1900.[29]

The marine cycle of mercury has been reviewed by Fitzgerald.[30] It is thought that there are three potentially important chemical forms of dissolved mercury in natural waters. These are elemental mercury Hg(0), divalent mercury (Hg^{2+}), and methylmercury (CH_3); although ethylmercury can also occur.

Four factors strongly influence the environmental effects of mercury. (i) Mercury has a particularly high affinity for organic species, which results in its accumulation in marine biota. (ii) Inorganic mercury can undergo bio-mediated transformation into 'alkylated' forms (methyl and dimethylmercury) which are particularly toxic species of the element. In the environment methylmercury compounds can be synthesized by a variety of microorganisms. This occurs almost entirely under aerobic conditions in the marine environment (see *e.g.* Landner[31]). The principal form of mercury in fish is methylmercury. (iii) Mercury is non-conservative in estuaries where it tends to be accumulated in fine-grained near-shore sediments from which it can be remobilized (see *e.g.* Campbell *et al.*[32]). (iv) In the particulate form, mercury has a strong association with organic particles, *via* which it is readily transmitted to biota.

The residence time of mercury in the oceans has been estimated to be ~ 350 y,[30] and indicates a high degree of biogeochemical activity and a rapid removal from the water column. Open-ocean concentrations of reactive mercury (which includes 'labile' organo-Hg associations, but not the more stable organo-Hg associations) are around ~ 2–~ 10 pmol l^{-1} (see *e.g.* Bruland[33] and Fitzgerald[30]), although concentrations can rise to ~ 100 pmol l^{-1} in polluted coastal waters.[34] The WHO maximum tolerable limits for mercury consumption are 0.3 mg per week, of which not more than 0.2 mg should be as methylmercury. These levels are not likely to be exceeded except under 'extreme' circumstances where seafood from highly contaminated waters is consumed. Such 'extreme' circumstances occurred at Minimata in Japan, and in Minimata Bay concentrations of as high as 10 nmol l^{-1} were reported (see *e.g.* Bishop[35]).

In what Goldberg[36] described as a typical example of the 'surprise factor' in environmental problems, outbreaks of methylmercury poisoning at Minimata were first discovered during the 1950s. By the mid-1970s around a thousand cases had been diagnosed with many more suspected, and 41 deaths had occurred; there was also evidence of the long-term accumulation of mercury in the human brain from eating seafood caught in Minimata Bay.[37]

The Minimata incident provides a prime example of how a pollutant released into the environment can move along the food chain to affect humans. Mercury compounds were released into the waters of Minimata Bay from a factory using mercury(II) chloride in the synthesis of vinyl chloride and mercury(II) sulfate in the production of acetaldehyde.[36] According to Waldron[38] there is some disagreement over the form in which the mercury initially entered the environment. It was claimed that mercury was released by the factory in an inorganic form, and that this was subsequently methylated by bacteria in the sediments of Minimata Bay. However, the rate of conversion is probably too slow to have accounted for the high levels of methylmercury in the waters of the bay. The methylmercury was markedly accumulated by fish and shellfish and appeared in man from the consumption of the seafood.

2.3.4.2 Lead. The global cycle of lead has been strongly perturbed by anthropogenic effects, and the metal is an example of the input of a contaminant to the natural environment mainly through release into the atmosphere. For example, Niriagu[39] estimated that the global anthropogenic emission of Pb to the atmosphere was ($\sim 332 \times 10^6$ kg y^{-1}), which exceeded the natural emissions ($\sim 12 \times 10^6$ kg y^{-1}) by a factor of ~ 28.

Lead is used in a number of industrial applications, but the main anthropogenic input to the environment is *via* the combustion of fossil fuels, especially from the use of lead alkyls (tetraethyl and tetramethyl lead) as an anti-knock additive in combustion engine fuel. However, the use of lead in fuels has decreased markedly over the past few years with the reduction of the anti-knock additive in Europe and North America; for example, the maximum lead concentrations in gasoline in the US fell from 2.5 g US gal^{-1} in 1970 to 0.5 g US gal^{-1} in 1979.[40]

Three factors strongly influence the distribution of lead in the marine environment. (i) Its global cycle is strongly perturbed by anthropogenic inputs. (ii) It is a 'scavenging-type' trace metal which is rapidly removed from surface waters, and which has an oceanic residence time of only a few hundred years. (iii) It has a large atmospheric signal to surface waters and because it is emitted into the atmosphere from anthropogenic sources as small particles (see *e.g.* Arimoto and Duce[41]) it undergoes long-range transport. As a result, both the open-ocean and the coastal cycles of lead have been strongly perturbed by man's activities. However, the major effects of Pb contamination are found in the coastal environment, and Pb can be used to illustrate the accumulation of a pollutant trace metal in the sediment reservoir.

In coastal regions accumulation rates are fast enough for sediments to record anthropogenic inputs. For example, with respect to Pb, Chow *et al.*[42] reported data on the distribution of the metal in dated sediment cores from a number of basins off the coast of southern California. Aluminium, which has a mainly natural origin, was used as a normalizing element and Pb:Al ratios were used to establish the background lead levels in the sediments. The data showed that in the sediments of some basins which received anthropogenic inputs there was an increase in both the concentrations of lead and the Pb:Al ratios in sediments deposited after ~ 1940. In addition, lead isotopic ratios indicated that the source of the 'excess' lead had been influenced by gasoline emissions. The study provides an example of how natural lead cycles in the marine environment can be perturbed by anthropogenic inputs. However, according to Bernhard[43] the marine environment represents only a negligible source of lead to humans. Nonetheless, situations can arise where the sudden release of lead into the marine environment is potentially harmful. There have also been examples of an increase in alkyl lead compounds in biota and sediments from areas where there has been no evidence of abnormal discharges of alkyl lead to the environment. For example, mortality among seabirds has been attributed to poisoning from trialkyl lead, a stable decomposition product of tetra-alkyl lead, in the River Mersey estuary, which receives a number of industrial effluent discharges.[44]

2.3.4.3   Tin.   Tin is now recognized as being potentially a very serious marine
pollutant and is an example of a substance introduced into the marine
environment for a specific purpose, rather than one which enters the system
either as a by-product of anthropogenic activity, or as a result of accidental
release.

Tin has a number of industrial uses, but it is the organotin compounds
(notably tributyl tin oxide – TBTO) which have given rise to the greatest
concern. The principal uses of organotin compounds are as stabilizers for PVC
and as biocides, especially anti-fouling paints used on ships hulls. Organotin
compounds in the marine environment derive from two main sources: (i) *via* the
bacterial methylation of inorganic tin, and (ii) by the sea water leaching of alkyl
and aryl tin from some types of anti-fouling paints which may affect non-target
species, especially in areas such as harbours, boat yards and marinas where
there are high concentrations of ships.[8]

TBTO can produce major harmful responses in a number of marine organisms, such as commercial Pacific oysters (*Crassostrea gigas*) and dogwhelks
(*Nucella lapillus*). In the latter, TBTO can cause 'imposex' – a phenomenon in
which females develop male sexual characteristics leading to sterility and
population collapse. Concern has also been expressed over TBTO in farmed
salmon, where it has been used on breeding cages (Davies and McKie;[45] Davies
et al.[46]). Other marine organisms affected by TBTO include phytoplankton and
zooplankton, and evidence suggests that the TBTO can have harmful effects at
concentrations as low as the ng l^{-1}-level. Waters in many regions associated
with commercial or recreation shipping may have TBTO concentrations which
exceed this level, and a North Sea Quality Status Report[15] identified numerous
areas around the North Sea where serious TBTO effects may occur; these
include the Solent, the Wash and regions adjacent to the major continental
European rivers such as the Elbe and the Rhine. Even remote areas, such as the
Shetland islands, may be affected because of the use of TBTO on fishing boats.
Although many countries have now banned the use of tin-based anti-fouling
paints, either entirely or at least on small-sized vessels, there is a legacy of old
paintwork that is likely to sustain elevated sea water concentrations of TBTO
for some time.

2.3.5   Radioactivity (see also Chapter 18)

The release, or even the potential release, of radioactive substances to the
environment is always a major issue, as was clearly demonstrated by the
decision of the French Government to resume weapons testing on Mururoa
Atoll in the southern Pacific and the subsequent actions of the Greenpeace
organization.

The subject of marine radioactivity has been extensively reviewed by Burton[47]
and Preston,[8] and only a brief description of some of the more important, or
recent, issues is included in this section.

2.3.5.1   The Natural Radioactivity of Sea Water.   Sea water and marine

Table 2.9 *The concentrations of some radionuclides in sea water*[48]

Radionuclide	Concentration (Bq kg^{-1})
Potassium-40	11.84
Tritium	0.022–0.11
Rubidium-87	1.07
Uranium-234	0.05
Uranium-238	0.04
Carbon-14	0.007
Radium-228	$0.0037–0.37 \times 10^{-2}$
Lead-210	$0.037–0.25 \times 10^{-2}$
Uranium-235	0.18×10^{-2}
Radium-226	$0.15–0.17 \times 10^{-2}$
Polonium-210	$0.022–0.15 \times 10^{-2}$
Radon-222	0.07×10^{-2}
Thorium-228	$0.007–0.11 \times 10^{-3}$
Thorium-230	$0.022–0.05 \times 10^{-4}$
Thorium-232	$0.004–0.29 \times 10^{-4}$

sediments are naturally radioactive (Table 2.9). Sea water itself has a radio-activity of around 12.6 Bq l^{-1}, with that of marine sands and muds being around 200–400 Bq kg^{-1} and 700–1000 Bq kg^{-1}, respectively. Most of the radiation comes from the isotope potassium-40, but there are numerous members of the uranium and thorium decay series present as well. In addition, the creation of lighter radioisotopes through the interaction of cosmic rays and atmospheric gases, with the products subsequently transferred to the ocean surface, also makes a contribution (Table 2.9). Chemically, radionuclides behave almost identically to their stable counterparts (where these exist).[49] They are, therefore, partitioned between water, sediments and biota according to their behavioural properties. Examples of this chemically determined behaviour include: (i) Cs-137, which is largely water soluble and behaves as a fairly conservative property of sea water moving with the prevailing currents, and (ii) plutonium-239/240 isotopes, which are highly non-conservative and form strong associations with fine-grained sediments.

One of the more important consequences of the natural radioactivity of marine systems is that marine organisms have evolved in a comparatively radioactive environment, so that the additional radiation introduced as a result of human activities is to a considerable extent a Type I rather than a Type II pollutant problem.

2.3.5.2 Radiation Releases from Weapons Testing Programmes. Between the first atomic weapon test explosion in 1945 and 1996, around 2000 tests have been carried out. Many of the early tests resulted in significant contamination of marine systems[50] although comparatively few of those conducted since 1963 have resulted in similar contamination. This is because, from 1963, the countries which have ratified the Partial Nuclear Test Ban Treaty confined their tests to underground sites; however, non-signatory nations have continued to carry out 'open' tests.

2.3.5.3   Routine Releases from Nuclear Power Plants.   Under normal operating conditions, power generating reactors release very small quantities of radioactivity to the environment. Such isotopes that are released include tritium, sulfur-35, zinc-65 and cobalt-60 which primarily derive from the neutron activation of cooling water or soluble species within it. In the UK, the station with the highest discharges is probably Heysham Station 1 which, in 1990, released 0.40 and 157 TBq of sulfur-35 and tritium respectively and 0.058 TBq of other activity.[51]

2.3.5.4   Releases from Nuclear Fuel Reprocessing Plants to Marine Systems. Over the years, the most important non-military source of routine radiation discharges to the marine environment has been the routine low-level discharges from nuclear fuel reprocessing plants such as those at Sellafield in the UK and Cap de la Hague in France. The main isotopes discharged in this manner include tritium, C-14, Co-60, Sr-90, Te-99, Ru-106, I-129, Cs-134 and 137, Ce-144, Pu-241 and Am-241.[52]

The consequences of the Sellafield discharges to radiation levels in the Irish Sea and human populations have been reviewed by Kershaw *et al.*[52] who report that between 1983 and 1988 collective radiation doses to the UK population have decreased from 70 to 40 man-Sievert with an increase in 1986 as a result of the Chernobyl accident (see below). The resultant individual doses to members of the local fishing community and "typical members of the fish-eating public consuming fish landed at Whitehaven/Fleetwood" are 0.11–0.16 and 0.004 mSv respectively compared to a recommended principal dose limit of 1 mSv a^{-1}.

2.3.5.5   Reactor Accidents on Land.   Reactor accidents on land have been rare, but the major incident at Chernobyl in the Ukraine (April, 1986) resulted in very widespread contamination of western Europe. In the UK contamination of both land and sea was considerable,[53] and whilst the aquatic burdens have now essentially reduced to previous levels the effects on upland pastures are still influencing the marketability of animals grazed on upland areas.

2.3.5.6   Dumping of Low Level Waste.   Disposal of low-level, packaged radioactive waste in the oceans was used by a number of countries until 1983 when the authorizing body, the London Dumping Convention (now the London Convention) took the political decision to suspend it. This decision was confirmed in 1995 when the members of the Convention voted to cease all marine waste disposal. The total radiation inventory dumped in the north-east Atlantic is given in Table 2.10.

2.3.5.7   Naval Sources of Radioactivity to the Oceans.   Since 1945 at least 50 nuclear warheads and nine reactors have been introduced to the world's oceans mainly as a result of accidents to submarines.[54] The total count is still rising as a result of the increased information flow from the former Soviet Union about disposal of radioactive wastes in the Russian Arctic. The incident of greatest

Table 2.10 *Amounts of intermediate and low level radioactive waste dumped in the Northeast Atlantic*[49]

Year	α emitters (TBq)	β/γ emitters (TBq)	Tritium (TBq)
1974	15.5	40 700[a]	
1975	28.9	1130	1100
1976	32.6	1200	775
1980	70.3	3075	3630
1981	77.7	2930	2750
1982	51.8	1830	2860
Total 1948–82	660	38 000	15 000

[a] Combined total of β/γ emitters and tritium.

concern appears to be the loss of a Soviet 'Mike Class' submarine containing two liquid metal nuclear reactors, and two nuclear armed torpedoes, which sank some 270 miles north of Norway in April 1989 after a fire on board. The proximity of this wreck to land, and the damage sustained by the vessel prior to sinking, must make the risks of radiation releases in the relatively near future fairly high. The same factors also make any kind of recovery operation potentially very hazardous.

2.3.5.8 The Effects of Artificial Radioactivity on the Marine Environment. Humans are the most sensitive of living organisms to the effects of radiation, and so all radiological population measures are designed to protect humans on the understanding that all other species will therefore automatically receive adequate protection.[55] Nevertheless, studies of the effects of anthropogenically derived radiation on marine organisms have been conducted.[52] For example, studies of the possible effects of Sellafield discharges on plaice eggs (the most sensitive developmental stage) indicate that the LD_{50} at metamorphosis to be 0.9 Gy, with local dose rates estimated at ~ 1 ng Gy h^{-1}.[56] The general conclusions are that the lowest dose rates at which minor radiation-induced disturbances of physiology or metabolism might be detectable are around 0.4–8 mGy h^{-1},[57] with the highest (mid-1970s) Sellafield dose rates being about two orders of magnitude below those likely to cause observable effects at the population level. Since then discharges have declined by several orders of magnitude.

2.4 CONCLUSIONS

The influence of human activities on the marine environment can now be detected in even the most remote regions. The main impacts are, not surprisingly, in coastal waters which are both closest to the sources of pollutants and the most physically, chemically and biologically active zones. In such coastal areas waste discharges combine with other pressures, such as coastal developments and fishing, to produce deleterious effects which may be environmentally significant. The situation in the more remote oceanic regions has not yet become as serious, and human activities are normally only detectable though the presence of artificially produced (Type II) chemicals and, with the exception of lead, all Type I pollutants normally fall within the range of natural variability.

Although the pressures on marine systems are now widely recognized there is still a considerable gap between the scientific recognition of actual, or potential, problems and the social and/or political willingness to take appropriate action. The future health of the oceans therefore depends on integrated management strategies designed to provide the maximum possible levels of environmental protection.

2.5 REFERENCES

1. R. Chester, 'Marine Geochemistry', 2nd Edn., Blackwell Scientific, Oxford, 1999, 506 pp.
2. J. H. Simpson, in 'Understanding the North Sea System', ed. H. Charnock, K. R. Dyer, J. M. Huthnance, P. S. Liss, J. H. Simpson and P. B. Tett, The Royal Society, Chapman and Hall, London, 1994, pp. 1–4.
3. Institute of Petroleum, 'Petroleum Statistics 1993', Institute of Petroleum, London, 1993, 2 pp.
4. GESAMP, Rep. Stud., GESAMP No. 50, 1993, 180 pp.
5. J. W. Doerffer, 'Oil Spil Response in the Marine Environment', Pergamon Press, Oxford, 1992, 391 pp.
6. P. Ambrose, *Mar. Poll. Bull.*, 1991, **22**, 262.
7. NAS, 'Oil in the Sea, Inputs, Fates and Effects', National Academy of Sciences, Washington, DC, 1985.
8. M. R. Preston, 'Marine Pollution', in 'Chemical Oceanography' ed. J. P. Riley, Academic Press, London, 1989, Vol. 9, pp. 53–196.
9. CONCAWE, 'A Field Guide to Coastal Oil Spill Control and Clean-up Techniques', Rep. No. 9/81, CONCAWE, Den Haag, 1981, 112 pp.
10. J. E. Smith, '"Torrey Canyon" Pollution and Marine Life', Cambridge University Press, Cambridge, 1970, 196 pp.
11. CONCAWE, 'A Field Guide to Inland Oil Spill Clean-up Techniques', Rep. No 10/83, CONCAWE, Den Haag, 1983, 104 pp.
12. MARPOL 73/78, International Maritime Organization, 1992, 485 pp.
13. ITOPF, 'Responses to Marine Oil Spills', International Tanker Owners Pollution Federation Ltd., London, 1987.
14. DoE, 'Digest of Environmental Statistics', No. 17, HMSO, London, 1995.
15. NSQSR, 'North Sea Quality Status Report 1993', Oslo and Paris Commissions, London, 1993, 132 pp.
16. NRA, 'The Mersey Estuary, a Report on Environmental Quality', HMSO, London, 1995, 44 pp.
17. M. Takizawa, W. L. Straube, R. T. Hill and R. Colwell, *Appl. Env. Micobiol.*, 1993, **59**, 3406–3410.
18. GESAMP, Rep. Stud., GESAMP No. 5, 1976, 23 pp.
19. D. Broman, 'Transport and Fate of Hydrophobic Organic Compounds in the Baltic Aquatic Environment', PhD thesis, University of Stockholm, 1990.
20. S. Tanabe, *Env. Poll.*, 1988, **50**, 5–28.
21. E. Atlas, T. Bidleman and C. S. Giam, in 'PCBs in the Environment', ed. J. S. Waid, CRC Press, Boca Raton, FL, Vol. 1, 1986.

22. E. Carlson, A. Giwercman, N. Keiding and N. Skakkebaek, *Br. Med. J.*, 1992, **305**, 609–612.
23. R. M. Sharpe and N. E. Skakkebaek, *Lancet*, 1993, **341**, 1392–1395.
24. W. R. Kelce, C. R. Stone, S. Laws, L. E. Gray, J. A. Kemppainen and E. M. Wilson, *Nature*, 1995, **375**, 581–585.
25. IEH, 'IEH Assessment on Environmental Oestrogens: Consequences to Human Health and Wildlife', Institute for Environment and Health, Leicester, 1995, 107 pp.
26. R. E. Hester and R. M. Harrison, 'Endocrine Disrupting Chemicals', Issues in Environmental Science and Technology, No. 12, Royal Society of Chemistry, Cambridge, 1999, 151 pp.
27. J. G. Vos, E. Dybing, H. A. Greim, O. Ladefoged, C. Lambre, J. V. Tarazona, I. Brandt and A. D. Vethaak, 'Health Effects of Endocrine-disrupting Chemicals on Wildlife, with Special Reference to the European Situation', *Crit. Rev. Toxicol.*, 2000, **30**, 71–133.
28. J. O. Nriagu, in 'The Bio-geochemistry of Mercury in the Environment', ed. J. O. Nriagu, Elsevier, North Holland, Amsterdam, 1979, p. 23.
29. A. Andren and J. O. Nriagu, in 'The Biogeochemistry of Mercury in the Environment', ed. J. O. Nriagu, Elsevier, Amsterdam, 1979, p. 7.
30. W. F. Fitzgerald, in 'Chemical Oceanography', ed. J. P. Riley, R. Chester and R. A. Duce, Academic Press, London, 1989, Vol. 10, p. 152.
31. L. Landner, *Nature*, 1971, **230**, 452.
32. J. A. Campbell, E. Y. Chan, J. P. Riley, P. C. Head and P. D. Jones, *Mar. Poll. Bull.*, 1986, **17**, 36.
33. K. W. Bruland, in 'Chemical Oceanography', ed. J. P. Riley and R. Chester, Academic Press, London, 1983, Vol. 8, p. 157.
34. C. W. Baker, *Nature*, 1977, **270**, 230.
35. P. L. Bishop, 'Marine Pollution and its Control', McGraw-Hill, New York, 1983, 357 pp.
36. E. D. Goldberg, in 'Chemical Oceanography', ed. J. P. Riley and G. Skirrow, Academic Press, London, 1975, Vol. 3, p. 39.
37. S. Smith, in 'Understanding Our Environment: An Introduction to Environmental Chemistry and Pollution', ed. R. M. Harrison, Royal Society of Chemistry, Cambridge, 1992, p. 244.
38. H. A. Waldron, in 'Pollution: Causes, Effects, and Control', ed. R. M. Harrison, Royal Society of Chemistry, Cambridge, 1990, p. 261.
39. J. O. Nriagu, in 'Control and Fate of Atmospheric Trace Metals', ed. J. M. Pacyna and B. Ottar, Kluwer Academic Publishers, Dordrecht, 1989, p. 3.
40. J. R. Ashby and P. J. Craig, in 'Pollution: Causes, Effects, and Control', ed. R. M. Harrison, Royal Society of Chemistry, Cambridge, 1990, p. 309.
41. R. Arimoto and R. A. Duce, *J. Geophys. Res.*, 1986, **92**, 2787.
42. T. J. Chow, K. W. Bruland, K. Bertine, A. Soutar, M. Koide and E. D. Goldberg, *Science*, 1973, **181**, 551.
43. M. Bernhard, in 'Lead in the Marine Environment', ed. M. Branica and Z. Konrad, Pergamon Press, Oxford, 1980, p. 345.
44. J. P. Riley and J. V. Towner, *Mar. Poll Bull.*, 1987, **15**, 153.
45. I. M. Davies and J. C. McKie, *Mar. Poll. Bull.*, 1987, **18**, 405.
46. I. M. Davies, S. K. Bailey and D. C. Moore, *Mar. Poll. Bull.*, 1987, **18**, 400.
47. J. D. Burton, in 'Chemical Oceanography', ed. J. P. Riley and G. Skirrow, Academic Press, London, 1975, Vol. 3, pp. 91–191.
48. R. B. Clarke, 'Marine Pollution', 3rd edn., Oxford University Press, Oxford, 1992, 172 pp.
49. IAEA, International Atomic Energy Agency, Vienna, TECDOC, 1985, 329, 183 pp.
50. A. B. Joseph, P. F. Gustafson, I. R. Russell, E. A. Schuert, H. L. Volchok and A. Tamplin, in 'Radioactivity in the Marine Environment', National Academy of Sciences, Washinton DC, 1971, pp. 6–41.

51. MAFF, 'Aquatic Environment Monitoring Report No. 29', MAFF Directorate of Fisheries Research, Lowestoft, 1992, 68 pp.
52. P. J. Kershaw, R. J. Pentreath, D. S. Woodhead and G. J. Hunt, 'Aquatic Environment Monitoring Report No. 32', MAFF Directorate of Fisheries Research, Lowestoft, 1992, 65 pp.
53. W. C. Camplin, N. T. Mitchell, D. R. P. Leonard and D. F. Jeffries, 'MAFF Aquatic Environment Monitoring Report', 1986, **15**, 49 pp.
54. M. Evans, *The Times*, 7th June 1989.
55. ICRP, 'Recommendations of the International Commission on Radiological Protection', Annals of the ICRP, ICRP Publication 26, Pergamon Press, 1977, 53 pp.
56. D. Woodhead, *Radiat. Res.*, 1970, **43**, 582–597.
57. IAEA, Tec. Report International Atomic Energy Agency, Vienna, 1976, 172, 131 pp.

CHAPTER 3

Drinking Water Quality and Health

J. K. FAWELL and G. STANFIELD

3.1 INTRODUCTION

An adequate supply of safe drinking water is one of the major prerequisites for a healthy life. The importance of clean water and the link between contaminated or putrid water and illness was recognized in the distant past, even though the actual cause of disease was not properly understood until the latter half of the 19th century. Herodotus who lived in the 5th century BC records that Cyrus the Great, who lived a century before, carried boiled water stored in flagons of silver when he travelled. This was sound practice since boiling would have killed all of the pathogenic microorganisms and silver ions would have acted as a bacteriostatic agent. Contaminated water was recognized by its foetid smell and unpleasant taste although we now know that sweet tasting water can still carry microorganisms.

The Romans developed a sophisticated water supply system bringing clean water from uncontaminated sources and also separating sewage disposal from drinking water supply. Sextus Julius Frontinus, who was in charge of Rome's water supply in the latter part of the 1st century AD, was the first well-known water engineer. He was certainly aware of the problems surrounding the provision of clean drinking water in sufficient quantity to meet demand in a large city. Most of the Roman knowledge of water supply was lost after the fall of the Roman Empire, although some remnants can be detected in certain monasteries. As cities grew the problems associated with water supply also grew and in the 19th century outbreaks of typhus and cholera brought misery and fear to many of the great cities of Europe and North America. Abstracting water upstream of where sewage entered rivers was a great benefit as was the introduction of filtration through gravel and sand. Attempts were made to introduce disinfection using chlorine and ozone in the 19th century. John Snow, who identified the Broadwick Street pump as the source of cholera in the outbreak in that part of London in 1854, is reported to have attempted to treat the source with calcium hypochlorite. However, it was not until the 20th century that disinfection with chlorine was introduced on a wide scale with the first permanent chlorination plant at Middelkerke in Belgium in 1902.

59

Chlorination spread rapidly through the United States but it was not until the 1937 Croydon typhoid outbreak and the advent of the Second World War that universal chlorination was introduced in the UK. The first drinking water standards related primarily to microbiological contamination, although a number of inorganic contaminants were recognized. It was not until the development of advanced analytical techniques, such as gas chromatography/ mass spectrometry in the 1970s, that organic contaminants in drinking water were recognized as an issue. This in turn resulted in a major change in the way that drinking water was regulated and provided the impetus for the development of new methods of water treatment.

Unfortunately, waterborne disease is still a major cause of death in many parts of the world, particularly in children, and it is also a significant economic constraint in many subsistence economies. In developed countries there have been enormous changes in our knowledge of water contaminants and in the way that water is treated and supplied. The major risks associated with significant microbiological contamination have largely been overcome with the advent of both sewage treatment and the ability to treat large quantities of drinking water to a high level of microbiological and chemical purity. However, contamination can still occur and there is a constant need to maintain a high level of vigilance.

3.2 DRINKING WATER SOURCES

Drinking water is derived from three basic sources: lowland rivers and reservoirs, upland reservoirs, and groundwater. All public water supplies in the UK receive treatment that reflects the quality of the raw water and the variability in that quality. Lowland surface water sources contain a high level of natural and anthropogenic organic and inorganic matter and are much more prone to pollution. They receive the highest level of treatment to remove microorganisms, particulate matter and chemical contaminants. Upland sources usually suffer less from anthropogenic contamination but they contain high levels of naturally occurring organic matter from decaying vegetation. This, in combination with iron, is the cause of the brown colouration seen in many upland streams. There can also, on occasion, be microbiological contamination from farm animals and wildlife. They are usually low in inorganic constituents and are, therefore, soft. Groundwaters are usually low in organic matter and are less vulnerable to both microbiological and chemical contamination. They frequently contain high levels of inorganic substances from the rocks through which the water percolates.

Drinking water treatment is intended to remove microorganisms and more commonly, depending on the circumstances, chemical contaminants. However, the process can in itself result in the formation of other contaminants, such as the trihalomethanes, from the reaction of chemical oxidants with naturally occurring organic matter. This requires a balance to be struck between the benefits of the chemical oxidants in destroying microorganisms and the potential risks from the by-products.

In the UK only a small minority of the population does not receive water through the public water supply. Approximately 4% of the population obtain their water from private supplies, which may be wells, springs or streams, or small reservoirs fed by streams or springs. They usually supply only one or a small number of households and are often quite remote. The level of treatment received by these supplies is very variable and is often minimal, while source protection is often poor.

3.3 DRINKING WATER TREATMENT

Drinking water treatment as applied to public water supplies, consists of a series of barriers in a treatment train which will vary according to the requirements of the supply and the nature and vulnerability of the source. Broadly these will be coagulation and flocculation, filtration and oxidation.

Coagulation and flocculation using a coagulant such as aluminium sulfate or ferrous sulfate will remove particulates, including microorganisms, and some dissolved organic matter. Filtration can consist of slow sand filtration which will remove particles, microorganisms and some organic matter, rapid gravity filtration which will remove particles and associated microorganisms, and granular activated carbon (GAC) which removes many dissolved organic substances by adsorption. Granular activated carbon may also be operated in a biological mode, when a significant biofilm develops and provides a means of enhanced biodegradation of contaminants. In this mode the carbon does not require frequent regeneration, in contrast to some applications in the adsorptive mode. GAC filtration also removes microorganisms, but because of its ability to adsorb organic material it can become a growth substrate for bacteria. As a consequence, the filtrate from GAC beds can contain more heterotrophs, including total coliform bacteria than the influent, particularly in the summer months. In combination with ozonation GAC filtration is often termed advanced water treatment. This enhanced filtration process removes by-product precursors and reduces the assimilable organic carbon content making the water more biostable, thereby reducing the potential for microbial aftergrowth in the distribution system.

Oxidants are used both as part of the process for removal of chemicals, particularly ozone, and as disinfectants, which include chlorine, chlorine dioxide, ozone and ultraviolet irradiation. However, chlorine and chloramine are the main disinfectants used in the UK since they are the only oxidants that can provide a residual through the distribution system. New advanced treatment processes are rapidly gaining ground, particularly membrane filtration, which provides a more comprehensive physical barrier, used for removal of microorganisms and a number of chemical contaminants. However, there are disadvantages to membrane systems which generate a significant reject stream and, where there is significant removal of minerals, may require remineralization to reduce the corrosivity of the water to the distribution system and domestic plumbing.

Water supplies in any one area can be quite complex with several sources providing water to discrete sub-areas. In some cases water from two or more sources may be blended to achieve a particular water quality or to dilute a contaminant, such as nitrate, which may be present in one of the sources. Sources may also be remote from the area supplied, requiring long distribution systems, and they are sometimes changed over time in response to demand, water quality problems or operational difficulties. As demand for water has increased, particularly where availability might be limited, efforts are being made to make supplies more flexible so that demand can be balanced over a wider area.

3.4 SOURCES OF CONTAMINATION

Water can be contaminated from both natural and anthropogenic sources. All surface waters will contain natural organic matter in the form of humic substances, large complex carbohydrate molecules. Some will contain inorganic substances dissolved from the minerals through which the water percolates and some will contain particles of clay and sand washed from the surrounding catchment, the banks and the bottom sediments. There are a number of possible sources of man made contaminants, some of which are more important than others. These fall into the categories of point and diffuse sources. Discharges from industrial premises and sewage treatment works are point sources and as such are more readily identifiable and controlled, in the UK, by the Environment Agency (see Chapter 21). However, run-off from agricultural land and from hard surfaces, such as roads, are not so obvious, or easily controlled. Such sources can give rise to a significant variation in the contaminant load over time.

There are other sources of contamination, some of which are tightly regulated. Drinking water treatment can remove many contaminants, but it can also change the contaminants present, *e.g.* phenol reacts with chlorine to give chlorophenols, or add other contaminants such as disinfection by-products, *e.g.* trihalomethanes or THMs. Contamination from distribution may arise from materials used in distribution such as iron which can corrode to release iron oxides, or from ingress into or diffusion through the distribution system. This latter phenomenon is a particular problem when the components of oil, spilt into the surrounding soil, diffuse through plastic pipes giving rise to taste and odour problems. Contamination can also take place in consumer's premises from materials used in plumbing or from back siphoning of liquids into the distribution system as a consequence of improper connections, in breach of by-laws. Such contaminants can be either chemical or microbiological. It is recommended that water is not drunk from taps supplied from storage tanks which may be open to the atmosphere because they can be easily contaminated, or from the hot water system which may increase dissolution of metals from the plumbing.

Contaminants may be present most of the time, in which case the issue for consumers is chronic exposure and chronic toxicity, or for short periods, such as in a spill penetrating treatment, in which case the issue is likely to be acute toxicity following short-term exposure. Although the average water usage for a family of two adults and two children is about 475 l per day, only a very small proportion of this is used for drinking or cooking. An adult drinks approximately 1 to 2 l per day, a significant proportion of which is boiled. Children drink smaller quantities of water but this can be greater than adults when considered in relation to bodyweight. Normally, in carrying out risk assessments for chemical contaminants in drinking water, 1 l is allowed for a 10 kg child while a bottle fed infant will ingest significantly greater quantities of up to 150 ml kg^{-1} bodyweight, all of which will be boiled. Exposure to contaminants can also arise as a consequence of washing and showering. In general it is considered that exposure to volatile contaminants can be approximately doubled by skin exposure and inhalation.

3.5 DRINKING WATER STANDARDS

The basis on which drinking water is judged in Europe is the Directive on the Quality of Water for Human Consumption,[1] which is transcribed into UK law in the Drinking Water Regulations.[2] The Directive lays down the minimum standards that must be met, but member states may extend the parameters covered or apply more stringent standards. The regulations cover a range of contaminants in public water supplies and specify sampling requirements for those contaminants. Similar regulations also apply to private supplies. There are separate regulations that apply to bottled waters and water that is classified as natural mineral water under different regulations.

Many of the parameters are based on WHO Guidelines[3] and the revision of the WHO guidelines for drinking water quality in 1993 provided the stimulus for a revision of the original directive.[4,5] The directive specifies three groups of parameters. These are microbiology, chemical contaminants and finally indicator parameters (Tables 3.1, 3.2 and 3.3), many of which are not directly related to health and which member states can apply in a more flexible way. There are only a limited number of contaminants covered by the drinking water directive because it is not possible, or desirable, to set legally binding standards for all possible contaminants. However, the WHO guidelines cover a much wider range of contaminants than the standards and are an important source of information on acceptable concentrations for drinking water contaminants. The parameters covered in the 1993 edition of the WHO guidelines are also supplemented by a series of revisions that will continue as a programme of rolling revision to take into account new data on toxicity and occurrence.

One of the most frequently encountered questions regarding drinking water quality relates to the consequences of a standard being exceeded. To make an appropriate assessment of the implications of a particular exceedence, it is important to know the basis on which the standard is set. For example, not all values in the standards are derived from data relating to health. Some, such as the standards for iron and aluminium, are based on preventing discolouration and dirty water problems. The current standard for polycyclic aromatic hydrocarbons (PAH) is, in part, based on an old WHO guideline which reflects what might be expected in an unpolluted supply.[3] The pesticide parameter of 0.1 μg l^{-1} is based on a political, precautionary approach that reflected the limit of analytical detection when the original directive was formulated in the early 1980s. When a health-based standard for a chemical is exceeded, the key questions that need to be asked are: how long will exposure to the high concentration last and what is the margin of safety in the guideline value or standard? Most standards are set on the basis of long-term exposure and some averaging of exposure is acceptable, while most contain a significant margin of safety, which will allow leeway in the case of an exceedence. However, this may not always be the case. It is, therefore, important not just to take standards and guidelines at their face value but to examine the rationale behind them.

3.6 MICROBIOLOGICAL QUALITY

One of the great scourges of cities in Europe and North America in the 19th century was the danger of outbreaks of waterborne diseases such as cholera and typhoid. The development of sewerage systems and drinking water treatment were major factors in breaking the cycle of contamination of drinking water sources by infected faecal matter and continuing infection of the population. This was undoubtedly one of the great public health triumphs of the 19th century and its importance is still being demonstrated by outbreaks of these diseases in developing countries, whenever drinking water treatment breaks down.

The problem arises as a consequence of contamination of water by faecal matter containing pathogenic organisms. However, since detection and enumeration of the pathogens is still extremely difficult, standards are stipulated in terms of microbiological indicators of faecal contamination (Table 3.1). The indicators used are *Escherichia coli* and faecal streptococci, which must not be detectable in a 100 ml sample. The assumption is that if the indicators are detected, the pathogens, including viruses, could also be present and therefore

Table 3.1 *Bacteriological Quality Standards and Guidelines, comparison of EC Drinking Water Directive and WHO Guidelines*

Parameter	EC Directive	WHO Guideline	Comments
E. coli	0 in 100 ml	0 in 100 ml	WHO – may use thermotolerant coliforms
Enterococci	0 in 100 ml	Not included	
Clostridium perfringens and spores	0 in 100 ml	Not included	Indicator parameter for surface waters triggers investigation
Colony count 22°	No abnormal change	Not included	Indicator parameter for distribution
Total coliforms	0 in 100 ml	0 in 100 ml	Indicator parameter for distribution

appropriate action is required. One further indicator parameter is enumeration of the spores of *Clostridium*, which are not as susceptible to disinfection as the other faecal indicators and provide a measure of the efficacy of filtration. In particular a change in numbers of *Clostridium* spores can indicate a problem has arisen in the filtration process.[4,6]

Drinking water is not, however, sterile and bacteria can be found in the distribution system and at the tap. Most of these organisms are harmless, but some opportunist pathogens such as *Pseudomonas aeruginosa* and *Aeromonas* spp. may multiply in distribution given suitable conditions. Currently there is considerable debate as to whether these organisms are responsible for a significant level of waterborne, gastrointestinal disease in the community. Research is continuing in this field, but there are many difficulties, including the level of reporting of minor gastrointestinal illness and the confirmation that drinking water is the source of an infection. This is further complicated by the potential for secondary spread of infection from person to person.

A number of organisms are emerging as potential waterborne pathogens and some have emerged as significant pathogens that do give rise to detectable waterborne outbreaks of infection. The most important of these is *Cryptosporidium parvum*, a protozoan, gastrointestinal parasite which gives rise to severe, self-limiting diarrhoea and for which there is, currently, no specific treatment. *Cryptosporidium* is excreted as oocysts from infected animals, including man, which enables the organism to survive in the environment until ingested by a new host. Although environmental stress will probably reduce the numbers of viable oocysts, this is a very uncertain process. Significant concentrations might be expected in farm slurry and in some wastewater discharges which will contaminate surface sources, and groundwater sources with close connection to surface water. This organism has given rise to a number of waterborne or water associated outbreaks in the UK, although the actual numbers of cases in any outbreak are usually relatively small. By contrast an outbreak of cryptosporidiosis in Milwaukee, in the United States, resulted in many thousands of cases, and probably a number of deaths among the portion of the population which were immunocompromised. However, water is not the only source of infection or, possibly not the major source of infection. It is most likely that person to person spread following contact with faecal matter from infected animals is the largest source and there have been outbreaks involving milk and swimming pools. The organism is resistant to concentrations of chlorine used in water treatment and the primary barriers are normally coagulation and flocculation, and filtration. It is detected by filtering samples, staining with an immunofluorescent tag and examining under a microscope. Unfortunately this method gives no indication whether the oocysts are viable or not. *Cryptosporidium* oocysts have been detected in water supplies without evidence of a detectable outbreak of clinical disease. This could be due to non-viable organisms or a high level of immunity in the exposed population. One of the great difficulties facing water suppliers and health professionals is what to do if small numbers of oocysts are detected in a supply in the absence of evidence of an outbreak of cryptosporidiosis in the community. Investigative action is an imperative, but the imposition of a 'boil water order' leads to a number of difficulties, not least of which is determining the conditions under which the order will be lifted.

In the UK the water supply regulations have been amended to include a standard for *Cryptosporidium*. These regulations are innovative in many ways. This is the first occasion on which a standard has been stipulated in terms of a pathogen rather than an indicator organism. Additionally the standard requires, again for the first time, continuous monitoring at treatment works which have been assessed as being at risk. The standard is not absolute in that it requires water to be filtered at a rate of 40 l per hour and that, on average, there must be less than one oocyst in 10 l.[7]

It is important to recognize *Cryptosporidium* as an emergent pathogen, rather than a new pathogen. Its existence was recognised in the latter part of the 19th century. However, its importance as a threat to the safety of water supplies has only been recognized through the development of improved diagnostic techniques and more sensitive epidemiological procedures. There has been no increased disease burden on drinking water consumers, indeed this continues to decline, but a better understanding of the causal agents. Whilst the use of chlorine and other oxidants has been of unquestionable value in the provision of safe drinking water, the emergence of *Cryptosporidium* as a waterborne pathogen demonstrates the potential folly of over-reliance on such treatments. The provision of safe drinking water can only be achieved through the use of a multiple barriers to infectious agents, since no single barrier can be expected to consistently provide adequate protection from the wide spectrum of pathogens that can be present. It is vital that water treatment continues to utilize the combined removal capacities of physical treatments such as coagulation and filtration and disinfection.

It is possible that as diagnostic techniques advance other pathogens will emerge and this will require the security of the treatment barriers to be reinforced. Any association with water is unlikely to come from direct monitoring of water supplies, but from surveillance data of diseases in the community, good laboratory diagnostics and epidemiological investigation. Risk assessments of pathogens can be valuable in identifying potential problem organisms, particularly where information on sources, environmental loading, size and resistance to disinfection are known. Through this technique other protozoan pathogens such as *Cyclospora* and *Microsporidia* have been highlighted as being of potential concern, particularly to the growing numbers of immunocompromised in the community.

Private supplies show evidence of a high frequency of microbiological contamination and this is clearly a significant concern for such supplies, particularly where treatment is minimal and there is a possibility of faecal contamination from livestock. These circumstances can give rise to contamination by some of the more virulent pathogenic organisms such as *Cryptosporidium* or pathogenic *E. coli*, e.g. *E. coli* 0157. This is of particular concern when groups of individuals who may be immunologically naive are exposed. A number of waterborne outbreaks of disease caused by *E. coli* 0157 due to undisinfected private supplies have occurred and have even resulted in deaths in the USA and Canada.

Although the common waterborne diseases of the 19th century are now almost unknown in developed countries, it is vital that vigilance is maintained at a high level because these diseases are still common in many parts of the world. The seventh cholera pandemic, which started in 1961, arrived in South America in 1991[8] and caused 4700 deaths in one year. There are still an estimated 12.5 million cases of *Salmonella typhi* per year and waterborne disease is endemic in many developing countries. In this age of rapid global travel the potential for the reintroduction of waterborne pathogens still remains. In addition, as our knowledge of microbiological pathogens improves, we are able to identify organisms that were only suspected previously. These include viruses such as the Norwalk-like viruses, named after a major waterborne outbreak in North America, and a range of emerging pathogens including *Campylobacter*, a major cause of food poisoning

3.7 CHEMICAL CONTAMINANTS

3.7.1 Inorganic Contaminants

3.7.1.1 Lead. Lead in drinking water arises from dissolution from plumbing materials and lead pipes in particular. Lead has been used in plumbing since Roman times and the word plumber even comes from the Latin for lead. It has also been recognized as neurotoxic since Roman times, and possibly even long before. The Romans were aware of the possible dangers of lead water pipes and Charles Penney expressed his concern about the action of soft water from Loch Katrine on lead pipes in Glasgow, and its consequences, in the 19th century. However, in more recent times the problem of frank lead poisoning has been addressed and concerns centre around more subtle effects resulting from low doses.

The most widely recognized of these effects is the potential for lead to adversely affect the neurological development of young children resulting in learning deficits as measured by IQ.[9] These effects in children of four years and above show a probable reduction in IQ by 1–3 points per 10 μg dl^{-1} increase in blood lead up to 25 μg dl^{-1}. Above that the relationship changes and below about 10 μg dl^{-1} blood lead it is difficult to detect any change because of background 'noise' from other factors such as nutrition, parental IQ, socio-economic status and the type of environment the child lives in.

Lead can also have quite broad biochemical effects, and although some can apparently be measured at blood lead levels as low as 5 μg dl^{-1}, the clinical significance at such low blood lead levels is extremely doubtful. More recently new evidence has been emerging on the adverse effects of blood lead on blood pressure. More particularly, a small increase in blood pressure appears to be associated with blood lead levels of greater than 7.5 μg dl^{-1}. Although there remain uncertainties, it is an additional factor that is being considered in national policies regarding the reduction in environmetal lead. Exposure to lead, of course, includes air, dust and food, as well as water. The efforts that have been made to reduce lead over the past two decades have resulted in a significant reduction in the average blood lead in the population, especially children, for whom average blood lead levels are now well below 5 μg dl^{-1}.[10]

The concentrations of lead in drinking water have also steadily fallen with time as a consequence of lead pipe removal and treatment to reduce lead levels. The majority of lead arises from contact with service connections to the mains and from lead plumbing, usually where the properties were built before 1960. Some may also arise from the use of lead-based solders, although this is no longer officially permitted, and from fittings containing a proportion of lead in the metal to improve milling properties. Where water is known to be plumbo-solvent, it has the propensity to dissolve lead from pipework; it is treated before entering supply, usually with small amounts of phosphate. This not only reduces the level of lead in the water but also probably reduces the bioavail-ability of the dissolved lead, and therefore, the amount absorbed into the body. In some circumstances minute particles of lead can be found in the water but the extent that these are bioavailable is uncertain. However, they can give rise to high concentrations in water samples, which may give a misleading picture of actual lead exposure in the property.

The standard for lead in the UK was 50 μg l^{-1} as a maximum in any sample. However, as a consequence of WHO introducing a guideline value for lead in drinking water of 10 μg l^{-1},[4] the European standard is also being reduced to 10 μg l^{-1} *via* an interim standard of 25 μg l^{-1} (Table 3.2).[1] The standard will be measured by a method which reflects average exposure over a period of about a week, since it is based on the WHO provisional tolerable weekly intake (PTWI) of 25 μg kg^{-1} of bodyweight for infants and children. This value in turn was based on an intake in infants that would not give rise to any accumulation of lead. WHO used an allocation of 50% of the PTWI to water, assuming a 5 kg bottle-fed infant drinking 0.75 l per day.

The groups most at risk from high lead levels are, therefore, the foetus, bottle-fed infants whose intake of water in relation to bodyweight is much greater than other groups, and children. However, care is needed in assessing the significance of any exceedences of the lead standard in a single sample and even quite large exceedences of the new standards may have relatively little impact on blood lead and therefore on potential health effects.

The difficulty in dealing with lead in drinking water is that high lead is associated with individual properties. It is, therefore possible to find widely differing lead concentrations in drinking water even from adjacent properties. Routine sampling can identify areas with the greatest problems, but solving the problem finally requires investigation at the household level. It is also important that when old lead pipe is replaced, or new copper pipe is installed, that old-style lead solder is not used. Although the use of lead solder in drinking water systems is not legal, there are still occasions when it is used and does give rise to elevated lead concentrations.

3.7.1.2   Nitrate.   Nitrate is found in both surface and groundwater sources for

Table 3.2   *EC Directive standards for chemical parameters and comparison with WHO guidelines*

Parameter	EC Directive	WHO Guideline	Comments
Acrylamide	0.1 μg l^{-1}	0.5 μg l^{-1}	Control through product spec
Antimony	5 μg l^{-1}	5 μg l^{-1}	WHO provisional (analysis)
Arsenic	10 μg l^{-1}	10 μg l^{-1}	WHO provisional (health data)
Benzene	1 μg l^{-1}	10 μg l^{-1}	WHO 10^{-5} cancer risk
Boron	1 mg l^{-1}	0.5 mg l^{-1}	WHO provisional (health data)
Bromate	10 μg l^{-1}	25 μg l^{-1}	WHO provisional (analysis)
Cadmium	5 μg l^{-1}	3 μg l^{-1}	
Chromium	50 μg l^{-1}	50 μg l^{-1}	WHO provisional (health data)
Copper	2 mg l^{-1}	2 mg l^{-1}	WHO provisional (health data). EC relates to plumbing
Cyanide	50 μg l^{-1}	70 μg l^{-1}	
1,2-dichloroethane	3 μg l^{-1}	30 μg l^{-1}	WHO 10^{-5} cancer risk
Epichlorohydrin	0.1 μg l^{-1}	0.4 μg l^{-1}	WHO provisional (health data) Control through product spec
Fluoride	1.5 mg l^{-1}	1.5 mg l^{-1}	
Lead	10 μg l^{-1}	10 μg l^{-1}	EC average over week at tap. Applies 2015, 25 μg l^{-1} 2003 to 2013. Relates to plumbing
Mercury	1 μg l^{-1}	1 μg l^{-1}	
Nickel	20 μg l^{-1}	20 μg l^{-1}	WHO provisional (health data). EC relates to plumbing
Nitrate	50 mg l^{-1}	50 mg l^{-1}	Applied as sum of ratio of value
Nitrite	0.5 mg l^{-1}	3 mg l^{-1}	against WHO GV for nitrate and nitrite. EC also 0.1 mg l^{-1} nitrite ex-treatment works
Pesticides each	0.1 μg l^{-1}	Individual pesticides	EC is precautionary value not based on health, includes
Total	0.5 μg l^{-1}		'relevant' degradation products
Polycyclic aromatic hydrocarbons	0.1 μg l^{-1}	Not included	EC four specified PAH based on early WHO GV. WHO state value for fluoranthene unnecessary
Benzo[*a*]pyrene	0.01 μg l^{-1}	0.7 μg l^{-1}	EC based on 1984 WHO GV
Selenium	10 μg l^{-1}	10 μg l^{-1}	
Tetrachloro- and trichloroethene	10 μg l^{-1}	40 μg l^{-1} & 70 μg l^{-1}	EC precautionary value
Trihalomethane total	100 μg l^{-1}	Individual THMs	EC, 150 μg l^{-1} 2003 to 2008. WHO applied as sum of ratios of individual THMs
Vinyl chloride	0.5 μg l^{-1}	5 μg l^{-1}	WHO 10^{-5} cancer risk. Applied through product spec

drinking water, primarily as a consequence of agricultural activity. This is not simply due to the use of excessive amounts of artificial fertilizers but the presence of excess nitrate from any source when there is insufficient plant growth to take up that excess and when rainfall is present to cause leaching from the soil. Indeed one of the major sources of nitrate in some groundwaters was the ploughing of ancient grassland following the Second World War. While the appearance of nitrate in surface water following a leaching event can be quite rapid, it may take many years for nitrate to reach some groundwaters, which can mean that that, even when surface inputs stop, the levels in groundwater will continue to rise for some time after. In some cases this can be more than 10 years.

For most individuals, vegetables are the main source of exposure to nitrate, although significant levels may be found in some cured meats. Vegetables and fruit usually contain between 200 and 2500 mg kg^{-1} nitrate with vegetables such as beetroot, lettuce and spinach being particularly high in nitrate. The intake of nitrate from the diet varies from about 40 to 130 mg day^{-1} so water with nitrate levels above the drinking water standard of 50 mg l^{-1} can make a significant contribution to daily intake.

Nitrate is often cited as a contaminant of concern for health but, although it has been widely studied, the only health effect which has been proved to be a consequence of high nitrate levels is methaemoglobinaemia or blue-baby syndrome in bottle-fed infants.[4] Even this condition appears to be significantly influenced by the simultaneous presence of microbiological contamination. Most cases occur at nitrate concentrations in excess of 100 mg l^{-1} but some cases do occur between 50 and 100 mg l^{-1} and even below 50 mg l^{-1} in unusual circumstances with significant microbiological contamination. Nitrate is reduced to nitrite in the body and this subsequently reduces haemoglobin to methaemoglobin with a consequential reduction in oxygen carrying capacity. When the proportion of methaemoglobin rises from normal levels of <1 to 3% to about 10%, clinical symptoms are observed. Bottle-fed infants under three months of age are particularly vulnerable to this condition for four reasons: they have a high intake of water in relation to bodyweight, they have a high stomach pH which allows growth of bacteria particularly efficient at converting nitrate into nitrite, foetal haemoglobin which is present in the first few months of life is more readily oxidized to methaemoglobin and there is a deficiency of the enzymes which convert methaemoglobin back into haemoglobin. The condition is now unusual and there has not been a case associated with a public supply recorded in the UK since 1972. However, in some countries nitrate is still a significant issue, particularly in relation to shallow wells in agricultural areas.

There has been considerable interest in the possibility that nitrate, reduced to nitrite, can react with secondary amines in the stomach of adults, giving rise to the formation of *N*-nitroso compounds. Some of this family of compounds have been shown to be potent carcinogens in a range of laboratory animals and so concern was raised that high nitrate levels could be a contributory cause of cancer. However, in spite of many studies, WHO concluded that there is no convincing evidence of a relationship between nitrates in drinking water and cancer in exposed populations.[4,5] This conclusion is supported by the majority of more recent epidemiological studies.[11]

There have been a number of additional theories as to the possible contribution of nitrate in drinking water to other human diseases and conditions. One of the recent suggestions has been that there is an association with childhood diabetes mellitus. However, there were flaws in the initial studies, particularly in the measurement of exposure and subsequent studies in the UK and the Netherlands have failed to observe any significant association.[12,13]

Nitrate is secreted in human saliva and there is now evidence, from studies of the metabolism and enterosalivary circulation of nitrate in mammals, that nitrate, converted into nitrite in the oral cavity, is a key part of an important resistance mechanism against infectious disease through the production of oxides of nitrogen. There is also evidence that this process may help prevent the formation of carcinogenic nitrosamines.

3.7.1.3 Aluminium. Aluminium has been widely used in drinking water treatment as a coagulant and flocculating agent to remove particulate matter, including microorganisms, and soluble organic matter. It is also found extensively in raw waters although the chemical form will vary and will on occasion be found as inert aluminosilicates in clay particles. In the 1970s it was noted that some patients receiving kidney dialysis developed a severe form of dementia and that this was associated with high aluminium levels in the fluid used in the dialysis procedure. When the levels of aluminium were maintained at less than about 50 μg l^{-1}, dialysis dementia did not occur. At the same time it was shown that soluble aluminium salts injected into the brains of cats caused severe neurotoxicity. When aluminium was found in the lesions of patients, in parts of the Pacific, with Parkinsonism Dementia and Amyotrophic Lateral Sclerosis, concern was expressed that aluminium could also be associated with Alzheimer's Disease (AD), a major cause of pre-senile and senile dementia in Europe and North America. Aluminium was also found in the lesions characteristic of AD and a number of ecological type epidemiological studies reported an association between aluminium in drinking water and AD.[14] However, exposure through drinking water is a very small part of daily exposure and the biological plausibility of the hypothesis was uncertain. A number of subsequent epidemiological studies, including a well conducted case control study, found no association between aluminium in drinking water and AD. However, studies in human volunteers using radiolabelled aluminium and advanced analytical techniques indicate that aluminium in drinking water is of similar low bioavailability to aluminium in food.[15] In addition drinking water contributes only a very small proportion of daily aluminium intake compared to food. Since the first studies on aluminium and AD, there has been an increase in understanding of AD and there is extensive evidence that genetic factors play a major role in susceptibility to the formation and development of the neurological lesions characteristic of the condition. Research is continuing into the potential association between aluminium in drinking water and AD but the majority view of scientists working in this field is that aluminium in drinking water is unlikely to play a primary causal role in the development of AD. WHO concluded "There is no evidence to support a primary causal role of aluminium in AD, and aluminium does not induce AD pathology *in vivo* in any species, including humans."[5] In spite of this the issue continues to be controversial.

3.7.1.4 Fluoride. Fluoride is found in water sources in many parts of the world, sometimes in concentrations of up to 10 mg l^{-1} and higher. It is also added to drinking water in a number of countries to supplement natural fluoride, which has been shown to be beneficial in preventing dental caries at a concentration of about 1 mg l^{-1}. However, above this concentration there is an increasing risk of adverse cosmetic effects in teeth and, at high concentrations, of skeletal fluorosis. This condition arises from increasing density of the bones, which can eventually give rise to severe skeletal deformation. There are many factors that appear to influence the risk of such adverse effects, including intake from other sources, volume of drinking water and nutritional status. In general, significant effects are not seen below about 4 mg l^{-1}, which is the health based maximum contaminant level goal set for drinking water by the United States Environmental Protection Agency.[16] However, water supplies with high fluoride concentrations are still a major problem in some countries, particularly in parts of Africa and the Indian sub-continent.

3.7.1.5 Arsenic. Arsenic occurs in drinking water primarily as a consequence of dissolution from arsenic bearing rocks into groundwater. Although problems in the UK are very limited, arsenic is a major problem in many parts of the world including the Indian sub-continent, South America, the far east and the United States. Arsenic is the only contaminant that has been shown to be the cause of human cancers as a consequence of exposure through drinking water. Arsenic gives rise to a range of adverse effects besides cancer of the skin and probably liver, including hyperkeratosis and peripheral vascular disease. In some countries, arsenic contamination ranks second only to microbiological contamination in its importance. However, the epidemiological data also demonstrate that many local factors are important, including nutritional status and there is little evidence of any health problem in developed countries where the concentrations are between 50 and 100 μg l^{-1}, although the margin of safety appears to be small.[17] WHO have set a provisional guideline value of 10 μg l^{-1} based on the then practical limit of detection and on the JECFA evaluation, but the data are insufficiently precise to be able to distinguish whether a standard of 5, 10, or 15 μg l^{-1} is the most appropriate.[4] Since arsenic removal at these concentrations is difficult, it is necessary to achieve some sort of compromise and this is the reason that the European Commission have adopted the WHO provisional guideline.

3.7.1.6 Water Hardness. The hardness of water is associated with a number of polyvalent metallic ions but is dominated by calcium and magnesium. The most common measure of hardness is in milligrams of calcium carbonate equivalent per litre. Hardness in drinking water generally ranges from about

10 to 500 mg calcium carbonate 1^{-1} and waters with concentrations of less than about 60 mg 1^{-1} are considered to be soft water. In spite of the fact that hardness is caused by cations, anions also play an important part in determining temporary (bicarbonate) hardness and permanent (non-carbonate) hardness. The former is removable but the latter, consisting primarily of sulfates, is not. In the UK hardness tends to follow a north to southeast gradient with the softest waters being encountered in Scotland and the North of England, although parts of Wales and the Southwest also have soft water. This is because the softest waters come from upland supplies with little contact with rocks such as limestone and chalk. The hardest waters tend to be in the South East and East Anglia, although parts of the west of England have some of the hardest waters in the UK. Calcium and magnesium are both essential elements and are found in a variety of foods. The importance of calcium and magnesium in drinking water to dietary intake is a subject of some debate.[4,6] However, in individuals who are marginal in calcium or magnesium, it is reasonable to suppose that hard drinking water could provide a significant source of these minerals.

Mortality from cardiovascular disease also tends to follow a southeast to north trend with the highest rates lying in the north where the water is softest. A number of epidemiological studies, some of which were very large like the British Regional Heart Study, have shown a significant inverse relationship between hardness and mortality from cardiovascular disease. This significant association remains even when socio-economic factors are taken into account. The reason for the inverse association between cardiovascular mortality and drinking water hardness remains uncertain and there has been no serious suggestion that soft water supplies should be hardened and, indeed, some studies have failed to find any association. In spite of this, research in the field has continued, extending to examination of cerebrovascular mortality. A study from Sweden concluded that magnesium and calcium in drinking water were important protective factors for death from heart attacks among women[18] and a study from Taiwan[19] concluded that magnesium in drinking water was protective against cerebrovascular disease, although they found no significant association with calcium.

There is continuing interest in water hardness and health and there is now some evidence that hard water may exacerbate eczema in school children.

3.7.1.7 Other Inorganic Contaminants. There are of course many other inorganic contaminants that can reach drinking water.[20] Most of these are present in trace amounts, but some can be found in substantial quantities. Copper from domestic plumbing is usually present at a few tens of micro-grammes, but on occasion in aggressive waters with a long time in contact with the pipes it can be present at much higher concentrations. The primary concern, apart from the aesthetic impact of discoloured fittings, is acute gastrointestinal irritation, which although not life threatening is very unpleasant. It would appear that concentrations in excess of about 3 mg 1^{-1} can cause this in some individuals. In some parts of the world high selenium concentrations from seleniferous rocks has given rise to clinically detectable adverse effects, particularly skin lesions. Iron is found in many waters and can also give rise to dirty water as a consequence of disturbing the sediment from old cast iron water mains. The iron may naturally be present in the ferrous form but in drinking water the presence of oxygen rapidly oxidizes the iron into the ferric form. This is brown in colour and the primary problem with iron is discolouration of water, which can be detectable at iron concentrations as low as 0.4 mg l^{-1}. A similar problem exists with manganese, which can also give rise to black deposits of oxides in distribution and very unpleasant dirty water problems. Such problems are not only unpleasant for the consumer but they can potentially mask more serious problems and can undermine the confidence of consumers in the safety of the supply.

3.7.2 Organic Contaminants

3.7.2.1 Introduction. There are many organic substances that can reach drinking water sources, either as a consequence of discharges of wastewater or run-off from urban and agricultural areas.[21] Most are present in extremely small concentrations of no more than a few microgrammes per litre but occasionally much higher concentrations can be encountered as a consequence of spills. Organic contaminants are found in the greatest number in surface water but some can, and do, reach groundwater. It is not possible or productive to consider most of the potential organic contaminants but WHO (1993, 1996) have reviewed the occurrence and effects of a substantial number.

3.7.2.2 Disinfection By-products. The use of chemicals, such as chlorine, as disinfectants in water treatment has significantly contributed to the microbiological safety of drinking water this century. For example, eight out of ten waterborne outbreaks, between 1937 and 1986, involving public water supplies were associated with defective chlorination. All of the 13 outbreaks involving private supplies over the same period were also a result of problems with the disinfection process.[22] However, chlorine and similar disinfectants are reactive chemicals and will react with organic material such as humic and fulvic acids. This reaction results in the formation of unwanted by-products or disinfection by-products (DBPs), which are present in low concentrations. It was not until 1974 that Rook[23] demonstrated the presence of chloroform in the drinking water of Rotterdam. Since then intensive research has resulted in the identification of many more by-products of chlorination and, more recently, of ozone, some of which are present at concentrations of only a few nanograms per litre. The discovery of chloroform and other trihalomethanes in drinking water also coincided with the discovery that chloroform increased liver and kidney tumours in laboratory studies with rats and mice. A number of epidemiological studies were carried out, mainly in the USA, which found a weak association between chlorinated drinking water and cancer of the rectum, colon and, latterly, bladder. The majority of these studies are of the ecological type, analysing data that characterize a population rather than individuals, but there are also a number of case control studies, which compare individuals with and without the specific cancer under study. Where positive associations were found, they were weak to moderate with most relative risks below 2.0 and the mean below 1.5. The data relating to bladder cancer appear to be consistent between studies, but on closer examination the consistency between sub-groups within the studies is much less consistent. The data on colon and rectum are lacking in consistency. One case control study found a positive association with brain cancer, but there have been no other studies that have examined this site. All of the studies have some limitations in design and/or methodology. One problem in this respect is comparing chlorinated surface waters with unchlorinated groundwater, where there are other differences than chlorination by-products. Another is the problem of measuring exposure, particularly as considerable efforts have been made to reduce the concentrations of DBPs in drinking water so that current levels of exposure may give a misleading picture of exposure in the past.[24]

Chloroform has been shown to increase the number of liver and kidney tumours in laboratory rats and mice given high doses for their lifetime. Most of the studies used oral dosing in corn oil but subsequent studies using water as a dosing vehicle resulted in much lower numbers of tumours or no tumours at all. Other THMs have also been studied and of these the most convincing evidence for tumourigenicity is associated with bromodichloromethane which caused an increase in tumours of the liver, kidney and, in rats, the large intestine. Bromoform also gave rise to a small increase in large intestinal tumours in rats but chlorodibromomethane only caused an increase in liver tumours in mice, which are not usually considered to be relevant for man, particularly in the absence of tumours in a second species. There is considerable debate about the mechanism by which the THMs cause tumours since they are not mutagenic (chloroform) or only weak mutagens in *in vitro* studies, but are largely negative *in vivo*.[25] Several other important DBPs have been tested for carcinogenicity but although some, such as di- and trichloroacetic acid, cause increases in liver tumours in rats, they are not mutagenic and only cause tumours at extremely high doses. One DBP which is different is 3-chloro-4-dichloromethyl-5-hydroxy-2(5H)-furanone (MX) which is a very potent mutagen in bacterial tests. It is reported to induce a range of tumours in laboratory rodents but its mutagenic potency is not replicated *in vivo* and the estimated risks associated with the nanogram concentrations found in drinking water are extremely low.[6]

More recently concern has centred on the possibility of adverse birth outcomes being associated with drinking chlorinated water. In theory the problem of measuring exposure is smaller since the possible effects are unlikely to be related to paternal exposure and the critical exposure period will be up to nine months. The outcomes of interest which have emerged so far are low birth weight, preterm delivery, spontaneous abortions, stillbirth and various birth defects including defects of the central nervous system, heart, oral cleft, respiratory system and neural tube. The strongest evidence is for an association between low birth weight and total THMs but this evidence is not conclusive, partly because all of these studies suffer from methodological problems. In addition, the toxicological data do not, so far, provide evidence for a biologically plausible mechanism for the THMs or other DBPs to cause these effects. However, there are a number of other studies in progress and it will be some time before the weight of evidence is sufficient to determine whether these associations stand up to closer scrutiny and whether the relationship is likely to be causal.[26]

WHO have considered all of the evidence and have determined that there is insufficient evidence to conclude that there is a causal relationship between DBPs and either cancer or adverse birth outcomes in humans.[27] They have proposed guideline values for a range of DBPs in drinking water but they emphasize that effective disinfection should never be compromised in meeting guideline values for these substances.[4]

3.7.2.3 Pesticides.   One of the most common groups of contaminants cited by the media, and various pressure groups, as a problem in drinking water is pesticides. This concern arose from exceedences of the European drinking water standard of 0.1 μg l^{-1} for any pesticide and 0.5 μg l^{-1} for total pesticides.[1] This standard is a political rather than a scientifically based standard and, therefore its relevance to public health is limited. Pesticides are a very diverse group of substances with widely differing chemical characteristics and toxicity. WHO[4,5] and the USEPA[28] are two organizations that have promulgated health-based guidelines or standards for pesticides in drinking water and these values have been rarely exceeded in drinking water in the UK. Advanced water treatment processes have been installed in part to combat this problem with great success. Some of the major contaminating pesticides were herbicides, which will often tend to be of a higher water solubility than insecticides, particularly atrazine and simazine. These compounds were approved for use on hard standing to control weeds on the edges of roads, between paving and on other non-crop land such as railway lines because they were persistent. Unfortunately their persistence combined with water solubility resulted in significant movement to drains and to both ground and surface water. Some years ago the approval for use on non-crop land was withdrawn which has resulted in a steady decline in problems with these herbicides.

3.7.2.4 Endocrine Disrupters (see also Chapters 1 and 19).   Endocrine disrupting chemicals have become a general concern in the light of possible reductions in sperm quality over time and the increase in a number of cancers which are potentially influenced by endocrine activity, such as testicular cancer and breast cancer. Much of the evidence for endocrine disruption comes from aquatic wildlife. TBTO causes imposex in Dog Whelks and some other molluscs at very low concentrations. The mechanism is suspected to be by disruption of the endocrine system but there is still uncertainty as to the exact mechanism. Fish, living in streams affected by pulp and paper mill effluent, have been found to show masculinization of females although the actual substances responsible have not been properly identified. Male roach living in lagoons containing sewage effluent were found to exhibit ovarian tissue in their testis. Subsequently male trout exposed to sewage effluent were shown to produce vitellogenin, an egg yolk protein under the control of oestrogen, normally only observed in breeding females. In Lake Apopka in Florida where there had been a spill of the pesticide Dicofol which was heavily contaminated with DDE, the alligator population collapsed. It was shown that males had low serum testosterone and a reversal of the testosterone/oestrogen ratio. They also exhibited reduced phallus and seminiferous tubules while the females showed signs of super-feminization. Eggs dosed with the pesticide mixture showed similar developmental effects in the embryos but the mechanism appears to be through an antiandrogen effect rather than oestrogenicity. Similar effects were noted in Red Eared turtles in the lake.[29]

A number of chemicals have been shown to bind or block oestrogen and androgen receptors in yeast and cell culture assays. Some have also been shown to possess activity in whole animals but they are very much less active than natural hormones. In some cases, such as the phthalates, there is evidence for species differences in metabolism, which imply that they will not possess the same activity in primates as in rodents. Studies by the Environment Agency on fractionating sewage effluent showed that the primary cause of the activity in fish was the natural hormones, oestrone and 17β-oestradiol. These are excreted as the water soluble glucuronides and sulfates but these compounds are broken down by microbiological activity in sewage treatment to yield the parent hormone, some of which will also be broken down. In some cases there appeared to be a contribution from the synthetic steroid ethinyl oestradiol and, in one instance, alkyl phenols. In this case on the River Aire in Yorkshire there are a significant number of wool scourers which discharge to the sewage treatment works and they were using alkylphenol ethoxylate based industrial detergents. These were breaking down in treatment to give smaller molecules with shorter ethoxylate chains and the parent alkylphenols. When most of these companies switched to alternative detergents the vitellogenin response in trout was also reduced.[30]

Most of these substances are lipophilic and are readily removed in water treatment. Laboratory studies have also been shown that water treatment techniques remove or break down these substances fairly readily. These findings have been confirmed by specific chemical analysis at very low limits of detection. Epoxy resins used in relining cast iron water mains have also been examined for the presence of bisphenol A and F as contaminants since these have been proposed as possible endocrine disrupters, but, so far, none has been found in water as it is supplied.[31] An additional pointer for drinking water which is reassuring is that two of the countries reporting the greatest effects (increased testicular cancer and decreased sperm quality) in man, Scotland and Denmark, do not practise re-use of waste-water and so no sewage effluent is in the water used for water supply. In addition Scotland uses primarily upland surface water while Denmark uses clean groundwater.[32]

3.7.2.5 PAH. Polycyclic aromatic hydrocarbons are a substantial family of substances found widely in the environment as a consequence of incomplete combustion. In 1676 Percival Potts studied chimney sweeps in London who at that time were exposed to soot all over the body, particularly as boys were sent up chimneys to sweep them. Soot is high in many PAH including benzo[a]-

pyrene. The sweeps showed a significant level of scrotal cancer, which is a very rare cancer in other circumstances. Subsequently some of the PAH, including benzo[a]pyrene were shown to induce skin cancer in mice. Up to the 1950s, cast iron water mains were often lined with coal tar to protect them against the corrosive effect of the water. The lining was hard and shiny but only in very clean waters was it particularly durable. The most important PAH which leaches from coal tar linings is fluoranthene and this was included, with benzo[a]pyrene, in a group of six PAH to comprise the standard for drinking water of 200 ng l^{-1}. However, only fluoranthene is frequently observed at up to 1 μg l^{-1}, causing a breach of the standard. WHO have examined the toxicity of fluoranthene and the evidence indicates that although it is mutagenic in bacteria it is not carcinogenic in animals. WHO indicated that there was no need to propose a guideline value for fluoranthene, which has been removed from the current standard in the directive. Where problems arise with PAH in drinking water the cause is usually the disturbance of sediment in water mains and is, therefore, a consequence of particulate matter. In these circumstances the water would be discoloured and, possibly, would also give rise to taste and odour problems. The problem can be rapidly identified and exposure through drinking water can be minimized.[4,5]

3.7.2.6   Tri- and Tetrachloroethene.   These two substances which were widely used as chlorinated solvents have caused significant problems in some groundwaters. They are not found at significant concentrations in surface waters because they are rapidly lost to the atmosphere. However, they are poorly adsorbed to soil and are only degraded extremely slowly so when spilt on the ground they can move down to groundwater where they are very persistent. Improved controls on the way these substances are used and increased awareness by users has significantly reduced the amount of new contamination and the problem is now largely as a consequence of historical pollution. Both of these substances have shown evidence of increasing the number of tumours in laboratory animals, although they are, at worst, only weakly genotoxic. The International Agency for Research on Cancer (IARC)[33] re-evaluated the data on the epidemiology of renal cancer associated with exposure to both tri- and tetrachloroethene and concluded that there was some evidence to support a causal association. However, other groups would not agree with this conclusion and consider that the totality of the epidemiological data does not support a causal association. WHO, in 1993,[4] proposed guideline values of 30 μg l^{-1} and 70 μg l^{-1} for tri- and tetrachloroethene respectively on the basis of cancer in laboratory animal studies but caused by a non-genotoxic mechanism. However, the EU has introduced a more stringent precautionary standard of 10 μg l^{-1} for the two substances in total.[1]

3.8   CONCLUSIONS

Clean and wholesome drinking water is an essential requirement for a healthy life. There are many potential sources of contamination with both pathogens and toxic chemicals and the control and treatment of drinking water has evolved to combat these threats. The primary requirement for clean drinking water is that of microbiological safety. Our knowledge about waterborne pathogens has increased significantly and we now face pathogens that are new or newly recognized, such as *Cryptosporidium*. These have required changes in the approaches to monitoring drinking water and in drinking water treatment, particularly for those supplies derived from surface waters.

In general the concentrations of chemical contaminants are very small but there appears to be a higher perception of chemical contaminants by consumers than of microbiological contaminants. There are many possible contaminants, both inorganic and organic, that can arise naturally or from man's activities. However, in developed countries, the contaminants of greatest concern have been those arising from domestic plumbing, such as lead, and arising as a consequence of the use of chemical disinfection processes, such as chlorination. In the developing world there are other substances of concern, in particular, arsenic, fluoride, and nitrate. There are now rigorous standards for these and other contaminants and WHO are continuing to consider chemical and

Table 3.3 *Comparison of EC Directive indicator parameters and WHO Guidelines*

Parameter	EC Directive	WHO Guidelines	Comments (WHO only set GVs on basis of health)
Aluminium	200 μg l^{-1}	No GV	Prevents discolouration
Ammonium	0.5 mg l^{-1}	No GV	Indicator of possible bacterial, sewage and animal waste pollution
Chloride	250 mg l^{-1}	No GV	Corrosion and taste
Colour	Acceptable to consumers & no abnormal change	No GV	Indicator of pollution and operational problems
Conductivity	2.500 μS cm^{-1} at 20 °C	No GV	Corrosion
pH	$\geqslant$6.5 & $\leqslant$9.5	No GV	Operational parameter
Iron	200 μg l^{-1}	No GV	Discolouration. 2 mg l^{-1} not hazard to health (WHO 1993)
Manganese	50 μg l^{-1}	500 μg l^{-1}	Discolouration
Odour	Acceptable to consumers & no abnormal change	No GV	Indicator of pollution and operational problems
Oxidizability	5 mg l^{-1} O$_2$	No GV	Operational parameter not needed if Total Organic Carbon (TOC) measured
Sulfate	250 mg l^{-1}	No GV	Corrosion, taste
Sodium	200 mg l^{-1}	No GV	Taste
Taste	Acceptable to consumers & no abnormal change	No GV	Indicator of pollution or operational problems
Total Organic Carbon (TOC)	No abnormal change	No GV	Indicator of pollution and disinfectant demand
Turbidity	Acceptable to consumers & no abnormal change	No GV	Indicator of pollution and operational problems

microbiological contaminants in a rolling revision of their Guidelines for Drinking Water Quality. These guidelines have been, and will continue to be, the primary basis for drinking water standards in Europe and their revision in the future.

3.9 REFERENCES

1. Council Directive 98/83/EC on the quality of water intended for human consumption.
2. 'The Water Supply (Water Quality) Regulations 1989', Statutory Instrument 1147.
3. 'World Health Organization, Guidelines for Drinking-Water Quality' Volume 1, 'Recommendations', WHO, Geneva, 1984.
4. 'World Health Organization, Guidelines for Drinking-Water Quality', 2nd edn., Volume 1, 'Recommendations', WHO, Geneva, 1993.
5. 'World Health Organization, Guidelines for Drinking-Water Quality', 2nd edn., 'Addendum to Volume 1, Recommendations', WHO, Geneva, 1998.
6. 'World Health Organization, Guidelines for Drinking-Water Quality', 2nd edn., Volume 2, 'Health Criteria and Other Supporting Information', WHO, Geneva, 1996.
7. 'The Water Supply (Water Quality) (Amendment Regulations) 1999', *Cryptosporidium* in Water Supplies.
8. E. Salazar-Lindo, M. Alegre, M. Rodriguez, P. Carrion and N. Razzeto in 'Safety of Water Disinfection: Balancing Chemical and Microbial Risks', ed. G. F. Craun, ILSI Press, Washington, D.C., 1993, 401.
9. Joint FAO/WHO Expert Committee on Food Additives, 'Evaluation of Certain Food Additives and Contaminants: Forty-first Report', WHO Technical Report Series, No. 837, WHO, Geneva, 1993.
10. Institute for Environment and Health, 'Recent UK Blood Lead Surveys', Report R9, IEH, Leicester, 1998.
11. National Research Council, 'Nitrate and Nitrite in Drinking Water', National Academy Press, Washington, D.C., 1995.
12. J. M. Van Maanen, H. J. Albering, S. G. Van Breda, D. M. Curfs, A. W. Amberg, B. H. Wolffenbuttel, J. C. Kleinjans and H. M. Reeser, *Diabetes Care*, 1999, **22**, 1750.
13. A. Casu, M. Carlini, A. Contu, G. F. Bottazzo and M. Songini, *Diabetes Care*, 2000, **23**, 1043.
14. 'International Programme on Chemical Safety, Aluminium', EHC 194, WHO, Geneva, 1997.
15. N. D. Priest, R. J. Talbot, D. Newton, J. P. Day, S. J. King and L. K. Fifield, *Hum. Expt. Toxicol.*, 1998, **17**, 296.
16. National Academy of Sciences, 'Health Effects of Ingested Fluoride', National Academy Press, Washington, D.C., 1993.
17. National Research Council, 'Arsenic in Drinking Water', National Academy Press, Washington, D.C., 1999.
18. E. Rubenowitz, G. Axelsson and R. Rylander, *Epidemiology*, 1999, **10**, 31.
19. C. Y. Yang, *Stroke*, 1998, **29**, 411.
20. J. K. Fawell, *Ann. Ist. Super. Sanita.*, 1993, **29**, 293.
21. M. Fielding, T. M. Gibson, H. A. James, K. McLoughlin and C. P. Steel, 'Organic Micropollutants in Drinking Water', Technical Report TR159, Water Research Centre, Marlow, 1981.
22. N. J. Galbraith and R. Stanwell-Smith. *JIWEM*, 1987, **1**, 7.
23. J. J. Rook, *Water Treatment Exam.*, 1974, **23**, 234.
24. G. F. Craun, in 'Safety of Water Disinfection: Balancing Chemical and Microbial Risks', ed. G. F. Craun, ILSI Press, Washington, D.C., 1993, 277.

25. J. K. Fawell, *Food Chem. Toxicol.*, 2000, **38**, suppl. 1, S91.
26. M. J. Nieuwenhuijsen, M. B. Toledano, N. E. Eaton, J. Fawell and P. Elliott, *Occ. Env. Med.*, 2000, **57**, 73.
27. International Programme on Chemical Safety, 'Disinfectants and Disinfectant By products', EHC 216, WHO, Geneva, 2000.
28. http://www.epa.gov/safewater/mcl.htm
29. Institute for Environment and Health, 'Environmental Oestrogens: Consequences to Human Health and Wildlife. Assessment A1', IEH, Leicester, 1995.
30. J. K. Fawell, D. Sheahan, H. A. James, M. Hurst and S. Scott, *Water Res.*, 2001, **35**, 1240.
31. J. K. Fawell and J. K. Chipman, *J. CIWEM.*, in press.
32. J. K. Fawell and M. Wilkinson, *J. Water. SRT-Aqua.*, 1994, **43**.
33. International Agency for Research on Cancer, 'IARC Monograph', Vol. 63, IARC, Lyon, 1995.

CHAPTER 4

Water Pollution Biology

C. F. MASON

4.1 INTRODUCTION

Pollution can be defined as "the introduction by man into the environment of substances or energy liable to cause hazards to human health, harm to living resources and ecological systems, damage to structure or amenity, or interference with legitimate uses of the environment".[1] Pollutants are therefore chemical or physical in nature and can be measured more or less accurately in water. The measured quantities can then be compared with standards of allowable concentrations. Why then do we need to undertake quantitative studies of organisms and biological communities when we know that they can be defined with much less precision than chemical or physical parameters? There are a number of reasons why biological studies are important.

Firstly, the definition of pollution given above includes the adverse effects on living resources and ecological systems, so that these impacts need quantifying. People, of course, by drinking water, eating fish, and using freshwaters as recreation areas, are also linked to the aquatic environment. Indeed, in many developing countries, the river is often central to the life of the community. We can consider the effects of pollutants we record on aquatic life as an early warning of potential harm to ourselves.

Secondly, animal and plant communities respond to intermittent pollution which may be missed in a chemical surveillance programme. A chemical survey will indicate to the water manager what is present in a sample at a particular moment in time. However, a plug of pollution could have passed down a river before the sample was taken, or between sampling occasions, while a disreputable factory owner or farmer may well know which day the river inspector takes his sample so that he can discharge his toxic waste accordingly. The pollution will kill the most vulnerable members of the aquatic community, the most sensitive species acting as *indicators* of pollution. The amount of change in the community will be related to the severity of the incident. Because the community can only be restored to its former diversity by reproduction and immigration, its recovery is likely to be slow. If the intermittent pollution occurs with some frequency, the community will remain impoverished. The biologist will be able to detect such damage and suggest a more detailed surveillance programme, both biological and chemical, to find the source of pollution.

Thirdly, biological communities may respond to unsuspected or new pollutants in the environment. There are well in excess of 1500 potential pollutants discharged to freshwaters, while only some 30 determinands will routinely be tested for. It is financially prohibitive to look for more compounds. If there is a change in the biological community, however, then a wider screening for pollutants can be initiated. For example, male rainbow trout (*Oncorhynchus mykiss*), caged in the River Lea, north of London, below the point of discharge of effluent from a sewage treatment works, exhibited marked increases in vitellogenin levels.[2] This compound is normally produced in the liver of female fish in response to the hormone oestradiol and is incorporated into the yolk of developing eggs. Clearly, some chemical in the effluent was behaving like a female hormone and the male trout provided an early warning of a potential problem requiring urgent investigation. Nonylphenols (widely used as antioxidants) were suspected of stimulating vitellogenin production in the trout. The effect has since been demonstrated in fish from a number of other rivers. Laboratory experiments using yeast cells that have been genetically modified to respond to chemicals with oestrogenic activity have identified three such active compounds in treated sewage effluent, two naturally occurring hormones and an active ingredient of birth control pills.[3] There are many other substances, including DDT and some PCBs, which exert similar effects. For example, abnormalities in the eggs and hatchlings of snapping turtles (*Chelydra serpentina*) from the Great Lakes region of North America were considered to be caused by polychlorinated aromatic hydrocarbons.[4] Increases in testicular cancer and falling sperm counts in the human male may be related to oestrogen-mimics released into the environment,[5] and these *endocrine disruptors* are now a major cause of concern and research effort.

Finally, some chemicals are accumulated in the bodies of certain organisms, concentrations within them reflecting environmental pollution levels over time. In any particular sample of water, the concentration of a pollutant may be too low to detect using routine methods, but nevertheless will be gradually accumulated within the ecosystem to levels of considerable concern in some species. Heavy metals, organochlorine pesticides, and PCBs have caused particular problems in aquatic habitats and are potential threats to human health. Eels (*Anguilla anguilla*), for example, have been used to detect mercury below discharges of sewage effluent.[6]

Pollutants having biological impacts may derive from *point sources*, often discharges known to the authorities, which can be effectively controlled if resources are available. However, the example of endocrine disruptors above emphasizes that we are not always aware of the effects on the ecology of a receiving river of compounds coming from apparently well-regulated sewage effluent discharges. Alternatively, sources of pollution may be *diffuse*, entering watercourses from land drainage and run-off, fertilizers and pesticides applied to crops providing examples. Much pollution is *chronic*, the watercourse receiving discharges continuously or regularly, and such pollution can generally be reduced to acceptable levels given the right regulatory framework. A greater problem is *episodic* or accidental pollution, which is unpredictable in space or time. A recent example is the spillage of five million cubic metres of toxic waste when a dam burst at a mine in southern Spain in April 1998. The site was upstream of the Coto Doñana National Park, one of Europe's largest and most important wetlands. The acid water and its burden of toxic metals killed all life, including some 30 000 tons of fish, in its path as it headed towards the park.[7] It is too early to know the long-term effects but one year after the incident half of the wintering greylag geese (*Anser anser*), and many other wetland birds, were considered to have near lethal concentrations of metals in their tissues. Accidents do happen but many, like this dam burst, could be avoided if thorough risk assessments are carried out.

This review will describe the effects of major types of pollutants on aquatic life, though it must be remembered that any particular source of pollution may contain a range of compounds, such that the effects on ecosystem processes are likely to be complex. The review will conclude with a description of some of the methods of assessing the biological impacts of pollution.

4.2 ORGANIC POLLUTION

Organic pollution results when large quantities of organic matter are discharged into a watercourse to be broken down by microorganisms which utilize oxygen to the detriment of the stream biota. A simple measure of the potential of organic matter for deoxygenating water is given by the *biochemical oxygen demand* (BOD) which is determined in the laboratory by incubating a sample of water for five days at 20 °C and determining the oxygen used.

One of the major sources of organic pollution are effluents from sewage treatment works. In the United Kingdom, such effluents are supposed, as a minimum requirement, to meet the Royal Commission Standard, allowing no more than 30 mg l^{-1} of suspended solids and 20 mg l^{-1} BOD (a 30:20 effluent). The effluent should be diluted with at least eight volumes of clean river water, having a BOD of no more than 2 mg l^{-1}, to ensure that there is minimal impact on the aquatic environment. Unfortunately, the design capacities of many sewage treatment works are below the population they are now having to serve (Figure 4.1). This may cause chronic pollution of rivers or result in periodic flushes of poor quality water which damages the aquatic community. In the majority of poorer countries of the world there are few, or indeed often no, sewage treatment facilities and the faecal contamination of water results in many parasitic infections and waterborne diseases such as dysentery, cholera, and poliomyelitis. One billion people do not have access to a safe water supply and two billion have no safe sanitation. Contaminated water supplies are the cause of one third of all deaths in the developing world, children being especially susceptible. Countless millions more are made ill by waterborne diseases.

Other sources of organic pollution include industries such as breweries, dairies, and food processing plants. Run-off from hard surfaces and roads of towns, especially during storm conditions, can be very polluting. Farm effluents

Figure 4.1 *A poor quality effluent discharged from a sewage treatment works*

have become a particular problem over the last two decades, with the intensification of livestock rearing and the overwintering of animals in confined buildings, as well as the increased use of silage to feed them. Silage effluent can be 200 times as strong as settled sewage, measured in terms of BOD. In Southwest England, a predominantly rural area, many rivers are failing to comply with their long-term quality objectives, agricultural pollution being the major reason for the decline.

Figure 4.2 outlines the general effects of an organic effluent on a receiving stream.[1] At the point of entry of the discharge there is a sharp decline in the concentration of oxygen in the water, known as the *oxygen sag curve*. At the same time there is a large increase in BOD as the microorganisms added to the stream in the effluent and those already present utilize the oxygen as they break down the organic matter. As the organic matter is depleted, the microbial populations and BOD decline, while the oxygen concentration increases, a process known as *self-purification*, assisted by turbulence within the stream and by the photosynthesis of algae and higher plants.

The effluent will also contain large amounts of suspended solids which cut out the light immediately below the discharge, thus eliminating photosynthetic organisms. Suspended solids settle on the stream bed, altering the nature of the substratum and smothering many organisms living within it.

Under conditions of fairly heavy pollution, *sewage fungus* develops. This is an attached, macroscopic growth containing a whole community of microorganisms, dominated by *Sphaerotilus natans*, which consists of unbranched filaments of cells enclosed in sheaths of mucilage, and by zoogloeal bacteria. Sewage fungus may form a white or light brown slime over the surface of the

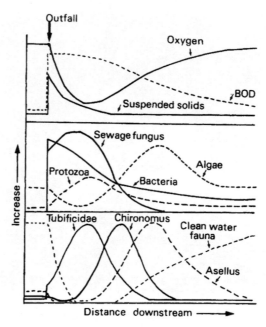

Figure 4.2   *Changes in water quality and populations of organisms in a river below a discharge of an organic effluent*

substratum, or it may exist as a fluffy, fungus-like growth with long streamers trailing into the water.

Protozoans are chiefly predators of bacteria and respond to changes in bacterial numbers. Attached algae are eliminated immediately below the outfall due to the diminished penetration of light, but they gradually reappear below the zone of gross pollution, *Stigeoclonium tenue* being the initial colonizer. With the decomposition of organic matter large quantities of nitrates and phosphates are released, stimulating algal growth and resulting in dense blankets of the filamentous green *Cladophora* smothering the stream bed. Similarly, higher aquatic plants (macrophytes) may respond to the increased nutrient concentration, though only *Potamogeton pectinatus* is very tolerant of organic pollution.

Heavy organic pollution affects whole taxonomic groups of macroinvertebrates, rather than individual sensitive species and it is only in conditions of mild pollution that the tolerances of individual species within a group assume significance. The groups most affected are those which thrive in waters of high oxygen content and those which live on eroding substrata, the most sensitive being the stoneflies (*Plecoptera*) and mayflies (*Ephemeroptera*). The differences in tolerances of groups of macroinvertebrates form the basis of methods for monitoring, as will be described later in this chapter.

In the most severe pollution the tubificid worms, *Limnodrilus hoffmeisteri* and/or *Tubifex tubifex*, are the only macroinvertebrates to survive. The organic effluent provides an ideal medium for burrowing and feeding, while in the absence of predation and competition, the worms build up dense populations, often approaching one million individuals per square metre of stream bed. These worms contain the pigment haemoglobin, which is involved in oxygen transport, and they can survive anoxic conditions for up to four weeks. As conditions improve slightly downstream, the larvae of the midge *Chironomus riparius*, which also contains haemoglobin, thrives in dense populations and, as the water self-purifies, other species of this large family of flies appear, the proportion of *Chironomus riparius* gradually declining. Below the chironomid zone, the isopod crustacean *Asellus aquaticus* becomes numerous, especially where large growths of *Cladophora* occur. At this point, molluscs, leeches and the predatory alder fly (*Sialis lutaria*) may also be present in some numbers. As self-purification progresses downstream the invertebrate community diversifies, though some stonefly and mayfly species, which are sensitive even to the mildest organic pollution, may not recolonize the stream.

Fish are the most mobile members of the aquatic community and they can swim to avoid some pollution incidents. In conditions of chronic organic pollution they are absent below the discharge, reappearing in the *Cladophora/Asellus* zone, the tolerant three-spined stickleback (*Gasterosteus aculeatus*) being the first to take advantage of the abundant invertebrate food supply. Organic pollution is usually most severe in the downstream reaches of rivers and may prevent sensitive migratory species, such as salmon (*Salmo salar*) and sea trout (*S. trutta*), from reaching their pollution-free breeding grounds in the headwaters.

The impact of organic pollution may extend to organisms living outside of the river channel. Pipistrelle bats (*Pipistrellus pipistrellus*) have been observed to forage significantly less downstream of sewage effluent discharges than in adjacent stretches upstream of the discharge,[8] presumably because of a reduction in emerging insects.

It must be remembered that many toxic compounds, such as ammonia, may occur in organic effluents, particularly those emanating from sewage treatment works. These make the prediction of the effects on the aquatic community of any particular discharge rather difficult.

4.3 EUTROPHICATION

It has already been described how the release of nutrients during the breakdown of organic matter stimulates the growth of aquatic plants. The addition of nutrients to a waterbody is known as eutrophication. Other important sources of nutrients include phosphorus-containing detergents (much of these entering the river in sewage effluent), agricultural run-off and leaching of artificial fertilizers, the washing of manure from intensive farming units into water, the burning of fossil fuels which increases the nitrogen content of rain, the felling of forests which causes increasing erosion and run-off, and erosion resulting from recreational boating.

Nitrogen and phosphorus are the two nutrients most implicated in eutrophication and, because growth is normally limited by phosphorus rather than nitrogen, it is the increase in phosphorus which stimulates excessive plant production in freshwaters. Nitrate is highly soluble and fertilizers form the main source of nitrogen input to rivers, accounting for some 70% of the annual mass flow of nitrogen in East Anglian rivers. Some 50% of the nitrogen applied to crops is lost to water. Phosphorus is largely insoluble, so erosion is the main route of the element from land to water; 90% of the phosphorus entering East Anglian rivers is in the effluent from sewage treatment works.

The concentrations of nitrate and phosphate in the water of Ardleigh Reservoir, a eutrophic waterbody in East Anglia, are shown in Figure 4.3.[9] Note that the concentration of nitrate increases during the late winter when fertilizer is applied to growing crops and is washed in large amounts into streams which feed the reservoir. By contrast, the concentration of phosphate peaks in late summer, when low flows in the feeder streams consist largely of treated sewage effluent while, at this time of year, much phosphate is released from the reservoir sediments into the water.

Table 4.1 lists the guidelines for assessing the trophic status of a waterbody. Peak phosphorus concentrations in Ardleigh Reservoir were some 250 times the minimum concentration for assigning a waterbody as eutrophic, while peak concentrations of nitrogen were 10 times the minimum.

One of the major biological effects of eutrophication is the stimulation of algal growth. As eutrophication progresses, there is a decline in the species diversity of the phytoplankton and a change in species dominance as overall populations and biomass increase. Figure 4.4 illustrates the seasonal changes in biomass of the dominant groups of algae in Ardleigh Reservoir.[9] Typical of

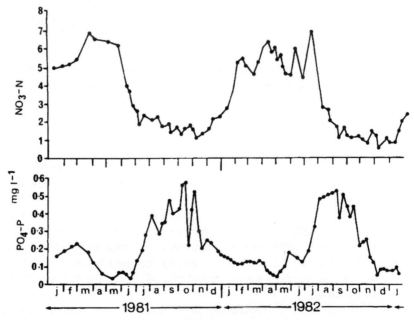

Figure 4.3 *Concentrations* (mg l^{-1}) *of nitrate and phosphate in Ardleigh Reservoir, eastern England, over two years*

Table 4.1 *Eutrophication survey guidelines for lakes and reservoirs*

	Oligotrophic	Mesotrophic	Eutrophic
Total phosphorus ($\mu g\ l^{-1}$)	< 10	10–20	> 20
Total nitrogen ($\mu g\ l^{-1}$)	< 200	200–500	> 500
Secchi depth (m)	> 3.7	3.7–2.0	< 2.0
Hypolimnetic dissolved oxygen (% saturation)	> 80	10–80	< 10
Chlorophyll-a ($\mu g\ l^{-1}$)	< 4	4–10	> 10
Phytoplankton production ($g\ C\ m^{-2}\ d^{-1}$)	7–25	75–250	350–700

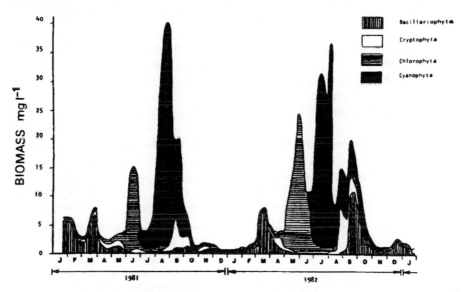

Figure 4.4 *Seasonal variation in phytoplankton composition and total biomass* (mg l^{-1} *wet weight) in Ardleigh Reservoir, eastern England, over two years*

temperate lakes is the early peak of diatoms (*Bacillariophyta*), followed by a late spring peak of green algae (*Chlorophyta*). Eutrophic lakes are characterized by enormous summer growths of cyanobacteria, in the case of Ardleigh Reservoir mainly *Microcystis aeruginosa*.

Large populations of algae, and in some cases of the zooplankton which they support, may block the filters in the water treatment works. The reservoir may then have to be taken out of service, sometimes for several weeks, because the water becomes untreatable, resulting in considerable extra cost in providing alternative sources of potable water. Some of the smaller algae may pass through the treatment process altogether, to decompose in the water main, giving drinking water an unpleasant taste and odour. The crash of an algal bloom in the reservoir and its subsequent decomposition may also result in a potable water so unpleasant to taste that it is almost undrinkable.

Drinking water which is high in nitrates presents potential health problems. In particular, babies under six months of age who are bottle fed may develop *methaemoglobinaemia* (blue baby syndrome), the nitrate in their milk being reduced to nitrite in their acid stomachs and then oxidizing ferrous ions in the haemoglobin of their blood, so lowering its oxygen carrying capacity. There have been several thousand cases world-wide, many of them fatal, but none in the United Kingdom since 1972. The disease only occurs when bacteriologically impure water, containing nitrate levels approaching 100 mg l^{-1}, is supplied. Treated, piped water is safe, the problem being associated with water from wells. There are also claims, so far unsubstantiated, that nitrates may convert into carcinogenic nitrosamines in the adult stomach. The European Health Standard for drinking water recommends that nitrates should not exceed 50 mg NO_3 l^{-1}, but concentrations in some lowland areas of Britain frequently exceed 100 mg NO_3 l^{-1}.

The algal blooms associated with excessive amounts of nutrients have other consequences for the aquatic ecosystem. The macrophyte swards of many lakes have been eliminated as the light is reduced on the lake bed, preventing photosynthesis of germinating plants. Growths of epiphytic algae on the leaf surface may also restrict light uptake by aquatic plants so that they become scarce in the lake. Zooplankton use macrophyte canopies as refuges from fish. Without macrophytes, they are very vulnerable and this itself can accelerate the eutrophication process as the grazing pressure on phytoplankton is reduced in the absence of zooplankton, allowing denser algal blooms to develop.

The Norfolk Broads of eastern England are a series of shallow lakes, formed during medieval times when peat workings became flooded, and famous for their rich flora of *Charophytes* (stoneworts) and aquatic angiosperms, supporting a diverse assemblage of invertebrates. During the 1960s a rapid deterioration set in as nutrient concentrations increased and a survey of 28 of the main broads in 1972–73 revealed that eleven were devoid of macrophytes and only six had a well developed aquatic flora. The invertebrate fauna was similarly impoverished. Since then the situation has deteriorated further with additional losses of macrophytes and the Norfolk Broads now have some of the highest total phosphorus concentrations of world freshwater lakes.[10]

It appears that shallow lakes have two alternative stable states over a range of nutrient concentrations.[11] They may either have clear water, dominated by aquatic vegetation, or turbid water with high algal biomass. The clear water, macrophyte community may be stabilized by luxury uptake of nutrients, making nutrients unavailable to plankton, by secreting chemicals to prevent plankton growth, and by sheltering large populations of grazing zooplankton which eat algae. Fish predation of zooplankton is reduced in the structured macrophyte sward.[12] The turbid water phytoplankton community may be stabilized by an early growing season, shading the later germination of macrophytes, and by producing large, inedible algae, which also acquire carbon dioxide more easily, especially at high pH. An absence of macrophytes increases the vulnerability of herbivores to predation by fish in the unstructured environment, while the predatory pike (*Esox lucius*) declines without cover for ambushing its prey, allowing even larger populations of planktivorous fish to survive.[13]

In many lakes, including the Norfolk Broads, the stability of the macrophyte community has been broken. The causes of the switch may have included the widespread introduction of insecticides and herbicides into agriculture in the 1950s. Zooplankton are especially vulnerable to insecticides, while the cocktail of herbicides running off from land into watercourses could have weakened the growth of macrophytes. The loss of zooplankton would have allowed the build-up of large algal populations, switching the lakes into the turbid alternative stable state.

Fish communities also change as oligotrophic lakes become more eutrophic. Cold water fish with high oxygen requirements, such as salmonids and white-fish, are replaced by less demanding cyprinids, the commercial value of the fishery declining. Some algae at high densities produce toxins which kill fish. In the Norfolk Broads *Prymnesium parvum* has caused several large fish kills over the last four decades. The sudden collapse of an algal bloom may result in rapid deoxygenation of the water and large kills of fish. Cyanobacteria may also produce potent poisons which can induce rapid and fatal liver damage at low concentrations. Livestock, dogs and wildlife have been killed. Toxins do not always occur in blooms and can be highly variable with time, making them difficult to predict, detect and monitor.[14]

Problems of eutrophication are not restricted to standing waters. Nutrient rich streams support dense growths of aquatic weeds which impede the flow of water, increasing the risk of severe flooding following summer storms. The weeds have to be cut, often twice a year, at great expense and the situation has been made worse by the clearance in the interests of 'river improvement' of bankside trees, which otherwise shade the stream and reduce weed growth.

It is possible to reverse eutrophication though, in many cases, expensive remedial measures have proved much less successful than was hoped.[15] Methods include 'bottom-up' (nutrient control) and 'top-down' (biomanipulation) approaches. Bottom-up approaches involve controlling inputs of phosphorus because this element is normally limiting to plant growth and much of it in freshwaters is derived from point sources (*e.g.* sewage treatment works), whereas nitrogen enters the aquatic ecosystem diffusely, *via* land drainage. Modifications in agricultural practices may reduce erosion and the loss of phosphorus to water, while the development of buffer strips of grassland, woodland or marshland along the sides of streams may help to filter out nutrients. Phosphorus can be removed from sewage effluent at the treatment works by chemical flocculation (tertiary treatment) or other techniques and the phosphorus recovered is potentially a valuable renewable resource.[16] Phosphorus may also be precipitated before water enters a lake or reservoir, in pre-treatment lagoons.[17] However, the reduction in phosphorus in the inflow water may be counterbalanced by the regeneration of phosphorus from the sediments, so-called *internal loading*, which may be as high as 278 mg P m^{-2} d^{-1} in the Norfolk Broads.[18] To reduce internal loading the sediment must be removed, or phosphorus can be precipitated out of the water column and sealed in the sediment with chemicals such as iron or aluminium sulfate,[17] the floc of ferric hydroxide which forms acting as a barrier to phosphorus release.

Biomanipulation normally requires the management (or complete removal) of planktivorous fish, allowing the re-establishment of populations of the larger herbivorous zooplankton, which graze algae.[19,20] Fish removal has led to increases in water clarity, encouraging the growth of macrophytes. Once the macrophyte stable state has developed it should be possible, indeed it is desirable, to reintroduce fish, along with the predatory pike. Biomanipulation technology, however, is still rather new and the clear water phase may only be temporary, several lakes showing increased turbidity after the first few years of management. Some lakes have not responded as predicted and it appears that an increase in herbivorous wildfowl prevent the development of macrophytes.[21] The control of phosphorus is in most cases a necessary prerequisite to biomanipulation, which is likely to be most effective in shallow waters where macrophytes are a major component of the ecosystem.

4.4 ACIDIFICATION

Acid rain and acidification of freshwaters has received considerable recent attention but the problem is not new, for there were observations of lakes losing their fish populations in Scandinavia as early as the 1920s. Studies of the diatom remains in cores of sediment from lochs in South-west Scotland have indicated that progressive acidification began around 1850, with an increase in *Tabellaria binalis* and *T. quadriseptata*, species characteristic of acid waters.[22,23] However, there has certainly been an acceleration of the process in the last three decades. For example, of 87 lakes in southern Norway surveyed in both the periods 1923–49 and 1970–80, 24% had a pH below 5.5 in the earlier period compared to 47% in the second period.[24] The major source of acid is undoubtedly the burning of fossil fuels, releasing oxides of sulfur and nitrogen to the atmosphere. Electricity generating stations produce much of these polluting gases but domestic and industrial sources are also significant, as are the exhausts of vehicles.

The acids either fall directly into waterbodies in precipitation or are washed in from vegetation and soils within the catchment. Three broad categories of water which differ in acidity can be recognized:

(i)   Those which are permanently acid, with a pH less than 5.6, low electrical conductivity, and an alkalinity close to zero. Such conditions occur in the headwaters of streams and in lakes, where the soils are strongly acid, or in the outflows of peat bogs.

(ii) Those which are occasionally acid, where pH is normally above 5.6 but because they have low alkalinity (usually less than 5.0 mg l^{-1} $CaCO_3$) the pH may drop below 5.6 periodically. These include streams and lakes in upstream areas of low conductivity on rocks unable to neutralize acid quickly. Such waters may show episodes of extreme acidity, for instance during snowmelt or following storms. These may be very damaging to aquatic life but the infrequency of acid events makes the problem difficult to detect.

(iii) Those which are never acid, the pH never dropping below 5.6 and the alkalinity always above 5 ml^{-1} CaCO$_3$.

Much of northern and western Britain has a solid geology consisting of granites and acid igneous rocks; there is little or no buffering capacity. The situation is exacerbated in those catchments which have been extensively planted with conifer forests. The sulfate ion is very mobile and transfers acidity very efficiently from soils to surface waters. The nitrate ion behaves similarly but is normally quickly taken up by plant roots.

In Canada, one experimental lake was artificially acidified, from 1975, by adding sulfuric acid and the pH dropped from 6.8 to around 5.0 over an eight year period. Among the first animals to disappear from the lake were shrimps and minnows. Within a year, at pH 5.8 opossum shrimps (*Mysis relicta*), with an estimated population of almost seven million before the experiment, had all but disappeared. Fathead minnow (*Pimephales promelas*) failed to reproduce and died out a year after the shrimps. Young trout were failing to appear and many of their food items were killed at pH 5.8 and as the pH fell still further cannibalism was noted in trout. At pH 5.6 the exoskeleton of crayfish (*Oronectes virilis*) began to lose its calcium and the animals became infested with a microsporozoan parasite. Some species, such as chironomid midges, did well as they were released from the pressure of predation in the simplified ecosystem. Since no species of fish reproduced at values of pH below 5.4, it was predicted that the lake would become fishless within ten years, based on knowledge of the natural mortalities of long-lived species. Some observations were unexpected in this experiment. There was no decrease in primary production, in the rate of decomposition, or in nutrient concentrations. The changes in the lake were caused solely by changes in hydrogen ion concentration and not by any secondary effects such as aluminium toxicity.[25]

Table 4.2 provides a generalized summary of the sensitivity of aquatic organisms to lowered pH based on studies in Scandinavian lakes. Changes in the community begin at pH 6.5 and most species have disappeared below pH 5.0 leaving just a few species of tolerant insects and some species of phyto and zooplankton.

Considerable research has been directed towards the effects of acidification on fish because of their economic and recreational importance. Figure 4.5 shows

Table 4.2 *Sensitivities of aquatic organisms to lowered pH*

pH 6.0	Crustaceans, molluscs, *etc.* disappear. White moss increases
pH 5.8	Salmon, char, trout and roach die
	Sensitive insects, phytoplankton and zooplankton die
pH 5.5	Whitefish, grayling die
pH 5.0	Perch, pike die
pH 4.5	Eels, brook trout die

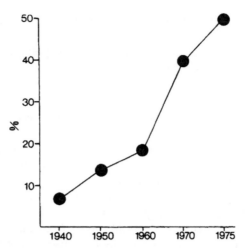

Figure 4.5 *Percentage of lakes* (n = 2850) *in Scandinavia which have lost their population of brown trout*

the proportion of lakes in southern Norway which have lost their brown trout (*Salmo trutta*) over the 35 years since 1940. By 1975 over half a total of 2850 lakes were without trout. The effect of acidity on fish is mediated *via* the gills. The blood plasma of fish contains high levels of sodium and chloride ions and those ions which are lost in the urine or from the gills must be replaced by active transport, against a large concentration gradient, across the gills. When calcium is present in the water it reduces the egress of sodium and chloride ions and the ingress of hydrogen ions. The main cause of mortality in acid waters is the excessive loss of ions such as sodium which cannot be replaced quickly enough by active transport. When the concentrations of sodium and chloride ions in the blood plasma fall by about a third, the body cells swell and extracellular fluids become more concentrated. To compensate for these changes, potassium may be lost from the cells, but, if this is not eliminated quickly from the body, depolarization of nerve and muscle cells occurs, resulting in uncontrolled twitching of the fish prior to death.

Aluminium has been shown to be toxic to fish in the pH range 5.0–5.5 and, during episodes of acidity, aluminium ions are frequently present in high concentrations, especially where water is draining conifer plantations. Aluminium ions apparently interfere with the regulation by calcium of gill permeability so enhancing the loss of sodium in the critical pH range. They also cause clogging of the gills with mucus and interfere with respiration. The developmental stages of fish are particularly sensitive to acidification[26,27] and it is thought that aluminium may interfere with the calcification of the skeleton of fish fry resulting in a failure of normal growth. This failure in recruitment results in a gradual decline of the fish population to extinction.

Many invertebrate species, such as crustaceans, molluscs and caddis are absent from acidified streams even though their food supply may be present. Physiological stress is the most likely cause. Aquatic invertebrates need to actively take up sodium, chloride, potassium and calcium ions for survival. Uptake is dependent on external concentrations. In acid waters ion concentrations may be too low while hydrogen and aluminium ions become dominant in the water. These ions are small and mobile and may be transported in instead of essential ions, upsetting the normal ionic balance, leading to death. Aluminium in acidic conditions has be has been shown to damage the ion-regulatory organs of caddis larvae, disrupting osmoregulation and leading to increased mortality. Sensitivity to aluminium differed between species.[28]

The simplification of the aquatic ecosystem due to acidification may also affect higher levels of the food chain, such as birds and mammals. The dipper (*Cinclus cinclus*) lives along swiftly flowing streams where it feeds mainly on aquatic invertebrates which it collects by 'flying' underwater or foraging on the stream bottom. Dippers in Wales are scarcest along streams with low mean pH (less than 5.5–6.0) and elevated concentrations of aluminium (greater than 0.80–1.0 g m^{-3}). A decline in pH of 1.7 units on one river over the period 1960–1984 resulted in a 70–80% decline in the dipper population. On acidic streams nesting started later and clutch sizes were smaller than in those dippers breeding elsewhere.[29] If fish populations are eliminated from headwaters of streams by acidification, then the piscivorous otter (*Lutra lutra*) may not use them. In general, however, acidification results in a reduction in the carrying capacity for otters rather than a decrease in distribution.[30]

The process of acidification has been reversed by liming, enabling normal plant and animal communities to be maintained. In some cases a single treatment of acidified headwater streams has proved successful and relatively inexpensive.[31] However, due to the remoteness of most localities and distance from sources of limestone the technique is generally very expensive and, because of the vast number of lakes and streams which have lost their fish, it is hardly realistic on a large scale.

Re-acidification may occur, resulting in high mortality of stocked trout.[32] For long-term effects, liming of catchments rather than water might prove more effective. Clearly the control of emissions at power stations and from vehicle exhausts is a more sensible alternative. Emissions of sulfur dioxide in the UK have been halved since 1980. During the 1980s it was predicted that a 30% reduction in sulfur emissions in Europe would reduce the number of acid lakes in southern Norway by only 20%, while a 50% reduction would reduce the number by only 35%.[33] In South-west Scotland mathematical modelling has indicated that a 50% reduction in acid emissions would be necessary to prevent further increases in stream acidity on moorlands, more for afforested catchments.[34] Despite the marked reductions of acidifying emissions, it has been predicted that by 2010 the critical load for sulfur in Sweden, for example, will still be exceeded in 1.6 million hectares, concentrated in the south and southwest of the country. It is considered that the acidification of freshwaters will remain a problem long after emissions have been reduced because of the large amount of sulfur deposited in soils.

It is necessary briefly to mention the problems of acid mine drainage which, for example, affects some 19 000 km of streams and 73 000 ha of lakes in the

USA alone.[35] Chemosynthetic bacteria in mine water oxidize the mineral iron pyrite to ferric hydroxide with the formation of sulfuric acid, the water flowing from mines having a pH often below 3.5, resulting in the extermination of much of the biota in the watercourse into which it flows. The problem becomes much worse when mines are abandoned and pumping ceases, as is currently being observed in many areas of Great Britain following the demise of the mining industry. Pumping during the active life of a mine keeps the water table several hundreds of metres below ground but when pumping stops the water levels begin to rise. The acidic conditions bring metals, from ores exposed by the mining activity, into solution so that the effluent from the mine is also highly toxic. The closure of the Wheal Jane tin mine in South-west England led to the release of more than 45 million litres of toxic water, loaded with cadmium, arsenic, zinc, and iron into the neighbouring river and down to coastal waters. Ferric hydroxide in mine waters precipitates out on to the stream bed covering the substratum with a brown slime and smothering benthic organisms and macrophytes. Many mines are situated in rural localities so aquatic ecosystems in otherwise pristine environments may be destroyed. Gold mining in the Amazon region of Brazil is resulting in widespread contamination with mercury, used to extract gold, in fish, livestock and the human residents of the region.[36]

4.5 TOXIC CHEMICALS

Some aspects of toxic pollution have already been mentioned but it is now appropriate to describe the modes of action of toxic chemicals. The major types of toxic compounds are:

(i)   metals, such as zinc, copper, mercury, cadmium;
(ii)   organic compounds, such as pesticides, herbicides, polychlorinated biphenyls (PCBs), phenols;
(iii) gases, such as chlorine, ammonia;
(iv)   anions, such as cyanide, sulfide, sulfite;
(v)   acids and alkalis.

There are a number of terms in regular use in the study of toxic effects:

acute – causing an effect (usually death) within a short period;
chronic – causing an effect (lethal or sub-lethal) over a prolonged period of time;
lethal – causing death by direct poisoning;
sub-lethal – below the level which causes death but which may affect growth, reproduction or behaviour so that the population may eventually be reduced;
cumulative – the effect is increased by successive doses.

A typical example of a toxicity curve is given in Figure 4.6. Note that both axes are on a logarithmic scale. The median periods for survival are plotted

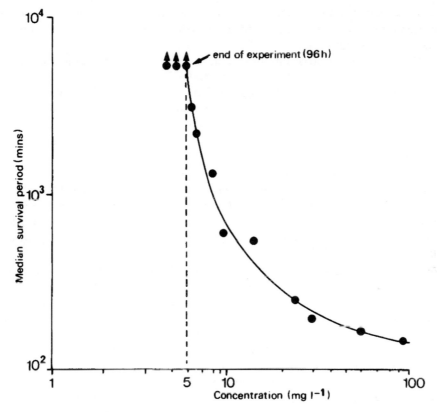

Figure 4.6 *A typical toxicity curve*

against a range of concentrations. In many cases the relationship between survival time and concentration is curvilinear and there is a level below which the organism is likely to survive for long periods (5 mg l^{-1} in Figure 4.6), a point known as the incipient LC_{50}. The lethal concentration (LC) is used where death is the criterion of toxicity. The number indicates the percentage of animals killed at that concentration and it is also usual to indicate the time of exposure. Thus 96 hour LC_{50} is the concentration of toxic material which kills fifty per cent of the test organism in ninety six hours. The incipient level is usually taken as the concentration at which fifty per cent of the population can live for an indefinite period of time. Where effects other than death are being sought, for example respiratory stress or behavioural changes, the term used is the effective concentration (EC) which is expressed in a similar way, *e.g.*, 96 hour EC_{50}.

Effluents are frequently complex mixtures of poisons. If two or more poisons are present together they may exert a combined effect on the organism which is additive, for example mixtures of zinc and copper or ammonia and zinc. In other cases, there may be antagonism, the overall toxicity being less than when compounds are present alone; calcium, for example, antagonizes the toxic effect of lead and aluminium. In some cases the overall effect of mixtures of poisons on an organism may be more than additive (synergistic), for example, mixtures of nickel and chromium.

There has been a large amount of data collected on the acute toxicity of chemicals, especially to fish and invertebrates,[37,38] and this has undoubtedly been of great value in elucidating the mechanisms of toxicity. The value of these data to the river manager is more questionable. Incidents resulting in large mortalities of fish and other organisms are usually accidents over which the river manager has no control. He can merely assess the damage and perhaps restock when conditions improve. In addition, detailed information on toxicity of a range of compounds is available for only a few test organisms, such as rainbow trout or *Daphnia*, and it is well known that even closely related species may show very different responses to particular pollutants. It is the sub-lethal effects of pollutants which are of particular concern in many field situations, for low levels of pollutants may result in the gradual loss of populations, without any overt signs of a problem.

Experiments on sub-lethal effects are more difficult to carry out because they invariably take longer and individuals under test may respond very differently to low levels of pollution. Furthermore, the reaction to pollutants may vary over the lifetime of an organism, developmental stages often being more susceptible. It is therefore necessary to study an organism under experimental conditions over its lifetime to find the weak link in its response to pollution and such long-term experiments, possibly over several generations, are essential to discover any carcinogenic, teratogenic or mutagenic effects of pollutants. Sub-lethal effects may be manifested at the biochemical, physiological, behavioural or life cycle level.[1] Although it is possible in the laboratory to demonstrate small effects at very low levels of pollution, for example in biochemistry or on growth, it is essential to show that these are likely to reduce the fitness of an organism in its natural environment and are not merely within the organism's range of adaptation. Nevertheless these sub-lethal effects may be quite subtle and can be measured long before any outward toxic effects are manifested. They can be used as *biomarkers*, to show that an organism has been exposed to contaminants at levels which exceed the normal detoxification and repair capabilities.[39,40] A biomarker can be defined as 'an xenobiotically-induced variation in cellular or biochemical components or processes, structure, or functions that is measurable in a biological system or sample'. Biomarkers can be used to predict what concentrations of a pollutant are likely to cause damage, rather than merely to measure concentrations when damage has been noted.

Porphyrins, involved in the synthesis of haem, have been used as biomarkers. Concentrations of porphyrins in the livers of herring gulls (*Larus argentatus*) from various sites in the Great Lakes of North America were compared with samples from the Atlantic coast. Samples from Hamilton Bay, Lake Ontario and Green Bay, Michigan, had porphyrin levels respectively 38 times and 28 times those in livers from the Atlantic coast, although it was not possible to determine which compound was responsible for the increase; gull livers contain many xenobiotics.[40]

Those liver enzymes which metabolize xenobiotics (mixed-function oxidases) are often used as biomarkers. In the Belgian River Meuse a once common fish, the barbel (*Barbus barbus*), is now scarce. The activities of three liver enzymes were strongly correlated with concentrations of PCBs in the liver. At high PCB levels there were marked alterations in the liver ultrastructure, with a change in mitochondrial membranes and an excessive growth of rough endoplasmic reticulum.[41] High PCB concentrations also reduce reproductive success in barbels,[42] which would explain the decline of fish populations in the Meuse.

Organisms which are regularly subjected to toxic pollutants may develop tolerance to them. This may be achieved either by functioning normally at high concentrations of pollutants or by metabolizing and detoxifying pollutants. Algae living in streams receiving mine drainage are highly tolerant to metals and this adaptation has been shown to be genetically determined. Metal tolerance has also been observed in invertebrates from metal contaminated streams, and in fish. Exposure stimulates the production of metallothioneins, low molecular weight proteins containing many sulfur-rich amino acids which bind and detoxify some metals.

Of particular concern to environmental toxicologists are those compounds which accumulate in tissues, especially the heavy metals and organochlorines (pesticides and PCBs). From often undetectable concentrations in water, organisms may accumulate levels of biological significance. Furthermore, these compounds are passed along the food chain so that top carnivores feeding on contaminated prey may accumulate enormous concentrations of pollutants. The biomagnification factor of PCBs from water to top carnivores may be as high as twenty-five million times.[43]

Figure 4.7 illustrates the frequency distribution of concentrations of total mercury in the muscle of eels (*Anguilla anguilla*) and roach (*Rutilus rutilus*) collected from waters in five regions of Great Britain.[44] It is likely that, in the vast majority of rivers and lakes from which the fish originated, mercury concentrations in water would have been below the limit of detection. Eels are probably more contaminated than roach because they spend much of the winter buried within sediments which also accumulate heavy metals. The EC directive for mercury levels in fish flesh taken for consumption should not exceed 300 μg kg^{-1}. None of the roach exceeded this standard but 25% of the eels, a species which is commercially exploited, especially for the export market, contained more than 300 μg kg^{-1} Hg. Most of the fish contained lead and cadmium also, a good proportion of them above recommended standards for consumption. Eels contained more cadmium than roach but roach were more heavily contaminated with lead. It has been cogently argued that we have little knowledge of the long-term, low level exposure of human populations to toxic metals and over ten billion individuals are being unwittingly exposed to elevated levels of toxic metals and metalloids in the environment.[45]

Wildlife species which feed predominantly on fish may also be at risk from heavy metals. Chicks of little blue herons (*Egretta caerulea*) exposed to cadmium in Louisiana wetlands had slower growth rates than non-exposed chicks. Exposure to lead resulted in increased mortality.[46]

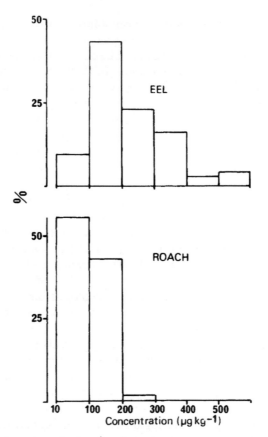

Figure 4.7 *Concentration (μg kg^{-1}) of total mercury in 85 eels and 79 roach from freshwaters in Great Britain*

Much of the evidence of the effects of PCBs on wild populations of vertebrates has come from the Great Lakes region of North America. In Green Bay, Lake Michigan, Forster's Terns (*Sterna forsteri*) exhibit impaired reproduction. The incubation is extended, few eggs hatch, the chicks have lower body weight and their livers are larger than normal. They show a high incidence of congenital deformities. The parents are inattentive at nesting and this further reduces their reproductive success. It was concluded, following a detailed toxicological analysis, that those PCB congeners which induce the enzyme aryl hydrocarbon hydroxylase (AHH) were the only contaminants present in sufficient amounts to cause the observed effects on eggs and chicks. More than 90% of the effect could be explained by two pentachlorobiphenyls. The behavioural abnormalities in adults were caused by total PCBs.[47] In addition to the Forster's Tern, six other species of fish-eating bird from the Great Lakes have exhibited growth deformities and physiological defects.[48] These symptoms have only been observed in the past two decades and it is considered that, in earlier years, symptoms were masked by the effects of DDE, which thinned eggs to such an extent that they did not survive long enough for abnormalities to be expressed.

Recent studies on humans also give cause for concern. Behavioural and neurological disorders have been found in the children of mothers who eat modest amounts of fish caught in the Great Lakes. Parallel studies on rats have shown that these behavioural changes are still measurable two generations after exposure.[43]

There has been considerable, detailed research on PCBs in the Great Lakes but the problem, of course, is not confined to that region. For example, evidence suggests that PCBs have been largely responsible for the decline of the otter over large areas of Western Europe, in some places to extinction. PCBs are known to affect the reproductive success of mammals, being powerful endocrine disruptors. There is a strong relationship between the mean amount of PCBs in tissues of otters and the extent of the population decline, relationships which do not hold for other contaminants.[49] Using vitamin A in liver as a biomarker, a strong negative correlation was found between concentrations of this vitamin and PCB concentrations, which coincided with the incidence of disease,[50] the PCBs presumably affecting the immune system. A relationship between PCB concentrations and the incidence of disease has also recently been reported in harbour porpoises (*Phocoena phocoena*).[51]

With the restrictions placed on the manufacture and use of PCBs, concentrations have declined in tissues of otters in Britain to average levels in the early 1990s which are unlikely to have adverse health effects[52] (Figure 4.8). The species has shown a considerable increase in range in rivers in lowland England from where it had been absent for four decades, and otters are now beginning to penetrate into the city of London. However, much of the PCBs in the environment eventually ends up in the marine ecosystem and, because of their persistence and the transfer across the generations, concentrations in top carnivores there are unlikely to decrease significantly for some years to come. This should be a matter of great concern in view of the dependence of many human communities on marine resources.

4.6 THERMAL POLLUTION

Cooling water discharges from electricity generating stations are the main sources of pollution by heat. Such effluents also contain a range of chemical contaminants which, though small in relation to the volume of cooling water, may in fact have a greater impact on the ecology of the receiving stream.[53] An increase in temperature alters the physical environment, in terms of both a reduction in the density of the water and its oxygen concentration, while the metabolism of organisms increases. Cold water species, especially of fish, are very sensitive to changes in temperature and they will disappear if heated effluents are discharged to the headwaters of streams. As the temperature increases, the oxygen consumption and heart rate of a fish will increase to obtain oxygen for increased metabolic processes but, at the same time, the oxygen concentration of the water is decreased. For example, at 1°C a carp

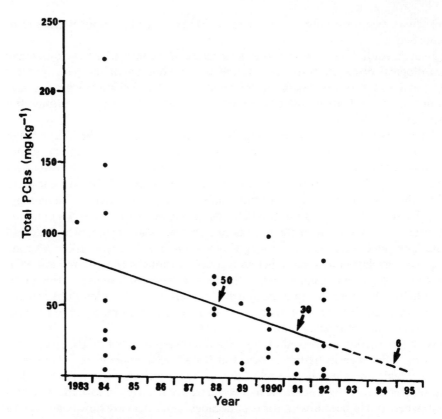

Figure 4.8   *PCB concentrations* (mg kg^{-1} lipid) *in tissues of otters collected in England and Wales between 1984 and 1992. The arrow at 50 mg kg^{-1} illustrates the concentration of PCBs causing reproductive failure in mink* (Mustela vison), *that at 30 mg kg^{-1} is a more stringent standard based on the precautionary principle. The more stringent standard of 6 mg kg^{-1} is based on dose effects levels on vitamin A deficiency*

(*Cyprinus carpio*) can survive in an oxygen concentration as low as 0.5 mg l^{-1}, whereas at 35 °C the water must contain 1.5 mg l^{-1}. The swimming speeds of some species declines at higher temperatures, *e.g.*, trout at 19 °C, making them less efficient predators. Resistance to disease may also change. The bacterium *Chondrococcus columnaris* is innocuous to fish below 10 °C but it invades wounds between 10 and 21 °C while it can invade healthy tissues above 21 °C.

However, it must be remembered that temperature changes are a feature of natural ecosystems so that organisms have the ability to adapt to the altered conditions provided by thermal effluents and, although much research has been carried out into thermal pollution, it is now considered to be of little importance compared to other sources of pollution.[53] Indeed, there are some benefits of heated effluents, for growth and productivity may be enhanced.

Global warming could have a major impact in future on freshwaters. One problem is the increased frequency of unpredictable events, such as storms, leading to widespread flooding and financial loss. Summer droughts, by contrast, lead to the drying out of many river headwaters with severe consequences for the biota. Surface water temperatures may rise so that the effects described above for thermal discharges may become general. They could, for example influence the physiology, reproduction and survival of some fish species differentially, leading to a change in the community within rivers.[54] In a eutrophic German lake, Heiligensee, an increase in temperature of slightly less than 3 °C over 21 years has led to a shift in phytoplankton dominance from diatoms and cryptophytes to one of cyanobacteria, while the zooplankton has shifted from large-bodied *Daphnia galeata* to small-bodied forms.[55] In the drought of 1995/1996, the first toxic blooms of algae were recorded in British rivers.[56] Effluents already make up a substantial proportion of the low summer flow of many rivers and, should these become more concentrated, only the most resilient aquatic species may survive. Higher concentrations of bioaccumulating compounds (*e.g.* metals, PCBs) could have ramifications for the food-chain far beyond the boundaries of the river. Finally, with predicted rises in sea-level, many coastal freshwaters may become brackish, while salt intrusions may push far inland up rivers.

4.7 RADIOACTIVITY

Chemically, radionuclides behave in the same way as their non-radioactive isotopes but, if they accumulate up the food chain the radioactive isotopes have much greater significance. Radionuclides come mainly from fall-out from weapons testing and the effluent from nuclear power stations. Because ionizing radiation is highly persistent in the environment, causing cancer and genetic disorders in humans, it has always attracted special concern and the release of radionuclides is strictly monitored and controlled.

Until recently, it was thought that the natural environment was little affected by radionuclides, which are discharged very locally. However, the explosion at the nuclear power station at Chernobyl in the Ukraine in April, 1986 and the subsequent spread and deposition of radionuclides over large areas of western Europe emphasized how potentially damaging such pollution can be. There were major effects on the terrestrial ecosystem for up to 40 km from the reactor, but not apparently on the local aquatic ecosystem.[57]

In Sweden, levels of radioactivity in freshwaters rose quickly but fish species varied in the amount and rate of accumulation of radionuclides. In one lake, bream (*Abramis brama*), which feed on bottom-living invertebrates had accumulated 1000 Bq kg^{-1} of caesium-137 within two weeks. By July accumulation had increased to 3800 Bq kg^{-1}. Perch (*Perca fluviatilis*), which feed on zooplankton, accumulated caesium more slowly but to higher levels, 8000 Bq kg^{-1} by July. Concentrations in the fish-eating pike (*Esox lucius*) were little higher than those in bream in July.[58] Between 4000 and 7000 Swedish lakes had fish with a Cs-137 load more than the consumption guideline of 1500 Bq kg^{-1} and activity levels in pike could exceed this value well into the 21st century.[59]

Because wild caught freshwater fish feature little in the diet of the vast majority of the human population it was considered that the Chernobyl accident presented no cause for concern to human health away from the vicinity of the reactor. However, other species within the aquatic ecosystem are more exclusively piscivorous. Samples of otter faeces (spraints) were collected from river banks in central Wales and South-west Scotland. Central Wales was outside of the main deposition area for Chernobyl fallout. In the September following the accident, radioactivity in otter spraints was more than double that in a sample collected some fifteen months prior to the accident and stored in a deep freeze. By the following January levels had returned to normal. By contrast, South-west Scotland, which is an area of soft waters, received large quantities of Chernobyl fall-out and average radioactivity was more than six times that in Wales (13 000 Bq kg^{-1}, with a maximum of 79 500 Bq kg^{-1} in one sample).[60] Unfortunately, there was no pre-Chernobyl sample but levels were still high in the following January, as has been found for other biological materials from this area. Whether or not such high levels of radioactivity could adversely affect otters is unknown and it would be impossible to disentangle any mortality caused by radiation from other mortality factors, a major problem with many ecotoxicological studies.

4.8   OIL

Compared to the marine situation comparatively little work has been done on the effects of oil in freshwater ecosystems. Nevertheless the chronic pollution of freshwaters with hydrocarbons is widespread. Much of it derives from petrol and oil washed from roads together with the illegal discharge of engine oil. Other sources include boats and irrigation pumps, while accidents involving transporters and spillages from storage tanks are also significant. In 1998, a typical year, 30% of all reported river pollution incidents in England and Wales were caused by oils. These incidents would seem minor compared with the pollution of rivers and wetlands caused by unregulated oil exploitation in Siberia.[61]

The water soluble components of crude oils and refined products may prove toxic to freshwater animals though the prediction of toxic effects is rather difficult owing to the complex chemical nature of discharges. Eggs and young stages of animals are especially vulnerable. In a study using the tree frog *Hyla cinerea*, hatching success of eggs was not influenced by the presence of oil but growth rates and metamorphosis were.[62] A study of pearl dace (*Margariscus margarita*) in pools contaminated with diesel fuel demonstrated severe pathological changes and a failure in reproduction which would eventually lead to the extermination of this fish population.[63]

In general terms, the aliphatic compounds of oils are relatively innocuous while the aromatic hydrocarbons, such as benzene, xylene and toluene are highly toxic. There are also marked species differences in the susceptibility to these compounds, further adding to the difficulties of making predictions about toxicity. Some components of oils, such as PCBs and lead, will accumulate in tissues. Emulsifiers and dispersants, used to clean up spillages, are themselves often highly toxic. The surface active agents which they contain make membranes more permeable and increase the penetration of toxic compounds into animals. In this way mixtures of oils and dispersants are often more toxic than either applied separately.

The physical properties of floating oil are a special threat to higher vertebrates, especially aquatic birds because contamination reduces buoyancy and insulation, while the ingestion of oil, frequently the result of attempts to clean the plumage, may prove toxic. A further problem is the tainting of flesh, especially of fish, which is detectable to the human palate at very low levels of contamination and renders fish inedible. The major sources of taint are light oils and the middle boiling range of crude oil distillates but there are a number of other sources, such as exhaust from outboard motors, waste from petrochemical factories, refinery wastes, and all crude oils.

4.9 BIOLOGICAL MONITORING OF POLLUTION IN FRESHWATERS

The biological assessment of pollutants includes both laboratory and field techniques.[1] Two widely used laboratory methods are biostimulation tests and toxicity testing. Biostimulation tests are used mostly for evaluating the nutrient status of water bodies and are normally carried out with algae, the most frequently used species including *Selenastrum capricornutum*, *Asterionella formosa*, and *Microcystis aeruginosa*. The Standard Bottle Test measures either the maximum specific growth rate or the maximum standing crop.[1] Chlorophyll-a fluorescence of *Selenastrum* has also been used in toxicity studies.[64]

The chief uses of toxicity tests are for a preliminary screening of chemicals, for monitoring effluents to determine the extent of risk to aquatic organisms, and, for those effluents which are toxic, to determine which component is causing death so that it can receive especial treatment. The simplest type of test is the static test in which an organism is placed in a standard tank in the water under investigation for 48–96 h. There are normally a series of tanks with test water of different dilutions. More sophisticated techniques involve the periodic replacement of test water or indeed continuous flow systems. Fish have traditionally been used as test organisms. In the United States, the main test species have been fathead minnows and bluegill sunfish (*Lepomis macrochirus*), though the tendency now is to use a range of test animals. Much toxicity work in Britain has been with rainbow trout but the tropical harlequin (*Rasbora heteromorpha*) has become increasingly popular because it is smaller and has a similar sensitivity to pollutants. Fish require large volumes of clean water for maintenance and, because tests need replicating, much space is needed. Furthermore there are growing ethical objections to using vertebrates for routine toxicological assessments. There is therefore increasing interest in developing other test organisms. The planktonic crustacean *Daphnia* is widely used for it is easily cultured, has a high reproductive rate and is sensitive to a range of pollutants.[65] Other invertebrates that have been used include tubificid worms,[66] zebra mussels (*Dreissena polymorpha*), whose water filtration rates are measured,[67] and the flagellate *Euglena gracilis*, in which movement is the measure of toxicity.[68] Bacterial tests are also being developed, of which one, the Microtox test, is commercially available. It utilizes the bioluminescence of the marine *Photobacterium phosphoreum*, the reduction in light output being the measure of toxicity. The test is sensitive, precise and reproducible. It has been used, for example, to demonstrate that the quality of water in the River Meuse is, over most of its length, below environmentally acceptable standards for a range of organic compounds.[69]

Fishes show distinct physiological and behavioural responses to low levels of pollutants and studies have attempted to use these to devise automatic alarm systems. Automatic fish monitors should provide rapid indications of a deteriorating water quality and have potential uses for monitoring river waters and raw waters which are abstracted for potable supply and for monitoring effluents from treatment plants. Fish alarm systems have monitored movement and respiration. Movements are monitored with photosensitive cells and respiration by changes of potential between two electrodes.[70] Multiple species monitors, using invertebrates as well as fish, are also being developed; two or more species are more likely to respond to a wider range of pollutants than a single species and they may be deployed in the laboratory or field.[71,72]

It has already been shown that pollutant levels in water are frequently below the limits of detection but that some organisms can accumulate large quantities of some contaminants which reflect the pollution level over periods of time. Because the degree to which organisms concentrate pollutants varies, for large-scale surveys a single, widespread species is needed. Care must be taken in the interpretation of results because many factors may influence the total pollutant content and concentration. These include the age of the organism, its sex, size and weight as well as the time of year, sampling position, and the relative levels of other pollutants in the tissues. In addition to sampling organisms from the field to measure pollutant levels, they can also be placed in the field in cages so that uptake rates can be measured over defined periods of time.

Fish have frequently been used as biomonitors because they can be considered as a *critical material*, forming part of the diet of man. Eels occur in most freshwater habitats and are hence particularly useful for large-scale surveys. Molluscs and macrophytes have also been used locally while mosses have proved especially valuable for monitoring pollution caused by metals.[73,74]

The biological assessment of water quality in the field may involve a number of levels of effort: survey, surveillance, monitoring, or research.[1] It is usually impossible to investigate the entire fauna and flora in a pollution study because of constraints of time and the wide range of sampling methods required for different groups of organisms. A survey or monitoring programme should therefore be based on those organisms which are most likely to provide the right information to answer the particular question being posed. To be suitable, the chosen group must meet a number of criteria. For example, the presence or absence of organisms must be a function of water quality rather than other ecological factors and the group must be relatively easy to sample and identify.

Diatoms are used to monitor eutrophication[75] and bryophyte communities and attached algae have been employed to monitor general water quality.[76,77] In Great Britain the most favoured group for routine biological surveillance is the macroinvertebrate assemblage living in the stream bed.

To sample sites for macroinvertebrates a pond net (mesh size 900 μm) is used. The net is placed on the stream bed, facing downstream, and the feet are shuffled on the substratum, animals being dislodged and floating into the net. As much of the stream bed as possible is sampled and additional habitats, such as aquatic macrophytes or marginal vegetation, are also sampled by sweeping the net through them. The sampling period is three minutes and the objective is to obtain a comprehensive list of species with the minimum of sampling effort.[78] These samples are normally collected from eroding substrata (rocks, gravels, *etc.*). If there are long stretches of stream consisting of depositing substrata (silts and muds) or which are too deep to wade then sampling may have to be restricted to the margins. Alternatively artificial substrata, which are colonized by invertebrates, can be left *in situ* for later collection and analysis.[1]

Once the invertebrates have been identified and counted, there are two commonly used methods for data presentation and interpretation, diversity indices and biotic indices. Diversity indices take into account the number of species within the collection (species richness) and the relative abundance of species within the collection (evenness). It is argued that a community from an unstressed, *i.e.*, pollution free, environment will contain a large number of species (high richness), many at fairly low densities (high evenness), so that the calculated diversity index will be high. As pollution stress increases, species will gradually decline in number and disappear (low richness), while a few tolerant species will build up big populations in the absence of predation and competition (low evenness) resulting in a low diversity index. Diversity indices take no account of the tolerances of individual species to pollution. Many diversity indices have been devised.[79]

Biotic indices take account of the sensitivities of different species to pollution, that is, species are used as indicators of water quality. Those species which are sensitive to pollution (such as stoneflies) are given a high score, tolerant species (such as tubificid worms) are given a low score. In Britain, the biotic index most in favour is the Biological Monitoring Working Party (BMWP) Score (Table 4.3). Identification is necessary only to the family level. Score values for individual families reflect their pollution tolerance based on current knowledge of distribution and abundance. Each family is given a score depending on its position in Table 4.3 and a site score is calculated by summing the individual family scores. This total score can then be divided by the number of families recorded in the sample to derive the average score per taxon (ASPT), which is less sensitive to sample size and sampling effort, and hence gives more information for less effort. High ASPT scores characterize clean, upland sites which contain large numbers of high scoring taxa such as stoneflies, mayflies and caddisflies. Lower ASPT values are obtained from slow-flowing, lowland sites which are dominated by molluscs, chironomids and tubificid worms.

Table 4.3   *The BMWP score system*

Families	Score
Siphlonuridae, Heptageniidae, Leptophlebidae, Ephemerellidae, Potamanthidae, Ephemeridae, Taeniopterygidae, Leuctridae, Capniidae, Perlodidae, Perlidae, Chloroperlidae, Aphelocheiridae, Phryganeidae, Molannidae, Beraeidae, Ondontoceridae, Leptoceridae, Goeridae, Lepidostomatidae, Brachycentridae, Sericostomatidae	10
Astacidae, Lestidae, Agriidae, Gomphidae, Cordulegasteridae, Aeshnidae, Corduliidae, Libellulidae, Psychomyiidae, Philopotamidae	8
Caenidae, Nemouridae, Rhyacophilidae, Polycentropidae, Limnephilidae	7
Neritidae, Viviparidae, Ancylidae, Hydroptilidae, Unionidae, Corophiidae, Gammaridae, Platycnemididae, Coenagriidae	6
Mesoveliidae, Hydrometridae, Gerridae, Nepidae, Naucoridae, Notonectidae, Pleidae, Corixidae, Haliplidae, Hygrobiidae, Dytiscidae, Gyrinidae, Hydrophilidae, Clambidae, Helodidae, Dryopidae, Elminthidae, Chrysomelidae, Curculionidae, Hydropsychidae, Tipulidae, Simuliidae, Planariidae, Dendrocoelidae	5
Baetidae, Sialidae, Piscicolidae	4
Valvatidae, Hydrobiidae, Lymnaeidae, Physidae, Planorbidae, Sphaeriidae, Glossiphoniidae, Hirudidae, Erpobdellidae, Asellidae	3
Chironomidae	2
Oligochaeta (whole class)	1

Multivariate techniques (TWINSPAN and DECORANA) have been used to classify the running waters of Great Britain on the basis of their macroinvertebrate fauna. Substratum type and alkalinity explain much of the variation between invertebrate groups, long-term average flow and distance from source also being important.[80] From this study it has become possible to predict the probability with which a given species or family will be captured at a site, using environmental data, and a software package, RIVPACS (River Invertebrate Prediction and Classification System), has been developed.[81] The predicted target assemblage of macroinvertebrates can be used to generate expected BMWP or ASPT scores against which to assess the results of field surveys. The technique has been compared with alternative approaches using large sets of data and has been found to be extremely robust.[82]

4.10   CONCLUSIONS

The river catchment is subject to use by a wide range of activities, many of which are increasing. These include, amongst others, the abstraction of water for agriculture, industry and potable supply, the discharge of waste products, angling and other forms of recreation, and the conservation of wildlife. It is clear that there are many areas of potential conflict between river users. The

Environment Agency aims to minimise such conflicts and to achieve and maintain a sustainable river environment by planning at the catchment level. This is done by catchment based Local Environment Agency Plans (LEAPs), which provide a local agenda for integrated actions leading to environmental improvements. It allows stakeholders in the catchment the opportunity to express opinions about the environmental needs of the local area and on the way that it is being managed.

Water Quality Objectives are used as the basis for setting consents for the discharge of effluents to rivers and preventing pollution. This is done by a River Ecosystem Classification, a national scheme which places each river into a category depending on concentrations of organic matter and other common pollutants which influence the communities of animals and plants which live there. The measured determinands are dissolved oxygen, BOD, total ammonia, unionized ammonia, pH, copper and zinc. A stretch of river is given a quality target according to the use to which that stretch is put (*e.g.* fisheries, water supply, public amenity). The LEAP is a mechanism to ensure that targets are achieved or maintained. There are five river ecosystem (RE) classes. RE1 and RE2 rivers are of very good or good water quality capable of supporting all fish species, including the pollution sensitive trout and salmon. RE3 stretches have fair quality water supporting high quality coarse fish populations, RE4 stretches also hold these species. RE5 stretches have poor quality water which limit fish populations. In addition stretches may be unclassified if the water is so bad that no fish population can be supported, or if there are insufficient data to enable a classification to be made.

In addition to these targets, the Environment Agency produces general quality assessments of biological and chemical (ammonia, dissolved oxygen and BOD concentrations) conditions, based on routine monitoring, which provide an annual statement of river quality. They allow trends to be examined.

The biologist works within a multifunctional team in formulating and acting upon LEAPs and quality assessments. However there are two areas of river management in which the biologist has the pre-eminent role – fisheries and wildlife conservation. Angling remains the largest participatory sport in Great Britain and one aim of the Environment Agency is to ensure that all waters in England and Wales will be capable of sustaining healthy and thriving fish populations. The 1992 Convention on Biodiversity has resulted in the development of Local Biodiversity Action Plans for conserving threatened species and habitats. Those related to the aquatic environment will keep many biologists in the water industry fully employed for the foreseeable future.

Despite the vast amount of scientific information gathered on pollution over the past four decades, many of our freshwaters still suffer from poor water quality, with severe consequences for the flora and fauna they support. Scientific knowledge is subverted by economic and political constraints, and in many instances by careless accidents within the catchment. A continuing programme of surveillance and monitoring is therefore essential. Newly synthesized materials are also constantly being added to our waterways as traces or in effluents and the long-term effects of these are largely unknown.

Constant vigilance is required to protect our water resources and the biologist will continue to have a central role in the management team.

4.11 REFERENCES

1. C. F. Mason, 'Biology of Freshwater Pollution', 4th edn., Pearson, Harlow, 2002.
2. C. E. Purdom, P. A. Hardiman, V. J. Bye. N. C. Eno, C. R. Tyler and J. P. Sumpter, *Chem. Ecol.*, 1994, **8**, 275.
3. Environment Agency, 'Endocrine-disrupting Substances in the Environment: What Should be Done?', Environment Agency, Bristol, 1998.
4. C. A. Bishop, P. Ng, K. E. Pettit, S. W. Kennedy, J. J. Stegeman, R. J. Norstrom and R. J. Brooks, *Environ. Pollut.*, 1998, **101**, 143.
5. R. M. Sharpe and N. E. Skakkebaek, *Lancet*, 1993, **341**, 1392.
6. C. F. Mason and N. A.-E. Barak, *Chemosphere*, 1990, **21**, 695.
7. J. O. Grimalt, M. Ferrer and E. Macpherson, *Sci. Total Environ.*, 1999, **242**, 3.
8. N. Vaughan, G. Jones and S. Harris, *Biol. Conserv.*, 1996, **78**, 337.
9. M. M. Abdul-Hussein and C. F. Mason, *Hydrobiologia*, 1988, **169**, 265.
10. B. Moss, *Biol. Rev.*, 1983, **58**, 521.
11. M. Scheffer, S. H. Hosper, M.-L. Meijer, B. Moss and E. Jeppeson, *Trends Ecol. Evol.*, 1993, **8**, 275.
12. P. Schriver, J. B. Øgestrand, E. Jeppeson and M. Søndergaard. *Freshwater Biol.*, 1995, **33**, 255.
13. M. Klinge, M. P. Grimm and S. H. Hosper, *Water Sci. Tech.*, 1995, **31**, 207.
14. L. A. Lawton and G. A. Codd, *J. IWEM*, 1991, **5**, 460.
15. P. Cullen and C. Forsberg, *Hydrobiologia*, 1988, **170**, 321.
16. G. K. Morse, S. W. Brett, J. A. Guy and J. N. Lester, *Sci. Total Environ.*, 1998, **212**, 69.
17. P. Daldorph and R. Price, *Arch. Hydrobiol.*, 1994, **40**, 231.
18. G. Phillips, R. Jackson, C. Bennett and A. Chilvers, *Hydrobiologia*, 1994, **275/276**, 445.
19. G. L. Phillips, M. R. Perrow and J. Stansfield in 'Aquatic Predators and their Prey', eds S. P. R. Greenstreet and M. L. Tasker, Fishing News Books, Oxford, 1996, p. 174.
20. B. Moss, J. Madgwick and G. Phillips, 'A Guide to the Restoration of Nutrient-enriched Shallow Lakes', Broads Authority, Norwich, 1996.
21. E. Bergman, L.-A. Hansson, A. Persson, J. Strand, P. Romare, M. Enell, W. Granéli, J. M. Svensson, S. F. Hanrin, G. Cronberg, G. Andersson and E. Bergstrand, *Hydrobiologia*, 1999, **404**, 145.
22. R. J. Flower, R. W. Battarbee and P. G. Appleby, *J. Ecol.*, 1987, **75**, 797.
23. R. W. Battarbee, *Hydrobiologia*, 1994, **274**, 1.
24. A. Wellburn, 'Air Pollution and Climate Change: the Biological Impact', Longman, London, 1994.
25. D. W. Schindler, *Science*, 1988, **239**, 149.
26. M. Appelberg, E. Degerman and L. Norrgren, *Finnish Fish. Res.*, 1992, **13**, 77.
27. B. W. Stallsmith, J. P. Ebersole and W. G. Haggar, *Freshwater Biol.*, 1996, **36**, 731.
28. K.-M. Vuori, *Freshwater Biol.*, 1996, **35**, 179.
29. S. J. Tyler and S. J. Ormerod, *Environ. Pollut.*, 1992, **78**, 49.
30. C. F. Mason and S. M. Macdonald, *Water, Air, Soil Pollut.*, 1989, **43**, 365.
31. D. M. Downey, C. R. French and M. Odom, *Water, Air, Soil Pollut.*, 1994, **77**, 49.
32. B. T. Barlaup, Å. Åtland and E. Kleiven, *Water, Air, Soil Pollut.*, 1994, **72**, 317.
33. A. Henriksen, L. Liem, B. O. Rosseland, T. S. Traaen and I. S. Sevaldrud, *Ambio*, 1989, **18**, 314.
34. P. G. Whitehead, C. Neal and R. Neale in 'Reversibility of Acidification', ed. H. Barth, Elsevier, London, 1987, p. 126.

35. R. L. P. Kleinmann and R. Hedin in 'Tailings and Effluent Management', eds M. E. Chalkly, B. R. Conrad, V. I. Lakshmanan and K. G. Wheeland, Pergamon, New York, 1990, p. 140.
36. D. Palheta and A. Taylor, *Sci. Total Environ.*, 1995, **168**, 63.
37. J. M. Hellawell, 'Biological Indicators of Freshwater Pollution and Environmental Management', Elsevier Applied Science, London, 1986.
38. R. Lloyd, 'Pollution and Freshwater Fish', Fishing News Books, Oxford, 1992.
39. D. Peakall, 'Animal Biomarkers as Pollution Indicators', Chapman and Hall, London, 1992.
40. G. A. Fox, *J. Great Lakes Res.*, 1993, **19**, 722.
41. J. L. Hugla, J. C. Philippart, P. Kremers and J. P. Thome, *Netherlands J. Aquat. Ecol.*, 1995, **29**, 135.
42. J. L. Hugla, J. P. Thome and J. C. Philippart, *Cah. Ethol.*, 1993, **13**, 155.
43. T. Colborn, D. Dumanoski and J. P. Myers, 'Our Stolen Future', Dutton, New York, 1996.
44. C. F. Mason, *Chemosphere*, 1987, **16**, 901.
45. J. O. Nriagu, *Environ. Pollut.*, 1988, **50**, 139.
46. S. A. Spahn and T. W. Sherry, *Arch. Environ. Contam. Toxicol.*, 1999, **37**, 377.
47. T. J. Kubiak, H. J. Harris, L. M. Smith, D. L. Starling, T. R. Schwartz, J. A. Trick, L. Sileo, D. E. Docherty and T. C. Erdman, *Arch. Environ. Contam. Toxicol.*, 1989, **18**, 706.
48. J. P. Giesy, J. P. Ludwig and D. E. Tillitt, *Environ. Sci. Technol.*, 1994, **28**, 128A.
49. S. M. Macdonald and C. F. Mason, 'Status and Conservation Needs of the Otter (*Lutra lutra*) in the Western Palearctic', Council of Europe, Nature and Environment 67, 1994.
50. A. J. Murk, P. E. G. Leonards, B. van Hattum, R. Luit, M. E. J. van der Weiden and M. Smit, *Environ. Toxicol. Pharmacol.*, 1998, **6**, 91.
51. P. D. Jepson, P. M. Bennett, C. R. Allchin, R. J. Law, T. Kuiken, J. R. Baker, E. Rogan and J. K. Kirkwood, *Sci. Total Environ.*, 1999, **243/244**, 339.
52. C. F. Mason, *Chemosphere*, 1998, **36**, 1969.
53. T. E. Langford, 'Ecological Effects of Thermal Discharges', Chapman and Hall, London, 1990.
54. C. M. Wood and G. McDonald, 'Global Warming: Implications for Freshwater and Marine Fish', Cambridge University Press, Cambridge, 1997.
55. R. Adrian and R. Deneke, *Freshwater Biol.*, 1996, **36**, 757.
56. R. Everard, *Freshwater Forum*, 1996, **7**, 33.
57. Z. A. Medvedev, *Trends Ecol. Evol.*, 1994, **9**, 369.
58. R. C. Petersen, L. Landner and H. Blanck, *Ambio*, 1986, **15**, 327.
59. L. Lundgren, *Ambio*, 1993, **22**, 369.
60. C. F. Mason and S. M. Macdonald, *Water, Air, Soil Pollut.*, 1988, **37**, 131.
61. F. Pearce, *New Scientist*, 27 November, 1993, p. 28.
62. P. A. Mahaney, *Environ. Toxicol. Chem.*, 1994, **13**, 259.
63. R. A. Khan, *Bull. Environ. Contam. Toxicol.*, 1999, **62**, 638.
64. J. A. van der Heever and J. U. Grobbelaar, *Arch. Environ. Contam. Toxicol.*, 1998, **35**, 281.
65. R. Baudo, *Mem. Ist. Ital. Idrabiol.*, 1987, **45**, 461.
66. M. Leynen, T. Van den Berckt, J. M. Aerts, B. Castelein, D. Berckmans and F. Ollevier, *Environ. Pollut.*, 1999, **105**, 151.
67. M. H. S. Kraak, F. Kuipers, H. Schoon, C. J. de Groot and W. Admiraal, *Hydrobiologia*, 1994, **294**, 13.
68. H. Tahedl and D. Häder, *Water Res.*, 1999, **33**, 426.
69. H. J. G. Polman and D. de Zwart, *Water Sci. Technol.*, 1994, **29**, 253.
70. D. Gruber, C. H. Frago and W. J. Rasnake, *J. Aquat. Ecosystem Health*, 1994, **3**, 87.
71. J. Cairns, P. V. McCormick and B. R. Niederlehner, *Ergemn. Limnol.*, 1994, **42**, 267.
72. R. Schulz and M. Liess, *Environ. Toxicol. Chem.*, 1999, **18**, 2243.

73. J. Lopez, M. D. Vasquez and A. Carballeira, *Freshwater Biol.*, 1994, **32**, 185.
74. I. Bruns, K. Friese, B. Markert and G.-J. Krauss, *Sci. Total Environ.*, 1997, **204**, 161.
75. M. G. Kelly, *Water Res.*, 1997, **32**, 236.
76. A. Vanderpoorten, *Environ. Pollut.*, 1999, **104**, 401.
77. C. Vis, C. Hudon, A. Catteneo and B. Pinet-Alloul, *Environ. Pollut.*, 1998, **101**, 13.
78. M. T. Furse, J. F. Wright, P. D. Armitage and D. Moss, *Water Res.*, 1981, **15**, 679.
79. A. E. Magurren, 'Ecological Diversity and its Measurement', Croom Helm, London, 1988.
80. J. F. Wright, P. D. Armitage, M. T. Furse and D. Moss, *Regul. Rivers*, 1989, **4**, 147.
81. J. F. Wright, M. T. Furse and P. D. Armitage, *Eur. Water Pollut. Control*, 1993, **3(4)**, 15.
82. D. Moss, J. F. Wright, M. T. Furse and R. T. Clarke, *Freshwater Biol.*, 1999, **41**, 167.

CHAPTER 5

Sewage and Sewage Sludge Treatment

J. LESTER and D. EDGE

5.1 INTRODUCTION

It is estimated that the volume of water used daily in England and Wales (exclusive of water abstracted for cooling purposes) amounts to 5000×10^6 gal (23×10^6 m^3) or approximately 95 gal (430 l) per capita per day. Domestic use accounts for nearly 1800×10^6 gal (8×10^6 m^3) of this average daily total. Nearly all of the water used domestically and approximately 1500×10^6 gal (6.8×10^6 m^3) of the water used by industry each day is discharged to the sewers, yielding a total sewage flow of 3100×10^6 gal (14.1×10^6 m^3) or about 60 gal (275 l) per capita per day.

The sewage from approximately 44 million people in England and Wales is treated by conventional wastewater treatment processes and discharged to inland waters. That from about a further six million people is discharged to the sea with full or partial treatment (only that from small communities of less than 10 000 population equivalents (pe) is discharged to sea with no treatment). Some one to two million people are not connected to the sewerage system and are served by septic tanks, cess-pits and rotating biological contactors (rbc). To achieve this degree of wastewater treatment requires some 5000 sewage treatment works serving populations in excess of 10 000; these are distributed throughout the ten Water Utilities (plcs), which have been privatized, in England and Wales. The sewerage systems which carry the sewage to the site of treatment, or point of discharge, are of two types. Foul sewers carry only domestic and industrial effluent. In areas serviced in this way there are entirely separate systems for the collection of stormwater which is discharged directly to natural water courses. However, in older towns and cities considerable use has been made of combined foul and stormwater systems. The use of combined sewerage systems leads to very significant changes in the flow of sewage during storms. However, even in foul sewers significant changes in the flow occur due to variations in the pattern of domestic and industrial water usage which is essentially diurnal, and at its greatest during the day. Infiltration will also influence the flow in the sewage system. Although a properly laid sewer is watertight when constructed, ground movement and aging may allow water to enter the sewer if it is below the water table. The combined total of average daily flows to a sewage treatment works is called the dry weather flow (*DWF*). The *DWF* is an important value in the design and operation of the sewage treatment works and other flows are expressed in terms of it. *DWF* is defined as the daily rate of flow of sewage (including both domestic and trade waste), together with infiltration, if any, in a sewer in dry weather. This may be measured after a period of seven consecutive days during which the rainfall has not exceeded 0.25 mm.

The *DWF* may be calculated from the following equation:

$$DWF = PQ + I + E$$

where

P = population served
Q = average domestic water consumption $(l\,d^{-1})$
I = rate of infiltration $(l\,d^{-1})$
E = volume (in litres) of industrial effluent discharged to sewers in 24 hours

5.1.1 Objectives of Sewage Treatment

Water pollution in the United Kingdom was already a serious problem by 1850. It is probable that the early endeavours to control water pollution were considerably stimulated by the state of the lower reaches of the River Thames which, at the point where it passed the Houses of Parliament, was grossly polluted. An early solution to these problems was sought through the construction of interceptor sewers. These collected all the sewage draining to the River Thames and carried it several miles down the river before discharging it to the estuary on the ebb tide. From this it moved towards the sea and in so doing received greater dilution. Despite these measures, and the passing in 1876 of the first Act of Parliament to control water pollution, the situation continued to deteriorate. The requirement for, and the objectives of, sewage treatment were first outlined by the Royal Commission on Sewage Disposal (1898–1915). The objectives of sewage treatment have developed significantly since this report; however, the standards described then are still applicable in many areas and this report provided the framework around which the United Kingdom wastewater industry has developed.

Originally the objective of sewage treatment was to avoid pestilence and nuisance (disease and odour) and to protect the sources of potable supply.

During sewage treatment disease-causing organisms may be destroyed or concentrated in the sludges produced; similarly, offensive materials may be concentrated in the sludges or biodegraded. As a consequence the quantities of these agents present in the sewage effluent is much less than in the untreated sewage and their dilution in the receiving water far greater. The benefits of sewage treatment are not limited to greater dilution, however, since each receiving water has a certain capacity for 'self-purification'. Providing sewage treatment reduces the burden of polluting material to a value less than this capacity then the ecosystem of the receiving water will complete the treatment of the residual materials present in the sewage effluent. Thus sewage treatment in conjunction with the selection of appropriate points for sewage effluent discharge has resulted in the elimination of waterborne disease in the UK and many other advanced countries. However, as the population has expanded and become urbanized with a concomitant development of water-consuming industries an additional requirement has been placed upon sewage treatment.

It is now the objective of sewage treatment in many parts of the UK to produce a sewage effluent which after varying degrees of dilution and self-purification is suitable for abstraction for treatment to produce a potable supply. This indirect re-use affects some 30% of all water supplies in the UK.

5.1.2 The Importance of Water Re-use

That the United Kingdom practises indirect re-use to a greater extent than most other countries may appear surprising given the annual rainfall. Indeed, that re-use should be important in global terms given the abundance of water on the earth's surface may also be considered improbable in all but the most arid regions. However, two important factors readily explain this situation; firstly a vast amount of the available water is too saline to be used as a potable supply (the salinity is too costly to remove in all but the most extreme cases) and secondly the non-uniform distribution of the population and the available water supply. The available water supply is determined by the rainfall, the ability of the environment to store water (essentially the size of lakes and rivers, which are small in the United Kingdom) and their location, *e.g.* Wales has an abundance of suitable water supplies, but limited population, whilst South East England has a large population with limited water resources.

It has been estimated that of the water falling on the United Kingdom 50% is not available for use as a result of run-off to the sea. Of the remainder approximately 17% was utilized in the 1960s and this doubled by the 1990s, when consumption became stable. Thus the potential reserves are very limited. However, because demand and supply are not geographically proximate re-use is already essential. As a consequence the traditional concept of water supply employing single-purpose reservoirs impounding unused river water has been abandoned in favour of multi-purpose schemes designed to permit repeated use of the water before it reaches the sea. In these schemes sewage treatment plays a vital role in addition to being an integral part of the hydrological cycle.

5.1.3 Criteria for Sewage Treatment

Sewage is a complex mixture of suspended and dissolved materials; both categories constitute organic pollution. The strength of sewage and the quality of sewage effluent are described in terms of their suspended solids (SS) and biochemical oxygen demand (BOD); these two measures were either proposed or devised by the Royal Commission (1898–1915). The SS are determined by weighing after the filtration of a known volume of sample through a standard glassfibre filter paper, the results being expressed in mg l^{-1}.

Dissolved pollutants are determined by the BOD they exert when incubated for five days at 20 °C. Samples require appropriate dilution with oxygen-saturated water and suitable replication. The oxygen consumed is determined and the results again expressed in mg l^{-1}.

The two standards for sewage effluent quality proposed by the Royal Commission were for no more than 30 mg l^{-1} of suspended solids and 20 mg l^{-1} for BOD, the so called 30:20 standard. The Royal Commission envisaged that the effluent of this standard would be diluted 8:1 with clean river water having BOD of 2 mg l^{-1} or less. This standard was considered to be the normal minimum requirement and was not enforced by statute because the character and use of rivers varied so greatly. It was intended that standards would be introduced locally as required. For example, a river to be used for abstraction of potable supplies would require a higher standard such as the 10:10 standard imposed by the Thames Conservancy. Whilst other countries which are members of the European Union have adopted 'uniform emission standards', that is, the same quality of effluent regardless of the state or use of the river, the United Kingdom has continued with its pragmatic approach whereby effluent standards are set depending on the 'water quality objectives' of the river, which in turn is determined by its function or use. In the 1970s, with the reorganization, the water industry's reliance solely on the 30:20 standard was abandoned, although this standard is probably still the most commonly applied. Sewage treatment now attempts to consistently produce an effluent with a quality superior to its 'Legal Consent' and attempts to achieve an 'Operating Target', frequently half the Legal Consent. In addition considerable importance has been placed upon the concentration of ammonia in the effluent. In the case of a works attempting to nitrify the effluent (see Section 5.2.3.8) the ammonia concentration is frequently limiting. Typical Legal Consent and Operating Targets are outlined in Table 5.1. It is evident that the Operating Targets included in Table 5.1 are the same as the Royal Commission 30:20 standard.

Table 5.1 *Legal Consent and Operating Target values for a conventional two-stage sewage treatment works*

Parameter	Legal consent value (mg l^{-1})	Operating target value (mg l^{-1})
SS	50	30
BOD	35	20
Ammonia	25	12

5.1.4 Composition of Sewage

Domestic sewage contains approximately 1000 mg l^{-1} of impurities of which about two-thirds are organic. Thus sewage is 99.9% water and 0.1% total solids upon evaporation (see Figure 5.1). When present in sewage approximately 50% of this material is dissolved and 50% suspended. The main components are: nitrogenous compounds – proteins and urea; carbohydrates – sugars, starches and cellulose; fats – soap, cooking oil and greases. Inorganic components include chloride, metallic salts and road grit where combined sewerage is used. Thus sewage is a dilute, heterogeneous medium which tends to be rich in nitrogen.

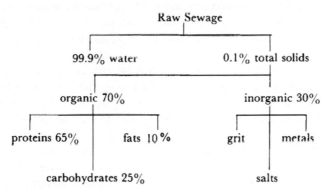

Figure 5.1 *Composition of a typical raw sewage*

5.2 SEWAGE TREATMENT PROCESSES

Conventional sewage treatment is a three-stage process, including preliminary treatment, primary sedimentation and secondary (biological) treatment; these are presented schematically in Figure 5.2. In addition some form of sludge treatment facility is frequently employed, typically anaerobic digestion (see Section 5.3.1).

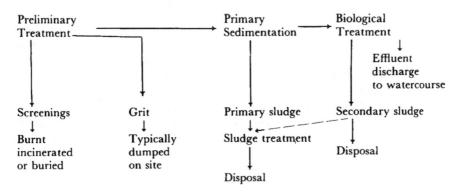

Figure 5.2 *Flow diagram of a conventional sewage treatment works*

5.2.1 Preliminary Treatment

These treatment processes are intended to remove the larger floating and suspended materials. They do not make a significant contribution to reducing the polluting load, but render the sewage more amenable to treatment by removing large objects which could form blockages or damage equipment.

Floating or very large suspended objects are frequently removed by bar screens. These consist of parallel rods with spaces between them which vary from 40 to 80 mm, through which the influent raw sewage must pass. Material which accumulates on the screen may be removed manually with a rake at small works, but on larger works some form of automatic raking would be used. The material removed from the screens contains a significant amount of putrescible organic matter which is objectionable in nature and may pose a disposal problem. Typically the material is buried, incinerated and less frequently subject to open burning.

If screens have been used to remove the largest suspended and virtually all the floating objects, then it only remains to remove the small stones and grit, which may otherwise damage pumps and valves, to complete the preliminary treatment. This is most frequently achieved by the use of constant velocity grit channels. The channels utilize differential settlement to remove only the heavier grit particles whilst leaving the lighter organic matter in suspension. A velocity of 0.3 m s^{-1} is sufficient to allow the grit to settle whilst maintaining the organic solids in suspension. If the grit channels are to function efficiently the velocity must remain constant regardless of variation of the flow to the works (typically between 0.4 and 9 *DWF*. This is achieved by using channels with a parabolic cross section controlled by venturi flumes. The grit is removed from the bottom of the channel by a bucket scraper or suction, and organic matter adhering to the grit is removed by washing, with the wash water being returned to the sewage. Small sedimentation tanks from which the sewage overflows at such a rate that only grit will settle out may also be used. These are compact and by the introduction of air on one side a rotary motion can be induced in the sewage which washes the grit *in situ*. However, these tanks do not cope with the variation in hydraulic load in such an elegant and effective manner as the grit channel.

To avoid the problems associated with the disposal of screenings, comminutors are frequently employed in place of screens. Unlike screens, which precede grit removal, the comminutors are placed downstream of the grit removal process. The comminutors shred the large solids in the flow without removing them. As a result they are reduced to a suitable size for removal during sedimentation. Comminutors consist of a slotted drum through which the sewage must pass. The drum slowly rotates carrying material which is too large to pass through the drum towards a cutting bar upon which it is shredded before it passes through the drum.

The total flow reaching the sewage treatment works is subjected to both these preliminary treatment processes. However, the works is only able to give full treatment up to a maximum flow of 3 *DWF*. When the flow to the works exceeds this value the excess flows over the weir to the storm tanks which are normally empty. If the storm is short, no discharge occurs and the contents of the tanks are pumped back into the works when the flow falls below 3 *DWF*. If the storm is prolonged then these tanks will begin to discharge to a nearby watercourse, inevitably causing some pollution. However, this excess flow has been subjected to sedimentation which removes some of the polluting material. Moreover, as a consequence of the storm, flow in the watercourse will be high, giving greater dilution.

5.2.2 Primary Sedimentation

The raw sewage (containing approximately 400 mg l^{-1} SS and 300 mg l^{-1} BOD) at a flow rate of 3 *DWF* or less and with increased homogeneity as a result of the preliminary treatment process enters the first stage of treatment which reduces its pollutant load, primary sedimentation, or mechanical treatment. Circular (radial flow) or rectangular (horizontal flow) tanks equipped with mechanical sludge scraping devices are normally used (see Figure 5.3). However, on small works hopper bottom tanks (vertical flow) are preferred; although more expensive to construct these costs are more than offset by savings made as a result of eliminating the requirement for scrapers (see Figure 5.3).

Removal of particles during sedimentation is controlled by the settling characteristics of the particles (their density, size, and ability to flocculate), the retention time in the tank (h), the surface loading ($m^3 m^{-2} d^{-1}$) and to a very limited degree the weir overflow rate ($m^3 m^{-1} d^{-1}$). Retention times are generally between 2 and 6 h. However, the most important design criterion is the surface loading; typical values would be in the range 30 to 45 $m^3 m^{-2} d^{-1}$. The surface loading rate is obtained by dividing the volume of sewage entering the tank each day ($m^3 d^{-1}$) by the surface area of the tank (m^2). The retention time may be fixed independently of the surface loading by selection of the tank depth, typically 2 to 4 m, which increases the volume without influencing the surface area. Because they strongly influence the value for surface loading selected, the nature of the particles in the sewage is one of the most important factors in determining the design and efficiency of the sedimentation tank. Of the three factors mentioned before, flocculation is perhaps the most significant.

Four different types of settling can occur:

Class 1 Settling: settlement of discrete particles in accordance with theory (Stokes' Law).
Class 2 Settling: settlement of flocculant particles exhibiting increased velocity during the process.
Zone Settling (Hindered Settlement): at certain concentrations of flocculant particles, the particles are close enough together for the interparticulate forces to hold the particles fixed relative to one another so that the suspension settles as a unit.
Compressive Settling: at high solids concentrations the particles are in contact and the weight of the particles is in part supported by the lower layer of solids.

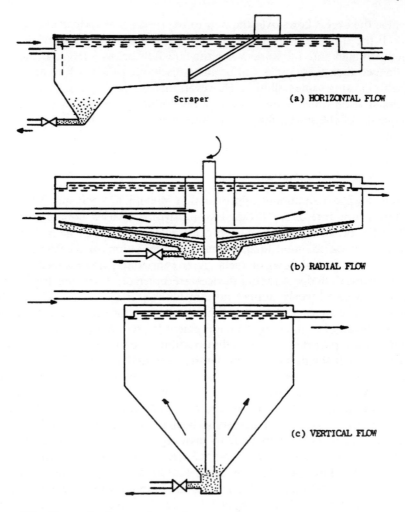

Scraper   **(a) HORIZONTAL FLOW**

(b) RADIAL FLOW

(c) VERTICAL FLOW

Figure 5.3   *Types of sedimentation tank*

During primary sedimentation settlement is of the Class 1 or 2 types. However, in secondary sedimentation (see Section 5.2.4) zone or hindered settlement may occur. Compressive settlement only occurs in special sludge thickening tanks.

Primary sedimentation removes approximately 55% of the suspended solids and because some of these solids are biodegradable the BOD is typically reduced by 35%. The floating scum is also removed and combined with the sludge. As a result the effluent from the primary has an SS of approximately 150 mg l^{-1} and a BOD of approximately 200 mg l^{-1}. This may be acceptable for discharge to the sea or some estuaries without further treatment. The solids are concentrated into the primary sludge which is typically removed once a day under the influence of hydrostatic pressure.

5.2.3 Secondary (Biological) Treatment

There are two principal types of biological sewage treatment:

(i) The percolating filter (also referred to as a trickling or biological filter).
(ii) Activated sludge treatment.

Both types of treatment utilize two vessels, a reactor containing the microorganisms which oxidize the BOD, and a secondary sedimentation tank, which resembles the circular radial flow primary sedimentation tank, in which the microorganisms are separated from the final effluent.

The early development of biological sewage treatment is not well documented. However, it is established that the percolating filter was developed to overcome the problems associated with the treatment of sewage by land at 'sewage farms', where large areas of land were required for each unit volume of sewage treated. It was discovered that approximately 10 times the volume of sewage could be treated in a given area per unit by passing the sewage through a granular medium supported on underdrains designed to allow the access of air to the microbial film coating the granular bed.

The origins of the percolating filter are present in land treatment and its development was an example of evolution. The second and probably predominant form of biological sewage treatment, the activated sludge process, arose spontaneously and represents an entirely original approach. This process involves the aeration of freely suspended flocculant bacteria, 'the activated sludge floc' in conjunction with settled sewage which together constitute the 'mixed liquor'. Activated sludge treatment continues the trend established by the change from land treatment to the percolating filter in that, at the expense of higher operating costs, it is possible to treat very much larger volumes of sewage in a smaller area.

The activated sludge process is probably the earliest example of a continuous bacterial (microbial) culture deliberately employed by man, and certainly the largest used to date. Development of the activated sludge process was announced by its originators Fowler, Ardern and Lockett in 1913, based upon their research at the Davyhulme Sewage Treatment Works, Manchester. These scientists very generously did not patent the process to facilitate its rapid and widescale application.

Development of these two forms of biological sewage treatment has been largely empirical and undertaken without the benefit of information about the fundamental principles of continuous bacterial growth, which began to be developed from the late 1940s when Monod published his work on continuous bacterial growth, although the relevance was not perceived until approximately 10 years later. This lack of microbiological knowledge is highlighted by the fact that the role of microorganisms in the activated sludge process was not fully accepted until after 1931; prior to this it was accepted by several workers that coagulation of the sewage colloids was the principal mechanism in the activated sludge process, although in the USA the role of bacteria in percolating filters was first recognized in 1889.

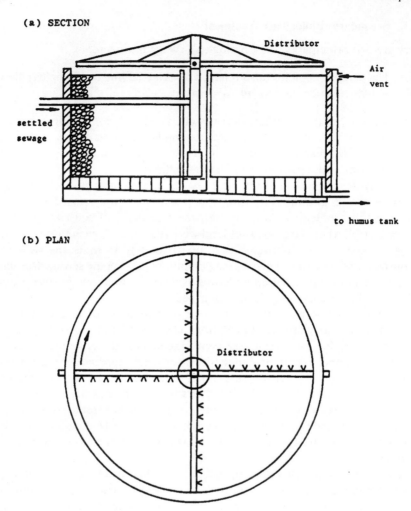

Figure 5.4   *A percolating filter*

5.2.3.1   *Percolating Filter.* These units consist of circular or rectangular beds of broken rock, gravel, clinker or slag with a typical size in the range of 50–100 mm. The beds are between 1.5 and 2.0 m deep and of very variable diameter or size depending on the population to be served. The proportion of voids (empty spaces) in the assembled bed is normally in the range 45 to 55% (see Figure 5.4). The settled sewage trickles through interstices of the medium which constitutes a very large surface area on which a microbial film can develop. It is in this gelatinous film containing bacteria, fungi, protozoa and on the upper surface algae that the oxidation of the BOD in the settled sewage takes place. The percolating filter is in fact a continuous mixed microbial film reactor. Settled sewage is fed onto the surface of the filter by some form of distributor mechanism. On circular filters a rotation system of radial sparge pipes is used which are usually reaction jet propelled although on larger beds they may be electrically driven. With rectangular beds electrically powered rope hauled arms are used.

The microorganisms which constitute the gelatinous film appear to be organized, at least near the surface of the filter where algae are present, into three layers (see Figure 5.5). The upper fungal layer is very thin (0.33 mm), beneath it the main algal layer is approximately 1.2 mm and both are anchored by a basal layer containing algae, fungi and bacteria of approximately 0.5 mm. However, algae do occur to some extent in all three layers. Beneath the surface, where sunlight is excluded and as a consequence the algae are absent, this structure is significantly modified, probably into a form of organization with only two layers. It has been calculated that photosynthesis by algae could provide only 5% or less of the oxygen requirements of the microorganisms in the filter. Furthermore, photosynthesis would only be an intermittent source of oxygen since it would not occur in the dark and algae are often present only in the summer months. Carbon dioxide generated by other organisms in the filter might, however, increase the rate of photosynthesis. It has been proposed that algae derive nitrogen and minerals from the sewage and that some may be facultative heterotrophs. The nitrogen fixing so-called 'blue–green algae', really bacteria, are frequently present in filters.

Whilst fungi are efficient in the oxidation of the BOD present in the settled sewage they are not desirable as dominant members of the microbial community. They generate more biomass than bacteria, per unit of BOD consumed, thus increasing the sludge disposal problem. Moreover, an accumulation of predominantly fungal film quickly causes blockages of the interstices of the filter bed material, impeding both drainage and aeration. The latter may result in a reduction in the efficiency of treatment which is dependent upon the metabolic activity of aerobic microorganisms.

Protozoa and certain metazoans (macrofauna) play an important role in the successful performance of the biological filter, although the precise nature of

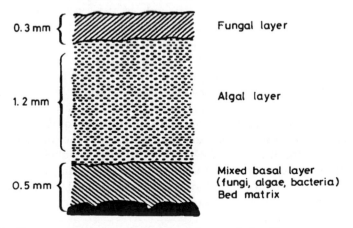

Figure 5.5 *Cross-section of the surface layers of a percolating filter*

this role is dependent on the extrapolation of observations made in the activated sludge process, which is more amenable to study. However, the similarity in the distribution of organisms within the two processes suggests strongly that their roles are the same in both. The protozoa in particular remove free-swimming bacteria, thus preventing turbid effluents since freely suspended bacteria are not settleable. Certain metazoans may also ingest free-swimming bacteria, but their most important function is to assist in breaking the microbial film which would otherwise block the filter. This film is 'sloughed off' with the treated settled sewage. Protozoa (principally ciliates and flagellates) tend to dominate in the upper layers of the filter, whilst the macrofauna (nematodes, rotifers, annelids and insect larvae) dominate the lower layers.

If film is not removed satisfactorily, frequently as the result of excessive fungal growth, the condition known as 'ponding' develops. In this condition the surface of the filter is covered in settled sewage, air flow ceases, treatment stops and the bed becomes anaerobic. Ponding may also be caused by the growth of a sheet or felt of large filamentous algae, principally *Phormidian* sp., on the face of the filter. To minimize film production recirculation of treated effluent is often employed. This reduces film growth by dilution of the settled sewage, improves the flushing action for the removal of loose film, and promotes a more uniform distribution of the film with depth.

Treated sewage is subject to secondary sedimentation which is similar to primary sedimentation as a result of which the suspended sloughed off film is consolidated into humus sludge and the final effluent discharged to the receiving water.

5.2.3.2 Activated Sludge. In the activated sludge process the majority of biological solids removed in the secondary sedimentation tank are recycled (returned sludge) to the aerator. The feedback of most of the cell yield from the sedimentation tank encourages rapid adsorption of the pollutants in the incoming settled sewage, and also serves to stabilize the operation over a wide range of dilution rates and substrate concentrations imposed by the diurnal and other fluctuations in the flow and strength of the sewage. Stability is also provided by the continuous inoculation of the reactor with microorganisms in the sewage and airflows, which are ultimately derived from human and animal excreta, soil run-off, water and dust. The reactor of the activated sludge plant is usually in the form of long deep channels. Before entering these channels the returned sludge and settled sewage are mixed thereby forming the 'mixed liquor'. The retention time of the 'mixed liquor' in the aerator is typically three to six hours; during this period it moves down the length of the channel before passing over a weir, prior to secondary sedimentation. The sludge which is not returned to the aerator unit is known as surplus activated sludge and has to be disposed of. In practice the conditions in the aeration unit diverge from the completely mixed conditions commonly used for industrial fermentations and it may be best described as a continuous mixed microbial deep reactor with feedback.

The design of the concrete tanks which form the reactor is strongly influenced by the type of aeration to be employed. Two types are available, compressed (diffused air) (see Figure 5.6) and mechanical (surface aeration) (see Figure 5.7).

In the diffused air system much of the air supplied is required to create turbulence, to avoid sedimentation of the bacteria responsible for oxidation. Surface aeration systems introduce the turbulence mechanically and only provide sufficient air for bacterial oxidation. Both types of system aim to maintain a dissolved oxygen concentration of between 1 and 2 mg l^{-1}.

In the diffused air system the air is released through a porous sinter at the base of the tank and this system is characterized by long undivided channels which may be quite narrow (see Figure 5.6). Mechanical aeration utilizes rotating paddles to agitate the surface thereby incorporating air and creating a rotating current which maintains the bacterial flocs in suspension. Each paddle is located in its own cell which has a hopper shaped bottom, this gives the plant the appearance of a square lattice (see Figure 5.7). However, beneath the face of the

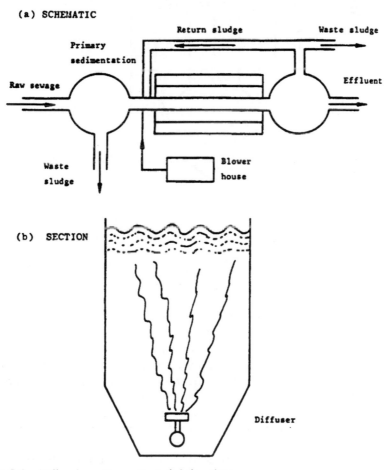

Figure 5.6   *Diffused aeration activated sludge plant*

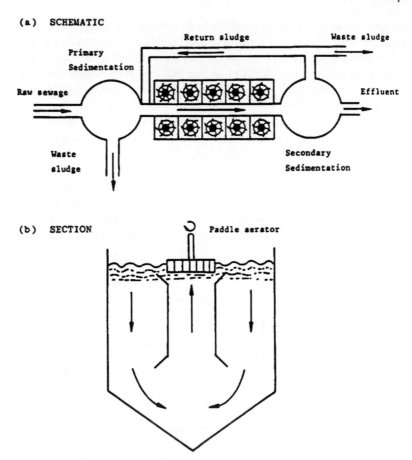

Figure 5.7 *Mechanically aerated activated sludge plant*

mixed liquor all the cells are connected, forming a channel. In both systems the channels are 2–3 m deep and 40–100 m long.

The success of the activated sludge process is dependent on the ability of the microorganisms to form aggregates (flocs) which are able to settle. It is generally accepted that flocculation can be explained by colloidal phenomena and that bacterial extracellular polymers play an important role, but the precise mechanism is not known. The significance of flocculation to the success of the process is not the only characteristic to distinguish it from other industrial continuous cultures. There are four additional and very significant differences: it utilizes a heterogeneous microbial population, growing in a very dilute multi-substrate medium, many of the bacterial cells are not viable and finally the objectives of the process, which are the complete mineralization of the substrates (principally carbon dioxide, water, ammonia and/or nitrate) with minimal production of both biomass and metabolites are also unique.

The heterogeneous population present in activated sludge includes bacteria, protozoa, rotifers, nematodes and fungi. The bacteria alone are responsible for the removal of the dissolved organic material, whilst the protozoa and rotifers 'graze', removing any 'free-swimming' and hence non-settleable bacteria; the protozoans and rotifers being large enough to settle during secondary sedimentation. The role of protozoa in activated sludge has been extensively studied; there are three groups involved: the ciliates, flagellates and amoebae. It is probably the ciliates (*Ciliophora*), which constitute the greatest number of species with the greatest number present in each species, that play the major role in the clarification process. The effect on effluent quality as a consequence of grazing by protozoa is summarized in Table 5.2. Not only do the protozoa remove free-swimming activated sludge bacteria but they play an important role in the reduction of pathogenic bacteria, including those which cause diphtheria, cholera, typhus and streptococcal infections. In the absence of protozoa approximately 50% of these types of organisms are removed while in their presence removals rise to 95%. Nematodes have no significant role in the process, whilst the effects of the fungi are generally deleterious and contribute to or cause non-settleable sludge known as 'bulking'. Members of the following bacterial genera have been regularly isolated from activated sludge, *Pseudomonas, Acinetobacter, Comamonas, Lophomonas, Nitrosomonas, Zoogloea, Sphaerotilus, Azotobacter, Chromobacterium, Achromobacter, Flavotacterium, Alcaligenes, Micrococcus* and *Bacillus*. Attributing the appropriate importance to each genus is a problem which confounds bacteriologists.

Of the principal groups of substrates listed in Section 5.1.4, only one single substrate (cellulose) was included. Each of the groups includes many substrates; for example the 'sugars' identified in sewage include glucose, galactose, mannose, lactose, sucrose, maltose and, arabinose, whilst the nitrogenous compounds include proteins, polypeptides, peptides, amino acids, urea, creatine and amino-sugars. Since bacteria normally only utilize a single carbon substrate or at the most two, this diversity of substrates in part explains the numerous genera of bacteria isolated from activated sludge because each substrate under most conditions will sustain one species of bacterium. Moreover, as a consequence of the large number of substrates present in the settled sewage the concentration of individual substrates is far less than the 200 mg l^{-1} of BOD present, perhaps 20–40 mg l^{-1} for the most abundant and less than 10 mg l^{-1} for the less common ones. The concentration of each substrate is further reduced in the aeration tank by dilution with the returned activated sludge which is typically mixed 1:1 with settled sewage resulting in a 50% reduction in substrate concentration.

Table 5.2 *Importance of ciliated protozoa in determining effluent quality*

Effluent property	Ciliates absent	Ciliates present
Chemical Oxygen Demand (mg l^{-1})	198–254	124–142
Organic Nitrogen (mg l^{-1})	14–20	7–10
Suspended Solids (mg l^{-1})	86–118	26–34
Viable bacteria (10^7 ml^{-1})	29–42	9–12

The low substrate concentration means that the bacteria are in a starved condition. As a consequence many of them are 'senescent', *i.e.* in that phase between death, as expressed by the loss of viability, and breakdown of the osmotic regulatory system (the moribund state); thus the bacterium is a functioning biological entity incapable of multiplication. That bacteria could exist in this condition was established at an early stage in a series of inspired experiments by Wooldridge and Standfast who published their results in 1933. They determined the dissolved oxygen concentrations and bacterial numbers (by viable counts) in a series of biochemical oxygen demand bottles containing diluted raw sewage on a daily basis. The viable count reached a maximum on the second day and thereafter fell rapidly. However, the consumption of oxygen increased by equal amounts until the fourth day and fell to a negligible value on the fifth day. There was no obvious relationship between viability and oxygen consumption. They tested experimentally the hypothesis that non-viable bacteria were apparently capable of oxygen uptake by destroying the capacity for division without significantly diminishing enzyme activity. Treatment of *Pseudomonas fluorescens* with a 0.5% formaldehyde solution prevented division but these bacteria exhibited vigorous oxygen uptake in both sewage and other media. Subsequently they were able to determine the presence of active oxidase and dehydrogenase enzymes in these non-viable bacteria. The effects of low substrate concentration on the viability of the bacteria are compounded by their specific growth rate. It is intended that biological wastewater treatment should result in the production of a final effluent containing negligible BOD. The biochemically oxidizable material in the effluent is composed of compounds originally present in the settled sewage, which have not been completely biodegraded, and bacterial products. Moreover this is to be achieved with the minimal production of biomass. These twin objectives are concomitant with the utilization of a bacterial population with a very low specific growth rate.

Unlike the percolating filter, bacterial growth in the activated sludge process is amenable to the type of description used by bacteriologists for conventional continuous cultures. However, although it is amenable to this type of treatment it inevitably appears to be very different from all other continuous cultures. The dilution rates (rate of inflow of settled sewage/aeration tank volume) used are invariably low by the standards of industrial fermentations, typically $0.25 \, h^{-1}$, *i.e.* one quarter of the aeration tank volume is displaced every hour, therefore the *hydraulic retention time* is four hours. Although in the conventional single pass reactor the dilution rate and the specific growth rate (time required for a doubling of the population) are identical; that is the state in which the rate of production of cells through growth equals the rate of loss of cells through the overflow. In the activated sludge process, because of the recycling of the biomass, the specific growth rate is very much lower than the dilution rate, typically in the range 0.002–$0.007 \, h^{-1}$. Since, under steady-state conditions, the bacteria are only able to grow at the same rate as they are lost from the system, recycling them dramatically lowers their specific growth rate and allows it to be controlled independently of the dilution rate. Under steady-state conditions the specific growth rate is equivalent to the specific rate of sludge wastage (mass of suspended solids lost by sludge wastage and discharged in the effluent in unit time as a proportion of the total mass in the plant) which is the reciprocal of the 'sludge age' or mean cell retention time which is typically 4–9 days. Thus, whilst the retention of the aqueous phase in the system is only four hours, the retention of the bacterial cells or sludge age is several days. The sludge age (θc) is a value which describes a great deal about the type of activated sludge plant; its purpose, quality of effluent and the bacteriological and biochemical states are all summarized by this item.

The activated sludge process may have up to four phases:

(i)   clarification, by flocculation of suspended and colloidal matter;
(ii) oxidation of carbonaceous matter;
(iii) oxidation of nitrogenous matter (see Section 5.2.3.8);
(iv) auto-digestion of the activated sludge.

The occurrence of these four phases is directly dependent on increasing sludge age. Those processes which operate at low sludge ages give rapid removal of BOD per unit time, but the effluent is of poor quality. Plants which have high sludge ages give good quality effluents but only a slow rate of removal. Low sludge ages result in actively growing bacteria, and consequently high sludge production, whilst bacteria grown at high sludge ages behave conversely. Figure 5.8 illustrates the relationship between the growth curve of the bacterial culture

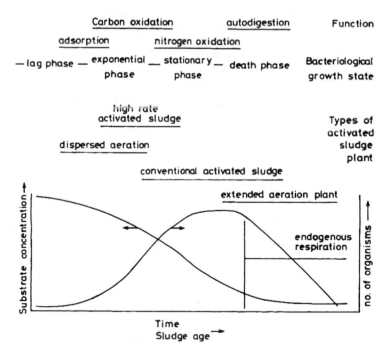

Figure 5.8 *Relationship between batch culture and type of activated sludge plant*

and the type of activated sludge plant. By operating continuously the activated sludge process functions only over a small region of the batch growth curve; this region is determined by the specific sludge wastage rate. The region selected determines the type of plant and its performance. These are summarized in Figure 5.8.

5.2.3.3   Dispersed Aeration.   This type of process is rarely used and is not applicable to the treatment of municipal sewage but may be of use in the preliminary treatment of some industrial wastes. The bacteria are growing rapidly (exponential phase), thus the process has the ability to remove a large quantity of BOD per unit of biomass and as a consequence a small reactor may be used which is cheap to construct. However, because of their high rate of growth, the bacteria convert much of the BOD into biomass, causing a sludge disposal problem, flocculation is limited so additional treatment is essential to remove solids. Furthermore, although BOD removal per unit biomass is high, the effluent BOD is also high.

5.2.3.4   High-rate Activated Sludge.   This shares many features of the previous process; however, flocculation proceeds satisfactorily and secondary sedimentation will remove the solids effectively. The growth rate of the bacteria is still high, but only carbonaceous material will be oxidized. However, some 60 to 70% of the influent BOD will be removed with a hydraulic retention time of approximately two hours. This type of process is probably most frequently used for industrial wastes prior to discharge to the sewers, although it is also used for domestic sewage treatment, perhaps most appropriately where effluents are to be discharged to estuarine waters where standards are less stringent.

5.2.3.5   Conventional Activated Sludge.   The two previous processes utilize actively growing bacteria in the exponential phase of growth. They achieve the oxidation of carbon compounds utilizing an exclusively heterotrophic bacterial population. Conventional activated sludge plants operate in the stationary or declining growth phases utilizing senescent bacteria. This very slow growth results in very low residual substrate concentrations and hence low values for effluent BOD. In addition, plants operating at sludge ages towards the upper end of this range contain autotrophic nitrifying bacteria. These organisms convert ammonia into nitrite and nitrate. This further improves the quality of the effluent since ammonia can exert an oxygen demand but nitrate cannot. In addition to maximizing effluent quality, conventional activated sludge plants limit the production of new cells. Bacteria which are growing slowly use much of the organic matter available in the maintenance of their cells rather than in the production of new cells. These features have made conventional activated sludge the most widely adopted biological sewage treatment process for medium and large communities. The rate of oxidation is highest at the inlet of the tank and it can be difficult to maintain aerobic conditions. Two solutions to this problem have been adopted. With *tapered aeration* rather than supplying air uniformly along the length of the tank the air is concentrated at the beginning of the tank and progressively reduced along its length. The volume of air supplied remains unchanged but it is distributed according to demand. Alternatively *stepped loading* may be utilized. This aims to make the requirement for air uniform by adding the settled sewage at intervals along the tank, thus distributing the demand.

5.2.3.6 Extended Aeration. This process operates at very high sludge ages, exclusively in the declining phase of growth. The retention time in the aeration tank is between 24 h and 24 days. As a consequence, the available substrate concentration is low and the bacteria undertake endogenous respiration (see Figure 5.8), that is respiration after the consumption of all available extracellular substrate. The result of utilizing endogenous materials is the breakdown of the sludge, sometimes referred to as auto-digestion. By this means sludge production is minimized and the small amount of material that must be disposed of is highly mineralized and inoffensive. This type of treatment has been extensively used for small communities; whilst capital costs of such plants are high, operating and sludge disposal costs are very low.

5.2.3.7 Contact Stabilization. The contact stabilization process is a variation of conventional activated sludge used for treating wastes with a high content of biodegradable colloidal and suspended matter. The process utilizes the adsorptive properties of the sludge to remove the polluting material very rapidly (0.5–1 h) in a small aeration tank. The mixed liquor is then settled and passed into a second aeration tank and aerated for a further 5 to 6 h, during which period the adsorbed material is oxidized. After this the sludge with its adsorptive capacity restored is returned to the contact basin. Although this process requires two aeration tanks, the two are very much smaller than the equivalent single tank, since the mixed liquor suspended solids in the contact basin are typically 2000 mg l^{-1} and in the second tank (digestion unit) they are about 20 000 mg l^{-1}.

5.2.3.8 Nitrification. The production of a final effluent with the minimum BOD value is dependent upon the complete nitrification of the effluent, which involves the conversion of the ammonia present into nitrate. This is a two-stage process undertaken by autotrophic bacteria principally from the genera *Nitrosomonas* and *Nitrobacter*. Nitrification occurs in percolating filters and activated sludge plants operated in a suitable manner. The first stage, sometimes referred to as 'nitrosification', involves the oxidation of ammonium ions to nitrite and follows the general equation:

$$NH_4^+ + 1.5O_2 \xrightarrow{\textit{Nitrosomonas}} NO_2^- + 2H^+ + H_2O$$

In the second stage nitrite is oxidized to nitrate:

$$NO_2^- + 0.5O_2 \xrightarrow{\textit{Nitrobacter}} NO_3^-$$

The overall nitrification process is described by the equation:

$$NH_4^+ + 2O_2 \longrightarrow NO_3^- + 2H^+ + H_2O$$

Two important points are evident from this last equation. Firstly, nitrification requires a considerable quantity of oxygen. Secondly, hydrogen ions are formed and hence the pH of the wastewater will fall slightly during nitrification.

The settled sewage is effectively self-buffering but a fall of 0.2 of a pH unit is frequently observed at the onset of nitrification. In this autotrophic nitrification process, ammonia or nitrite provide the energy source, oxygen the electron acceptor, ammonia the nitrogen source and carbon dioxide the carbon source. The carbon dioxide is provided by the heterotrophic oxidation of carbonaceous nutrient, by reaction of the acid produced during nitrification with carbonate or bicarbonate present in the wastewater, or by carbon dioxide in the air. Whereas for carbonaceous removal the oxygen requirement is roughly weight for weight with the nutrients oxidized, in the case of ammonia removal by nitrification requires approximately seven times as much oxygen as is required to achieve the removal of the same quantity of nutrient.

Nitrification significantly increases the cost of sewage treatment since more air is required. Furthermore, because these autotrophic organisms grow only slowly, longer retention periods are also required, resulting in higher capital costs. Nor does nitrification result in the production of an entirely acceptable sewage effluent. In areas where water re-use is practised the concentration of nitrate in river waters causes concern. There exists a limit on the concentration of nitrate in drinking water to avoid the occurrence of methaemoglobinaemia (so called 'blue baby' syndrome). As a consequence denitrification is now practised after nitrification in some activated sludge treatment plants. In this anoxic heterotrophic bacterial process, nitrite and nitrate replace oxygen in the respiratory mechanism and gaseous nitrogen compounds are formed (nitrogen gas, nitrous and nitric oxides). However, this procedure is not part of conventional sewage treatment practice at present.

5.2.4 Secondary Sedimentation

Both types of biological treatment require sedimentation to remove suspended matter from the oxidized effluent. Tanks similar to those normally employed for primary sedimentation are generally employed, although at a higher loading of approximately $40 \, m^3 \, m^{-2} \, d^{-1}$, at 3 *DWF*. Because of the lighter and more homogenous nature of secondary sludge, simple sludge scrapers are possible and scum removal is not necessary. The association of primary sedimentation tanks and a biological process for secondary treatment, results in a sewage treatment works, as opposed to sewage farms where only land treatment was (is) employed. As an awareness of environmental pollution, in addition to public health, developed in the 1950s and 1960s, the term water pollution control works was introduced to describe sewage treatment works, although this change of terminology was merely cosmetic. With the recognition of the importance of water re-use the term water reclamation works has found favour in some areas. Such works frequently apply additional tertiary treatment processes.

Sewage treatment results in the production of a final effluent suitable for discharge in the selected receiving water, and one or more sludges which may require treatment prior to disposal.

5.3 SLUDGE TREATMENT AND DISPOSAL

Sludge treatment and recycling or disposal is a significant element of the cost of wastewater treatment, accounting for some 40% of all site costs. The collection, handling, processing and disposal of sludge is an operational chain of finite capacity which must be managed as a single process to ensure that reliable capacity and consistent products are provided. Often given insufficient attention in the past, it is the decision as to how the final product will be disposed that will influence design and process options for the entire treatment facility. Disposal options have been reduced by the ban on sea dispersal,[11] the planned controls on landfill and incineration, and strict conditions for agricultural recycling. In the UK 1.12 million tonnes of dry solids are disposed of annually (Table 5.3).[12] This represents some 43 million tonnes of sludge production. Sewage sludge is the combination of the product of primary sedimentation of sewage and the by-product of secondary and tertiary treatment processes. Primary sludges are odourous and liable to become putrescent with the potential to cause odour nuisance. Secondary and tertiary sludge consist largely of bacterial solids. When fresh it is much less offensive than primary sludges but is still liable to putresce. These sludges are usually combined for treatment, though co-settlement of surplus activated sludges in primary settlement tanks is to be avoided as the mixture can be less easy to settle and can increase process odour.

The purpose of sludge treatment is to secure the reliable, safe, sustainable and economic disposal of sludge. When recycled to agriculture, sludge must be treated to reduce pathogens and to stabilize the product to ensure minimum odour. When sludge is to be incinerated, treatment aims to achieve the maximum dry solids content to safeguard energy efficient combustion.

Table 5.3 *UK sludge disposal routes 2000/01 (estimated)*[12]

Route	000s tonnes dry solids	% of total
Agriculture	846	60
Forestry, horticulture	55	4
Incineration	262	19
Other, including energy recovery	130	9
Landfill	65	5
Land reclamation	49	3
Total	1407	

5.3.1 Sources

The production of sludge can be estimated from the contributing population and the type and extent of the treatment process. The higher the quality of effluent from a site, the greater the degree of treatment, the more sludge is produced. The primary settlement tanks will produce some 40 g capita^{-1} day^{-1} at about 5% dry solids content. Secondary treatment processes will add 20–50 g capita^{-1} day^{-1} of bacterial solids at 1–3% dry solids. Where chemical phosphorus stripping is practised then total sludge production will be increased by 10–25%. Settlement processes are managed so that the dry solids content of sludge is optimized to reduce volumes to be processed without being to thick for easy handling or so stale that it becomes septic and odourous. Sludge treatment is usually centralized into few processing centres. This requires that untreated sludges are imported from many smaller satellite treatment works.

As treatment processes will be designed with a maximum hydraulic capacity it is important to ensure reliable management of dry solids content. The volume of sludge at 2.5% dry solids is twice that at 5% dry solids and therefore would incur twice the transport or pumping costs and could overload processing capacity. The reliability of the treatment chain protects sewage treatment processes by ensuring that sludge is regularly removed from settlement tanks. In addition to managing volume, the quality of the sludge must be protected from damaging pollutants. Effective trade effluent control will safeguard biological treatment processes, agricultural recycling, and combustion.

5.3.2 Recycling and Disposal Options

The decision as to which treatment is appropriate is dependent on the disposal option proposed for the site. It is based on a combination of economic, social and environmental considerations. There are three long term options, nutrient recycling, energy recovery, and materials recovery. For the next few years landfill will remain an option but this will be ended by the implementation of the EC Landfill Directive. Nutrient recycling includes agricultural recycling, forestry, land reclamation and is subject to the controls of the Sludge (Use in Agriculture) Regulations 1989 (to be revised in 2002) or the controls on waste disposal in the 1990 Environmental Protection Act.

Treated sludge beneficially recycled in agriculture is described as 'biosolids'. Agricultural recycling is, in many cases, the Best Practicable Environmental Option (BPEO). Biosolids contain useful amounts of nutrients, trace elements, and organic matter. Farmers value the product as a cost effective source of nutrient and soil conditioning (Table 5.4). However, farmers are customers for the products and must be assured that the product delivers real benefits, that soil quality is protected, and that food grown on treated fields is safe. The increase in public concern about food safety, arising from several food related health incidents in the 1980s and 1990s has affected biosolids recycling. Untreated recycling to land used for food production was ended after 1999 as the result of

Table 5.4 *Typical nutrient content and value of biosolids (20% dry solids digested cake)*

Nutrient	Content (kg tonne^{-1})	Est. value of available nutrients (£ ha^{-1})
Nitrogen (total)	8	10
Total phosphate (P$_2$O$_5$)	12	65
Total potash (K$_2$O)	0.4	2
Sulfur	1.5	3
Magnesium	0.75	6

an agreement between the UK water industry and the British Retail Consortium.

The concerns and constraints surrounding agricultural recycling and the ending of disposal at sea have increased interest in energy recovery options. It is expected that by 2005 40% of sludge will be disposed by an energy recovery route.[12] Incineration has been the disposal choice for large conurbations where access to farm land has been not been practicable because of the transport distances and potential traffic impacts. In the future smaller scale operations are likely to be more common. The main energy recovery options are incineration, gasification/pyrolysis, and co-fuelling of cement/aggregate kilns or power stations. The ash from energy processes, generally landfilled, can be incorporated into building materials. Landfill void space will be more scarce and expensive in the future as environmental taxes impose additional costs on landfill sites. The EC Landfill Directive has set targets for the reduction of landfilling of degradable material which will end landfilling as a practical disposal option for sludges. While the combustion options are a practical technical solution, there is significant public opposition to the construction of new incinerators. This public concern is reflected in the standards proposed in the Waste Incineration Directive and the time required for the planning process. The proposed emission standards will require substantial investment in clean-up technology, adding to the costs of combustion. Preparation of sludge prior to incineration requires dewatering to >35% dry solids to ensure that the process is autothermic. Alternatively sludge may be dried to +90% dry solids prior to processing.

5.3.3 Pre-treatment Handling

Sludge treatment processes cannot be considered in isolation. The process comprises the preparation and storage at satellite sites; transport, reception and processing at the treatment centre; and final disposal on farms. The capacity of the system to handle sludges is finite and bottlenecks in the system will cause delays in moving and processing sludge which can affect the efficient operation of the sewage treatment processes.

Thickening removes water from sludge to reduce volumes transported or processed. The process should produce a 6% dry solids product. Gravity thickening uses settlement to separate the water. A tank is equipped with valves at several levels in the tank to enable water layers to be identified and drained. Stirred settlement tanks, picket fence thickeners, avoid the problems of the development of layers of water within a tank by slow stirring. The thickened product is removed from the bottom of the tank and the water removed by overflow weirs at the top of the tank. Typical retention of a picket fence thickener is two days. For activated sludges mechanical thickening is preferred. It can be achieved with centrifuges, filter belt presses and gravity belt presses. The thickening is achieved with the help of polyelectrolytes to aid floc formation. Polymer is added at a rate of 4–10 kg of active ingredient per tonne dry solids of sludge processed.

Dewatering removes water from sludge to achieve a cake of 20–40% dry solids. Dewatering may be placed before or after the main treatment process. It can be achieved with centrifuges, filter belt presses or plate presses. Polyelectrolytes are used to improve dewaterability of the sludge. The selection of equipment is usually based on pilot trials as the response of sludges can vary according to the process used. The amount of polymers used will vary from 4–8 kg active ingredient per tonne of dry solids. A key process parameter is the quality of returned liquors (centrate or filtrate); these can add significant loads to the treatment process. Filter presses use large volumes of washwater to keep the belts clean. This washwater is usually the treated effluent containing significant numbers of pathogens. Where a pasteurized product is being dewatered, centrifuges or plate presses should be used to avoid recontamination of product by the washwater.

5.3.4 Treatment Processes

Treatment for agricultural recycling requires that the product achieves the pathogen reduction targets, is stable and, as far as is possible, is odour free. The critical control points for each process must be constantly monitored and recorded. The balance and availability of nutrients is different in different products. Generally drier and more processed product have lower agronomic values, but are often more acceptable products to the customer.

5.3.4.1 Digestion. During digestion the organic matter present in the sewage sludge is biologically converted into a gas typically containing 70% methane and 30% carbon dioxide. The process is undertaken in a closed reactor usually equipped with a separate gas holder. The methane produced is used for maintaining the digester temperature, and power production by combustion in gas engines.

Methane production is only significant at elevated temperatures, when 1 m^3 of methane is produced for every 3 kg of BOD degraded. Digesters are characterized by the temperature at which they operate, those in which gas production is optimum at 35 °C are described as 'mesophilic' whilst those at

55 °C are 'thermophilic', these terms describing the temperature preferences of the bacteria undertaking the process. Thermophilic digestion will achieve a greater pathogen reduction than mesophilic processes and can be classed as an enhanced treatment process.

Heat exchangers are used to transfer heat from the treated sludge to the influent sludge. The additional heat is provided by the combustion of methane. For efficient operation the digester requires a robust mixing system which may be mechanical or utilize the gas produced in the process to provide turbulence. A conventional anaerobic digester is illustrated in Figure 5.9. The result of anaerobic digestion is to reduce the volatile solids present in the original sludge by 50% and the total solids by 30%. In addition the unpleasant odour associated with the raw sludge is drastically reduced. During the 12 to 20 days required for digestion the sludge is stabilized and emerges with a slightly tarry odour.

Anaerobic digestion can be considered as a three-stage process (Figure 5.10). The first, involving the hydrolysis of the fats, proteins and polysaccharides present in the sludge, produces long chain fatty acids, glycerol, short chain peptides, amino acids, monosaccharides and disaccharides. The second step (acid formation) involves the formation of a range of relatively low molecular weight materials, including hydrogen, formic and acetic acids, other fatty acids, ketones and alcohol. It is now recognized that only hydrogen, formic acid and acetic acid can be utilized as substrates by the methanogenic bacteria. Thus in the third step compounds other than hydrogen, formic acid and acetic acid are converted by the obligatory hydrogen producing acetogenic (OHPA) bacteria. Some bacteria are able to undertake both steps one and two and produce hydrogen, formic acid and acetic acid and which therefore do not require step three. Once in operation, with reasonable retention times and volatile solids loadings, the routine operation of digesters must include careful monitoring of certain parameters which are used to indicate whether the process is about to

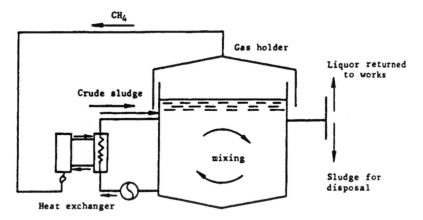

Figure 5.9 *Schematic diagram of an anaerobic digester*

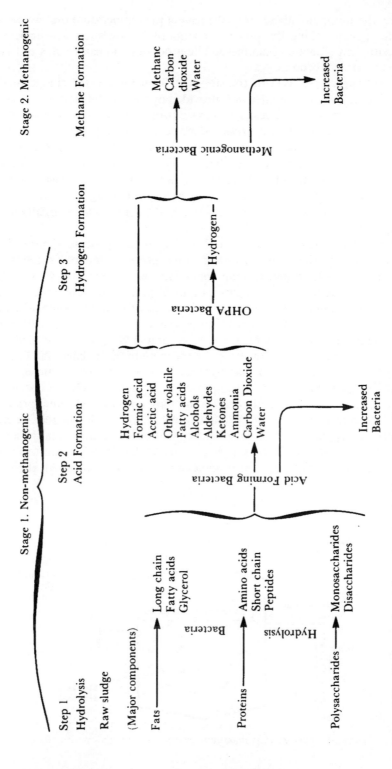

Figure 5.10 *Biochemical transformations involved in anaerobic digestion*

fail. The main parameters are volatile acids and hydrogen ion concentration (pH).

Anaerobic digestion is quite sensitive to fairly low concentrations of toxic pollutants, such as heavy metals and chlorinated organics, and to variations in loading rates and other operational aspects. If the balance of the process is upset it is most likely that the methanogenic organisms become inhibited first. This results in a build-up of the intermediate compounds, volatile fatty acids, at the stage immediately prior to methane formation. These compounds include formic, acetic and butyric acids and can be monitored to determine the state of the process. The volatile acids are important because of their acidic nature. Normally digesters operate in the pH range of neutrality (6.5–7.5). They also have some resistance to pH change. High concentrations of the volatile acids can cause a reduction in pH sufficient to inhibit bacterial activity to the extent where irreversible failure of the process occurs. Because of their capacity to resist changes in pH, volatile acid concentrations can build up to significant levels before pH change occurs. Therefore they can act as an early warning indicator of impending process failure. Normal levels of volatile acids are 250–1000 mg l^{-1}. If they exceed 2000 mg l^{-1} this could lead very quickly to failure, if they exceed 5000 mg l^{-1} failure is almost inevitable.

To ensure that mesophilic digestion process achieves the necessary 99% (or 2 log) reduction in indicator pathogens, a secondary storage phase is normally needed. Batch storage as liquid or cake for 14 days is normally sufficient but each site must be evaluated to ensure process standards can be met. Where an enhanced treated product is required mesophilic digestion may be preceded by a thermal treatment stage, pre-pasteurisation. Here batches of sludge are held at high temperature for a period of time to eliminate pathogens. Enhanced treatment process should achieve a 99.9999% (or 6 log) reduction in indicator pathogens. The thermal hydrolysis process combines high temperatures ($>135\,^{\circ}$C) and pressure (>8 bar) to pasteurize and hydrolyse sludge before digestion. This improves the efficiency of digestion, increases gas production, and improves dewaterability of the final product. Drying of sludge to achieve a $>90\%$ dry solids granule or pellet will ensure an enhanced treated product and a reduced transport cost. However, drying is a complex, energy intensive process that requires careful control.

5.3.4.2 Lime Stabilization. The use of lime for sludge treatment has increased recently as a short term solution to the ban on untreated recycling. Lime can be added to liquid sludges or to cakes. The effect is to raise the pH and because the reaction is exothermic it will also raise the temperature (Equation 1). Where a temperature effect is required then mixing with cake is preferred to reduce the mass of water to be heated.

$$CaO + H_2O \longrightarrow Ca(OH)_2 + heat \qquad (1)$$

Mixing is critical to ensure a complete exposure of the sludge mass to the pH and temperature effects. An enhanced treated product is produced when the appropriate time temperature rate is observed, supplementary heat can be provided to reduce the cost of lime. In storage there will be a slow reduction in pH as atmospheric CO_2 reacts with the calcium hydroxide to produce calcium carbonate (Equation 2). This, depending on the structure of the product, may take many weeks.

$$Ca(OH)_2 + CO_2 \longrightarrow CaCO_3 + H_2O \qquad (2)$$

5.3.4.3 Composting. This is an aerobic stabilization process utilizing both thermophilic and mesophilic bacteria to break down organic matter in sludge. Sludge is generally composted in a ratio of 1:2 with straw or green waste to produce an enhanced product which is very acceptable to customers. Composting can be conducted in static piles, turned windrows or in contained systems. The optimum operating temperature is in the range 45–55 °C. The use of imported wastes to provide the admixture will trigger the need for a waste management licence to operate the site. The product can achieve the enhanced product standards referred to in Section 5.3.5.

5.3.4.4 Combustion. Where agricultural recycling is not practical, energy recovery by thermal treatment is the usual alternative. Dried sludge has a similar calorific value to brown coal at 12–18 MJ kg^{-1}. The energy recovery options are incineration, gasification/pyrolysis or co-combustion. For the process to be energy positive then the recovered energy must exceed that used to dewater or dry the sludge prior to the thermal process. Incineration is the combustion in air of the organic content of sludge producing heat and flue gas. The residual ash, which can comprise 30% of the sludge mass must be landfilled or can be incorporated into building materials. Gasification is a partial oxidation process in a low oxygen environment at temperatures of 900–1100 °C. The organic content of sludge is converted into gas comprising CO, H_2, and CH_4, leaving a residual ash. Pyrolysis is the thermal degradation of the organic content in the absence of oxygen at 400–800 °C. In addition to the gas the process produces an oil residue and a char. The gas from these processes can be burned to generate energy or used as a chemical feedstock. Co-fuelling, or the combustion of sludge in coal fired power stations and cement kilns, provides a practical disposal option.

5.3.5 Controls

The practice of recycling sludge to farms has been carried on for as long as sewage treatment has been practised. Controls have evolved over the years in response to the need to protect the environment and food quality. Those controls have increased in recent years from both EU Directives and UK Regulations. In 1986 the EU produced Directive 86/278/EEC on protection of the soil when sludge used in agriculture.[13] The UK implemented this Directive in the Sludge (Use in Agriculture) Regulations 1989 and the associated Code of Practice.[14] The Regulations implemented limits on metals applications to soil and set soil quality standards (Table 5.5). They also limited the use of untreated

Table 5.5 *Maximum permissible concentrations of potentially toxic elements (PTE) in soil after biosolids application and maximum annual application rates*

PTE	Maximum permissible concentration of PTE in soil (mg kg^{-1} dry solids)				Maximum permissible average annual rate of PTE addition over a 10 year period (kg ha^{-1})
	pH 5.0 < 5.5	pH 5.5 < 6.0	pH 6.0 < 7.0	pH > 7.0 and CaCO$_3$ > 5%	
Zinc	200[1]	250[1] 200[3]	300[1] 200[3]	450[1] 300[3]	15
Copper[1]	80	100	135	200	7.5
Nickel[1]	50 for pH 5.0 and above	60	75	110	3
Cadmium[1]	3				0.15
Lead[1]	300				15
Mercury[1]	1				0.1
Chromium[2]	400				15
Molybdenum[3]	4				0.2
Selenium[3]	3				0.15
Arsenic[3]	50				0.7
Fluoride[3]	500				20

1. Statutory limit values.
2. Provisional limit value.
3. Advisory limit value.

sludge to protect food safety, requiring that untreated sludge be injected or immediately ploughed in and not be applied within 10 months of planting of crops which may be eaten raw.

The Urban Waste Water Treatment Directive[11] stopped the sea disposal of sludge at the same time as increasing production, especially from coastal communities. The recycling practice was reviewed by the Royal Commission on Environmental Pollution in their 1996 report.[15] They reported that there had been no instance in the UK in which a link has been established between the controlled application of sewage sludge to agricultural land and disease in the general population through food or water contamination. However, they recommended that all sludge applied to farm land should be treated. This recommendation was reinforced by the House of Commons, Environment, Transport and Regional Affairs Committee in their 1998 report on sewage treatment and disposal.[16] They concluded that all sludge recycled to land should be stabilized and pasteurized. At the same time, the public concerns about food safety, prompted by the BSE incidents and a number of food poisoning outbreaks led to a review of practices by the water industry in partnership with the British Retail Consortium. The outcome was the 'Safe Sludge Matrix',[17] an agreement to end untreated recycling and enhance controls on post application cropping (Table 5.6).

The UK regulations are to be revised in 2002 to give statutory effect to the matrix and to provide a statutory definition of treated and enhanced treated

Table 5.6 *The Safe Sludge Matrix*

Crop type	Untreated sludges	Treated sludges	Enhanced treated sludges
Fruit	✗	✗	✔
Salads	✗	✗ (30 months harvest interval applies)	✔
Vegetables	✗	✗ (12 month harvest interval applies)	✔
Horticulture	✗	✔	✔
Combinable and animal feed crops	✔ Target end date 31/12/99	✗	✔
Grass grazing	✗	✔ (Deep injected or ploughed down only)	✔
Grass and maize silage	✗	✔	✔

Note: ✔ All applications must comply with the current DoE Code of Practice.
  ✗ Applications not allowed (except where stated conditions apply).

products. The following definitions are expected to be applied to sludge used in agriculture:

- treated sludge has undergone a treatment processes so as to significantly reduce its fermentability and to achieve a minimum of $2 \log_{10}$ or 99% reduction of the indicator organism *Escherichia coli* and where the final product meets a 10^5 colony forming units (CFU) per gram dry solids standard in 90% of samples with no single sample exceeding 10^7 CFU per gram dry solids (until 31 December 2003). From 01 January 2004 all sludge samples must achieve the 10^5 CFU per gram dry solids standard;
- enhanced treated sludge has undergone a treatment processes so as to significantly reduce its fermentability and to achieve a minimum of $6 \log_{10}$ or 99.9999% reduction of *Escherichia coli* and where the final product contains no *Salmonella* spp. in 2 grams dry solids and contains less than 103 CFU *Escherichia coli* per gram dry solids.

Suitable processes will not be defined in law but it will be for each operator to demonstrate the efficacy of any treatment and identify the design criteria or critical controls for the process which assure the appropriate level of treatment. The operator shall maintain operational records of the performance of the sludge treatment process in terms of the critical controls. The EC Directive on sludge used in agriculture[13] is in the process of revision and amendment. The revisions are likely to introduce additional controls on organic contaminants that may be found in sludge and revise the existing metals standards for soil and sludge quality.

While the main purpose of the current Regulations has been control of contaminants, and protection of long term soil quality, it is the controls on nutrient inputs to the environment that limit the amount of biosolids that can be applied to farm land. The Nitrate Directive[18] established nitrate vulnerable zones (NVZs) within which nitrogen inputs to agriculture are controlled. In addition the MAFF Water Code[19] provides guidance on nutrient inputs for biosolids, manures and slurries in respect of nitrogen and phosphorus. The Code recommends that not more than 250 kg N ha^{-1} of liquid biosolids should be applied in one year or 500 kg N ha^{-1} of cake every other year. For phosphate it recommends that, for land with adequate soil reserves of phosphate, inputs of total phosphorus should not exceed crop offtake in the rotation.

The combustion technologies are licensed by the Environment Agency under the Integrated Pollution Control (IPC) requirements of the Environmental Protection Act 1990. This requires the application of Best Available Techniques Not Entailing Excessive Cost (BATNEEC) so emission standards tend to be process specific. In future, minimum standards will be set by the proposed EU Waste Incineration Directive and by the Integrated Pollution Prevention Control (IPPC) Directive which will use the Best Available Techniques (BAT) approach to emission control.

5.4 REFERENCES

1. C. R. Curds and H. A. Hawkes (eds.), 'Ecological Aspects of Used-Water Treatment', Volume 1, 'The Organisms and their Ecology', Academic Press, London, 1975.
2. Government of Great Britain, 'Water Pollution Control Engineering', HMSO, London, 1970.
3. Government of Great Britain, Department of the Environment, 'Taken for Granted, Report of the Working Party on Sewage Disposal', HMSO, London, 1970.
4. J. W. Birkett and I. N. Lester, 'Microbiology and Chemistry for Environmental and Public Health Engineers', 2nd edn., E. and F. N. Spon, London, 1999.
5. Metcalf and Eddy Inc., 'Wastewater Engineering: Treatment, Disposal, Re-use', 3rd edn., ed. G. Tchobanoglous and F. L. Burton, McGraw-Hill, New York, 1991.
6. J. N. Lester (ed.), 'Heavy Metals in Wastewater and Sludge Treatment Processes', Volume II, 'Treatment and Disposal', CRC Press, Boca Raton, FL, 1987.
7. J. N. Lester and R. M. Sterritt, 'Water Pollution Control, Module 3, Unit 2: The Basic Principles of Biological Wastewater Treatment', Manpower Services Commission, Sheffield, 1988.
8. R. M. Sterritt and J. N. Lester, 'Water Pollution Control, Module 3, Unit 4: The Treatment and Disposal of Sludge', Manpower Services Commission, Sheffield, 1988.
9. N. F. Gray, 'Water Technology. An introduction for Environmental Scientists and Engineers', Arnold, London, 1999.
10. F. Gaudy and E. T. Gaudy, 'Microbiology for Environmental Scientists and Engineers', McGraw-Hill, New York, 1980.
11. EC 91/271/EC Council Directive of 21 May 1991 concerning urban wastewater treatment.
12. A. Gendebien, C. Carlton-Smith, M. Izzo and J. Hall, 'UK Sewage Sludge Survey 1996/97', Environment Agency Report No. P165.

13. EC 86/278/EEC Council Directive of 12 June 1986 on the protection of the environment, and in particular the soil, when sewage sludge is used in agriculture.
14. Department of the Environment, 'Code of Practice for the Agricultural Use of Sewage Sludge', 1996.
15. Nineteenth Report of Royal Commission on Environmental Pollution, 'Sustainable Use of Soil', 1996.
16. House of Commons Environment, Transport and Regional Affairs Committee (1998), 'Sewage Treatment and Disposal', second report.
17. ADAS, 'The Safe Sludge Matrix', 1999.
18. EC 91/676/EEC Council Directive of 12 December 1991 concerning the protection of waters against pollution caused by nitrates from agricultural sources.
19. MAFF, 'Code of Good Agricultural Practice for the Protection of Water', 1998.

CHAPTER 6

The Treatment of Toxic Wastes

A. JAMES

6.1 INTRODUCTION

The presence of toxic substances in wastewater has always been a matter for concern. This concern has become much more pressing with the intentional or unintentional release of an ever larger variety of substances into the environment.[1]

A whole spectrum of difficulties has arisen in attempting to control the toxicity problem. The major issues may be summarized as follows:

(a) Assessment of environmentally safe concentrations – most toxicity testing measures lethal concentrations but subtle sub-lethal effects may impair ecological success at concentrations well below those causing death. Chronic effects due to prolonged exposure or bio-accumulation, toxic interactions and geo-chemical cycles involving toxins all add further complications.[2]

(b) Assessment of biodegradability persistence of toxins in the environment is clearly undesirable but there are still difficulties in designing suitable tests to assess the biodegradability of newly synthesized compounds.

These areas of uncertainty and lack of information have led to the development of hazard identification, risk assessment and risk management. Virtually all chemicals have the potential to cause harm. The hazard presented in each case is assessed by laboratory toxicity tests for humans and by ecotoxicity tests for the environment.[3]

Risk assessment is the process of estimating the likelihood of the hazards being realized. This obviously has a large stochastic element, both in terms of exposure and effect. The risk assessment needs to include not only the risks from the intended pattern of production and use but also the risks arising from accidental release.[4]

Risk management usually involves control through statutory regulations by the appropriate government agency but increasingly there is a role for economic pressures by consumers or non-governmental organizations.

The treatment of wastes containing toxic material may be technically difficult and/or expensive. Manufacturers or privatized water utilities may be reluctant to make large capital investments in wastewater treatment facilities, especially in periods of economic downturn.[5] For all these reasons the current emphasis in this field is for clean technology, to avoid or reduce the production of toxic waste (see also Chapter 16). The presence of an end of pipe treatment plant should not be taken as an acceptable reason for not minimizing or eliminating losses at source.[6,7]

Planning for the treatment and disposal of wastes containing toxic materials needs to be carried out in discussion with the appropriate government agencies. This will ensure that the most reliable methods of treatment are used, that any laboratory or pilot-scale testing is well planned and that the need for training is not overlooked. The agencies can also advise on contingency planning, for although main systems are now achieving around 98% reliability, some provision for a back-up may be required.

Because of the doubts and uncertainties the treatment and disposal of toxic wastewaters has remained mainly on an empirical level. This is reflected in the following notes which cover:

(a) Sources of Toxic Wastewaters
(b) Toxicity Problems in the Collection System
(c) Pretreatment
(d) Primary arid Secondary Treatment
(e) Passive Treatment
(f) Disposal
(g) Sludge Treatment and Disposal
(h) Case Studies

6.2 SOURCES AND TYPES OF TOXIC WASTES

Toxic substances are primarily associated with industrial wastes but may be found in all types of wastewaters as shown below:

(a) Domestic Wastewaters – these wastes contain ammoniacal nitrogen in concentrations up to 50 mg l^{-1} and when septic may contain sulfide at levels up to 50 mg l^{-1}. Both of these can cause damage to aquatic fauna unless diluted and dispersed. The former is particularly damaging to fresh water fish (Median Toxic Level for Rainbow Trout is around 2 mg l^{-1}) and the latter acts as an enzyme inhibitor in a wide variety of aquatic organisms at levels of a few mg per litre.

(b) Stormwater – the composition of stormwater is much more varied than domestic wastes and is influenced by the nature of the drainage area and the frequency of storms. But the toxic potential is mainly associated with heavy metals like zinc and lead and is invariably found in the first flush of run-off from a storm.

(c) Agricultural Wastes – these contain a wide variety of materials that are used for fertilizers and pest control. Fertilizers containing oxidized nitrogen can cause human toxicological problems. Pesticides and herbicides are the most potent aquatic toxins. Because of their toxicity and persistence in the environment some of these substances like eldrin have had to be prohibited. Wastes from animal husbandry can also be toxic, especially from silage.

(d) Industrial Wastes – the range of toxic substances present in industrial wastes is too wide to catalogue but Table 6.1 gives an indication of the main types of toxic industrial waste and the toxins they contain.

(e) Leachates – a wide range of toxic substances can leach out of sites used for dumping solid wastes, particularly where surface water or groundwater has access. This particularly applies to the dumping of wastes from

Table 6.1 *Sources of some common toxins*

Toxin	Sources
Acids – mainly inorganic but some organic causing pH < 6	Acid Manufacture Battery Manufacture Chemical Industry Steel Industry
Alkalis – causing pH > 9	Breweries Food Industry Chemical Industry Textile Manufacture
Antibiotics	Pharmaceutical Industry
Ammonia	Coke Production Fertilizer Manufacture Rubber Industry
Chromium	Metal Processing Tanneries
Cyanide	Coke Production Metal Plating
Detergents	Detergent Manufacture Textile Manufacture Laundries Food Industry
Herbicides and Pesticides	Chemical Industry
Metals	Metal Processing and Plating Chemical Industry
Phenols	Coke Production Oil Refining Wood Preserving
Solvents	Chemical Industry Pharmaceuticals

industry. The problems caused by contaminated land have become a major concern, leading to a whole new field of bioremediation.

See Watts[3] and UNIDO[8] for further details.

6.3   TOXICITY PROBLEMS IN THE COLLECTION SYSTEM

The cost of constructing a wastewater collection system in an urban area is extremely high, often accounting for 70% of the total cost for treatment and disposal. Damage to the fabric of the sewer is therefore to be avoided and strict controls are usually imposed on substances that may be discharged to the sewer. As shown in Table 6.2 these controls are also intended to control the discharge of substances like cyanides or sulfide that may give rise to poisonous gases which could damage the health of sewer workers. Levels of HCN and H_2S of 0.03% in the atmosphere are toxic and with H_2S there is an additional problem of anaesthesia which makes detection difficult.

Some organic chemicals may cause similar difficulties. They tend to be immiscible in water, volatile and intoxicating and may also form explosive mixtures.[9]

Table 6.2   *Typical consent conditions for discharge to sewers*

Parameter	*Consent condition*
Maximum temperature	40–45 °C
pH	6–10
Substances producing inflammable vapours	Nil
Cyanide concentration	5–10 mg l^{-1}
Sulfide concentration	1 mg l^{-1}
Soluble sulfates	1250 mg l^{-1}
Synthetic detergents	30 mg l^{-1}
Free chlorine	100 mg l^{-1}
Mercury	0.1 mg l^{-1}
Cadmium	2 mg l^{-1}
Chromium	5 mg l^{-1}
Lead	5 mg l^{-1}
Zinc	10 mg l^{-1}
Copper	5 mg l^{-1}
Zinc equivalent (Zn + Cd + 2Cu + 8Ni)	35 mg l^{-1}
Total non-ferrous metal	30 mg l^{-1}
Total soluble non-ferrous metal	10 mg l^{-1}

Note: There are also large numbers of other toxic substances whose discharge to sewers is controlled.

6.4   PRE-TREATMENT OF TOXIC WASTES

In general, industrial wastewaters are most readily and most economically treated in admixture with domestic wastewaters rather than in isolation. Many benefits of scale, balancing nutient supplementation as well as skilled operation, may be obtained by discharging the industrial wastewater to a sewer.[10] But there are a number of occasions when this is not possible or desirable:

(a) Rural areas without sewerage
(b) By-product recovery is economically and technically feasible
(c) Domestic effluent is used ultimately in irrigation
(d) Industrial wastewater does not meet consent conditions for discharge to a sewer
(e) Where industrial wastes form too high a proportion of the combined waste from a community
(f) Land reclamation sites where the products of remediation need to be treated on site or tankered away.

Under these circumstances some form of pre-treatment is needed to render the wastewater suitable for discharge, further treatment or disposal. The main advantages of treatment on site are the possibilities of recovering specific substances in an uncontaminated condition and economies which result from treatment at higher temperatures or concentrations. There may be an important additional advantage; that is the avoidance of contamination of a much larger wastewater stream which would cause difficulties in disposal. Where toxic materials are organic in nature there is often a problem in treatment due to inhibition of bacterial growth. It is often easier and cheaper to develop the necessary bacterial flora in an on-site treatment plant.

This to some extent depends upon the concentration and toxicity of the substances concerned. In some cases dilution of the wastes by admixture with sewage reduces the toxic inhibition making it preferable to treat the industrial waste and sewage together. Also many industrial wastes are deficient in some nutrients such as nitrogen or phosphorus. The desirable ratio of biochemical oxygen demand (BOD):N:P is 100:5:1 and the ratio in domestic waste is commonly 100:18:2.5 so that deficiencies in industrial wastes can be balanced.

There are other considerations in deciding for or against pre-treatment such as

(a) Availability of space – the site may be too restricted or land may be too valuable to be used for a treatment plant.
(b) Availability of expertise – the company may not wish to get involved in effluent treatment.
(c) Sludge and/or odour production may create a nuisance.
(d) Possibility for the introduction of clean technologies.
(e) Economic uncertainties may deter investment in non-profit making wastewater treatment.

Even where it is decided to carry out pre-treatment of the toxic waste by chemical or biological methods it is often useful to install devices to improve the effluent quality by simple physical means. These include some form of screening, coarse or fine, to reduce solids. Also some form of balancing to reduce variations in concentration, flow, pH, *etc.*, some traps to prevent the escape of oil and grease and some grit arrestors.

Every attempt should be made to minimize the quantity of material discharged through good housekeeping. This can take the form of any or all of the following techniques:

(a) Extending the life of process solutions by filtration, topping up, adsorption, *etc.*

(b) Altering the production process to use less toxic compounds, *e.g.* substituting copper pyrophosphate for copper cyanide in electroplating solutions.

(c) Dry cleaning prior to wash-down, which can remove a large proportion of the pollutant in solid form.

(d) Evaporation of strong organic liquors, which can often produce a burnable product.

(e) Minimizing and segregating any flows which contain toxic materials. In some cases it is necessary to separate wastes for safety reasons, *e.g.* cyanides or sulfides and acid wastes, trichlorethylene and alkaline wastes. In other cases it may be desirable to segregate for treatment reasons. However, segregation can be very expensive.

These and related matters are discussed in more detail in Rossiter,[11] Martin and Bostock[12] and Rhyner *et al.*[13]

In Table 6.3 the first five items are common methods of physical treatment employed by a wide range of industrial processes while the others are more costly and complex, require more skilled operation and are therefore less common forms of pre-treatment. Nevertheless these techniques can be extremely useful such as the recovery of specific non-polar organics using macroreticular ion exchange resins. Also membrane processes are finding an increasing role for clean-up prior to recycling in specialist areas where the problems of membrane fouling have been overcome. A relatively new membrane technique which has proved useful is pervaporation. This uses polymeric membranes for the removal of volatile organics from water. They are separated by evaporation through a membrane under vacuum. The separated compound is recovered by condensation, *e.g.* solvent recovery from process waters.[14]

Having minimized so far as possible the types, quantities and concentration of any toxic wastes, it may still be necessary to treat them prior to discharge either to a sewer or a water course. The processes which are used may be classified as physical, chemical and biological. The physical processes are summarized in Table 6.3. Where the toxic wastes contain or are composed of organic materials, it may also be necessary to provide some biological treatment especially if the effluent is to be discharged directly into a watercourse.[15] Many different types of process are used but the following are the most popular:

Table 6.3 *Physical methods of pre-treatment*

Process	Aim	Examples
Screening	Removal of coarse solids	Vegetable canneries, paper mills
Centrifuging	Concentration of solids	Sludge dewatering in chemical industry
Filtration	Concentration of fine solids	Final polishing and sludge dewatering in chemical and metal processing
Sedimentation	Removal of settleable solids	Separation of inorganic solids in ore extraction, coal and clay production
Flotation	Removal of low specific gravity solids and liquids	Separation of oil, grease and solids in chemical and food industry
Freezing	Concentration of liquids and sludges	Recovery of pickle liquor and non-ferrous metals
Solvent extraction	Recovery of valuable materials	Coal carbonizing, plastics manufacture
Ion exchange	Separation and concentration	Metal processing
Reverse osmosis	Separation of dissolved solids	Desalination of process and wash water
Adsorption	Concentration and removal	Pesticide manufacture, dyestuffs removal

(a) High-rate filtration using plastic media and very high rates of recirculation.

(b) Activated sludge using contact stabilization.

Like all biological processes these can suffer from toxicity problems, especially where the concentration of toxin is not constant. In general terms it is easier for bacteria and other microorganisms to adapt to toxic substances than for organisms like worms, fly larvae, *etc.* For this reason conventional percolating filters have not proved successful – the lack of grazing fauna has led to persistent ponding.

Biological processes may be either aerobic or anaerobic. The latter have proved especially popular for treating high strength industrial wastes. As aerobic and anaerobic processes utilize different groups of bacteria, a waste that is toxic to aerobic organisms may not have the same effect on anaerobes.

Due to a combination of high organic strength and inhibition from toxic substances it is unusual to obtain complete treatment of toxic industrial wastes by conventional primary and secondary treatment. The effluent from high-rate filters often has a BOD and COD (chemical oxygen demand) similar to settled

Table 6.4   *Chemicals used in industrial waste treatment*

Chemicals	Purpose
Calcium hydroxide	pH adjustment, precipitation of metals and assisting sedimentation
Sodium hydroxide	Used mainly for pH adjustment in place of lime
Sodium carbonate	pH adjustment and precipitation of metals with soluble hydroxide
Carbon dioxide	pH adjustment
Aluminium sulfate	Solids separation
Ferrous sulfate	Solids separation
Chlorine	Oxidation
Anionic polyelectrolytes	Enhance coagulation and flocculation

sewage and is suitable either for discharge to a sewer or for further biological treatment on site.

Chemical treatment of industrial wastes may be used in addition to, and to some extent in place of, biological treatment. The aims are somewhat different since biological treatment is mainly a way of oxidizing organic matter or a way of converting it into a settleable form. Chemical treatment is used only for oxidizing particular compounds, like cyanide, since it is expensive and liable to lead to the production of undesirable chlorinated organics. It is mainly used for pH correction and improving the removal of solids. The commonest chemicals in use are shown in Table 6.4.

6.5   PRIMARY AND SECONDARY TREATMENT

Provided that the pre-treatment of toxic industrial wastes is successful then no difficulties should be encountered in subsequent treatment. However, no pre-treatment system is perfect and malfunction will occasionally occur, mostly due to variations in the manufacturing process. As a result toxic material together with possible overload of organics and solids may be passed on to the subsequent treatment stages.

Wastewater treatment is conventionally divided into preliminary, primary and secondary treatment (plus tertiary treatment if necessary for some special purpose). The following notes omit any consideration of preliminary treatment as this usually consists of screening and grit removal which have little effect on toxic materials.

The effect of toxic materials on primary sedimentation is insignificant since this is a purely physical process of sedimentation and flocculation. However, the effect of primary sedimentation on toxic wastes can be very important. Toxic materials in suspension such as particulate metals are effectively removed. Also flocculant material has a great capacity for adsorption, removing the majority of dissolved metals, pesticides and other toxic organics. In one respect this is beneficial since it renders the waste material less inhibitory for biological treatment but it selectively concentrates the toxins in the sludge and may give rise to problems in digestion and in sludge disposal. Some indication of the removal of metals during primary treatment is given in Table 6.5.

Chemicals may be used to enhance the effectiveness of primary sedimentation, in some cases removing additional material by precipitation at the same time. Chemical addition can be expensive, often requires pH correction, and may produce large quantities of sludge with a disposal problem. For these reasons chemical enhancement of primary sedimentation is rarely practised at plants treating domestic wastes. Nevertheless for many toxic industrial wastewaters this is an attractive treatment option since it enables industry to avoid secondary biological treatment and enables the waste to be discharged to a sewer, estuary or the sea.[16]

Where chemically enhanced sedimentation is used the main aim is generally to increase removal of solids, but, since many toxins such as metals and chlorinated organics adsorb strongly, their removal is also increased to levels similar to combined primary and secondary treatment. The material employed for enhancement is lime or less frequently aluminium salts sometimes supplemented by polyelectrolytes. Some indications of metal removal achievable by sedimentation, with and without lime, are given in Table 6.6.

Whatever the form of primary treatment employed, further treatment is generally brought about by biological processes, either aerobic or anaerobic. The key to successful secondary treatment of wastewaters containing toxins is the adaption of the microorganisms to the presence of the toxin. Bacteria, and to a lesser extent protozoa, show a remarkable ability to acclimatize to the presence of toxic substances and a great adaptability in degrading new synthetic organic compounds. Metazoa are less adaptable so forms of treatment that rely on metazoa are best avoided in dealing with toxic wastes.[17]

It is important in the biological treatment of toxic waste that a microbial population is developed which is acclimatized to the presence of the toxin and, in the case of degradable toxins, it is essential that it contains sufficient numbers of organisms which can metabolize the toxins.[18]

These twin aspects of acclimatization require great care in the start-up operation and may need a period of several months before successful operation is achieved. Even after start-up is complete, particular processes involving sensitive bacterial species like nitrification or methane production can be easily disrupted by shock loads.

Biological processes for treating wastewaters may be divided into aerobic and anaerobic and each division may be subdivided into dispersed growth and fixed film. The resulting four categories of treatment process have advantages and disadvantages for the treatment of toxic wastewaters as shown in Table 6.7. The tolerance level of the processes are difficult to define precisely but some indication is given in Tables 6.8 and 6.9.

The possibilities of microbiological involvement in the treatment of toxic wastes have recently been extended in two important directions:

(a) Xenobiotic substances have been shown in some cases to be susceptible to microbiological breakdown, *e.g.* white rot fungus is capable of degrading

Table 6.5 Amounts of heavy metal ions removed from sewage by sludges

Heavy metal ion	Primary sedimentation		Percolating filter treatment		Activated-sludge process	
	Metal concentration in crude sewage (mg l⁻¹)	Proportion removed by treatment (%)	Metal concentration in crude sewage (mg l⁻¹)	Proportion removed by treatment (%)	Metal concentration in crude sewage (mg l⁻¹)	Proportion removed by treatment (%)
Copper	Up to 0.8	45	Up to 0.44	20	0.4	54
Copper					Up to 0.44	60
Copper	Up to 5	12			0.4–25	50–79
Copper					28	90–93
Copper					(as Cr)	
Dichromate	(as Cr) Up to 1.2	28	(as Cr) Up to 0.86	32	Up to 0.86	67–70
Dichromate					4.0	6.3
Dichromate					0.5–2	ca. 100
Dichromate					5	50
Dichromate					50	10
Iron (Ferric)	3–9	40	1.8–5.4	Nil	1.8–5.4	80
Lead	0.3–0.9	40	0.18–0.54	30	0.18–0.54	90
Nickel	0.1–0.3	20	0.08–10	40*	0.08–0.24	30
Nickel					2.0	31
Nickel					2.5–10	30
Zinc	0.7–1.6	40	0.4–1.0	30	0.4–1.0	60
Zinc					2.5	90
Zinc					2.5	95
Zinc	Up to 5	12			7.5	100
Zinc					15	78
Zinc					20	74

Table 6.6 *Metal removed by sedimentation*

Metal	Concentration in wastewater (mg l^{-1})	% Removal by sedimentation	% Removal (with lime) by sedimentation
Iron	6.3	48	80
Copper	0.6	28	60
Chromium	0.34	40	58
Lead	0.12	33	55
Mercury	0.028	15	50
Nickel	0.08	15	15
Zinc	0.7	38	70

Table 6.7 *Biological processes for treating toxic wastewaters*

Aerobic/ Anaerobic	Reactor type	Advantage/disadvantage	Example
Aerobic	Dispersed growth	Tend to be completely mixed therefore dilutes toxin but affects whole biomass. Liable to settling problems	Activated sludge
Aerobic	Fixed film	Plug flow so no dilution unless recirculate. Biomass more robust but metazoa more sensitive	High rate filters
Anaerobic	Dispersed growth	Tend to be completely mixed so suffer washout of methanogens	UASB
Anaerobic	Fixed film	Plug flow but attachment can be a problem. Need recirculation to dilute toxins	Anaerobic filters

Table 6.8 *Toxic levels in aerobic biological treatment*

Toxin	Significant level
Hydrogen ions	pH <6 or >9
Phenols	50–100 mg l^{-1}
Ammoniacal-N	500–1000 mg l^{-1}
Zinc	10–50 mg l^{-1}
Chromium	5–20 mg l^{-1}
Lead	5–30 mg l^{-1}
Alkyl Benzene Sulfonates	3–20 mg l^{-1}
Sulfide	5–50 mg l^{-1}

Table 6.9 *Toxic effects in anaerobic treatment*

Toxin	Inhibitory concentration (mg l^{-1})	
	In sewage	*In sludge*
Chromium	–	2
Cadmium	2	2
Copper	1.5	–
Iron	10	–
Lead	100	–
Nickel	80	–
Zinc	50	–
Detergent	–	2% of Suspended Solids
Benzene	–	50–200
Chloroform	–	0.1
Toluene	–	430–860

PCBs.[19] The range of this is being greatly extended by gene transfer techniques.

(b) Microorganisms in the form of dead biomass have been shown to be capable of adsorbing heavy metals from dilute waste streams. The adsorbtion is reversible and the metals can be recovered by electrolysis from acid rinsing of the biomass. Metals such as cadmium and copper can be removed down to microgram levels.[20]

Following primary and secondary treatment it may be desirable in some cases to employ tertiary treatment. For example with metals it is possible to ensure almost complete removal by using activated carbon columns as a tertiary stage. The carbon can be regenerated provided that they are acid washed before use.[21]

6.6 PASSIVE TREATMENT

Wastewater issuing from a mining operation, either current or abandoned, creates a peculiarly difficult treatment problem. The flow of the wastewater is largely determined by local hydrological conditions, which may, in times of storm or snow melt, generate enormous volumes of relatively dilute material. This can result in the oxidation and subsequent precipitation of many tonnes of iron and manganese (in the form of hydrated oxides) in the stream bed.

Such discharges are often outside the acceptable pH range for streams (generally 6–9), thus causing more damage to the aquatic community, and in extreme cases the pH may fall below 4, allowing aluminium to come into solution with consequent toxicity problems.

Unlike other industries the flow of contaminated water from a mine does not cease with the end of mineral extraction. Acid and alkaline mine drainage may continue for decades if not indefinitely. This places a severe financial constraint on the treatment options as often there is no industrialist to help finance the continuing treatment costs.

As a result of these financial pressures a number of relatively cheap alternative treatment methods have been developed. Since they involve little or no mechanical equipment or use of chemicals, they are often referred to as passive treatment. The main methods employed may be summarized as follows:

(a) Anoxic Limestone Drains. These consist of underground channels filled with crushed limestone. They add alkalinity and provide a reducing environment.
(b) Aerobic wetlands. These consist of shallow beds of reeds in which oxidation and precipitation of iron and manganese can occur.
(c) Anaerobic Wetlands. These consist of beds of peat or other organic material in which anaerobic conditions are established with the aim of reducing the acidity.
(d) Successive Alkali Producers. These are in effect a combination of an aerobic wetland and an anoxic limestone drain.

As with conventional treatment plants it is customary to use a combination of two or more of these processes in sequence to obtain a satisfactory effluent. The sequence often contains a sedimentation tank in addition to the other processes.

For further details of passive treatment see Younger.[22]

6.7 SLUDGE TREATMENT AND DISPOSAL

This topic is discussed in Chapter 5 so the remarks below are limited to the treatment of toxic sludges. These are generated almost exclusively by industrial sources although sludges from agriculture may cause difficulties. Industrial sludges are generally formed during primary sedimentation and tend to contain high levels of toxins and persistent organics which are adsorbed onto the solids. Where domestic wastes are treated along with industrial discharges a similar problem can arise.

The treatment of toxic sludges may be divided into a number of different processes:

(a) Dewatering
(b) Biological treatment
(c) Detoxification

Attempts to develop a process for sludge detoxification have not been successful, so some technique of dewatering like incineration, which considerably reduces the volume for disposal, is a useful alternative.

There are two possible strategies for disposal, namely dispersion or containment. Where the industrial wastewater contains substances which are not permitted in the environment, the overall strategy is to segregate the waste into a small volume of hazardous material for containment in a special site. Where the toxic substances are less dangerous, the usual aim is dispersion to reduce the concentration to below toxic levels.

Examples of these processes are as follows:

(a) Containment of radioisotopes by acid extraction and precipitation, followed by vitrification for long-term storage.[23]
(b) Dispersion of metals by spraying the sludge on to farm land. (Note that this practice is now much less common – see Chapter 5.)

6.8   DISPOSAL OF TOXIC WASTES

Ultimately toxic wastes have to be disposed of and this disposal must be in a manner which does not present a short term or long term hazard to man or the environment. Some hazardous wastes may be rendered innocuous by treatment prior to disposal. Incineration is frequently used for decomposition of organic toxins.[24] However, care is required particularly in dealing with halogenated materials since irritant corrosive gases may be produced. There is also a danger that the treatment plant may become too complex. Examples of the successful use of incineration may be found in the treatment of toxic sulfide liquor from the Kraft process for making paper pulp and in the treatment of steel pickle liquor. Both of these have the added advantage of regenerating useful compounds.[25]

The toxic wastewaters that give rise to the most serious problem are metallic wastes and radioactive wastes. These types of waste have common characteristics in that they contain hazardous elements which cannot be broken down (or not for many decades) and there appear to be less toxic substances that could be used to replace them.[26]

The two major alternatives for disposal of toxic waste may be described briefly as follows:

(a) To land – where wastewaters contain human toxins great care is required to avoid contamination of groundwater. Where aquifers are at great depth and not directly connected to streams or underground supplies, *e.g.* some Middle Eastern countries, then land disposal may be preferable to using shallow semi-landlocked seas.

    The usual disposal arrangement for less hazardous wastes is a lagoon, which may have a connection with a watercourse, but which also permits infiltration (plus some evaporation and possibly some degradation). In the long term, swelling and blinding of the soil may reduce the infiltration capacity.

    For more hazardous wastewaters the land disposal policy is segregation followed by long term containment of the hazardous material in impervious disposal sites.[27]

(b) To sea – the sea has an almost unlimited capacity for dilution but the corollary is that it has an almost infinite retention time. Disposal to sea is therefore capable of diluting acute toxins below their toxic threshold but problems may arise with substances that accumulate due to geochemical or biochemical mechanisms.

A further complication with marine disposal is the international aspect. Waste material discharged into the ocean may be transported around the world. Two means of disposal are commonly practised:

(i)   Discharge by pipeline to inshore waters. Dispersion in the buoyant jet can give adequate initial dilution although inshore areas are particularly sensitive to pollution, being used as shellfisheries and recreational zones.

(ii) Deep sea disposal of toxic wastes has in recent years been the subject of several international agreements as the result of which the volume of hazardous disposal has declined and the nature of the waste has changed. Substances like organohalogens, carcinogenic substances, mercury and cadmium compounds and plastics are all banned. Deep sea disposal of less hazardous material still takes place but in packaged form and only in deep sea.

Solidification can be used to render a variety of hazardous wastes suitable for disposal, including oily wastes, sludges contaminated with PCBs and fly ash contaminated with heavy metals.[23]

The technique relies upon reducing the mobility of the hazardous constituents by binding them into a solid matrix which has low permeability and is therefore resistant to leaching.

The mechanism of binding depends upon the agent employed which may be cement based, pozzolanic or silicate based, thermoplastic based or organic polymer based. The cheaper agents like cement, asphalt and pozzolanic based have been most widely used. Solidification has given promising results in short term tests but the longer term is less certain except for vitrification. This technique is only financially possible for nuclear waste – and apart from cost has the additional disadvantage of being virtually irreversible.

It should be appreciated that disposal of toxic wastes remains an intractable problem. Techniques of dispersion like spreading sludge on land have created build up of metals in the soil, so even sludge from domestic sources is increasingly placed in sanitary landfill, whilst sludges of industrial origin are often taken to special sites.

The overall strategy in many countries has shifted to trying to make the industrialist or the consumer responsible for waste disposal. All wastes for disposal are classified and may only be taken to appropriate sites. All waste disposal operations involve a series of transactions, with a corresponding paper record of the transfer, so that the waste can be tracked. Although not totally free from abuse this arrangement seems to be reasonably successful.

The pressures on finding sites for disposal of hazardous waste continue to increase and this has led to interest in processes for detoxification. A number of plants have been produced that can degrade organics like PCBs using different combinations of high temperature and chemicals like sodium carbonate. These processes have the attraction of immediate decomposition whereas the micro-biological alternatives, at present, have timescales in years.

In addition to the search for new sites for disposal there is a great backlog from decades of less controlled dumping. It is clear that long term disposal will remain a long term problem.[6]

6.9 INDUSTRIAL WASTE TREATMENT – CASE STUDIES

The foregoing notes have dealt in general with the problem of treating toxic wastes but it is not possible to explain the complexities in so general a discussion. The following notes therefore describe in more detail two examples of industries that produce toxic wastes and the methods used in treating them.

6.9.1 Tannery Wastes

Tanning is the process by which hides are converted into leather. The hides, after removal of flesh and fur, are treated with chemicals which cross-link the microscopic collagen fibres to form a stable and durable material. After tanning the hides will usually be further processed according to their intended end use. This will consist of trimming, drying, buffing and surface coating.

Wastes will arise from surplus, spent or washed-out chemicals used in the process; some chemical constituents may be toxic, while others are powerful pollutants in water and soil. The release of volatile sulfides gives rise to toxic and obnoxious odours. Certain solvent vapours can have adverse health effects after prolonged exposure. There will also be residues from operations such as cleaning, scraping, splitting and trimming. Each of these generates waste products which must be disposed of or reused. Solid waste products of animal origin are powerful pollutants in water, and are also highly odorous when they decompose in their solid forms.

A schematic diagram of the tanning process is presented in Figure 6.1, and this indicates the type of waste stream generated.

The composition of a combined tannery effluent that has not been treated is characterized by a high oxygen demand and a high fat content, and is strongly alkaline. It also contains a high level of suspended solids and possibly a persistent high load of chrome. Levels experienced in an actual tannery will vary from these values depending for example on water use, *etc.* The generally accepted range of water use is 25-80 m^3 t^{-1} of hide processing. Table 6.10 presents the chemical characteristics of a typical untreated combined tannery effluent.

Effluents from tanneries may be disposed of in one of four ways:

(a) Discharge to sewer
(b) Discharge to river or estuary
(c) Discharge to sea
(d) Discharge to underground

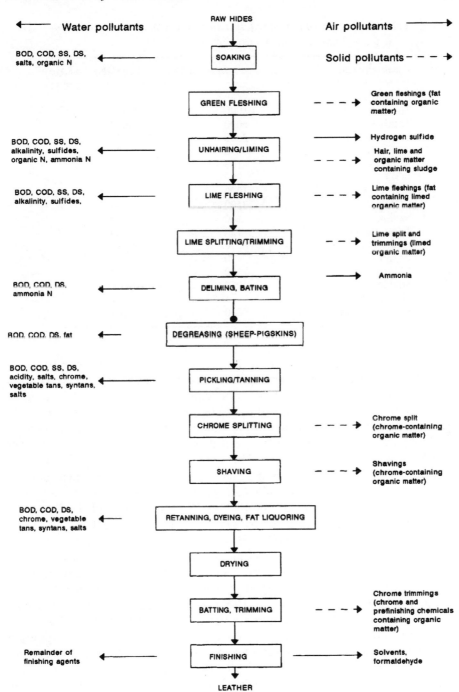

RAW HIDES

Water pollutants ←	Air pollutants →

Solid pollutants – – – →

BOD, COD, SS, DS, salts, organic N ← **SOAKING**

GREEN FLESHING – – – → Green fleshings (fat containing organic matter)

BOD, COD, SS, DS, alkalinity, sulfides, organic N, ammonia N ← **UNHAIRING/LIMING** → Hydrogen sulfide

– – – → Hair, lime and organic matter containing sludge

BOD, COD, SS, DS, alkalinity, sulfides, ← **LIME FLESHING** – – – → Lime fleshings (fat containing limed organic matter)

LIME SPLITTING/TRIMMING – – → Lime split and trimmings (limed organic matter)

BOD, COD, DS, ammonia N ← **DELIMING, BATING** → Ammonia

BOD, COD, DS, fat ← **DEGREASING (SHEEP-PIGSKINS)**

BOD, COD, SS, DS, acidity, salts, chrome, vegetable tans, syntans, salts ← **PICKLING/TANNING**

CHROME SPLITTING – – – → Chrome split (chrome-containing organic matter)

SHAVING – – – → Shavings (chrome-containing organic matter)

BOD, COD, DS, chrome, vegetable tans, syntans, salts ← **RETANNING, DYEING, FAT LIQUORING**

DRYING

BATTING, TRIMMING – – – → Chrome trimmings (chrome and prefinishing chemicals containing organic matter)

Remainder of finishing agents ← **FINISHING** → Solvents, formaldehyde

LEATHER

Figure 6.1 *Schematic diagram of tanning process*

Table 6.10   *Chemical characteristics of a typical untreated combined tannery effluent*[8]

Parameter (mg l^{-1})	Chrome tannage	Vegetable tannage
pH (units)	9	9
Total solids	10 000	10 000
Total ash	6000	6000
Suspended solids	2500	1500
Ash in suspended solids	1000	500
Settled solids (2 h)	100	50
BOD	900	1700
KMnO$_4$ value	1000	2500
COD	2500	3000
Sulfide	160	160
Total nitrogen	120	120
Ammonia nitrogen	70	70
Chrome	70	–
Chloride	2500	2500
Sulfide	2000	2000
Phosphorus	1	1
Ether extractable	200	200

The discharge of tannery waste to a river or estuary is usually subject to severe constraints on solids, BOD, toxins, pH, *etc*. Few tanneries have access to the sea and although underground disposal is popular in the USA it is not favourably regarded in the UK. Whether effluents are being discharged to sewers or to rivers the form of pre-treatment is often similar. In cases where space is limited a mechanically brushed perforated screen is all that is installed but wherever possible a balancing tank is added. This will reduce the need for pH correction, reduce the load of BOD and SS and enable any subsequent treatment units to operate continuously. Mechanical rotating screens which are self-cleaning have worked well within the industry; up to 30–40% of total suspended solids in the raw waste stream can be removed by a properly designed and operated screen. As a significant part of the COD load of the raw wastewater is due to organic solids, a preliminary settling operation can remove up to 30% of this COD, saving flocculating chemicals and reducing sludge volume in later treatment.

If the effluent is to be discharged to sewers some reduction in sulfide and metals will often be necessary. As H$_2$S gas is liberated when sulfide liquors become neutralized, these should be separated and treated. Three methods have been used: aeration in the presence of manganese catalyst; direct precipitation using ferrous sulfide or ferric chloride; and oxidation by chlorine. The relative economics of these processes tend to favour the aeration technique. The precipitation of chrome is relatively simple. The success of the operation depends on the ability to collect the major chrome-bearing liquors for treatment. Precipitation is achieved by raising the pH above 8.0 by the addition of lime, followed by the addition of aluminium salts (200 ppm) and anionic polyelectrolytes (5 ppm) to give a fast-settling flocculant.

Where tannery effluents have been discharged to sewers there have been some reports of difficulties at the municipal treatment plants. These difficulties seem to have arisen mainly because of the increase in organic concentration rather than any effect due to toxins. As undiluted tannery wastes may have BOD values around 2000 mg l^{-1} they require a minimum of 3–4 times dilution to avoid problems of oxygen transfer (care is needed in the use of COD data since the ratio of COD:BOD is often twice that found in raw sewage). The sulfide present in tannery waste may have an adverse effect on biological treatment in admixture with sewage, this is unlikely to be serious at concentrations of < 10 mg l^{-1}, even concentrations around 25 mg l^{-1} can be tolerated by the activated sludge process and even loads of up to 50 mg l^{-1} can be withstood for short periods of time. Percolating filters are very resistant to sulfide and concentrations of up to 100 mg l^{-1} can be treated successfully. Metal toxicity from tannery wastes has not been found to be a problem as 80% of the chrome is removed during primary sedimentation. The resulting sludges do not give rise to difficulties in digestion provided that the retention time exceeds 21 days.

Experience of treating tannery waste on its own has confirmed the general rule that it is better treated along with domestic waste. Anaerobic treatment has not been too successful, with moderate BOD removals and some difficulty in treating the resultant liquor.

Activated sludge appears to have good potential for treating tannery waste although this has not been exploited. Loading rates of 0.5–10 kg BOD kg^{-1} MLSS day^{-1} can be applied with BOD removals > 90%. The oxidation ditch, a low-load activated sludge system which is relatively cheap to install, with low maintenance and a long retention time, is well able to withstand the variable character of effluent and the shock loads experienced in the tanning industry.

Biological filters are often discounted as it is now generally recognized that a tannery waste treatment system must be robust enough to withstand the occasional shock load and any other operational irregularities.

6.9.2 Metal Processing Wastes

Discharge of metals to the aquatic environment has been a major cause of concern and the treatment of these wastes has consequently attracted considerable attention. Wastes containing metals may arise from a variety of industrial and agricultural operations including tanneries, paint manufacture, battery manufacture, pig wastes, *etc.*, but the main source is from metal processing. The wastes from metal processing may be classified as follows:

(a) Mining – ore production and washing – also contains inert SS.
(b) Ore processing – smelting, refining, quenching, gas, scrubbing, *etc.* – also contains sulfides, ammonia and organics.
(c) Machining – metal particles from machining usually mixed with lubricants.
(d) Degreasing – metals mostly in solution with cyanides, alkalis and solvents.

(e) Pickling – acids with metals and metallic oxides in solution.
(f) Dipping – alkalis with sodium carbonate, dichromate, *etc.*, plus metals.
(g) Polishing – particles of metals and abrasives together.
(h) Electrochemical or chemical brightening and smoothing – acids, mainly sulfuric, phosphoric, chromic and nitric with metals in solution.
(i) Cleaning – hot alkalis with detergents, cyanides and dilute acids plus metals in solution.
(j) Plating – acids, cyanides, chromium salts, pyrophosphates, sulfamates and fluoroborates plus metals in solution.
(k) Anodizing – chromium, cobalt, nickel and manganese – in solution.

The sources of wastes in metal processing are numerous and also extremely variable both in quantity and quality. Metals in the wastes occur in forms ranging from large particles of pure metal in suspension to metallic ions and complexes in solution. The most appropriate method of treatment depends upon the form of the metal, its concentration, pH, other constituents of the waste and the desired effluent standard. The technique most commonly employed in treating metal processing wastes is precipitation using pH adjustment. The optimum pH for precipitation varies depending on the particular metal and where several metals are involved a compromise pH is used. A typical value is in the range 8.0–9.0. With amphoteric metals, notably zinc, care must be taken to avoid too high a pH to prevent the formation of zincates. It should also be appreciated that other constituents of the waste, *e.g.* ammonia, can significantly affect the solubility of the metal hydroxides and it is therefore not possible to predict accurately the level of residual metal in the treated effluent.

Whilst the hydroxide precipitation method is satisfactory for most metals encountered in effluents both hexavalent chromium and lead are not precipitated in this way. Hexavalent chromium is present in wastes from metal plating and must first be reduced to the trivalent form before treatment with lime or caustic soda. The reducing agents commonly used are sodium bisulfate, sulfur dioxide and occasionally ferrous sulfate. The reduction is carried out under acid conditions and subsequent addition of alkali precipitates trivalent chromium hydroxide.

In the case of lead, the hydrated oxide formed when lime or caustic soda is added to the lead waste has an appreciable solubility and the resulting effluent after removal of solids would normally be unsatisfactory for discharge to sewer or watercourse. However, basic lead carbonate has a very low solubility and therefore sodium carbonate can be used in place of lime as the precipitating agent. Like zinc, lead is amphoteric and redissolves as plumbate at high pH, so careful pH control must be exercised. One additional method for removing lead is electrolytic. Addition of zinc to a waste containing lead will cause precipitation of the lead.

A particular type of precipitation system used in the metal plating industry is known as the Integrated Method of Treatment. The principal feature of this system is that the rinsing stage immediately after the metal plating stage is a chemical rinse which precipitates the metal from the liquid around the article being plated. A further water rinse is then required to wash off the treatment chemical. In the case of nickel plating the chemical rinse would contain sodium carbonate to precipitate nickel carbonate, whilst with chromium, a prior stage to effect reduction from hexavalent to trivalent form would be required. The integrated system has the advantages that water reuse can be readily practised and that the metals are not precipitated in a mixture and so can also be recovered. However, it is sometimes difficult to adapt the system to existing plating lines, since it necessitates the placement of an extra tank in the line.

Once the metals have been precipitated from solution, liquid and solid phases must be separated. The traditional method for this stage of treatment is settlement in either a circulate or rectangular tank. In small installations, where the effluent flow is less than say 25 m^3 day^{-1}, it is convenient to carry out the effluent treatment on a batch basis and to allow settlement to take place in the same tanks as that used for reaction. For larger installations a continuous flow system is required. The size of tank depends on both the maximum effluent flow rate and on the configuration adopted for the tank. The most common type of settlement vessel is of the vertical upward flow pattern having a central feed well, a peripheral collection launder, and a sludge cone at the bottom. Clarification of the effluent can be enhanced by the use of flocculating agents. Obviously the size and mode of operation of the precipitation system significantly affects quality of the effluent but typical figures for a well-designed, efficiently operated, settlement system for metal hydroxide precipitates would be in the range 10–30 mg l^{-1} suspended solids.

Where space is at a premium, a compact settling system utilizing parallel tilted plates or tubes can be used to perform the separation stage.

There are two factors which make this system efficient in terms of ground area used. These are:

(a) The distance through which a settling particle has to fall to become 'settled' is considerably reduced;
(b) The configuration produces laminar flow conditions which enhance the settling rate and overall efficiency.

Tilted plates can also be used to uprate settling tanks.

Flotation may be used as an alternative to settlement. This process, which is gaining in popularity, consists in the carrying of metal hydroxides and other particles in suspension to the surface of liquid in the flotation vessel by increasing particle buoyancy using gas bubbles which adhere to the particles. The scum containing the gas bubbles and separated solids is skimmed off. Variations in the process lie mainly in the method of producing the carrier gas bubbles. This may be done by injecting a super-saturated solution of air in water under pressure into the tank – dissolved air flotation – or by injecting air through a diffuser – dispersed air flotation – or by the electrolysis of water to yield fine bubbles of hydrogen and oxygen – electrolytic flotation. The gas bubbles produced in these processes are extremely small, normally in the range 70–150 μm.

The use of direct filtration appears to be a very attractive process for the phase separation but unfortunately is seldom appropriate, mainly because of the tendency for the filter media to blind (*i.e.* clog) rapidly. This tendency is largely due to the gelatinous nature of the metal hydroxide precipitates. Occasionally, where a more granular precipitate is obtained, direct filtration can be satisfactory and a high quality effluent can be obtained.

Whilst filtration has only limited application as the main means of solids removal it is frequently used to polish the effluent from a settlement or flotation system to produce a higher quality effluent.

Where the metal is substantially in solution there are various techniques for separation or concentration of the metal so that a high quality treated effluent may be obtained.

6.9.2.1 Ion Exchange. Ion exchange is a chemical treatment process used to remove dissolved ionic species from contaminated aqueous streams. Treatment for both anionic and cationic contaminants can be effected by ion-exchange processes. Ion exchangers are insoluble high molecular weight polyelectrolytes that have fixed ionic groups attached to a solid matrix. Both natural and synthetic ion exchangers are available. However, due to their greater stability, higher exchange capacity and greater homogeneity of their exchange properties, synthetic ion exchange materials are predominantly used today. Synthetic ion exchangers are generally polymeric materials (resins) that have been chemically treated to render them insoluble, and to exhibit ion exchange capacity. Often the exchanger is in the form of spherical resin beads, although ion-exchange membranes are also available. Various different polymers have been used as ion exchangers but copolymers of styrene and divinylbenzene (DVB) are the most common synthetic ion-exchange materials.

6.9.2.2 Evaporation. Evaporation is one of the most common methods used in industry for the concentration of aqueous solutions. Nevertheless, use of this process as a means of effluent treatment is rare and occurs only under special circumstances where the effluent contains a high concentration of a valuable material. The only application of note here is on the concentration of static rinses (drag out) from electroplating operations, especially chromium plating. In this application the rinse liquor is evaporated to a metal concentration which makes the concentrate suitable for direct reuse in the plating bath.

6.9.2.3 Molecular Filtration. Molecular filtration is divided into two categories: ultrafiltration (UF) and reverse osmosis (RO) processes. The differentiating characteristic between RO and UF is the molecular weight cutoff of the membrane, and corresponding pressure differentials required to achieve a given membrane flux. RO membranes operate with molecular cutoffs that are much smaller than UF membranes (100–200 Da compared with 2–1000 mDa), therefore the RO membrane will retain most organic materials, as well as many of the inorganic solutes. RO membranes operate with trans-membrane pressures of up to 500 psi, whereas UF membranes generally operate with pressure differentials only as high as 50 psi since most of the inorganics will pass through the UF membrane, minimizing the osmotic pressure resistance. This pressure difference has a significant economic implication. Often when reverse osmosis is used, upstream UF is provided as a pre-treatment for RO. The main operational problems associated with membrane processes are chemical and biological fouling of the membrane, and, particularly with RO, membrane deterioration. The RO process has been used on effluents from electroplating in the electronic components industry. The continuous development of the process and improved mechanical strength of the membranes will almost certainly increase the range of its applications.

6.9.2.4   Solvent Extraction.   In general, the solvents used in extraction operations are too expensive to be used just once, and furthermore the contaminants are highly concentrated in the extract. Therefore, the spent solvent from a liquid–liquid extraction operation needs to be further treated to reclaim the solvent for reuse and to reduce further the volume in which the contaminant is contained. Some solvent repurification sequences include the use of distillation or adsorption.

6.9.2.5   Electrodialysis.   Dissolved inorganics, the mineral content of a wastewater, can be removed by electrodialysis. When an inorganic salt is dissolved in a water solution it ionizes to produce positively charged cations and negatively charged anions. When an electrical potential is then passed through the solution, the cations migrate to the negative electrode and the anions to the positive electrode. Semi-permeable membranes are commercially available that allow the passage of ions of only one charge: cation-exchange membranes are permeable only to positive ions and anion-exchange membranes are permeable only to negative ions. When a series of these membranes is placed alternatively in a solution, and a voltage applied, the solution between one pair of electrodes becomes clarified as the ions concentrate in the solution in the adjacent compartments.

6.10   REFERENCES

1. P. Calow, 'Controlling Environmental Risks from Chemicals', Wiley, 1997.
2. J. F. Alabaster and R. Lloyd, 'Water Quality for Freshwater Fish', Butterworths, London, 1980.
3. R. J. Watts, 'Hazardous Wastes: Sources, Pathways and Receptors', Wiley, New York, 1990.
4. P. K. LaGoy, 'Risk Assessment: Principles and Applications for Hazardous Wastes', Noyes, New Jersey, 1994.
5. R. A. Selg (ed.), 'Hazardous Waste Cost Control', Marcel Dekker, New York, 1993.
6. H. M. Freeman (ed.), 'Standard Handbook of Hazardous Waste Management, Treatment and Disposal', McGraw Hill, 1998.
7. W. C. Blackmann, 'Basic Hazardous Waste Management', 2nd edn., CRC Lewis, New York, 1996.
8. UNIDO, 'Audit and Reduction Manual for Industrial Emmissions and Wastes', UNIDO Technical Report Series Number 7, 1990.
9. R. E. Train, 'Quality Criteria for Water', Castle House Publishers, 1979.

10. N. L. Nemerow and F. J. Agardy, 'Strategies of Industrial and Hazardous Waste Management', International Thompson Publishing, 1998.
11. A. P. Rossiter, 'Waste Minimization Through Process Design', McGraw Hill, 1995.
12. K. Martin and T. W. Bostock, 'Waste Minimisation: A Chemist's Approach', Royal Society of Chemistry, Cambridge, 1994.
13. C. R. Rhyner, L. J. Schwartz, R. B. Wenger and M. G. Kohrell, 'Waste Management and Resource Recovery', CRC Lewis, Ann Arbor, 1995.
14. WPCF, 'Hazardous Waste Treatment Processes. A Manual of Practice', Water Pollution Control Federation, FD-18, 1990.
15. R. E. Hinchee, G. D. Sayles and R. S. Skeen, 'Biological Unit Processes for Hazardous Waste Treatment', Batelle Press, Columbus, OH, 1995.
16. M. J. Hammer, 'Water and Wastewater Technology', 2nd edn., Wiley, New York, 1986.
17. R. C. Curds and H. A. Hawkes, 'Ecological Aspects of Used Water Treatment Processes and their Ecology', Acedemic Press, London, 1983.
18. S. J. Arcievala, 'Wastewater Treatment and Disposal Engineering in Pollution Control', Dekker, 1981.
19. G. R. Chaudry (ed.), 'Biological Degradation and Bioremediation of Toxic Chemicals', Chapman and Hall, London, 1994.
20. T. J. Butter, 'The Use of Biosorption, Elution and Electrolysis in the Removal and Recovery of Heavy Metals From Aqueous Solutions', PhD Thesis, University of Newcastle Upon Tyne, 1998.
21. G. Woodside, 'Hazardous Materials and Hazardous Waste Management', 2nd edn., Wiley, New York, 1999.
22. P. L. Younger, 'Minewater Treatment Using Wetlands, Proceedings of CIWEM Conference Newcastle Upon Tyne 1997', Chartered Institution of Water and Environmental Management, London, 1998.
23. J. L. Means, L. A. Smith, K. W. Nehring, S. E. Brauning, A. R. Gavasker and B. M. Sass, 'The Solidification and Stabilization of Waste Materials', CRC Lewis, Ann Arbor, 1995.
24. J. P. Lehmann (ed.), 'Hazardous Waste Disposal. Proceedings of NATO Symposium, Washington 1981', 1983.
25. P. N. Cheremisinoff and Y. C. Wu (eds.), 'Hazardous Waste Management Handbook', Prentice Hall, 1993.
26. A. Porteous (ed.), 'Hazardous Waste Management Handbook', Butterworths, London, 1985.
27. S. M. Testa, 'Geological Aspects of Hazardous Waste Management', CRC Lewis, Ann Arbor, 1994.

CHAPTER 7

Air Pollution: Sources, Concentrations and Measurements

R. M. HARRISON

7.1 INTRODUCTION

Before commencing the description of individual air pollutants, it is useful to start with a consideration of the terminology. Air pollutants may exist in *gaseous* or *particulate* form. The former includes substances such as sulfur dioxide and ozone. Concentrations are commonly expressed either in mass per unit volume (μg m^{-3} of air) or as a volume mixing ratio (1 ppm $= 10^{-6}$ v/v; 1 ppb $= 10^{-9}$ v/v). Particulate air pollutants are highly diverse in chemical composition and size. They include both solid particles and liquid droplets and range in size from a few nanometres to hundreds of micrometres in diameter; concentrations are expressed in μg m^{-3}. Gaseous and particulate air pollutants may be separated operationally by use of a filter.

Air pollutants emitted directly into the atmosphere from a source are termed *primary*. Thus, carbonaceous particles from diesel engine exhaust and sulfur dioxide from power stations are examples of primary pollutants. In contrast, *secondary* pollutants are not emitted as such, but are formed within the atmosphere itself. Thus, sulfuric acid and nitric acid, formed respectively from sulfur dioxide and nitrogen dioxide oxidation, are examples of secondary pollutants for which the atmospheric formation route far exceeds any primary emissions. The most commonly considered secondary pollutant is ozone, formed as a result of photolysis of molecular oxygen in the stratosphere, and of nitrogen dioxide in the troposphere (lower atmosphere).

This chapter will address the sources, concentrations and measurement methods for those pollutants considered most important in terms of human health effects and damage to crops and materials. Issues relating to the regional and global atmosphere will be considered in the following chapter.

7.2 SPECIFIC AIR POLLUTANTS

7.2.1 Sulfur Dioxide

The major source of sulfur dioxide is the combustion of fossil fuels containing sulfur. These are predominantly coal and fuel oil since natural gas, petrol and diesel fuels have a relatively low sulfur content. Until recently, emissions of sulfur dioxide from diesel engines led to a small but perceptible increment in sulfur dioxide alongside busy roads, but recent years have seen a very substantial reduction in the sulfur content of diesel fuel. Figure 7.1 shows in diagrammatic form the sources of sulfur dioxide emissions by source category for the United Kingdom.[1] Combustion of coal in power stations is far the most major single source of SO_2 emissions.

Over past years, two source categories, both associated with coal burning have tended to dominate the UK situation with respect to sulfur dioxide. Urban ground-level concentration of SO_2 fell rapidly between 1970 and 1990 largely due to a decline in the burning of coal in domestic fireplaces for home heating. Airborne concentrations fell much faster than total emissions of sulfur dioxide because over that period the reduction in emissions from power stations was quite limited. Since around 1990, urban and rural concentrations of sulfur dioxide have become almost indistinguishable because they have a common source in power station plumes superimposed on a low background from diffuse sources. Total UK emissions declined by more than half between 1990 and 1997 due largely to a cut in emissions from power stations effected by installation of flue gas desulfurization plant on some of the larger stations and a switch to electricity generation from combined cycle gas turbine plants burning natural gas. The main driver for this reduction has been international concern over acid rain problems (see Chapter 8) rather than domestic health concerns.

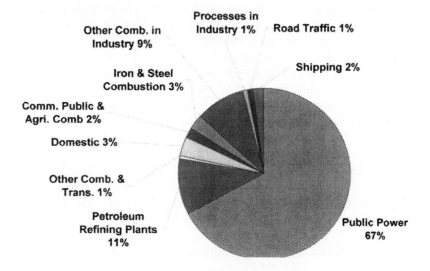

Figure 7.1   *Estimated UK emissions of sulfur dioxide by source category in 1998*[1]

7.2.1.1 Measurements of Sulfur Dioxide. In the UK two techniques are commonly used for the determination of sulfur dioxide in the atmosphere. One is a very simple method developed originally for use in the 'National Survey of Smoke and Sulfur Dioxide' which has now been discontinued, although some of the measurement sites still exist within the EC Directive monitoring network for sulfur dioxide, and the UK basic urban network. The method involves absorption of sulfur dioxide in hydrogen peroxide solution to form sulfuric acid. The resultant acid has traditionally been determined by acid–base titration which is subject to interference by other gaseous, acidic or basic compounds such as nitric acid or ammonia respectively. As sulfur dioxide levels have fallen, so the reliability of measurements made by titration has been reduced and many measurements are now made by determination of sulfate by ion chromatography, which yields a result specific to sulfur dioxide.

The most commonly used instrumental technique for measurement of sulfur dioxide is based upon measurement of fluorescence excited by radiation in the region of 214 nm. Commercial instruments are available, capable of measurement of sulfur dioxide to less than 0.1 ppb, as well as source instruments with ranges into the thousands of ppm. The method is potentially subject to interferences from water vapour, which quenches the SO_2 fluorescence, and hydrocarbons capable of fluorescence at the same wavelength as SO_2. Commercial instruments are generally equipped with diffusion dryers and hydrocarbon scrubbers to overcome these problems. The commonly used techniques for analysis of SO_2 and other pollutants are summarized in Table 7.1 and are described in more detail elsewhere.[2,3]

7.2.2 Suspended Particulate Matter

Airborne particles are very diverse in character, including both organic and inorganic substances with diameters ranging from less than 10 nm to greater than 100 μm. Since very fine particles grow rapidly by coagulation and vapour condensation, and large particles sediment rapidly under gravitational influence, the major part (by mass) generally exists in the 0.1–10 μm range. A schematic representation of the typical size distribution for atmospheric particles appears in Figure 7.2. There are three peaks, or modes, in the distribution. The smallest one relates to the transient nuclei, which are very tiny particles formed by condensation of hot vapours, or gas to particle conversion processes. Thus, primary particles from motor vehicle exhaust and sulfuric acid formed from SO_2 oxidation are initially in the transient nuclei mode. Such particles, when emitted, are present in very high numbers and are subject to rather rapid coagulation both with other fine particles and also with coarser particles already in the atmosphere. Through this mechanism they enter the accumulation range of particles typically with diameters between about 100 nm and 2 μm. Such particles are also capable of growth through the condensation of low volatility materials. Removal of the accumulation range particles by rainwater scavenging or dry deposition to surfaces is inefficient and such particles have a typical lifetime in the atmosphere of around one to two

Table 7.1 *Summary of commonly employed methods for measurement of air pollution*

Pollutant	Measurement technique	Sample collection period	Response time (continuous technique)[a]	Typical minimum detectable concentration
Sulfur dioxide	Absorption in H_2O_2 and titration or sulfate analysis	24 h		1 ppb
	Gas phase fluorescence		2 min	0.1 ppb
Oxides of nitrogen	Chemiluminescent reaction with ozone		1 s	0.1 ppb
Total hydrocarbons	Flame ionization analyser		0.5 s	10 ppb C
Specific hydrocarbons	Gas chromatography/ flame ionization detector	[b]		< 1 ppb
Carbon monoxide	Electrochemical cell		25 s	1 ppm
	Non-dispersive infrared		5 s	0.5 ppm
	Gas filter correlation		90 s	0.1 ppm
Ozone	UV absorption		30 s	1 ppb
Peroxyacetyl nitrate	Gas chromatography/ electron capture detection	[c]		< 1 ppb
Particulate matter	High volume sampler	24 h		5 μg m^{-3}
	TEOM	1 h		4 μg m^{-3}

[a] Time taken for a 90% response to an instantaneous concentration change.
[b] Samples of air concentrated prior to analysis.
[c] Instantaneous concentrations measured on a cyclic basis by flushing the contents of a sample loop into the instrument.

weeks. This renders them capable of very long range transport. The third mode, termed coarse particles, are mostly greater than 2 μm with sizes extending up to about 100 μm. This is comprised in the main of mechanically-generated particles such as wind-blown dust, sea spray and primary volcanic particles. These are formed by attrition of bulk materials and tend to be appreciably larger than transient nuclei or accumulation range particles.

The suspended particles in the atmosphere are typically referred to as the *atmospheric aerosol* and a number of other terms are used, which in the main are related to the method of collection or analysis.

In the UK, historically the largest set of measurements are *black smoke* which (see later) is a measurement related to the blackness or soiling capacity of the particles. Other measures of suspended particulate matter depend upon a gravimetric determination of particle mass. The term total suspended particu-

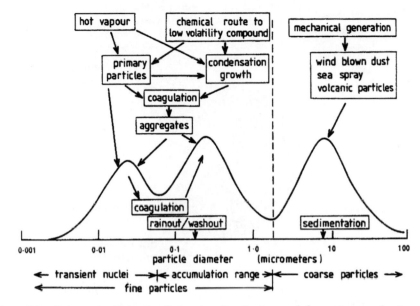

Figure 7.2 *Schematic diagram of the size distribution and formation mechanisms for atmospheric aerosols*
(adapted from Ref. 4)

late matter (TSP) has been used to describe the fraction of particles collected on a filter using the US high volume sampler. Since large particles, by virtue of their inertia, are not readily able to enter the inlet of such samplers, the measurement is dependent both on the orientation of the sampler with respect to the wind and the strength of wind, both of which influence the efficiency of particle aspiration.[2] To overcome this problem, the USEPA moved to use of a high volume sampler with a size selective inlet. This inlet has a 50% efficiency at 10 μm aerodynamic diameter and hence in simple terms may be considered to sample only those particle less than 10 μm in diameter, irrespective of orientation or wind speed. Measurements made using inlets meeting the EPA criterion are referred to as PM_{10}. A number of devices, other than the high volume sampler, are now available for PM_{10} measurement (see later). An inventory of UK emissions of primary particles as PM_{10} appears in Figure 7.3. This neglects secondary particles (which are formed in the atmosphere) and some diffuse sources such as resuspended soils and road dust which are hard to quantify. Inventories are also available for smaller particles such as $PM_{2.5}$ (*i.e.* particles below 2.5 μm diameter), $PM_{1.0}$ and even $PM_{0.1}$, termed the *ultrafine* particle fraction, which has stimulated much interest as it is believed to have exceptional toxicity per unit mass compared to coarser particles. The road traffic contribution to the smaller size fractions is greater than to PM_{10}, and is also larger in urban areas than in the UK as a whole.[4]

7.2.2.1 Black Smoke. Measurement of black smoke is still widely performed in the UK and many other countries around the world. Air is drawn through a

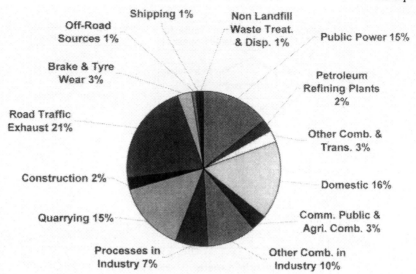

Figure 7.3   *Estimated UK emissions of primary PM$_{10}$ particulate matter by source category in 1998*[1]

cellulose filter upon which the particles are collected. At the end of the sampling period the ability of the filter surface to reflect light is determined quantitatively and the reflectance is related to a concentration of 'standard smoke' using a calibration graph constructed many years ago when urban particles were dominated by coal smoke. Nowadays, there are many other important sources of particles and the intrinsic darkness of urban particulate matter is much lower. Research has shown that the method measures elemental carbon, and thus the dramatic reduction in the inefficient combustion of bituminous coal in domestic grates over recent years has led to a large drop in the emissions of particles containing black elemental carbon. Whilst inventories of emission of black smoke are no longer routinely compiled, the major source of black particles in UK cities is now emissions from diesel road vehicles. Black smoke concentrations are appreciably elevated alongside major roads and black smoke can be a better indicator of vehicle emissions than gravimetrically determined particle concentrations.

7.2.2.2   Gravimetrically Determined Particulate Matter: PM$_{10}$.   PM$_{10}$ particulate matter is determined gravimetrically and therefore includes both primary and secondary particles. In UK urban areas two sources are dominant; one is road traffic emissions and the other is secondary particulate matter, mostly ammonium sulfate and ammonium nitrate particles. Whilst the former are black in colour, the latter are white and hence are not determined by the black smoke method. Other sources such as sea spray and wind-blown dust also contribute to urban PM$_{10}$, although their precise impact on airborne concentrations is more difficult to determine.[4,5] There are many other minor sources such as particles from building work and demolition, and biological particles such as pollens, spores and bacteria. The health effects of PM_{10} outlined in Chapter 11 do not appear to be a strong function of chemical composition.

In addition to adverse effects on human health, airborne particles are responsible for the soiling of buildings and loss of visibility. Airborne particles both scatter and absorb light and thus cause deterioration in the quality of image transmission through the atmosphere which manifests itself as a loss of visibility. Although the efficiency of light scattering and absorption per unit mass of particles is dependent upon the size distribution and chemical composition of the particles, the aerosol mass loading is a fairly good predictor of visibility impairment.[6]

Typical UK urban concentrations of PM_{10} together with those of other pollutants are shown in Table 7.2. These concentrations and those of black smoke are orders of magnitude below those associated with the major smog episodes of the 1950s and 1960s. The smogs were caused primarily by low level emissions from coal combustion during periods of meteorology unsuitable for effective pollutant dispersal (low windspeeds and shallow mixing depth). The combination with fog (smog = smoke + fog) led to dramatic losses in visibility. The smog of December 1952 is believed to have caused some 4000 premature deaths. For many years after, UK pollution control policy focused on smoke and sulfur dioxide, and it is only since the 1980s that the major focus has transferred to motor traffic as the major source of urban air pollution.

Measurement of PM_{10}. The simplest method for determination of PM_{10} involves use of a high volume air sampler capable of drawing air through a filter at a rate of about $1 \text{ m}^3 \text{ min}^{-1}$ through a 10 μm size selective inlet. The sampler is run for 24 h and the gain in mass of the filter, together with the volume of air passed, is used to calculate the airborne concentration of PM_{10} over the 24 h sampling period.

Measurements in pseudo-real time may be made using the Tapered Element Oscillating Microbalance (TEOM). In this device, particles are collected on a

Table 7.2 *Annual mean and recent trends in airborne pollutant concentration in Birmingham, UK*

Pollutant	Annual mean concn. (1998)	Max hourly concn. (1998)	Trend in annual mean
Benzene	0.85 ppb	14.2 ppb	$-0.04 \text{ ppb yr}^{-1}$
1,3-Butadiene	0.15 ppb	3.82 ppb	$-0.01 \text{ ppb yr}^{-1}$
Carbon monoxide	0.5 ppm	3.9 ppm	-0.2 ppm yr^{-1} [a]
Nitrogen dioxide	22 ppb	72 ppb	-0.7 ppb yr^{-1}
Ozone	18 ppb	53 ppb	Not significant
PM_{10}	19 μg m^{-3}	249 μg m^{-3}	$-1.0 \mu\text{g m}^{-3} \text{ yr}^{-1}$
Sulfur dioxide	5 ppb	159 ppb	-1.0 ppb yr^{-1}

[a] For CO, trend is in 98 percentile of hourly mean concentrations.
Note: Concentrations measured at a central urban background location. Measurements at roadside show higher concentrations of the traffic-generated pollutants and lower concentrations of ozone.

filter which is attached to a vibrating element whose vibrational frequency changes with the accumulation of particles on its tip. Measurement of the vibrational frequency is used to estimate the mass of particles collected. An example of averaged diurnal data for PM_{10} and carbon monoxide collected in Birmingham is shown in Figure 7.4 which clearly illustrates the influence of road traffic emissions on both PM_{10} and carbon monoxide concentrations.

Specific Components of Suspended Particulate Matter. Studies of airborne particles in the urban atmosphere of developed countries have shown that their composition can be described in terms of three major components.

- combustion particles comprising mainly fine (less than 2.5 μm diameter), particles of elemental and organic carbon derived predominantly from road vehicle traffic;
- secondary particles of mainly ammonium sulfate (or sulfuric acid) and ammonium nitrate predominantly within the fine (less than 2.5 μm diameter) size range;
- coarse particles arising largely from soil and road surface dust resuspended by traffic activity and made up mainly of inorganic mineral components.

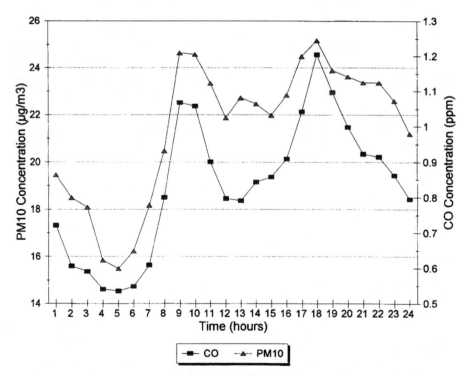

Figure 7.4   *Averaged diurnal variation of carbon monoxide and PM_{10} measured in central Birmingham, winter 1993*

Further discussion of the water soluble fraction of the particles is given in Chapter 8. Whilst these three categories generally make up the major part of the particle mass, it is trace components associated with the various source categories which often stimulate the greatest interest and concern.

The components of airborne particles which give rise to most interest in the context of air pollution tend to be trace metals and trace organic compounds. A summary of concentrations of trace metals measured in central London appears in Table 7.3. Also indicated are trends in concentration which in general have been clearly downwards for several years. Interest in trace metals relates primarily to their potential toxicity. Arsenic, chromium(VI) and nickel are believed to be genotoxic carcinogens and the World Health Organization has estimated unit risk factors for them.[7] Cadmium and mercury have also elicited a great deal of interest because of their high toxicity and there have been moves to control these elements throughout the developed world.

The trace metal which has stimulated by far the greatest interest in relation to public health has been lead. The main source of airborne lead is its combustion in leaded petrol (gasoline) to which it is added as an octane improver in the form of tetraalkyllead compounds, notably tetramethyllead, $(CH_3)_4Pb$ and tetra-ethyllead, $(C_2H_5)_4Pb$. Upon combustion in the engine, these are emitted predominantly as an aerosol of fine particles of inorganic lead. Developed countries have without exception introduced regulations to limit the lead content of gasoline, and the United States has been essentially lead-free for many years, with Europe becoming lead-free at the end of 1999. In Europe, low-lead gasolines were introduced in response to concerns over the health effects of lead, and there have been substantial fiscal incentives to the purchase of unleaded gasoline because of its requirement for use in vehicles fitted with catalytic converters, and in order to further reduce exposure of the general population to the metal. The maximum permitted level of lead in UK gasoline stood in 1972 at 0.84 g L^{-1}, although such high concentrations were not generally used. By 1981, this had been reduced to 0.40 g L^{-1}, and in January 1986 fell sharply to 0.15 g L^{-1} in line with most other western European countries. The decline in the use of lead in petrol between 1980 and 1998 is

Table 7.3 *Trace metals in air in central London in 1998*

Metal	Concentration in 1998 (ng m^{-3})	% Reduction since 1984
Cadmium	0.87	69
Chromium	2.9	77
Copper	17.7	34
Iron	832	15
Lead	38	93
Manganese	10.6	44
Nickel	8.1	52
Zinc	35	65
Vanadium	4.4	86

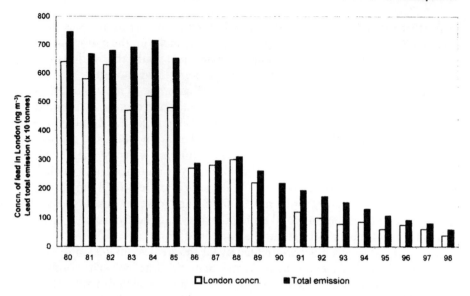

Figure 7.5   *Trends in emissions of lead from motor traffic and airborne concentrations in central London, 1980–1998*[1]

documented in Figure 7.5, which also shows the decline of lead in air concentrations in central London. In general, it may be seen that the two have proceeded very much in parallel with one another, an example of a 'linear rollback' in which a primary pollutant with a single major source declines in proportion to the reduction in emissions.

The abrupt fall in the lead emissions from road traffic at the beginning of 1986 was exploited as the basis for an experiment to evaluate its impact upon population blood leads. In the event, blood leads fell slightly between 1985 and 1986, but little more, if at all, than between 1984 and 1985, and between 1986 and 1987 (when petrol lead remained almost constant) and only slightly more for groups heavily exposed to airborne lead than for less exposed groups.[8]

Concentrations of lead and most other metals in air are determined by use of air filtration to collect a sample. Non-destructive analytical methods are available, and these include X-ray fluorescence (both wavelength dispersive and energy dispersive) and instrumental neutron activation analysis. More commonly, destructive methods are used which involve dissolving the sample in oxidizing acids and analytical procedures such as atomic absorption spectrometry, inductively coupled plasma (ICP) emission spectrometry, or ICP-mass spectrometry.

The organic components of atmospheric particles of most interest as air pollutants are the semi-volatile groups of compounds including the polynuclear aromatic hydrocarbons (PAH), polychlorinated biphenyls (PCB) and polychlorinated dibenzo-dioxins and dibenzo-furans (PCDD and PCDF). The sampling, measurement and environmental behaviour of such compounds is discussed in Chapter 17.

7.2.3 Oxides of Nitrogen

The most abundant nitrogen oxide in the atmosphere is nitrous oxide, N_2O. This is chemically rather unreactive and is formed by natural microbiological processes in the soil. It is not normally considered as a pollutant, although it does have an effect upon stratospheric ozone concentrations (see Chapter 9) and there is much evidence that use of nitrogenous fertilizers is increasing atmospheric levels of nitrous oxide.

The pollutant nitrogen oxides of concern are nitric oxide, NO and nitrogen dioxide, NO_2. By far the major proportion of emitted NO_x (as the sum of the two compounds is known) is in the form of NO, although most of the atmospheric burden is usually in the form of NO_2. The major conversion mechanism is the very rapid reaction of NO with ambient ozone,

$$NO + O_3 \longrightarrow NO_2 + O_2$$

The alternative third order reaction with molecular oxygen is relatively slow at ambient air concentrations, although it can be of importance when NO concentrations reach about 1 ppm.

$$2NO + O_2 \longrightarrow 2NO_2$$

This reaction can become important in severe winter pollution episodes when a catalytic cycle is established.[9]

The major source of NO_x is the high temperature combination of atmospheric nitrogen and oxygen in combustion processes, there being also a lesser contribution from combustion of nitrogen contained in the fuel. An emission inventory for the UK appears in Figure 7.6.

Typical hourly average air concentrations of NO_x are normally in the range 5–100 ppb in urban areas and less than 20 ppb at rural sites. The proportion present as the more toxic pollutant nitrogen dioxide in general becomes greater the lower the total NO_x concentration. Time trends in annual mean NO_x emissions from road traffic and NO_2 concentrations in central London appear in Figure 7.7. This can be contrasted with Figure 7.5 showing the linear rollback of lead in air concentrations as lead emissions from road traffic decreased. Over the period in question, road traffic will undoubtedly have been the main source of NO_x in the atmosphere of London and the apparent lack of relationships between the NO_x emissions and the NO_2 concentration is due to the atmospheric chemistry of nitrogen oxides. NO_x as noted above is emitted predominantly in the form of NO, depending upon reaction with ozone to convert it into NO_2. This reaction is limited mainly by the availability of ozone, and hence in an oxidant-limited situation, reductions in NO_x emissions lead to a reduction in NO_x concentrations, but an increase in the NO_2 to NO ratio. The typical relationship between hourly mean, nitrogen dioxide and NO_x appears in Figure 7.8 and it may be seen that over the predominant range of NO_x concentrations (NO_x between 40 ppb and 800 ppb), there is little change in NO_2 concentration.

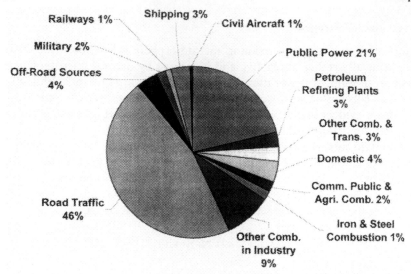

Figure 7.6   *Estimated UK emissions of oxides of nitrogen by source category in 1998*[1]

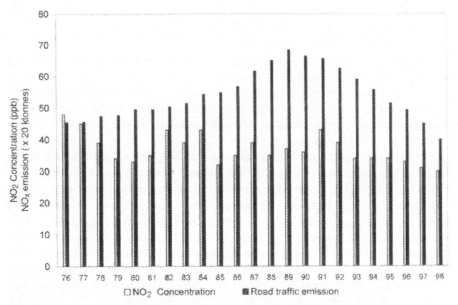

Figure 7.7   *Trends in emissions of NO_x from road traffic and annual mean airborne concentrations of nitrogen dioxide in central London, 1976–1998*

Thus, the annual mean concentration of NO_2, which is made up from many hourly concentrations largely within this range is relatively insensitive to changes in NO_x. The implication is that major reductions in NO_x emissions will be required to obtain modest reductions in NO_2 and hence to achieve the objectives the National Air Quality Strategy outlined below.

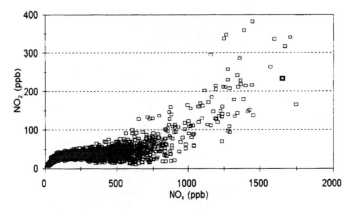

Figure 7.8 *Relationship between hourly concentrations of NO_2 and NO_x (i.e. the sum of NO and NO_2) at Cromwell Road, London in winter months*

7.2.3.1 Measurement of Oxides of Nitrogen. Instrumental analysers have been available for a good number of years. The currently favoured technique for determination of oxides of nitrogen is based upon the chemiluminescent reaction of nitrogen oxide and ozone to give an electronically excited nitrogen dioxide which emits light in the 600–3000 nm region with a maximum intensity near 1200 nm:

$$NO + O_3 \longrightarrow NO_2{}^* + O_2$$
$$NO_2{}^* \longrightarrow NO_2 + h\nu$$

In the presence of excess ozone generated within the instrument, the light emission varies linearly with the concentration of nitrogen oxides from 1 ppb to 10^4 ppm. The apparatus is shown schematically in Figure 7.9.

The method is believed to be free of interference for measurement of NO and may be used to measure NO_x by prior conversion of NO_2 into NO in a heated stainless steel or molybdenum converter. Dependent upon which converter is used, some interference in the NO_x mode is likely from compounds such as peroxyacetyl nitrate and nitric acid. Some instruments incorporate two reaction chambers, one running permanently in the NO mode, the other analysing NO_x after NO_2 into NO conversion. Thus, pseudo-real-time NO_2 concentrations may be measured.

A widely used inexpensive technique for measuring nitrogen dioxide is based upon the use of diffusion tubes. These are straight, hollow tubes of length about 7 cm and diameter 1 cm, sealed at one end which is placed upwards, and open at the other end. At the sealed end a metal grid is coated with triethanolamine which acts as a perfect sink for nitrogen dioxide. Access of nitrogen dioxide from ambient air to the triethanolamine is by molecular diffusion along the tube, and analysis of nitrite collected by the triethanolamine reagent and application of Fick's Law allow calculation of airborne concentrations of nitrogen dioxide. The tubes have a tendency to over-estimate nitrogen dioxide

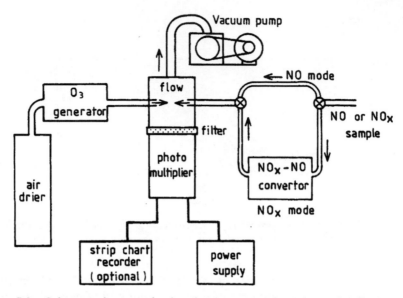

Figure 7.9 *Schematic diagram of a chemiluminescent analyser for oxides of nitrogen*

concentrations because wind-induced turbulence in the entry to the tube reduces the effective diffusion length, and due to chemical reactions converting NO into NO_2 within the tube. Nonetheless, diffusion tubes have been used very widely, and particularly in dense networks, to evaluate the spatial distribution of pollutants within an urban area.[10] Nitrogen dioxide has also been mapped across the entire United Kingdom on the basis of a network of diffusion tube samplers.

7.2.4   Carbon Monoxide

As exemplified by the inventories,[1] carbon monoxide is a pollutant very much associated with emissions from petrol vehicles. Within urban areas where concentrations tend to be highest, motor traffic is responsible for about 98% of emissions of carbon monoxide, and in the UK as a whole road traffic accounted for 73% of total emissions in 1998.

The major sink process is conversion into CO_2 by reaction with the hydroxyl radical (see Chapter 8). This process is, however, rather slow and the reduction in CO level away from the source areas is almost entirely a function of atmospheric dilution processes. Carbon monoxide can be a problem in heavily trafficked areas especially in confined 'street canyons' where concentrations may reach 50 ppm or more for short periods.

7.2.4.1   Measurement of Carbon Monoxide.   Non-dispersive infrared may be used to measure carbon monoxide in street air where levels encountered normally lie within the range 1–50 ppm. Using a long-path cell, the IR absorbance of polluted air at the wavelength corresponding to the C–O stretching vibration is continuously determined relative to that of reference air containing no CO. This is achieved without wavelength dispersion of the IR radiation by using cells containing CO at a reduced pressure as detectors for two beams which are chopped at a frequency of about 10 Hz and passed respectively through sample and reference cells. Absorption of radiation by the CO causes a differential pressure between the two detector cells which is sensed by a flexible diaphragm between them and used to generate an electrical signal. Because of partial overlap of absorption bands, carbon dioxide and water vapour interfere. The latter may be removed by passing the air sample through a drying agent, and the former interference by interposing a cell of carbon dioxide between the sample and reference cells and the detectors.

The other most common type of instrumental analyser is based upon gas filter correlation (see Figure 7.10). Infrared broad band radiation passes sequentially through gas cells containing carbon monoxide and molecular nitrogen contained within a spinning wheel, prior to passage of the radiation through a multi-pass optical cell through which ambient air is drawn. There are thus two beams separated in time but not space, one of which is absorbed by carbon monoxide in the ambient air passing through the sample cell; the other of which is already depleted in the wavelengths absorbed by carbon monoxide and is hence affected only by absorption by components other than carbon monoxide in the sample air. The difference in signal between the two beams is thus the result of absorption by carbon monoxide within the sample cell. Interferences from water vapour and carbon dioxide are claimed to be negligible. The instrument has a minimum detection limit of about 0.1 ppm, and reads up to 50 ppm.

An analyser for continuous determination of carbon monoxide at levels down to 1 ppm uses an electrochemical cell. Gas diffuses through a semi-permeable membrane into the cell and at an electrode CO is oxidized to CO_2 at a rate proportional to the concentration of CO in the air. The response time is fairly

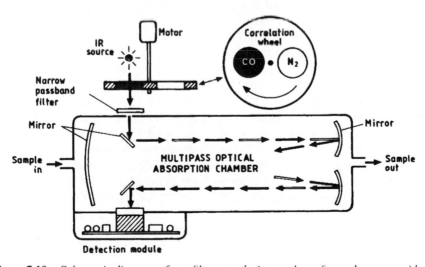

Figure 7.10 *Schematic diagram of gas filter correlation analyser for carbon monoxide*

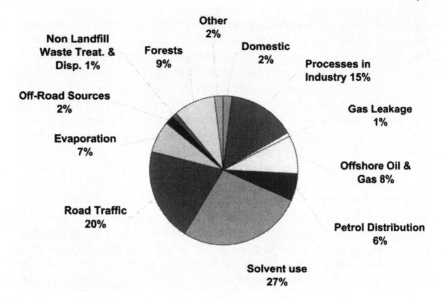

Figure 7.11   *Estimated UK emissions of volatile organic compounds by source category for 1998*[1]

short and interferences from other air pollutants at normally encountered levels are minimal. Analysers based upon electrochemical cells are also available for measurement of sulfur dioxide and oxides of nitrogen.

7.2.5   Hydrocarbons

The major sources of volatile organic compounds (which are mainly but not exclusively hydrocarbons) in the UK atmosphere are shown in Figure 7.11. It may be seen that these are rather more diverse than for many of the pollutants and include natural sources such as release from forest trees. In urban areas, road transport is probably the major contributor, although use of solvents, for example in paints and adhesives, can be a very significant source. Emissions from road transport include both the evaporation of fuels and the emission of unburned and partially combusted hydrocarbons and their oxidation products from the vehicle exhaust.

Many sources emit a range of individual compounds and careful analytical work has shown measurable levels of in excess of 200 hydrocarbons in some ambient air samples. In the UK the Hydrocarbon Network,[1] which makes automated hourly measurements of volatile organic compounds, reports data on some 25 individual hydrocarbons. Both benzene and 1,3-butadiene* are subject to regulation through the UK National Air Quality Strategy (Section 7.3); typical concentrations appear in Table 7.2. Methane, which is not often measured, far exceeds the other hydrocarbons in concentration. The Northern Hemisphere background of this compound is approximately 1.8 ppm and

_____
* The compound known in air pollution as 1,3-butadiene is properly named buta-1,3-diene.

elevated levels occur in urban areas as a result particularly of leakage of natural gas from the distribution system.

There are two major reasons for interest in the concentrations of hydrocarbons in the polluted atmosphere. The first is the direct toxicity of some compounds, particularly benzene and 1,3-butadiene, both of which are chemical carcinogens. The second cause of concern regarding hydrocarbons is due to their role as precursors of photochemical ozone (see Chapter 8). Compounds differ greatly in their potential to promote the production of ozone which has led to a system of classifying hydrocarbons according to their photochemical ozone creation potential (see also Chapter 8).

7.2.5.1 Measurement of Hydrocarbons. Determination of specific hydrocarbons in ambient air normally requires a pre-concentration stage in which air is drawn through an adsorbent such as a porous polymer or activated carbon, or a tube where freeze-out of the compounds by reduced temperature occurs, followed by injection into a gas chromatograph. Excellent separations of many compounds have been achieved, the best results coming from use of capillary columns. Detection may be by flame ionization or mass spectrometer (GC-MS), the latter technique allowing a more positive identification of individual compounds. Recent years have seen the development of automated systems for gas chromatographic measurements of hydrocarbon on a cyclic basis of about one hour to complete sampling and analysis. The UK Hydrocarbon Network was the first national network to adopt such instrumentation on a routine basis. It is now operated at some 13 sites. It is also possible to measure 'total hydrocarbons' by passage of a full air sample to a flame ionization detector. Results are reported as ppb C (parts per billion carbon) since the response of the FID is related closely to the rate of introduction of organic carbon atoms into the flame. Non-methane hydrocarbons may be determined with such instruments by alternate selective removal of hydrocarbons other than methane from the air stream prior to analysis and determination by difference.

7.2.6 Secondary Pollutants: Ozone and Peroxyacetyl Nitrate

7.2.6.1 Ozone. Atmospheric reactions involving oxides of nitrogen and hydrocarbons cause the formation of a wide range of secondary products. The most important of these is ozone. In severe photochemical smogs, such as occur in southern California, levels of ozone have in the past exceeded 400 ppb.

In Europe the classic Los Angeles type of urban smog is not experienced. Nonetheless, the same chemical processes give rise to elevated concentrations of ground-level ozone, often in a regional phenomenon extending over hundreds of kilometres simultaneously. Thus, hydrocarbon and NO_x emissions over wide areas of Europe react in the presence of sunlight causing large scale pollution, which is further extended by atmospheric transport of the ozone.[11] The phenomenon is crucially dependent upon meteorological conditions and hence, in Britain, is observed on only perhaps 10–30 days in each year on average. Concentrations of ozone measured at ground-level commonly exceed 100 ppb

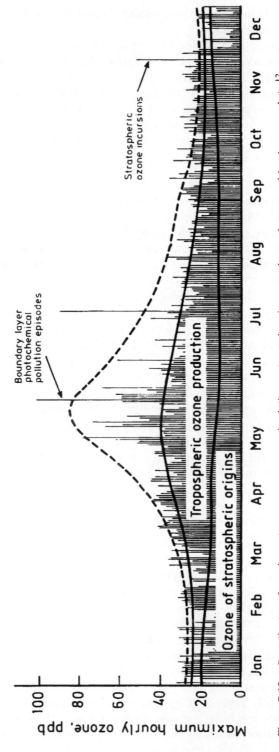

Figure 7.12   *Contributions from the major ozone sources to the daily maximum hourly mean observed at ground-level at a rural site*[12]

during such 'episodes' and have on one severe occasion been observed to exceed 250 ppb in southern England. These levels may be compared with a background of ozone at ground-level arising from downward diffusion of stratospheric ozone and general tropospheric production of 20–50 ppb. This is seen in Figure 7.12, showing measurement data from southern Scotland.[12]

The UK air quality standard for ozone of 50 ppb measured as a rolling eight-hour average[13] is currently exceeded several times each year in most parts of the United Kingdom; the greatest number of exceedences occurring in rural areas of the south-eastern UK. The least frequent exceedences occur in the urban sites where high levels of nitric oxide emissions from road traffic suppress ozone (see Chapter 8). For the rural sites, there is a general gradient in ozone with highest levels in the south and east of Britain and lowest in the north and west. Damage to crop plants can occur at ozone concentrations as low as 40 ppb; this is elaborated on in Chapter 12.

Measurement of Ozone.   The UV absorption of ozone at 254 nm may be used for its determination at levels down to 1 ppb. Interferences from other UV-absorbing air pollutants such as mercury and hydrocarbons may be minimized by taking two readings. The first reading is of the absorbance of an air sample after catalytic conversion of ozone into oxygen and the second is of an unchanged air sample, the difference in absorbance being due to the ozone content of the air. Available instruments perform this procedure automatically and give a read-out in digital form. Although truly continuous measurement of ozone levels is not possible, response is fast and readings may be taken at intervals of less than one minute. The instrument is shown diagrammatically in Figure 7.13.

7.2.6.2   Peroxyacetyl Nitrate (PAN).   PAN is a product of atmospheric photochemical reactions and is a characteristic product of photochemical smog (see Chapter 8).

$$CH_3-\underset{\underset{O}{\|}}{C}-O-O-NO_2 \quad PAN$$

Levels in southern California lie typically within the range 5–50 ppb on smoggy days. In Europe, the formation is far less favoured[11] and concentrations are more usually well below 10 ppb.

Chemical Analysis of Peroxyacetyl Nitrate.   PAN may be determined by long-path IR measurements, the greatest sensitivity being achieved when Fourier-transform methods are used. The most sensitive and specific routine technique involves gas chromatographic separation and detection of specific peroxyacyl nitrates by electron capture. The detection limit of below 1 ppb permits it to be used as a direct atmospheric monitor under circumstances of high pollution.

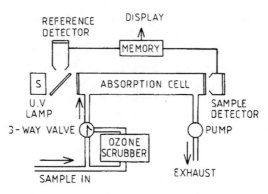

Figure 7.13 *Schematic diagram of a UV photometric analyser for ozone*

Lower concentrations may be measured by pre-concentration of PAN from the air.

7.3 AIR QUALITY MANAGEMENT

In recent years national and local government authorities have been bringing quite sophisticated approaches to bear on the subject of air quality management. The basic elements of the approach are illustrated in Figure 7.14. The first stage of the process is to carry out an air quality assessment. This entails bringing together the available air quality data, and where necessary making new measurements. These measurements are then assessed against an agreed set of air quality objectives. These air quality objectives are levels of air pollutant concentration which the strategy is aiming to achieve by a certain target date. If current concentrations exceed the air quality objectives, then model calculations are conducted to predict concentrations in the future target year taking account of measures already in place to reduce emissions within the intervening period. If concentrations in the target year are predicted to exceed the air quality objective, then an air quality management plan is devised and implemented in order to bring future concentrations into compliance with the objectives by the target year.

   Air quality objectives are normally derived from air quality standards. An air quality standard is a health-based guideline which is a concentration that, if achieved, will reduce the adverse effects of air pollution to a level which is zero or negligible at a population level. The latter refers to a concentration that may not be wholly safe for exquisitely sensitive individuals or which may imply a very small incidence of cancers as a result of breathing the pollutant. However, when viewed at the level of a large population, such effects are so small compared to the other risks of life as to be considered negligible. The air quality standards used in the UK which are derived from recommendations of the Expert Panel on Air Quality Standards (EPAQS), appear in Table 7.4. The UK National Air Quality Strategy[14] objectives as laid down in the year 2000 appear in Tables 7.5 and 7.6. The objectives are generally determined by what it is

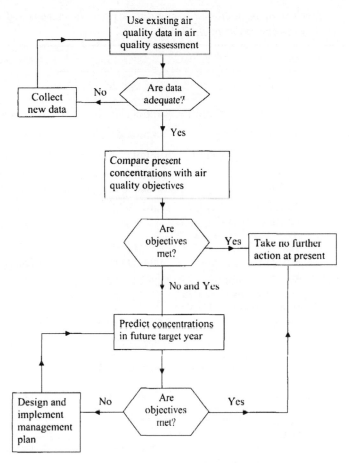

Figure 7.14 *Diagram illustrating the local air quality management process*

Table 7.4 *EPAQS recommended air quality standards*[14]

Pollutant	Concentration[a]	Standard measured as	EPAQS report
Benzene	5 ppb (16.25 μgm^{-3})	Running annual mean	1994
1,3-Butadiene	1 ppb (2.25 μg m^{-3})	Running annual mean	1994
Carbon monoxide	10 ppm (11.6 mg m^{-3})	Running eight hour mean	1994
Lead	0.25 μg m^{-3}	Annual mean	1998
Nitrogen dioxide	150 ppb (287 μg m^{-3})	One hour mean	1996
Ozone	50 ppb (100 μg m^{-3})	Running eight hour mean	1994
Particles (PM$_{10}$)	50 μg m^{-3}	Running 24 hour mean	1995
Sulfur dioxide	100 ppb (266 μg m^{-3})	15 minute mean	1995

[a] Conversion of ppb into μg m^{-3} and ppm to mg m^{-3} at 20 °C and 1013 mbar.

Table 7.5   *Objectives to be included in regulations for the purpose of local air quality management in the UK*[14]

Pollutant	Objective		Date to be achieved by
	Concentration[a]	Measured as[b]	
Benzene	16.25 μg m^{-3} (5 ppb)	Running annual mean	31 December 2003
1,3-Butadiene	2.25 μg m^{-3} (1 ppb)	Running annual mean	31 December 2003
Carbon monoxide	11.6 mg m^{-3} (10 ppm)	Running 8 hour mean	31 December 2003
Lead	0.5 μg m^{-3}	Annual mean	31 December 2004
	0.25 μg m^{-3}	Annual mean	31 December 2008
Nitrogen dioxide[c]	200 μg m^{-3} (105 ppb) not to be exceeded more that 18 times a year	1 hour mean	31 December 2005
	40 μg m^{-3} (21 ppb)	Annual mean	31 December 2005
Particles (PM$_{10}$)	50 μg m^{-3} not to be exceeded more than 35 times a year	24 hour mean	31 December 2004
	40 μg m^{-3}	Annual mean	31 December 2004
Sulfur dioxide	350 μg m^{-3} (132 ppb) not to be exceeded more than 24 times a year	1 hour mean	31 December 2004
	125 μg m^{-3} (47 ppb) not to be exceeded more than 3 times a year	24 hour mean	31 December 2004
	266 μg m^{-3} (100 ppb) not to be exceeded more than 35 times a year	15 minute mean	31 December 2005

[a] Conversions of ppb and ppm into μg m^{-3} and mg m^{-3} at 20 °C and 1013 mbar.
[b] How the objectives are to be measured is set out in Regulations.
[c] The objectives for nitrogen dioxide are provisional.

practicable to achieve on a given timescale. It will be seen from the tables that some objectives will lead to full achievement of compliance with the air quality standard, whereas others will lead only to a percentile compliance thus allowing some periods when the air quality standard is exceeded. Under UK legislation, for the majority of classical air pollutants excluding ozone (which is an international problem) responsibility is placed with local government to implement the objectives of the National Air Quality Strategy within their local boundaries.

Table 7.6 *National objectives not to be included in regulations for the purposes of local air quality management in the UK*[14]

Pollutant	Objective		Date to be achieved by
	Concentration[a]	Measured as[b]	
Objectives for the protection of human health			
Ozone[c]	100 μg m^{-3} (50 ppb) not to be exceeded more than 10 times a year	Daily maximum of running 8 hour mean	31 December 2005
Objectives for the protection of vegetation and ecosystems			
Nitrogen oxides[d]	30 μg m^{-3} (16 ppb)	Annual mean	31 December 2000
Sulfur dioxide	20 μg m^{-3} (8 ppb)	Annual mean	31 December 2000
	20 μg m^{-3} (8 ppb)	Winter average (1 October to 31 March)	31 December 2000

[a] Conversions of ppb and ppm into μg m^{-3} and mg m^{-3} at 20 °C and 1013 mbar.
[b] How the objectives are to be measured is set out in Regulations.
[c] The objective for ozone is provisional.
[d] Assuming NO_x is taken as NO_2.

7.4 INDOOR AIR QUALITY

Until rather recently, the emphasis in air quality evaluation has centred upon the outdoor environment. Recently, however, it has become clear that exposures to pollutants indoors may be very important (see also Chapter 19).

Pollutants with indoor sources may build up to appreciable levels because of the slowness of air exchange. An example is oxides of nitrogen from gas cookers and flueless gas and kerosene heaters which can readily exceed outdoor concentrations. Kerosene heaters can also be an important source of carbon monoxide and sulfur dioxide. Building materials and furnishings can also release a wide range of pollutants, such as formaldehyde from chipboard and hydrocarbons from paints, cleaners, adhesives, timber and furnishings. The tendency towards lower ventilation levels (energy efficient houses) has tended to exacerbate this problem.

Pollutants with a predominantly outdoor source may be reduced to rather low levels indoors due to the high surface area/volume ratios indoors leading to extremely efficient dry deposition of pollutants such as ozone and sulfur dioxide.

7.5 APPENDIX

7.5.1 Air Pollutant Concentration Units

Probably the most logical unit of air pollutant concentrations is mass per unit mass, *i.e.* μg kg^{-1} or mg kg^{-1}. This is, however, very rarely used. The commonest units are mass per unit volume (usually μg m^{-3}) or volume per unit volume, otherwise known as a volume mixing ratio (ppm or ppb). For particulate pollutants the volume mixing ratio is inapplicable.

Much confusion arises in the interconversion of μg m^{-3} and ppm. Whilst the volume mixing ratio is independent of temperature and pressure for an ideal gas (and air pollutant behaviour is close to ideal), the mass per unit volume unit is dependent on T and P conditions, and hence these will be taken into account.

7.5.1.1 Example 1.   Convert 0.1 ppm nitrogen dioxide into μg m^{-3} at 20 °C and 750 torr.

46 g NO_2 occupy 22.41 L at STP

46 g NO_2 occupy $22.41 \times \dfrac{293}{273} \times \dfrac{760}{750}$ L

   = 24.37 L at 20 °C and 750 torr

0.1 ppm NO_2 is 10^{-7} L NO_2 in 1 L, or

... 10^{-4} L NO_2 in 1 m^3

10^{-4} L NO_2 at 20 °C and 750 torr contain $46 \times \dfrac{10^{-4}}{24.37}$ g

   = 189 μg NO_2

   $\therefore$ NO_2 concentration = 189 μg m^{-3}

7.5.1.2 Example 2.   Convert 100 μg m^{-3} ozone at 25 °C and 765 torr into ppb.

48 g ozone occupy 22.41 L at STP

... occupy $22.41 \times \dfrac{298}{273} \times \dfrac{760}{765}$ L

   = 24.30 L at 25 °C and 765 torr

100 μg ozone occupy $24.30 \times \dfrac{100 \times 10^{-6}}{48}$ L

   = 50.6×10^{-6} L at 25 °C and 765 torr

   $\therefore$ Volume mixing ratio = 50.6×10^{-6} (L) $\div$ 1000 (L)

      = 50.6×10^{-9}

      = 51 ppb

(Note that for pressures quoted in millibars, the standard atmosphere, 760 torr = 1013 mbar = 1.013×10^5 Pa.)

The units ppm and ppb are sometimes designated ppmv and ppbv, as in Chapter 9.

7.6   REFERENCES

1. The UK National Air Quality Information Archive, URL: http://www.aeat.co.uk/netcen/airqual/
2. R. M. Harrison and R. Perry (eds.), 'Handbook of Air Pollution Analysis', 2nd edn., Chapman & Hall, London, 1986.
3. J. P. Lodge (ed.), 'Methods of Air Sampling and Analysis', 3rd edn., Lewis Publishers, Chelsea, Michigan, 1989.

4. APEG, 'Source Apportionment of Airborne Particulate Matter in the United Kingdom', The First Report of the Airborne Particles Expert Group, Department of Environment, Transport and the Regions, London, 1999.
5. Quality of Urban Air Review Group, 'Airborne Particulate Matter in the United Kingdom', QUARG, London, 1996.
6. R. J. Charlson, *J. Air Pollut. Control Assoc.*, 1968, **18**, 652.
7. World Health Organization, Air Quality Guidelines for Europe, URL: http://www.who.dk.envhlth/airqual.htm
8. Department of Environment, 'UK Blood Lead Monitoring Programme, 1984–1987. Results for 1987', Pollution Report No. 28, HMSO, London, 1990.
9. R. M. Harrison, J. P. Shi and J. L. Grenfell, *Atmos. Environ.*, 1998, **32**, 2769–2774.
10. Quality of Urban Air Review Group, 'Urban Air Quality in the United Kingdom', QUARG, London, 1993.
11. Photochemical Oxidants Review Group, 'Ozone in the United Kingdom', Fourth Report, DETR, London, 1997.
12. R. G. Derwent and P. J. A. Kaye, *Environ. Pollut.*, 1988, **55**, 191.
13. Expert Panel on Air Quality Standards, 'Ozone', HMSO, London, 1994.
14. Department of Environment, Transport and the Regions, 'The Air Quality Strategy for England, Scotland, Wales and Northern Ireland', HMSO, London, 2000.

CHAPTER 8

Chemistry and Climate Change in the Troposphere

R. M. HARRISON

8.1 INTRODUCTION

The atmosphere may conveniently be divided into a number of bands reflective of its temperature structure. These are illustrated by Figure 9.1 in Chapter 9. The lowest part, typically about 12 km in depth, is termed the troposphere and is characterized by a general diminution of temperature with height. The rate of temperature decrease, termed the lapse rate, is typically around 9.8 K km^{-1} close to ground level but may vary appreciably on a short-term basis. The troposphere may be considered in two smaller components: the part in contact with the earth's surface is termed the boundary layer; above it is the free troposphere. The boundary layer is normally bounded at its upper extreme by a temperature inversion (a horizontal band in which temperature increases with height) through which little exchange of air can occur with the free troposphere above. The depth of the boundary layer is typically around 100 m at night and 1000 m during the day, although these figures can vary greatly. The processes determining boundary layer height are introduced in Chapter 10. Pollutant emissions are generally into the boundary layer and are mostly constrained within it. Free tropospheric air contains the longer-lived atmospheric components, together with contributions from pollutants which have escaped the boundary layer, and from some downward mixing stratospheric air.

The average composition of the unpolluted atmosphere is given in Table 8.1. Some of the concentrations are very uncertain since:

(i)   Analytical procedures for some components have only recently reached the stage where good data can be obtained.
(ii) Some components such as CH_4 and N_2O are known to be increasing in concentration at an appreciable rate.
(iii) It is questionable whether any parts of the atmosphere can be considered entirely free of pollutants.

194

Table 8.1 *Average composition of the dry unpolluted atmosphere (based upon Seinfeld and Pandis[1] and Brimblecombe[2])*

Gas	Average concentration (ppm)	Approx. residence time
N_2	780 820	10^6 years
O_2	209 450	5000 years
Ar	9340	⎫
Ne	18	⎬ Not
Kr	1.1	⎭ cycled
Xe	0.09	
CO_2	360	100 years
CO	0.12 (N. Hemisphere)	65 days
CH_4	1.8	15 years
H_2	0.58	10 years
N_2O	0.31	120 years
O_3	0.01–0.1	100 days
NO/NO_2	10^{-6}–10^{-2}	1 day
NH_3	10^{-4}–10^{-3}	5 days
SO_2	10^{-3}–10^{-2}	10 days
HNO_3	10^{-5}–10^{-3}	1 day

Table 8.1 also includes estimates of the lifetime of the various components. Those with lifetimes of a few days or less are cycled mainly within the boundary layer. Components with longer lifetimes mix into the free troposphere more substantially and those with lifetimes of a year or more will penetrate the stratosphere to a significant degree. An indication of polluted air concentrations of some of these components is given in Chapter 7.

8.1.1 Pollutant Cycles

Pollutants are emitted from *sources* and are removed from the atmosphere by *sinks*. A typical cycle appears in Figure 8.1. Most pollutants have both natural and man-made sources; although the natural source is often of sizeable magnitude in global terms, on a local scale in populated areas pollutant sources are usually predominant.

Sink processes include both dry and wet mechanisms. Dry deposition involves the transfer and removal of gases and particles at land and sea surfaces without the intervention of rain or snow. For gases removed at the surface, dry deposition is driven by a concentration gradient caused by surface depletion; for particles this mechanism operates in parallel with gravitational settling of the large particles. The efficiency of dry deposition is described by the *deposition velocity*, V_g, defined as:

$$V_g(\text{m s}^{-1}) = \frac{\text{Flux to surface } (\mu g \text{ m}^{-2} \text{ s}^{-1})}{\text{Atmospheric concentration } (\mu g \text{ m}^{-2} \text{ s}^{-1})}$$

Table 8.2 *Some typical values of deposition velocity and corresponding lifetimes in boundary layers of 100 m and 1000 m in depth* (H)

Pollutant	Surface	Deposition velocity (cm s^{-1})	Lifetime ($H = 1000$ m)	Lifetime (h) ($H = 100$ m)
SO_2	Grass	1.0	28 h	2.8
SO_2	Ocean	0.5	56 h	5.6
SO_2	Soil	0.7	40 h	4.0
SO_2	Forest	2.0	14 h	1.4
O_3	Dry grass	0.5	56 h	5.6
O_3	Wet grass	0.2	5.8 days	13.9
O_3	Snow	0.1	11.6 days	27.8
HNO_3	Grass	2.0	14 h	1.4
CO	Soil	0.05	23 days	26
Aerosol ($<2.5\ \mu$m)	Grass	0.15	8 days	19

Some typical values of deposition velocity are given in Table 8.2. For a gas, such as sulfur dioxide which has a fairly high V_g, dry deposition has little influence upon near-source concentrations, but may appreciably influence ambient levels at large downwind distances.

Wet deposition describes scavenging by precipitation (rain, snow, hail, *etc.*) and is made up of two components, *rainout* which describes incorporation within the cloud layer, and *washout* describing scavenging by falling raindrops. The overall efficiency is described by the scavenging ratio, *W*, often rather misleadingly referred to as the Washout Factor.

$$W = \frac{\text{Concentration in rainwater (mg kg}^{-1})}{\text{Concentration in air (mg kg}^{-1})}$$

Typical values of scavenging ratio are given in Table 8.3. A large value implies efficient scavenging, perhaps resulting from extensive vertical mixing into the cloud layer, where scavenging is most efficient. Alternatively, it may be due to a

Table 8.3 *Typical scavenging ratios*

Species	W
Cl^-	600
SO_4^{2-}	700
Na	560
K	620
Mg	850
Ca	1890
Cd	390
Pb	320
Zn	870

large particle size (*e.g.* Ca), efficiently collected by falling raindrops. A related deposition process termed 'occult' deposition occurs when pollutants are deposited by fogwater deposition on surfaces. Pollutant concentrations in fogwater are typically much greater than in rainwater, hence the process may be significant despite the modest volumes of water deposited.

Another sink process involves chemical conversion of one pollutant into another (termed dry transformations in Figure 8.1). Thus atmospheric oxidation to sulfuric acid is a sink for sulfur dioxide. For many pollutants, a major sink is atmospheric reaction with the hydroxyl radical ($^{\bullet}$OH). Such reactions are described later in this chapter. Since pollutants are continually emitted into and removed from the atmosphere, they have an associated atmospheric lifetime or residence time defined below.

Many are chemically reactive and are transformed to other chemical species within the atmosphere. In some instances the products of such reactions, termed secondary pollutants, are more harmful than the primary pollutants from which they are formed. Thus an appreciation of atmospheric chemical processes is fundamental to any attempt to limit the adverse effects of air pollutant emissions.

Table 8.1 includes lifetimes (also termed residence times) of the atmospheric gases. In this context, lifetime, τ, is defined as:

$$\tau = \frac{A}{F}$$

where A = global atmospheric burden (Tg) (1 Tg = 10^{12} g)
F = global flux into and out of the atmosphere (Tg year $^{-1}$).

This treatment assumes a steady-state between the input and removal fluxes. In the case of a pollutant remote from its source, the lifetime corresponds to the

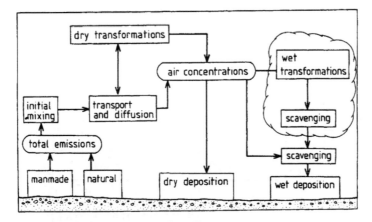

Figure 8.1 *Typical atmospheric cycle of a pollutant*
(Reprinted with permission from *Environ. Sci. Technol.*, 1988, **22**, 241, by permission of the American Chemical Society)

time taken to reduce its concentration to $\frac{1}{e}$, or approximately one third of its initial level, if it is not replenished.

Sources of trace gases are not normally evenly spaced over the surface of the globe. Thus spatial variability in airborne concentrations occurs. For gases with an atmospheric lifetime comparable with the timescale of mixing of the entire troposphere (a year, or more), there is little spatial variation in concentration, as mixing processes outweigh the local variability in source strengths. For gases with short lifetimes, atmospheric mixing cannot prevent a substantially variable concentration. High spatial variability will also be associated with high temporal variability at one point as differing air mass sources and different mixing conditions will advect different concentrations of trace gas to the fixed receptor. An example of a rather well mixed gas of long lifetime is carbon dioxide which shows little variation in concentration over the globe and only small fluctuations at a given site (except very close to major combustion sources). At the other extreme, ammonia, which is chemically reactive and subject to efficient dry and wet deposition processes, is highly variable on both spatial and temporal scales. The spatial variability may be described by the coefficient of variation (equal to the standard deviation of the mean concentration divided by the mean) which relates to residence time as shown in Figure 8.2. A process contributing to the flux of trace gases from the atmosphere is dry

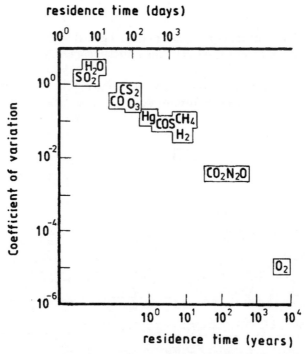

Figure 8.2 *The relationship between coefficient of variation in spatially averaged concentration and residence time for atmospheric gases*

deposition. Atmospheric lifetimes with respect to this process are equal to H/V_g, where H is the depth of the mixed boundary layer, and hence lifetimes are shortest when the boundary layer is very shallow. Values of atmospheric lifetime for boundary layer depths of 100 m and 1000 m appear in Table 8.2. These values apply only to the lifetime in the boundary layer and therefore tell us nothing about the lifetimes of substances within the free troposphere above, which will be determined by chemical reactions and mixing processes.

Early books on tropospheric chemistry tended to consider the cycle of one substance in isolation from those of others. It is now well recognized that many of the important atmospheric chemical cycles are closely interlinked and that a more integrated approach to study is appropriate. In this context, the hydroxyl radical has been recognized as having an immensely important role. It is responsible for the breakdown of many atmospheric pollutants, whilst its formation is dependent upon others. It is appropriate to commence a description of tropospheric chemistry with this short-lived free radical species.

8.2 ATMOSPHERIC CHEMICAL TRANSFORMATIONS

8.2.1 The Importance of the Hydroxyl Radical (OH)

The principal source of OH in the background troposphere is photolysis of ozone by light of short wavelengths ($\lambda < 315$ nm) to form singlet (excited state) atomic oxygen $O(^1D)$, which may either relax to the triplet (ground) state, $O(^3P)$ or may react with water vapour to form OH:

$$O_3 + h\nu \longrightarrow O(^1D) + O_2 \quad \lambda < 315 \text{ nm} \tag{1}$$
$$O(^1D) + M \longrightarrow O(^3P) + M \tag{2}$$
$$O(^1D) + H_2O \longrightarrow 2OH \tag{3}$$

(M is an unreactive third molecule such as N_2).

Minor sources are available also through reactions of $O(^1D)$ with CH_4 and H_2:

$$CH_4 + O(^1D) \longrightarrow CH_3 + OH \tag{4}$$
$$H_2 + O(^1D) \longrightarrow H + OH \tag{5}$$

Photolysis of both HONO and H_2O_2 produces OH directly:

$$HONO + h\nu \longrightarrow OH + NO \quad \lambda < 400 \text{ nm} \tag{6}$$
$$H_2O_2 + h\nu \longrightarrow 2OH \quad \lambda < 360 \text{ nm} \tag{7}$$

Formation from nitrous acid may be of significance in polluted air. The route from hydrogen peroxide is not likely to represent a net source of OH since the main source of H_2O_2 is from the HO_2 (hydroperoxy) radical:

$$HO_2 + HO_2 \longrightarrow H_2O_2 + O_2 \tag{8}$$

In polluted atmospheres, however, HO_2 is able to give rise to OH formation by a more direct route:

$$HO_2 + NO \longrightarrow NO_2 + OH \qquad (9)$$

A review of tropospheric concentrations of the hydroxyl radical[3] found considerable variations in concentrations estimated by direct spectroscopic measurement, indirect measurement and modelling. There are also genuine variations with latitude, season and the presence of atmospheric pollutants. The overall consensus was of tropospheric concentrations within the ranges 0.5–5×10^6 cm^{-3} daytime mean and 0.3–3×10^6 cm^{-3} 24 h mean. A seasonal variation of about threefold is suggested by model studies.

As the following sections will demonstrate, the hydroxyl radical plays a central role, *via* the peroxy radicals with which it is intimately related, in the production of ozone and hydrogen peroxide. It also itself contributes directly to formation of sulfuric and nitric acids in the atmosphere, as well as indirectly contributing *via* ozone and hydrogen peroxide. Thus atmospheric processes leading to ozone formation will also tend to favour production of other secondary pollutants, including the strong acids HNO_3 and H_2SO_4.

8.3 ATMOSPHERIC OXIDANTS

8.3.1 Formation of Ozone

Mid-latitude northern hemisphere sites show background ozone concentrations typically with the range 20–50 ppb. These concentrations were for many years attributed solely to downward transport of stratospheric ozone and the seasonal fluctuation, with a pronounced spring maximum and broad winter minimum, relating to adjustments in the altitude of the tropopause.

In the late 1970s, Fishman and Crutzen[4,5] showed that ozone could be formed from oxidation of methane and carbon monoxide in the troposphere in processes involving the hydroxyl radical:

$$CH_4 + OH \longrightarrow CH_3 + H_2O \qquad (10)$$
$$CH_3 + O_2 + M \longrightarrow CH_3O_2 + M \qquad (11)$$
$$CO + OH \longrightarrow CO_2 + H \qquad (12)$$
$$H + O_2 + M \longrightarrow HO_2 + M \qquad (13)$$

In the presence of NO:

$$CH_3O_2 + NO \longrightarrow CH_3O + NO_2 \qquad (14)$$
$$CH_3O + O_2 \longrightarrow CH_2O + HO_2 \qquad (15)$$
$$HO_2 + NO \longrightarrow OH + NO_2 \qquad (9)$$

Thus *via* reactions of the peroxy radicals HO_2 and RO_2 (R = alkyl), NO is converted into NO_2. Then:

$$NO_2 + h\nu \longrightarrow NO + O(^3P) \quad \lambda < 435\,nm \qquad (16)$$
$$O(^3P) + O_2 + M \longrightarrow O_3 + M \qquad (17)$$

The magnitude of this source of ozone is presently uncertain. However, there is much evidence to suggest that background northern hemisphere tropospheric ozone concentrations have approximately doubled since the turn of the century.[6] If this is indeed the case, the cause is almost certainly enhanced formation from the oxidation cycles of methane (whose concentration is known to be increasing), other less reactive hydrocarbons and carbon monoxide, involving anthropogenic nitrogen oxides. Such a source is also consistent with a spring maximum in ozone caused by reaction of hydrocarbons accumulated through the less reactive winter months. Figure 8.3 shows near-surface ozone concentrations measured in France and Italy in the nineteenth century compared with more recent data from Germany and Italy. These measurements provide strong evidence of an increase in ground-level ozone by a factor of approximately two over the past hundred years.

In polluted air there is abundant NO_2 whose photolysis leads to ozone formation. However, fresh emissions of NO lead to ozone removal and urban

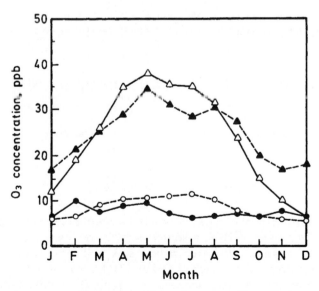

Figure 8.3 *Average monthly concentrations of ozone at Moncalieri near Turin (Italy), open circles 1868–1893, and at Montsouris (France), solid circles 1876–1886, and at Arkona (Germany), solid triangles 1983, and Ispra (Italy), open triangles 1986–1989*
(From Ref. 6)

concentrations of ozone are usually lower than those in surrounding rural areas; the full cycle of reactions is:

$$NO_2 + h\nu \xrightarrow{J_1} NO + O(^3P) \quad \lambda < 435 \text{ nm} \tag{16}$$
$$O(^3P) + O_2 + M \longrightarrow O_3 + M \tag{17}$$

$$NO + O_3 \xrightarrow{k_3} NO_2 + O_2 \tag{18}$$

J_1 and k_3 are the rate constants for the NO_2 photolysis and O_3 removal reactions respectively. All three reactions are rapid and an equilibrium is reached when the rate of ozone formation (equal to the rate of NO_2 photolysis if all $O(^3P)$ leads to O_3 formation) equals the rate of O_3 removal. Then:

$$J_1[NO_2] = k_3[O_3][NO] \tag{19}$$

$$\text{and} \quad [O_3] = \frac{J_1[NO_2]}{k_3[NO]} \tag{20}$$

This is termed the *photostationary state*, and thus the ozone concentration is determined by the value of J_1, highest at peak sunlight intensity, and the ratio of NO_2/NO. Hydrocarbons, and to a lesser extent CO, play a crucial role in producing HO_2 and RO_2 peroxy radicals which convert NO into NO_2 *without* consumption of O_3. For example, propene is attacked by hydroxyl:

$$CH_3CH=CH_2 + OH \longrightarrow CH_3CH-CH_2OH \tag{21}$$
$$CH_3CH-CH_2OH + O_2 \longrightarrow CH_3CH-CH_2OH \tag{22}$$
$$\qquad\qquad\qquad\qquad\qquad\qquad |$$
$$\qquad\qquad\qquad\qquad\qquad\qquad OO$$

$$CH_3CH-CH_2OH + NO \longrightarrow CH_3CH-CH_2OH + NO_2 \tag{23}$$
$$\quad\;\; |\qquad\qquad\qquad\qquad\qquad\qquad\; |$$
$$\quad\;\; OO\qquad\qquad\qquad\qquad\qquad\quad O$$

$$CH_3CH-CH_2OH \longrightarrow CH_2OH + CH_3CHO \tag{24}$$
$$\quad\;\; |$$
$$\quad\;\; O$$

$$CH_2OH + O_2 \longrightarrow HCHO + HO_2 \tag{25}$$

In this case atmospheric photochemistry acts as a source of O_3 and aldehydes. It also leads to formation of HO_2, a source of hydrogen peroxide. None of the above processes is an effective free radical sink and hence, during hours of daylight, reactive free radicals such as OH and HO_2 are constantly recycled.

In remote atmospheres, photochemistry is typically a sink, rather than a source of ozone. The ozone is photolysed to generate hydroxyl radicals (reactions (1) to (3)) leading to subsequent formation of peroxy radical such as methylperoxy (reactions (10) and (11)) or hydroperoxy (reactions (12) and (13)). Whereas in polluted atmospheres the peroxy radicals convert NO into NO_2

regenerating the OH radical (reactions (14), (15) and (9)), in clean atmospheres (less than about 20 ppt NO_x) there is insufficient NO and loss of peroxy radicals occurs through such processes as reaction (8) or (26) which destroys ozone:

$$HO_2 + O_3 \longrightarrow 2O_2 + OH \tag{26}$$

The apparent implication is that only massive cuts in NO_x emissions will reverse the increase in the northern hemisphere ozone background.

8.3.1.1 Polluted Atmospheres In heavily polluted urban areas, subject to photochemical air pollution (*e.g.* Los Angeles), the concentrations of NO, NO_2 O_3, and other secondary pollutants tend to follow characteristic patterns, illustrated in Figure 8.4. Primary pollutants NO and hydrocarbons tend to peak with heavy traffic around 7–8 am. This is followed by a peak in NO_2 some time later when the atmosphere has developed sufficient oxidizing capability to oxidize the NO emissions of motor vehicles. The NO_2/NO ratio is now high, favouring ozone production, and ozone peaks with peak sunlight intensity around noon, or a little later. In the United Kingdom, such diurnal profiles have been observed in urban areas during summer anticyclonic conditions, but peak ozone concentrations are generally rather later, around 3 pm. The situation is more complex, however, as in the UK much of the ozone is generated during long-range transport of air pollution, often from continental European sources.

Perhaps surprisingly, UK rural sites exhibit diurnal variations in ozone which are remarkably similar to those at urban sites. Figure 8.5 shows average variations at Stodday, near Lancaster, for various months in 1983. This diurnal change is not due to the same causes as that in urban areas, where fresh NO emissions destroy ozone in the night-time (reaction (18)). Measurement of

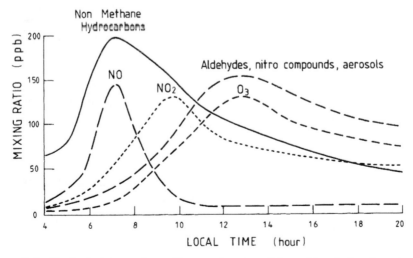

Figure 8.4 *Typical mixing ratio profiles, as a function of time of day, in the photochemical smog cycle*

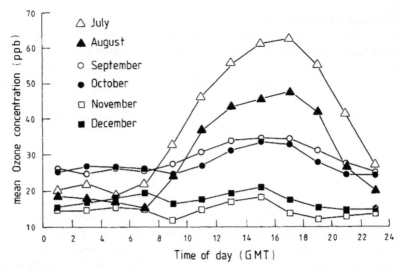

Figure 8.5   *Mean diurnal variations of ground-level ozone, July–December 1983 at Stodday, near Lancaster*[7]

vertical profiles of ozone and temperature (Figure 8.6) show that nocturnal depletion of ozone is a surface phenomenon due to dry deposition, with little change in airborne concentrations at 50 m to 100 m altitude. The temperature profiles show the depletion occurs from a non-turbulent stable layer at the surface indicated by temperature rising with height (termed a temperature inversion). During the winter months, little diurnal change in ozone is seen (see Figure 8.5) since the atmospheric temperature structure shows a less marked change between night and day.

As mentioned in Chapter 7, many volatile organic compounds contribute to ozone formation through the chemical reactions exemplified above by methane and propene. Their individual contributions depend upon their concentrations and reactivity, generally with OH. The contributions of individual compounds per unit mass of emissions, termed their photochemical ozone creation potential (POCP), can be determined from numerical models of atmospheric photochemistry. The twenty compounds contributing most to ozone formation in the UK atmosphere are listed in Table 8.4.

8.3.2   Formation of PAN

Peroxyacetyl nitrate (PAN) is of interest as a characteristic product of atmospheric photochemistry, as a reservoir of reactive nitrogen in remote atmospheres and because of its effects upon plants. The formation route is *via* acetyl radicals (CH_3CO) formed from a number of routes, most notably acetaldehyde oxidation.

$$CH_3CHO + OH \longrightarrow CH_3CO + H_2O \qquad (27)$$

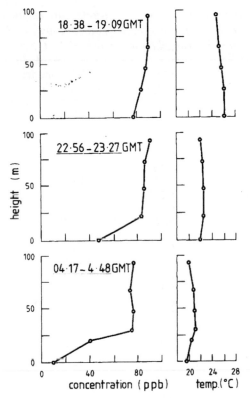

Figure 8.6 *Vertical profiles of ozone and temperature on 11–12 July 1983*[8]

$$CH_3CO + O_2 \longrightarrow CH_3C(O)OO \tag{28}$$

$$CH_3C(O)OO + NO_2 \longrightarrow CH_3C\!\!-\!\!OONO_2 \tag{29}$$
$$\underset{O}{\overset{\|}{}}$$
$$(PAN)$$

8.3.3 NO$_y$ Compounds

The term NO$_y$ refers to the sum of all odd nitrogen compounds in the atmosphere (*i.e.* not N$_2$O):

$$NO_y = \sum NO + NO_2 + HNO_3 + HONO + 2N_2O_5 + PAN$$

It thus contains not only the highly reactive species NO and NO$_2$ whose role in ozone formation is outlined above, but also products of NO$_x$ oxidation (HNO$_3$ and N$_2$O$_5$ – see Section 8.4.4) and reservoir species, HONO and PAN.

Table 8.4   *Mean rural summertime concentrations, rate coefficients for reaction with OH radicals and estimated contributions to O_3 formation for each individual hydrocarbon measured in the UK*

Species	Mean (ppb)	OH + species $k_{OH} \times 10^{12}$	O_3 prod. (ppb h^{-1})
i-Butene	0.21	51.4	0.773
Propylene	0.27	26.3	0.511
Ethylene	0.67	8.52	0.411
Isoprene	0.05	101	0.364
1,2,4-Trimethylbenzene	0.15	32.5	0.352
m- + *p*-Xylene	0.21	19	0.287
1,3,5-Trimethylbenzene	0.06	57.5	0.256
trans-But-2-ene	0.05	64	0.230
Toluene	0.46	5.96	0.197
trans-Pent-2-ene	0.04	66.9	0.193
Buta-1,3-diene	0.04	66.6	0.192
cis-But-2-ene	0.04	56.4	0.162
i-Pentane	0.52	3.9	0.146
o-Xylene	0.14	13.7	0.138
But-1-ene	0.05	31.4	0.113
3-Methyl-*cis*-pent-2-ene	0.02	86.7	0.112
n-Butane	0.59	2.52	0.107
m-Ethyltoluene	0.07	19.2	0.100
3-Methyl-*trans*-pent-2-ene	0.02	86.7	0.097
2-Methylpent-2-ene	0.02	89	0.097

Nitrous acid (HONO) is unimportant in clean air, but can play an important role as source of hydroxyl radical through photolysis in polluted atmospheres

$$HONO + h\nu \longrightarrow OH + NO \qquad (6)$$

The source of the HONO is currently not fully understood, but it appears to arise mainly from heterogeneous reactions of nitrogen dioxide and water on wet surfaces

$$2NO_2 + H_2O \longrightarrow HNO_3 + HONO \qquad (30)$$

Its photolysis (reaction (6)) can be an important source of hydroxyl radicals in the early morning within polluted atmospheres due to the build-up of HONO concentrations in the hours of darkness.

Peroxyacetyl nitrate is very important as a reservoir species of NO_x, as are its higher homologues such as peroxypropionyl nitrate (PPN). It is unstable with respect to dissociation (reverse of equation (29)):

$$\underset{\displaystyle \overset{|}{O}}{CH_3C}\text{-}OONO_2 \longleftrightarrow \underset{\displaystyle \overset{|}{O}}{CH_3C}\text{-}OO + NO_2 \qquad (31)$$

Cooler temperatures enhance the stability of PAN compounds and they can comprise a major component of NO_y in polar regions and in the mid and upper troposphere. When moved by the atmosphere to warmer locations, however, they can dissociate, releasing NO_2 and hence acting as a reservoir for NO_x.

8.4 ATMOSPHERIC ACIDS

8.4.1 Weak Acids

Well known weak acids in the atmosphere which may contribute to corrosion processes and influence the pH of precipitation are carbon dioxide and sulfur dioxide. Atmospheric CO_2 at a concentration of 340 ppm leads to an equilibrium pH of 5.6 at 15 °C in otherwise unpolluted rainwater. This is normally taken as the boundary pH below which rain is considered acid. Sulfur dioxide is a stronger acid than CO_2 and at a concentration of only 5 ppb in air will, at equilibrium, cause a rainwater pH of 4.6 at 15 °C.[2] In many instances, this pH is not attained due to severe kinetic constraints upon achievement of equilibrium, as is the case with many atmospheric trace gases. Dissolved SO_2 after oxidation may contribute appreciably to total sulfate and acidity in urban rainwater.

Organic acids such as formic acid may be formed in the atmosphere from oxidation of aldehydes, in this case formaldehyde (methanal):

$$HO_2 + HCHO \longleftrightarrow HOOCH_2O \longleftrightarrow OOCH_2OH \qquad (32)$$
$$OOCH_2OH + NO \longrightarrow OCH_2OH + NO_2 \qquad (33)$$
$$OCH_2OH + O_2 \longrightarrow HCOOH + HO_2 \qquad (34)$$

Carboxylic acids are believed to contribute significantly to rainwater acidity in remote areas, although their contribution is not at all well quantified. In polluted regions they are unlikely to be of much importance.

8.4.2 Strong Acids

Strong acids of major importance in the atmosphere are as follows: H_2SO_4, HNO_3, HCl and CH_3SO_3H (methanesulfonic acid).

8.4.3 Sulfuric Acid

Atmospheric oxidation of NO_2 proceeds *via* a range of mechanisms. Consequently, dependent upon the concentrations of the responsible oxidants, the oxidation rate is extremely variable with space and time but typically lies around $1\% \ h^{-1}$.

Several gas phase mechanisms have been investigated, including photooxidation, reaction with hydroxyl radical, Criegee biradical, ground state atomic oxygen, $O(^3P)$ and peroxy radicals. Table 8.5, based upon the treatment of Finlayson-Pitts and Pitts,[9] summarizes the rate data and clearly indicates the overwhelming importance of the hydroxyl radical reaction in the gas phase.

Table 8.5 *Homogeneous oxidation mechanisms for SO_2*
(based on Ref 9)

Oxidizing species	Concentration[a] (cm^{-3})	Rate constant $(cm^3\,molec^{-1}\,s^{-1})$	Loss of SO_2 $(\%\,h^{-1})$
OH	5×10^6	9×10^{-13}	1.6
O_3	2.5×10^{12}	$<8 \times 10^{-24}$	$<7 \times 10^{-6}$
Criegee biradical	1×10^6	7×10^{-14}	3×10^{-2}
$O(^3P)$	8×10^4	6×10^{-14}	2×10^{-3}
HO_2	1×10^9	$<1 \times 10^{-18}$	$<4 \times 10^{-4}$
RO_2	3×10^9	$<1 \times 10^{-18}$	$<1 \times 10^{-3}$

[a] Assuming a moderately polluted atmosphere.

The mechanism of the SO_2–OH reaction is as follows:

$$SO_2 + OH \longrightarrow HOSO_2 \tag{35}$$
$$HOSO_2 + O_2 \longrightarrow HO_2 + SO_3 \tag{36}$$
$$SO_3 + H_2O \longrightarrow H_2SO_4 \tag{37}$$

In the presence of water droplets, which can take the form of fogs, clouds, rain or hygroscopic aerosols, sulfur dioxide will dissolve, opening up the possibility of aqueous phase oxidation. Upon dissolution, the following equilibria operate.

$$SO_{2(g)} + H_2O \longleftrightarrow SO_2 \cdot H_2O \tag{38}$$
$$SO_2 \cdot H_2O \longleftrightarrow HSO_3^- + H^+ \tag{39}$$
$$HSO_3^- \longleftrightarrow SO_3^{2-} + H^+ \tag{40}$$

These equilibria are sensitive to pH, and HSO_3^- is the predominant species over the range pH 2–7. The other consequence of these equilibria is that the more acidic the droplet, the greater the degree to which the equilibria move towards gaseous SO_2 and limit the concentrations of dissolved S(IV) species. Some rate constants are pH-dependent in addition.

The major proposed oxidation mechanisms for SO_2 in liquid droplets include the following:

(i)   uncatalysed oxidation by O_2
(ii) transition metal-catalysed oxidation by O_2
(iii) oxidation by dissolved oxides of nitrogen
(iv) oxidation by ozone
(v)   oxidation by H_2O_2 and organic peroxides.

Relative rates for typical specified concentrations of reactive species are indicated in Table 8.6.

At high pH, all mechanisms in Table 8.6 are capable of oxidizing SO_2 at appreciable rates, generally far in excess of those observed in the atmosphere

Table 8.6 *Aqueous phase oxidation mechanisms for SO_2* (adapted from Ref. 9)

Oxidant	Concentration[a]	Oxidation rate (% h^{-1})	
		pH 3	pH 6
O_3	50 ppb (g)	3×10^{-2}	5×10^3
H_2O_2	1 ppb (g)	8×10^2	5×10^2
Fe-catalysed	3×10^{-7} M (l)	2×10^{-2}	5×10^1
Mn-catalysed	3×10^{-8} M (l)	3×10^{-2}	7
HNO_2	1 ppb (g)	5×10^{-4}	3

[a] (g) and (l) denote gas and liquid phase concentrations respectively. Conditions are 5 ppb gas phase SO_2 at 25 °C with no mass transfer limitations

over a significant averaging period. Slower rates are observed because of mass transfer limitations to the rate of introduction of SO_2 and oxidant to the water droplets and the previously noted effect of pH reduction (due to H_2SO_4 formation) on the solubility of SO_2. At pH 3, only the reaction with hydrogen peroxide is very fast, due to an increased rate constant at reduced pH compensating for the lower solubility of SO_2. Experimental studies indicate that this reaction can be very important in the atmosphere, but is limited by the availability of H_2O_2 which rapidly becomes depleted. The rate of formation of H_2SO_4 is therefore a function more of atmospheric mixing processes than of chemical kinetics in this case.

Several studies have emphasized the importance of SO_2 oxidation upon the surface of carbonaceous aerosols.[10,11] Other studies[12] have shown much slower rates of oxidation, and it remains unclear to what extent this mechanism contributes to the atmospheric oxidation of SO_2. It is likely to be limited rapidly by the ageing of the soot particles surfaces.

8.4.4 Nitric Acid

The main daytime route of nitric acid formation is from the reaction:

$$NO_2 + OH \longrightarrow HNO_3 \tag{41}$$

The rate constant is 1.1×10^{-11} cm^{-3} $molec^{-1}$ s^{-1} at 25 °C,[9] implying a rate of NO_2 oxidation of 19.8% h^{-1} at an OH concentration of 5×10^6 cm^{-3} (*cf.* Table 8.5). This is thus a much faster process than gas phase oxidation of SO_2. This process is not operative during hours of darkness due to near zero OH radical concentrations.

At night-time, reactions of the NO_3 radical become important which are not operative during daylight hours due to photolytic breakdown of NO_3. The radical itself is formed as follows:

$$NO_2 + O_3 \longrightarrow NO_3 + O_2 \tag{42}$$

It is converted into HNO_3 by two routes. The first is hydrogen abstraction from hydrocarbons or aldehydes:

$$NO_3 + RH \longrightarrow HNO_3 + R \qquad (43)$$

Typical reaction rates imply formation of HNO_3 at about 0.3 ppb h^{-1} in a polluted urban atmosphere by this route.[9] This is modest compared to daytime formation from NO_2 and OH.

The other night-time mechanism of HNO_3 formation is *via* the reaction sequence:

$$NO_3 + NO_2 \overset{M}{\longleftrightarrow} N_2O_5 \qquad (44)$$
$$N_2O_5 + H_2O \longleftrightarrow 2HNO_3 \qquad (45)$$

The reaction involving water is rate determining and may be fairly slow at low relative humidity, contributing HNO_3 at ~ 0.3 ppb h^{-1}.[9] As humidity increases, and especially in the presence of liquid water, more rapid reactions may be observed. This is presently supported only by indirect evidence and requires further experimental investigation.

Aqueous phase oxidation of NO_2 is of little importance due primarily to low aqueous solubility of NO_2.

8.4.5 Hydrochloric Acid

Hydrochloric acid differs from sulfuric and nitric acids in that it is emitted into the atmosphere as a primary pollutant and is not dependent upon atmospheric chemistry for its formation.

Some perspective of its significance as an acidic pollutant may be gained from Table 8.7 which shows the estimated emissions of SO_2, NO_x and HCl in the UK and the potential hydrogen ion equivalent. At only 2.6%, the HCl contribution appears small until it is realized that this is immediately available acidity. Using diurnally averaged concentrations of reactive species, oxidations of SO_2 and NO_2 may be proceeding at only $\sim 1\%$ and 10% h^{-1} respectively, or less, and thus, at smaller distances from a major source of all three pollutants (*e.g.* a power plant), HCl may be the predominant strong acid.

Table 8.7 *Total UK acid emission*
(1998 data derived from Ref. 13)

Species	*Emission* (kt a^{-1})	*Potential H^{+} equivalent* (kt a^{-1})	*% Total potential acidity*
SO_2	1615	50.5	55.5
NO_x (as NO_2)	1753	38.1	41.9
HCl	89	2.4	2.6

Another source, not considered in Table 8.6, may also contribute to HCl in the atmosphere. HCl is a more volatile acid than either H_2SO_4 or HNO_3 and thus may be displaced from aerosol chlorides, such as sea salt:

$$2NaCl + H_2SO_4 \longrightarrow Na_2SO_4 + 2HCl \qquad (46)$$
$$NaCl + HNO_3 \longrightarrow NaNO_3 + HCl \qquad (47)$$

This can be an appreciable source of HCl in areas influenced by maritime air masses.

8.4.6 Methanesulfonic Acid (MSA)

This strong acid is unlikely to be of importance in polluted regions but can contribute appreciably to acidity in remote areas. It is a major product of the oxidation of dimethyl sulfide, $(CH_3)_2S$. A likely mechanism[9] is:

$$CH_3SCH_3 + OH \xrightarrow{M} \underset{\underset{OH}{|}}{CH_3SCH_3} \qquad (48)$$

$$\underset{\underset{OH}{|}}{CH_3SCH_3} \longrightarrow CH_3SOH + CH_3 \qquad (49)$$

$$CH_3SOH + O_2 \xrightarrow{M} CH_3SO_3H \qquad (50)$$

At night-time, the main breakdown mechanism of dimethyl sulfide is reaction with NO_3 radicals, which can also lead to MSA formation. Dimethyl sulfide is a product of biomethylation of sulfur in seawater and coastal marshes and its production shows a strong seasonal pattern, with peak emissions in spring and early summer.

Oxidation of dimethyl sulfide can also lead to formation of sulfuric acid, both with and without the involvement of sulfur dioxide as an intermediate. Whether DMS oxidation gives MSA or sulfuric acid, the product is particulate and contributes to the number of cloud condensation nuclei. It has been postulated that biogenic dimethyl sulfide may act as a climate regulator through this mechanism.[14]

8.5 ATMOSPHERIC BASES

Carbonate rocks such as calcite or chalk, $CaCO_3$, or dolomite, $CaCO_3 \cdot MgCO_3$, exist in small concentrations in atmospheric particles and provide a small capacity for neutralizing atmospheric acidity. In western Europe, however, the major atmospheric base is ammonia. Although not emitted to any major extent by industry or motor vehicles, ammonia is arguably a man-made pollutant as it arises primarily from the decomposition of animal wastes; atmospheric concentrations relate closely to the density of farm animals in the locality. Release from chemical fertilizers can also be significant.[1]

In areas with a moderate or high ammonia source strength, ground-level atmospheric acidity is generally low. Sulfuric acid is present as highly neutralized $(NH_4)_2SO_4$, and HNO_3 and HCl predominantly as NH_4NO_3 and NH_4Cl. The latter two salts are appreciably volatile and may release their parent acids under conditions of low atmospheric ammonia, high temperature or reduced humidity.

$$NH_4NO_3 \longleftrightarrow HNO_3 + NH_3 \tag{51}$$
$$NH_4Cl \longleftrightarrow HCl + NH_3 \tag{52}$$

The relative neutrality of ground-level air at such locations may, however, not be reflective of far greater acidity at greater heights above the ground, as is shown from simultaneous sampling ground-level air and rainwater (see later). Additionally, once deposited in soils, ammonium salts are slowly oxidized to release strong acid in a process which may be represented as:

$$(NH_4)_2SO_4 + 4O_2 \longrightarrow H_2SO_4 + 2HNO_3 + 2H_2O \tag{53}$$

Thus the neutralization process has only a temporary influence and ultimately causes additional acidification.

Deposition of ammonia and ammonium may also contribute directly to damage to vegetation, which has proved to be a particular problem in the Netherlands. Since the reaction of ammonia with the OH radical is slow, the main sinks lie in wet and dry deposition processes.

8.6 ATMOSPHERIC AEROSOLS AND RAINWATER

8.6.1 Atmospheric Particles

A substantial proportion of the atmospheric aerosol over populated areas is termed secondary as it is formed in the atmosphere from chemical reactions affecting primary pollutants. In the UK, water-soluble materials account typically for about 60% of the aerosol mass[15] and are composed of nine or ten major ionic components:

Anions: SO_4^{2-}; NO_3^-; Cl^-; (CO_3^{2-})
Cations: Na^+; K^+; Mg^{2+}; Ca^{2+}; NH_4^+; H^-

If concentrations are expressed in gram equivalents per cubic metre of air, some interesting relationships appear:

Concentration $(g \, equiv \, m^{-3})$ = Concentration $(g \, m^{-3}) \times (Z/M)$

where Z = ionic charge and M = molecular weight.

This is an expression of charge equivalents and hence if all major components are accounted for:

$$\sum \text{anions} = \sum \text{cations}$$

In marine aerosol,

$$Na^+ + Mg^{2+} \approx Cl^- \tag{54}$$

when expressed in gram equivalents.

This is a relationship very similar to that pertaining in seawater, suggesting the latter as the major source of these components. The aerosol over the oceans also contains sulfate in excess of that present in the sea (termed non-sea salt sulfate). This is present as finer particles than the coarse sea salt sulfate, and arises from pollutant inputs of sulfur dioxide together with sulfate formed from the oxidation of dimethyl sulfide (Section 8.4.6). Marine aerosol can penetrate considerable distances inland over continental interiors.

Over the continents, particles derived from the earth's crust (soil and rock fragments) make up a major part of the aerosol.[16] Chemically these are composed largely of minerals such as clays (complex aluminosilicates), α-quartz and calcite. As well as these crustal particles, the UK atmosphere contains marine-derived particles, secondary aerosol and primary pollutant particles such as road vehicle emissions, comprising largely elemental and organic carbon (see also Section 7.2.2).

The inorganic secondary particles are composed mainly of ammonium nitrate, and sulfate as either ammonium sulfate or sulfuric acid. Hence, in gram equivalents:

$$SO_4^{2-} + NO_3^- \approx NH_4^+ + H^+ \tag{55}$$

In UK air, the ratio H^+/NH_4^+ is normally very low, although in other parts of the world it can be much higher. Close examination of a large data set[15] revealed that the NH_4^+, which is not accounted for by the above relationships, and Cl^-, not accounted for by association with Na^+ and Mg^{2+} in seawater, are approximately equal, indicating the presence of NH_4Cl.

K^+ and Ca^{2+} are mainly soil-derived, although some seawater contribution is likely, whilst CO_3^{2-}, not always observed, arises from carbonate minerals such as $CaCO_3$ in rocks and soils.

Recent years have seen a realization that aerosol particles can play a substantial active role in atmospheric chemistry, rather than being simply an inert product of atmospheric processes. They can act as a surface for hetero-geneous catalysis or, probably more importantly, can provide a liquid phase reaction medium. Soluble particles deliquesce at humidities well below satura-tion and many compounds, such as the abundant $(NH_4)_2SO_4$ probably exist predominantly in solution droplet form in humid climates such as that of the UK. As indicated above these droplets can form a medium for reaction of solutes. They can also provide a medium for reaction with gaseous components, for example, the dissolution of nitric acid vapour in sodium chloride aerosol, with release of hydrochloric acid vapour:

$$HNO_3 + NaCl \longrightarrow NaNO_3 + HCl \qquad (47)$$

One fascinating aspect of aerosol chemistry is the fact that at typical tropospheric temperatures both ammonium nitrate and ammonium chloride aerosols are close to dynamic equilibrium with their gaseous precursors:

$$NH_4NO_3 \text{ (aerosol)} \longleftrightarrow NH_3(g) + HNO_3(g) \qquad (51)$$
$$NH_4Cl \text{ (aerosol)} \longleftrightarrow NH_3(g) + HCl(g) \qquad (52)$$

The position of these equilibria may be predicted from chemical thermodynamics, both for crystalline particles, and at higher humidities for solution droplets.[17] Deviations from equilibrium probably arise from kinetic constraints upon the achievement of equilibrium. The implication for the UK, where ammonia levels are fairly high, is that the ammonia concentration at a given atmospheric temperature and relative humidity controls the nitric acid concentration *via* the ammonium nitrate dissociation equilibrium. It appears also that very similar considerations apply to the concentration of HCl vapour in the air which is subject to control by the ammonium chloride dissociation equilibrium.

Atmospheric aerosols are responsible for the major part of the visibility reduction associated with polluted air. In the UK and other heavily industralized regions, visibility reduction tends to correlate closely with the concentration of aerosol sulfate, a component tending to be present in sizes optimum for scattering of visible light. In very dry climates, visibility reduction is primarily associated with dust storms and is thus correlated with airborne soil.

8.6.2 Rainwater

Substances become incorporated in rainwater by a number of mechanisms. The major ones are as follows:

(a) particles act as cloud condensation nuclei, *i.e.* they act as centres for water condensation in clouds and fogs when the relative humidity reaches saturation;
(b) particles are scavenged by cloudwater droplets or falling raindrops as a result of their relative motion;
(c) gases may dissolve in water droplets either within or below the cloud.

As mentioned earlier, in-cloud scavenging is referred to as rainout, whilst below-cloud processes are termed washout. Incorporation in rain is a very efficient means of cleansing the atmosphere as evidenced by the substantial improvements in visibility occasioned by passage of a front.

The dissolved components of rainwater are the same nine or, at high rainwater pH, ten major ions listed earlier for atmospheric aerosol. There are also insoluble materials, again similar to the insoluble components of the atmospheric aerosol. When the composition of rainwater collected in the Lancaster, UK, area is compared with that of aerosols collected over the same periods close to ground-level, the average composition as percentages of total anion or cation load is very similar for many components, but is markedly different for some ions in aerosol and rainwater.[18] The most obvious difference is in the H^+/NH_4^+ ratio: rainwater has a far higher ratio and is thus much more acidic. This arises for two main reasons:

(a) The neutralizing agent, ammonia, has a ground-level source and is thus more abundant at ground-level where the aerosol is sampled than at cloud level where the major pollutant load is incorporated into the rain.

(b) In-cloud oxidation processes of SO_2 (see above) may lead to appreciable acidification of the cloudwater.

The regional distribution of rainwater pH over the UK is shown by Figure 8.7. The trend of increasing acidity from west to east arises from the prevailing

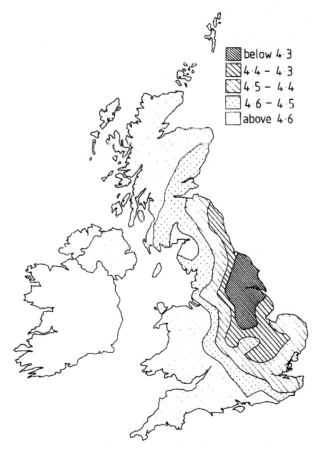

below 4·3
4·4 – 4·3
4·5 – 4·4
4·6 – 4·5
above 4·6

Figure 8.7 *Precipitation-weighted mean hydrogen ion concentration expressed as pH for 1986*[19]

south-westerly circulation bringing relatively clean air into the west of the country. Primary pollutants emitted over the country and the secondary pollutants formed from them lead to acidification which increases progressively as the air moves eastwards. The quantitative deposition of acidity in rain (as g H^+ m^{-2} per year) shows a rather different pattern which arises from the fact that the annual deposition field is derived from the product of the volume-weighted mean concentration and the annual rainfall. Thus areas of high rainfall (*e.g.* Scottish Highlands) exhibit high hydrogen ion inputs despite a relatively high pH. One feature of acidic inputs is the substantial influence of episodicity: a very small number of rainfall events in any year combining high acidity with a large rainfall amount can contribute a large proportion of the total annual hydrogen ion input.

8.7 ATMOSPHERIC COMPOSITION AND CLIMATE CHANGE

Some atmospheric gases, even at very low concentrations, can have important influences on the radiation budget of the atmosphere by absorbing infrared radiation emitted from the surface of the earth. As a consequence, more heat is retained in the lower atmosphere and thus there is a tendency towards gradual heating of the atmosphere, termed global warming. The gases with this property are termed *greenhouse gases* and the properties of the major contributors are listed in Table 8.8.

The characteristic which the gases have in common is that they are of rather long atmospheric lifetime, and all are currently increasing in concentration. However, as a result of the Montreal Protocol the emissions of the chloro-fluorocarbons have peaked (see Chapter 9). The relative impacts of the gases per unit mass are expressed in terms of Global Warming Potentials. These are estimates of the radiative forcing per kilogram of substance relative to the effect of 1 kilogram of carbon dioxide. Thus, gases, such as CFC-12, although present in low concentration, have a relatively large Global Warming Potential (GWP) and thus can impact considerably upon climate. Different time frames of effect applying to GWPs are 20, 100 and 500 years to account for the differences in atmospheric longevity of the various substances and the fact that their relative impacts vary according to time into the future for which the prediction is made. 'Indirect effects' upon climate occur due to, for example, depletion of strato-spheric ozone due to CFC emissions and enhancement of tropospheric ozone. Such effects, and those due to atmospheric aerosol, are particularly hard to quantify.

Figure 8.8 shows the contribution, globally averaged, of the various green-house gases from pre-industrial times to the present day to radiative forcing. The main effect is the direct infrared absorbing property of the greenhouse gases. The estimated impact through stratospheric ozone is opposite in sign, but relatively small, and the indirect greenhouse effect of tropospheric ozone is smaller in magnitude than the direct effect of the greenhouse gases. There is a negative influence on radiative forcing due to aerosol particles in the lower atmosphere. These exert a direct effect by back-scattering solar radiation to

Table 8.8 *A summary of key greenhouse gases affected by human activities*[20]

	CO_2	CH_4	N_2O	CFC-12	HCFC-22 (a CFC substitute)	CF_4 (a perfluorocarbon)
Pre-industrial concentration	280 ppmv	700 ppbv	275 ppbv	Zero	Zero	Zero
Concentration in 1992	355 ppmv	1714 ppbv	311 pptv	503 pptv	105 pptv	70 pptv
Recent rate of concentration change per year (over 1980s)	1.5 ppmv/yr 0.4%/yr	13 ppbv/yr 0.8%/yr	0.75 pptv/yr 0.25%/yr	18–20 pptv/yr 4%/yr	7–8 pptv/yr 7%/yr	1.1–1.3 pptv/yr 2%/yr
Atmospheric lifetime (years)	(50–200)[a]	(12–17)[b]	120	102	13.3	50 000

[a] No single lifetime for CO_2 can be defined because of the different rates of uptake by different sink processes.
[b] This has been defined as an adjustment time which takes into account the indirect effect of methane on its own lifetime
1 pptv = 1 part per trillion (million million) by volume

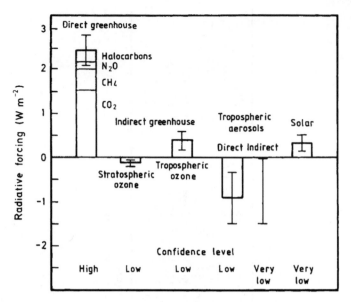

Figure 8.8   *Estimates of the globally averaged radiative forcing due to changes in green-house gases and aerosols from pre-industrial times to the present day and changes in solar variability from 1850 to the present day*[20]

space, and hence act to cool the lower atmosphere. There is also a very poorly quantified indirect effect which arises from the ability of water-soluble particles to act as cloud condensation nuclei, and in doing so to influence the albedo or reflectivity of clouds which can lead to reflection of solar radiation back to space. Currently, this effect is very poorly quantified indeed, and influences on cloud cover are one of the least well understood facets of climate change prediction. The final column in Figure 8.8 relates to changes in solar intensity between 1850 and the present day.

It is very difficult to discern from temperature records whether climate is actually changing. It is clear, however, that the last few decades have been notably warmer than the preceding 100 years and this is considered as evidence for global warming, although in the absence of a complete understanding of long, medium and short-term cycles in global climate, the interpretation cannot be certain. However, the evidence for global warming driven by human activity has strengthened considerably over the past decade.

8.8   REFERENCES

1. J. H. Seinfeld and S. Pandis, 'Atmospheric Chemistry and Physics', Wiley, New York, 1998.
2. P. Brimblecombe, 'Air Composition and Chemistry', 2nd edn., Cambridge University Press, Cambridge, 1996.
3. C. N. Hewitt and R. M. Harrison, *Atmos. Environ.*, 1985, **19**, 545.
4. J. Fishman and P. J. Crutzen, *J. Geophys. Res.*, 1977, **82**, 5897.
5. J. Fishman and P. J. Crutzen, *Nature*, 1978, **274**, 855.

6. D. Anfossi, S. Sandroni and S. Viarengo, *J. Geophys. Res.*, 1991, **96D**, 17 349.
7. I. Colbeck and R. M. Harrison, *Atmos. Environ.*, 1985, **19**, 1577.
8. I. Colbeck and R. M. Harrison, *Atmos. Environ.*, 1985, **19**, 1807.
9. B. J. Finlayson-Pitts and J. N. Pitts, Jr., 'Atmospheric Chemistry', Wiley, New York, 1986.
10. P. Middleton, C. S. Kiang and V. A. Mohnen, *Atmos. Environ.*, 1980, **14**, 463.
11. S. G. Chiang, R. Toosi and T. Novakov, *Atmos. Environ.*, 1981, **15**, 1287.
12. R. M. Harrison and C. A. Pio, *Atmos. Environ.*, 1983, **17**, 1261.
13. The UK National Air Quality Information Archive, URL: http://www.aeat./co.uk/netcen/airqual/
14. R. J. Charlson, J. E. Lovelock, M. O. Andreae and S. G. Warren, *Nature*, 1987, **326**, 655.
15. R. M. Harrison and C. A. Pio, *Environ. Sci Technol.*, 1983, **17**, 169.
16. R. M. Harrison and R. E. van Grieken, 'Atmospheric Particles', Wiley, Chichester, 1998.
17. A. G. Allen, R. M. Harrison and J. W. Erisman, *Atmos. Environ.*, 1989, **23**, 1591.
18. R. M. Harrison and C. A. Pio, *Atmos. Environ.*, 1983, **17**, 2539.
19. Warren Spring Laboratory, 'United Kingdom Acid Rain Monitoring', Stevenage, 1988.
20. Inter-Governmental Panel on Climate Change, 'Radiative Forcing of Climate Change – Summary for Policy Makers', WMO, UNEP, 1995.

CHAPTER 9

Chemistry and Pollution of the Stratosphere

A. R. MACKENZIE

9.1 INTRODUCTION

In the early 1970s the possibility of polluting the stratosphere was first raised. Concern centred on compounds capable of depleting the ozone layer and so allowing biologically harmful ultraviolet (UV) radiation to reach the ground. Because the chemical bonds in biological molecules – in DNA, for example – are broken down by UV radiation, increased UV radiation reaching the ground has potential adverse effects on human health and on aquatic and terrestrial ecosystems. Among the chemical agents considered harmful to ozone were nitrogen oxides, produced by the detonation of nuclear weapons or from the exhaust of supersonic aircraft, and chlorofluorocarbons (CFCs), especially $CFCl_3$, and CF_2Cl_2, which were used mainly as refrigerants and aerosol propellants. Public and scientific concern over ozone escalated abruptly in 1985 when Farman et al.[1] reported that the total ozone column over Antarctica in September and October had decreased by up to 40% over the past decade. No one had predicted such a dramatic change. The discovery of this 'ozone hole' over Antarctica precipitated a great deal of research on the chemistry and dynamics of the stratosphere in a variety of internationally supported scientific experiments involving satellites, aircraft, balloons and ground stations. The research has shown, amongst other important results, clear evidence that ozone destruction has occurred within the lower stratosphere over the Arctic too. Chemical compounds resulting from the breakdown of chlorine and bromine containing compounds are unambiguously implicated. For reasons outlined below, Arctic ozone depletion is much more variable than Antarctic ozone depletion. Some northern winters have resulted in ozone depletion as severe – in terms of the amount of ozone destroyed relative to the unperturbed background concentrations – as Antarctic winters of the mid-1980s, whilst other northern winters have resulted in almost no depletion. Variability notwithstanding, the fragility of the ozone layer has been amply demonstrated.

Ozone, particularly its absorption of solar ultraviolet radiation, has a

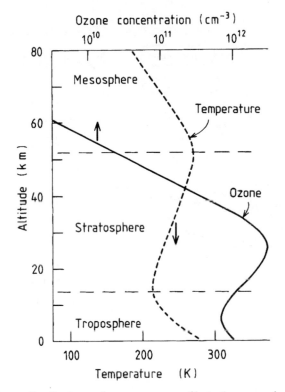

Figure 9.1 *Ozone distribution and temperature profile in the atmosphere*

profound impact on the structure of the atmosphere. Figure 9.1 shows a typical mid-latitude temperature profile. Most* ozone occurs in the stratosphere, and it is the presence of ozone that produces the characteristic temperature inversion and dynamic stability (*i.e.* hotter, lighter air above colder, denser air) of the stratosphere. The temperature structure, and implied stability, of the stratosphere distinguishes it from the troposphere below, where heating at the surface causes occasionally rapid, and often deep, vertical mixing. The stratosphere extends from a lower limit of between 8 km and 16 km, depending on latitude, to an upper limit of between 45 km and 60 km, again depending on latitude.

A meridional (*i.e.* north–south) cross-section of the circulation in the wintertime stratosphere is shown schematically in Figure 9.2. There is a slow overturning, caused by the interaction of internal, planetary-scale, waves on the mean stratospheric flow.[2] The meridional circulation is superimposed on rapid

---

* Three measures of local abundance are used in this chapter. Absolute abundances are reported as *number densities*, in molecules cm^{-3}, or as *partial pressures*, P_x, in nbar. These two concentration scales are related by the ideal gas law. Relative abundances are reported as *volume mixing ratios*, P_x/P_{air}, expressed as parts per million (ppmv, vmr = 10^{-6}) or parts per billion (ppbv, vmr = 10^{-9}). These are identical to the units ppm and ppb used in Chapters 7 and 8. Because pressure decreases exponentially with height, the absolute concentration of chemicals in air parcels moving upwards or downwards will change, but the relative abundance will be conserved in the absence of chemical reactions.

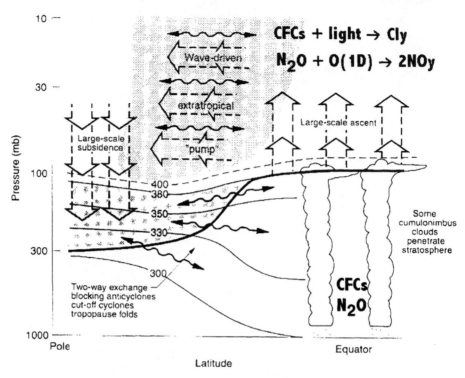

Figure 9.2 *Schematic of the mean diabatic circulation in the wintertime hemisphere (broad arrows). Source gases enter the stratosphere in the tropics and are transported slowly to high altitudes where they are broken down by short wavelength light or the electronically excited $O(^1D)$ atom. Descent into the polar regions brings down high mixing ratios of NO_y, Cl_y, and H_2O. Clouds in polar regions activate chlorine and so bring about ozone depletion. A strong jet of westerly winds is present at the edge of the polar vortex, preventing rapid horizontal mixing. A similar barrier to mixing exists in the subtropics. The tropopause is shown by the thick line. Thin lines are isentropic surfaces. The wiggly double-headed arrows denote transport across the tropopause by mid-latitude weather systems and associated features*

zonal (*i.e.*, east–west) flow: winds are westerly in the wintertime extra-tropics, easterly in the summertime extra-tropics, and either easterly or westerly in the tropics. The slow overturning brings air upwards, from the troposphere, in the tropics and downwards, from the mesosphere and upper stratosphere, at the pole. At a given altitude, therefore, air is "older", in the sense that it has been in the upper atmosphere for longer, near the winter pole and air is 'younger' near the equator. Since CFCs are destroyed in the upper stratosphere and meso-sphere, air near the winter pole is richer in the breakdown products of CFC photodissociation than is air near the equator. This gradient in the "age" of stratospheric air is particularly steep in the subtropics and at the edge of the polar vortex, where barriers to horizontal mixing exist. The barrier at the polar vortex edge is an important reason for the exacerbated ozone loss in the polar lower stratosphere. Between the mixing barriers, in the mid-latitude *surf zone*, quasi-horizontal stirring and mixing, on isentropic surfaces,* is rapid.

9.2 STRATOSPHERIC OZONE CHEMISTRY

Over 100 years ago, Hartley[3] pointed out that ozone is a normal constituent of the higher atmosphere and is in a larger proportion there than near the earth's surface. Fabry and Dobson pioneered the monitoring of atmospheric ozone, using spectroscopic techniques. We now know that ozone occurs in trace amounts throughout the atmosphere with a peak concentration in the lower stratosphere between about 20 and 25 km altitude (Figure 9.1). About 90% of all the ozone in the atmosphere resides in the stratosphere. Observations of the total amount of ozone in a vertical column have established the average column abundance and the general pattern of latitudinal, longitudinal, seasonal and meteorological variations.† Total or column ozone, per unit area, is usually stated in *Dobson Units* or matm cm ('milli-atmo centimetres'), related to the thickness of an equivalent layer of pure ozone at standard pressure and temperature. If the ozone in a column of the atmosphere were concentrated into a thin shell surrounding the earth at standard pressure (*i.e.*, 101.3 hPa) it would be about 3 mm thick, *i.e.* the average total amount of ozone is 300 DU. Seasonal and latitudinal variations at sub-polar latitudes are about $\pm 20\%$ of this value. The annual average total ozone is a minimum of approximately 260 DU at equatorial latitudes, and increases poleward in both hemispheres, to a maximum at sub-polar latitudes of about 400 DU. The high latitude maximum results from transport of ozone from the equatorial middle and upper stratosphere, the region of primary production, to the polar lower stratosphere (Figure 9.2).

9.2.1 Gas Phase Chemistry

How is this ozone produced? Chapman[4] proposed a dynamically static, pure oxygen, photochemical steady state model that agreed well with the observations available at the time. The reactions were

$$O_2 + h\nu(\lambda < 243 \text{ nm}) \longrightarrow O + O \qquad (1)$$
$$O + O_2 + M \longrightarrow O_3 + M \qquad (2)$$
$$O_3 + h\nu(\lambda < 1180 \text{ nm}) \longrightarrow O + O_2 \qquad (3)$$
$$O + O_3 \longrightarrow O_2 + O_2 \qquad (4)$$

---

* The stable stratification of the stratosphere is reflected in the fact that air tends to move on *isentropic* surfaces, *i.e.* surfaces of constant potential temperature. Potential temperature, θ, is defined as $\ln \theta = 0.286 \ln[T(1000/p)]$, where T is the temperature in Kelvin, and p the atmospheric pressure (in mbar), at the altitude of interest. θ is a measure of the entropy of air and so is conserved in adiabatic motion.
† It may appear strange, at first glance, to rely on an integrated measure of a chemical compound, but bear in mind that most of the ozone occurs in a layer, and that the UV-shielding effect of ozone relies on the total number of ozone molecules in the path sunlight takes to the earth's surface. For both these reasons, integrated measures of ozone are meaningful and extremely useful.

(a fifth reaction, the self-reaction of oxygen atoms, is too slow to be important in the stratosphere). Reaction (2) becomes slower with increasing altitude, since M, *i.e.* pressure, is decreasing, while Reactions (1) and (3) become faster, since the intensity of radiation, particularly short wavelength radiation, is increasing. Hence, O predominates at high altitudes and O_3 is favoured at lower altitudes. These two allotropes of oxygen – O and O_3 – are converted one into the other very rapidly, so that it is convenient to consider them together as *odd oxygen*. The rate of the loss of odd oxygen (Reaction (4)) was unknown at the time of Chapman's proposal. Laboratory measurements have since revealed that Reaction (4) is too slow to destroy ozone at the rate it is produced globally. Therefore, for ozone to attain a steady state close to the observations, there must exist chemical processes which provide more efficient routes for the loss of odd oxygen. The reaction sequence

$$X + O_3 \longrightarrow XO + O_2 \tag{5}$$
$$XO + O \longrightarrow X + O_2 \tag{6}$$
$$net\ O + O_3 \longrightarrow 2O_2$$

achieves the same result as reaction (4). The species X is a catalyst – it is not consumed in the process – and can effectively destroy many ozone molecules before being removed by some other chemical process. Several chemical species have been identified with X, including H, OH, NO, Cl and Br. The catalytic cycles are very fast and are able to compete with the direct reaction even when concentrations of X and XO are two or three orders of magnitude less than concentrations of the oxygen species. Because the interconversion of the various X/XO couples can be very rapid, it is often convenient to group the species together in *chemical families*. The various short-lived nitrogen compounds are grouped together as NO_x ($= N + NO + NO_2 + NO_3$, where the compounds are counted by number, *i.e.* moles, not mass), the various short-lived chlorine species are grouped together as ClO_x ($= Cl + ClO + 2Cl_2O_2$), short-lived hydrogen species are grouped as HO_x ($= H + HO + HO_2$), and so on. By extension, the total amount of potentially reactive nitrogen is known collectively as NO_y ($= NO_x + HONO + 2N_2O_5 + HNO_3 + HNO_4 + ClONO_2$), and the total amount of potentially reactive chlorine, the *inorganic chlorine*, is known collectively as Cl_y ($= ClO_x + HOCl + HCl + ClONO_2 + 2Cl_2 + BrCl$).

In addition to Reactions (5) and (6), ozone may also be destroyed in cycles not involving atomic oxygen. Such ozone-specific cycles are important in the lower stratosphere because concentrations of atomic oxygen are low there. Variations on the general form are (i) formation of an XO compound which itself reacts with ozone, for example:

$$OH + O_3 \longrightarrow HO_2 + O_2 \tag{7}$$
$$HO_2 + O_3 \longrightarrow OH + 2O_2 \tag{8}$$
$$net\ 2O_3 \longrightarrow 3O_2$$

and

$$NO + O_3 \longrightarrow NO_2 + O_2 \qquad (9)$$
$$NO_2 + O_3 \longrightarrow NO_3 + O_2 \qquad (10)$$
$$NO_3 + h\nu \longrightarrow NO + O_2 \qquad (11)$$
$$net\ 2O_3 + h\nu \longrightarrow 3O_2$$

or (ii) formation of a compound from two XO species, leading to the recombination of oxygen, for example, elimination of O_2 in a thermal reaction:

$$Br + O_3 \longrightarrow BrO + O_2 \qquad (12)$$
$$OH + O_3 \longrightarrow HO_2 + O_2 \qquad (7)$$
$$HO_2 + BrO \longrightarrow HOBr + O_2 \qquad (13)$$
$$HOBr + h\nu \longrightarrow OH + Br \qquad (14)$$
$$net\ 2O_3 + h\nu \longrightarrow 3O_2$$

or elimination of O_2 by photolysis of a different bond to the bond formed in the XO/YO combination:

$$Cl + O_3 \longrightarrow ClO + O_2 \qquad (15)$$
$$NO + O_3 \longrightarrow NO_2 + O_2 \qquad (9)$$
$$ClO + NO_2 + M \longrightarrow ClONO_2 + M \qquad (16)$$
$$ClONO_2 + h\nu \longrightarrow Cl + NO_3 \qquad (17)$$
$$NO_3 + h\nu \longrightarrow NO + O_2 \qquad (11)$$
$$net\ 2O_3 + 2h\nu \longrightarrow 3O_2$$

A very important example of this last kind of ozone-specific cycle is the *ClO dimer cycle* which is responsible for the majority of the lower stratospheric ozone loss inside the polar vortices (see below):

$$2(Cl + O_3 \longrightarrow ClO + O_2) \qquad (15)$$
$$ClO + ClO + M \longrightarrow Cl_2O_2 + M \qquad (18)$$
$$Cl_2O_2 + h\nu \longrightarrow Cl + ClOO \qquad (19)$$
$$ClOO + M \longrightarrow Cl + O_2 + M \qquad (20)$$
$$net\ 2O_3 + h\nu \longrightarrow 3O_2$$

Species X and XO may be involved in *null cycles* that do not remove odd oxygen. For X = NO we have

$$NO + O_3 \longrightarrow NO_2 + O_2 \qquad (9)$$
$$NO_2 + h\nu \longrightarrow NO + O \qquad (21)$$
$$net\ O_3 + h\nu \longrightarrow O_2 + O$$

This cycle is in competition with the catalytic cycle (Reactions (9), (10) and (11)) and the NO_x tied up in this cycle is ineffective as a catalyst. Together with Reaction (2), Reactions (9) and (21) make up the cycle that defines the

tropospheric photostationary state for ozone (see Chapter 8). The corresponding photostationary state for stratospheric ozone is more complicated, because
of the X/XO cycles. All the ozone destroying cycles above are in competition
with null cycles and termination reactions that occur simultaneously.

The termination of the catalytic and null cycles occurs by chemical conversion
of radicals into more stable oxidation products. For example

$$OH + NO_2 + M \longrightarrow HNO_3 + M \qquad (22)$$

The HNO_3 may be transported down into the troposphere and removed in rain
or may be photolysed to regenerate OH and NO_2. This latter process is
relatively slow, and HNO_3 is said to act as a *reservoir* of NO_x. Typically, in the
lower stratosphere, more than 90% of the stratospheric load of NO_y is stored in
this HNO_3 reservoir.[5] For ClO_x, about 40–80% of the stratospheric load is
sequestered into HCl[6] *via*

$$Cl + CH_4 \longrightarrow CH_3 + HCl \qquad (23)$$

The chlorine tied up as stable HCl may be released by

$$OH + HCl \longrightarrow H_2O + Cl \qquad (24)$$

or by heterogeneous reactions (see below). This ClO_x can then again participate
in catalytic cycles leading to the removal of ozone. Measurement and modelling
studies have emphasised the importance of more temporary reservoir species
such as $ClONO_2$, HOCl, N_2O_5 and HO_2NO_2, which act to lessen the efficiency
of ClO_x and NO_x species in destroying ozone in the lower stratosphere. These
reservoir species may be formed by

$$NO_2 + O_3 \longrightarrow NO_3 + O_2 \qquad (10)$$
$$NO_3 + NO_2 + M \longrightarrow N_2O_5 + M \qquad (25)$$
$$ClO + HO_2 \longrightarrow HOCl + O_2 \qquad (26)$$
$$HO_2 + NO_2 + M \longrightarrow HO_2NO_2 + M \qquad (27)$$
$$ClO + NO_2 + M \longrightarrow ClONO_2 + M \qquad (16)$$

These reactions emphasize the coupling between the HO_x, NO_x and ClO_x
families, and mean that the effects of the families are not additive. Members of
one family can react with members of another to produce null cycles, chiefly by
the production of NO_2, *e.g.*

$$NO + ClO \longrightarrow NO_2 + Cl \qquad (28)$$
$$NO + HO_2 \longrightarrow NO_2 + OH \qquad (29)$$

followed by Reaction (21). The recycling of HO_x, NO_x and ClO_x from the
reservoirs HNO_3, N_2O_5, HOCl, and $ClONO_2$ is by photolysis or heterogeneous

reaction. Removal of odd hydrogen in the lower stratosphere occurs mainly by:

$$OH + HNO_3 \longrightarrow H_2O + NO_3 \tag{30}$$
$$OH + HO_2NO_2 \longrightarrow H_2O + NO_2 + O_2 \tag{31}$$

These reactions also release NO_x from the reservoirs. The water formed may be physically removed from the stratosphere, or converted back into HO_x by reaction with electronically excited O atoms.

Termination reactions become less efficient with increasing molecular weight in the halogen series, *i.e.* from Cl to I. The reaction of Cl with CH_4 (Reaction (23)), to produce HCl, limits the abundance of active chlorine in the stratosphere, but the analogous reaction of Br with CH_4 is endothermic (*i.e.* not energetically favourable) and can be neglected. The temporary reservoirs for bromine – HOBr and $BrONO_2$ – are very photolabile. As a result, BrO is the major form of Br_y in the stratosphere. The result is that, on a molecule-for-molecule basis, bromine is about 50 times more efficient than chlorine at destroying O_3. Measurements of BrO show mixing ratios around 5×10^{-3} ppbv at 18 km. These are sufficient to account for about 25% of the ozone depletion in the Antarctic ozone hole,[7] and to make bromine cycles significant contributors to mid-latitude ozone loss. Reactions involving iodine radicals have also been suggested, but current upper limits to the concentration of IO in the stratosphere, and measurements of the relevant chemical reaction rates, suggest that iodine plays a very minor role in stratospheric chemistry.[8]

Figure 9.3 shows a model calculation of the ozone loss rate due to the various chemical families (HO_x, NO_x, ClO_x, BrO_x) as a function of altitude, for vernal equinox at 40° N.[9] These results show a smaller direct influence of NO_x-related cycles in the lower stratosphere than had been calculated before the inclusion of heterogeneous processes (see below), although the problem is raised that lower stratospheric NO_x concentrations in models are now lower than observed.

Hydrocarbon oxidation is closely related to all other reactive trace gas species and hence ozone photochemistry. Methane is the dominant stratospheric hydrocarbon and is a chemical source of water in the stratosphere, *via*

$$OH + CH_4 \longrightarrow CH_3 + H_2O \tag{32}$$

which is analogous to the Cl loss process (Reaction (23)) above. The CH_3 radical is further oxidized to CO_2 and another molecule of H_2O through intermediate products CH_3O_2, CH_3O, HCHO and CO. Below 35 km there is sufficient NO for CH_4 oxidation to be a net source of O_x ($O + O_3$) by the 'photochemical smog' reactions (see Chapter 8).

9.2.2 Heterogeneous Chemistry

Aerosol particles provide sites for reactions that would otherwise not occur in the stratosphere. There are many types of aerosol in the stratosphere, ranging

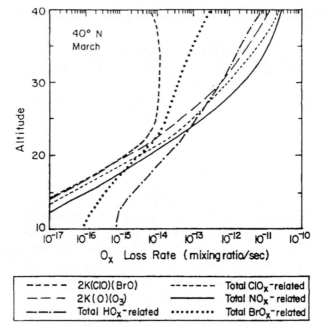

Figure 9.3   *Calculated 24-hour average O_x loss rates from various chemical cycles for 40°N in March for non-volcanic aerosol loading. In these circumstances the dominant ozone loss process below 22 km is due to reactions involving HO_x, with NO_x dominating between 23 km and 40 km. Under higher aerosol loading, coupled HO_x-halogen cycles become more important*
(From the two-dimensional model simulations of Garcia and Solomon[9])

from meteoritic mineral debris to water droplets.[10] Of primary importance in our current understanding of the chemistry of the stratosphere are the sulfuric acid aerosol and polar stratospheric clouds (PSCs).

Sulfuric acid aerosol occurs globally, in a wide altitude band between about 15 and 30 km. The aerosol is formed primarily by oxidation of SO_2 injected by volcanic eruptions, and its formation is therefore highly intermittent. An additional, very small but constant, source may be the percolation of COS, emitted by microbes, up to the stratosphere, with subsequent oxidation to sulfuric acid. The eruption of Mt. Pinatubo in June 1991 increased the mass of aerosol in the stratosphere by more than an order of magnitude (from a total loading of around 2 Tg (*i.e.*, 2 Mtonnes) to around 30 Tg). The timescale for removal of the aerosol is of the order of one year,[11] so that these volcanic injections can have a significant global effect (see below). Above temperatures of about 215 K, at 50 mbar, the aerosol exists as concentrated sulfuric acid droplets. Below this temperature solid sulfuric acid hydrates are the thermodynamically favoured phase, but there is a considerable kinetic barrier to freezing which enables liquid particles to remain until the temperature approaches the frost point,[12] taking up water and nitric acid as they cool. The frost point for stratospheric air at 50 mbar is about 188 K ($-85\,°C$).

Polar stratospheric clouds fall into two categories: type 1 PSCs are composed of nitric acid and water in volume ratios between 1:2 and 1:5, and form at temperatures about 2–5 K above the local frost point; type 2 PSCs are composed of water ice crystals and form, therefore, below the local frost point. PSCs evolve from the background sulfuric acid aerosol (note that, at PSC temperatures, the sulfuric acid seed or 'core' makes a very small contribution to the overall composition of the PSC particle). It is important to understand the details of particle evolution in the lower stratosphere because the different particles have different reactivities. The thermodynamically favoured solid phase for sulfuric acid aerosol is sulfuric acid tetrahydrate (SAT), and that for nitric acid aerosol is nitric acid trihydrate (NAT).

Reactions involving particles are more complicated to describe than the simple molecular collisions occurring in the gas phase. Diffusion is important, and reactants partition between the gas and condensed phases according to their solubilities. In liquid particles, if reaction is slow, the reactants will have time to continuously adjust to their equilibrium partitioning, and the rate of reaction has the form (*e.g.* Cox *et al.*[13])

$$-d[R_1]/dt = k[R_1][R_2]$$

where k is the rate of reaction in solution ($M^{-1} s^{-1}$) for reactants R_1 and R_2. By Henry's law and the ideal gas law

$$[R_n] = 10H^*.k_B.T.n_R$$

where R_n is either R_1 or R_2, H^* is the effective Henry's law coefficient ($M\,atm^{-1}$) defining the reactant's solubility, k_B is Boltzmann's constant ($J\,K^{-1}\,molec^{-1}$), T is the temperature (K), and n is the gas phase number density of the reactant (molec cm^{-3}). So the reaction rate can be given in terms of gas phase concentrations, which is important for coupling the heterogeneous chemistry to the gas-phase chemistry that is occurring simultaneously. If the rate coefficient for reaction is faster than about $10^5\,M^{-1}\,s^{-1}$, then the rate of reaction becomes limited by transport of reactants into the particle and the effective volume available for reaction is reduced.[14] In the limit, the rate is dependent on the aerosol surface area rather than the volume. For solid particles, reaction rates depend linearly on the particle surface area (but then, of course, surface area is not as easily measured or calculated as for spherical liquid drops).

Table 9.1 lists some of the most important reactions occurring on sulfuric acid aerosol and PSC particles. The overall effect of heterogeneous reactions is to convert the chlorine and hydrogen in reservoir compounds into more active forms, whilst converting the NO_x in temporary reservoir compounds into nitric acid. Many of the heterogeneous reactions in the sulfuric acid aerosol are strongly temperature dependent, due to the variation of aerosol composition, and thus Henry's law coefficients, with temperature. For aerosol loadings typical of a volcanically quiescent period, heterogeneous reactions begin to

Table 9.1 *Some heterogeneous reactions of importance in the stratosphere*

Reaction	Effect
$N_2O_5 + H_2O \rightarrow 2HNO_3$	Decreases NO_x concentration throughout lower stratosphere
$ClONO_2 + H_2O \rightarrow HOCl + HNO_3$	Chlorine activation at low temperatures, but HOCl less photolabile than Cl_2. Denoxification
$ClONO_2 + HCl \rightarrow Cl_2 + HNO_3$	Chlorine activation at low temperatures. Denoxification
$HOCl + HCl \rightarrow Cl_2 + H_2O$	Chlorine activation at low temperatures. Denoxification
$BrONO_2 + H_2O \rightarrow HOBr + HNO_3$	Indirectly affects ClO_x and HO_x concentrations
$HOBr + HCl \rightarrow BrCl + H_2O$	Indirectly affects ClO_x and HO_x concentrations

activate chlorine at temperatures below about 200 K. This activation is relatively slow, however. When aerosol loadings are much increased following a volcanic eruption, the temperature at which chlorine activation begins is increased by up to 5 K and aerosol reactions can effectively compete with reactions on type 1 PSCs. One heterogeneous reaction that is not temperature dependent is the hydrolysis of N_2O_5, and it is this reaction which has led to a revision of the importance of NO_x-related catalytic cycles in the lower stratosphere (Figure 9.3).

In the polar regions, when temperatures fall below about 195 K, PSCs can form. These have a much larger total volume than the background aerosol, and so can convert chlorine reservoirs much more rapidly. At PSC temperatures, Cl_2 partitions into the gas phase, whereas HNO_3 remains in the particles. This is known as *chlorine activation*. If the temperatures remain cold, the condensed phase can sediment out, leading to irreversible *denitrification* and *dehydration* of the air. Denitrification prevents the reformation of $ClONO_2$, and so increases the effective chain length of chlorine catalytic cycles. In sunlight, Cl_2 is photodissociated to release atoms that may then be converted into ClO. The small amount of NO_2, produced by photolysis of HNO_3, can convert the ClO back into $ClONO_2$. Hence we have

$$Cl_2 + h\nu \longrightarrow Cl + Cl \tag{33}$$
$$Cl + O_3 \longrightarrow ClO + O_2 \tag{15}$$
$$ClO + NO_2 + M \longrightarrow ClONO_2 + M \tag{16}$$

This chlorine nitrate is then available to react with any remaining HCl, to release more ClO_x until all the NO_x is removed. Hence within the polar vortices we would expect to find high concentrations of ClO (see below). Over the wintertime and springtime pole, methane concentrations are low as a result of downward transport from the methane sink region (Figure 9.2) and NO_x concentrations can be low as a result of denitrification. Hence the usual chain-breaking steps of Cl reaction with CH_4, or ClO reaction with NO_2 are particularly slow. In the Antarctic, the reduced rates of termination reactions, together with continued cold temperatures and PSCs, mean that all the ozone destroying catalytic cycles are able to continue throughout September. In the Arctic, however, sporadic cold temperatures and denitrification mean that there is more continuous competition between radical production and radical termination.

9.3 NATURAL SOURCES OF TRACE GASES

Nitrous oxide (N_2O) is the dominant precursor of stratospheric NO_x. It is emitted predominantly by biological sources in soils and water. Tropical forest soils are probably the single most important source of N_2O to the atmosphere. Tropical land-use changes and intensification of tropical agriculture indicate significant, growing sources of N_2O. The oceans, especially the upwelling regions of the Indian and Pacific Oceans, are also a significant source of N_2O. Natural sources of Cl_y, which could conceivably have a marked influence on stratospheric ozone, are few. Only methyl chloride* (CH_3Cl) and chloroform ($CHCl_3$) are thought to have substantial natural sources. Methyl chloride, with an atmospheric abundance of approximately 550 pptv, is the dominant halogenated species in the atmosphere. A production rate of around 3.5 Tg yr^{-1} is required to maintain this steady-state mixing ratio, most of which comes from biomass burning (not, of course, entirely a natural source: as well as lightning-induced natural fires, biomass burning is often associated with land clearance), with a smaller fraction from the oceans.[8] Another, as yet unknown, source is also supposed to exist, if the current estimates for the biomass burning and ocean sources are not grossly in error. Methyl chloride emission from wood-rotting fungi has been proposed, but the size of this source may not be large enough to completely bridge the current gap in the source inventory. Chloroform has sources from the ocean and from termite activity, both of which are as yet poorly quantified. Nevertheless, some substantial source, in addition to the known industrial sources, is required to balance the budget. Natural sources are the likely candidates.

There are many naturally-occurring bromine compounds in the atmosphere, most of which have appreciable tropospheric reactivities. This reactivity makes it difficult to measure or calculate the input of each organo-bromine compound, or the inorganic bromine compounds derived from each organo-bromine compound, into the stratosphere. The most abundant organo-bromine, methyl bromide (CH_3Br), has a very uncertain source inventory.[8] Soils and oceans are considered to act as net sinks for methyl bromide, but the best estimate of the emissions inventory requires an additional source to the known anthropogenic sources. Other naturally-occurring organo-bromine compounds are dibromomethane (CH_2Br_2), bromoform ($CHBr_3$), the bromomethane compounds (CH_2BrCl, $CHBr_2Cl$, $CHBrCl_2$), and ethylene dibromide ($C_2H_2Br_2$). These

---

* The nomenclature used to discuss organic compounds in the atmosphere is, perhaps unfortunately, a mixture of common and standard (IUPAC) names. Generally, when a compound has a large industrial source the common name is used.

reactive organo-bromine compounds are produced by marine algae. They are broken down rather quickly in the troposphere, so that measurements at the tropical tropopause show much lower concentrations than at the surface. Nevertheless, reactive organo-bromine compounds can contribute about 10% of the total bromine entering the stratosphere. This contribution can be either by the transport of the residual amount of organo-bromine remaining after tropospheric degradation, or by transport of the inorganic bromine formed by that degradation.

9.4 ANTHROPOGENIC SOURCES OF TRACE GASES

CFCs, unlike most other gases, are not chemically broken down or removed in the troposphere but rather, because of their exceptionally stable chemical structure, persist and slowly migrate up to the stratosphere. Depending on their individual structure, different CFCs can remain intact for decades or centuries. Once in the stratosphere, photodissociation of CFCs releases chlorine atoms that can take part in the catalytic destruction of ozone, *e.g.*

$$CF_2Cl_2 + h\nu \longrightarrow CF_2Cl + Cl \tag{34}$$

with the Cl released taking part in Reaction (5). CFCs were used in a wide variety of industrial applications including aerosol propellants, refrigerants, solvents and foam blowing. The CFCs that have attracted most attention in ozone depletion are CFC-11 ($CFCl_3$) and CFC-12 (CF_2Cl_2). Controls on their production and consumption (see Section 9.8) have slowed the rate of accumulation of CFCs in the atmosphere. Atmospheric concentrations of CFC-11 have levelled off and begun to decrease since about 1998, and concentrations of CFC-12 are rising more slowly than in previous decades. Emission rates of both compounds are in steep decline, but the longer atmospheric lifetime of CFC-12 means that its tropospheric concentration continues to grow, albeit at a slower rate than in previous decades. Total tropospheric chlorine peaked at 3.7 ppbv by 1994, with the stratospheric peak lagging by 3 to 6 years.[8]

Bromine is estimated to be about 50 times more efficient than chlorine in destroying stratospheric ozone on an atom-for-atom basis (see Section 9.2). Bromine is carried into the stratosphere in various forms such as halons and substituted hydrocarbons, of which methyl bromide is the predominant form. Three major anthropogenic sources of methyl bromide have been identified: soil fumigation; biomass burning; and the exhaust of automobiles using leaded petrol. Recent measurements have shown that there is three times as much methyl bromide in the Northern Hemisphere than in the Southern Hemisphere. Halon 1211 ($CBrClF_2$) and 1301 ($CBrF_3$) have been widely used in fire protection systems. Their production has now ceased but emissions from existing fire protection systems are expected to continue for decades. Global background levels are about 3.5×10^{-3} ppbv (H-1211) and 2.0×10^{-3} ppbv (H-1301). H-1211 concentrations continue to grow but H1301 concentrations have now begun to stabilize.[8]

Carbon dioxide concentrations are increasing largely as a result of burning fossil fuels.[15] Although not directly involved in ozone chemistry, carbon dioxide is a 'greenhouse gas', and so can affect ozone concentrations indirectly. Greenhouse gases warm the troposphere by trapping infrared radiation released by the earth's surface. Water vapour and carbon dioxide are the most important greenhouse gases, but others include methane, nitrous oxide, ozone, and the CFCs. Increased greenhouse warming in the troposphere leads to lower stratospheric temperatures, because less infrared radiation reaches the stratosphere. One consequence of stratospheric cooling is a slowing down of temperature-dependent (*e.g.* $O + O_3$, $NO + O$) ozone-destruction reactions. In the middle to upper stratosphere cooling would lead, therefore, to an increase in ozone. In the lower stratosphere, however, a decrease in temperature could increase the frequency and duration of PSCs, leading to more chlorine activation and hence more ozone depletion. In addition to its contribution as a greenhouse gas, methane has a direct chemical effect (see above). When methane is destroyed in the stratosphere, the chlorine and nitrogen cycles ozone-destruction cycles are suppressed – *i.e.* Reaction (23) and the 'smog' reactions of Chapter 8 compete more effectively with Reactions (5), (6), (9), (10), and (11), *etc.* – thus increasing stratospheric ozone abundances. The destruction process also produces water vapour, which can also indirectly destroy ozone through the production of OH. Current best estimates of the effects of increased greenhouse gases are that the recovery in polar ozone brought about by the Montreal Protocol will be delayed by 10–20 years relative to an atmosphere that is not experiencing climate change.

9.4.1 Direct Injection of Pollutants into the Stratosphere

In the 1970s, Johnston[16] postulated that NO_x, emitted in the exhausts of supersonic aircraft (SST), would result in large ozone depletions. There is once more increasing concern over the possible impact of aircraft flights, both supersonic and subsonic, on ozone levels in the lower stratosphere.[17] Production of a new generation of SSTs has not been ruled out by the aircraft industry. Total ozone changes, for cruise altitudes of 16 and 20 km, are calculated by models to be a few percent for reasonable estimates of the size of the supersonic fleet. Emissions higher in the stratosphere lead to larger local ozone losses since the NO_x emitted remains in the stratosphere for longer. Although changes in NO_x have the largest impact on ozone, the effects of H_2O emissions contribute about 20% to the calculated ozone change.

Many subsonic flights pass through the lowermost parts of the stratosphere, especially at high latitudes (recall Figure 9.2). Subsonic aircraft flying in the North Atlantic flight corridor emit 44% of their exhaust emissions into the stratosphere. Models predict an ozone decrease in the lower stratosphere of less than 1%, but modelling the lower stratosphere is particularly difficult since a number of chemical processes are of comparable importance to each other, and to the transport processes. Industry projections are for a 5% per year increase in air traffic from 1995 and 2015, so the impact of aircraft on the atmosphere is likely to grow significantly.

The exhaust products of rockets contain many substances capable of destroying ozone. Attention has focused on the chlorine compounds produced from solid fuel rockets. Rockets that release relatively large amounts of chlorine per launch into the stratosphere include NASA's Space Shuttle (68 tons per launch) and Titan IV (32 tons) rockets, and the European Space Agency's Ariane-5 (57 tons). But compared to the global release of CFCs, which is limited to 1×10^6 tons per year by the Montreal Protocol – and that is only 15% of the 1986 production – the chlorine source from rockets is tiny.

9.5 ANTARCTIC OZONE

The ozone hole continues to appear every austral spring, and was particularly severe in the 1990s. Figure 9.4 shows the monthly mean values of the ozone column over Halley Bay, Antarctica, for October. Prior to the mid-1970s the monthly mean column of ozone was approximately 300 DU. By 1985, the mean had fallen to below 200 DU and in the subsequent 5–6 years appeared to be

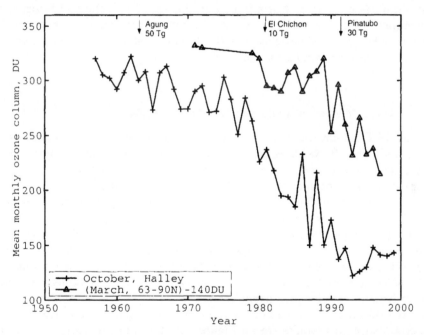

Figure 9.4   *Mean October value of column ozone over Halley Bay (76°S) since 1957 and mean March column ozone over northern polar regions (63–90°N). Note that 140 DU have been subtracted from the northern polar means to aid graph plotting. The dates of major volcanic eruptions which injected sulfur into the stratosphere are also marked*
[Ozone data courtesy of J. Shanklin (British Antarctic Survey, Cambridge)]

levelling off. However, the springs of 1992 and 1993 brought new record lows, with spot measurements below 100 DU for the first time, and the mean well below half that of the 1950s. The vertical profile of ozone above McMurdo station, Antarctica is shown in Figure 9.5(a). In mid-August the profile was near normal but in less than two months 97% of the ozone between 14 and 18 km in altitude has been destroyed.

The major chemical causes for the severe ozone depletion over Antarctica have now been established beyond reasonable doubt, and have been discussed in Section 9.2, above. The severe nature of polar depletion is caused by man-made chlorine pollution of the atmosphere in combination with unique wintertime meteorological conditions. The circulation of the winter stratosphere over the wintertime pole is dominated by the polar vortex (Figure 9.2). This is a region of very cold air surrounded by strong westerly winds. Air within the vortex is largely sealed off from that at lower latitudes and the chemistry occurs in near isolation (*e.g.* McIntyre;[18] Jones and MacKenzie[19]). In the lower stratosphere, temperatures within the vortex fall below 200 K; cold enough in the lower half of the stratosphere for heterogeneous chemistry on cold sulfuric acid aerosol, type 1 PSC, and type 2 PSCs. Chlorine activation, often accompanied by denitrification and dehydration, results.

Figure 9.6 shows the evolution of ozone and related compounds on a single isentropic surface (the 465 K surface, about 19 km) as measured by the Microwave Limb Sounder (MLS) satellite instrument.[20] In late autumn (28 April) descent of air at the pole has brought down higher mixing ratios of ozone, nitric acid and water vapour from the middle and upper stratosphere. Other measurements have shown that higher mixing ratios of Cl_y are also present in this air. The formation of the polar vortex isolates the polar air from its surroundings. PSC formation has yet to begin and so ClO mixing ratios in the vortex are the same as outside the vortex, near zero on this scale. By 2 June, temperatures have fallen below the threshold for type 1 PSC formation, as can be seen from the reduced mixing ratios of gas phase nitric acid in the MLS measurements. Chlorine activation takes place on the PSCs and the ClO formed is blown downwind, but there has not yet been enough exposure to sunlight for large-scale ozone depletion to be evident. By 17 August, temperatures are cold enough for type 1 and 2 PSCs to be present over large parts of the Antarctic continent, leading to reduced gas phase nitric acid and water vapours and to elevated ClO. Ozone depletion has begun on the sunlit outer rim of the vortex. By 1 November, temperatures are too high for PSC formation but nitric acid and water vapour are both reduced inside the vortex. This is due to denitrification and dehydration. ClO mixing ratios have returned to background levels almost everywhere since PSCs are no longer present and termination reactions have become effective. Ozone mixing ratios are now much reduced throughout the vortex, over an area larger than the Antarctic continent. During the 1990s, the area with ozone columns less than 220 DU approached 10 % of the area of the Southern Hemisphere.[11]

As the total inorganic chlorine loading increases, the rate of ozone depletion also increases. Hence, the lowest ozone values may be reached progressively

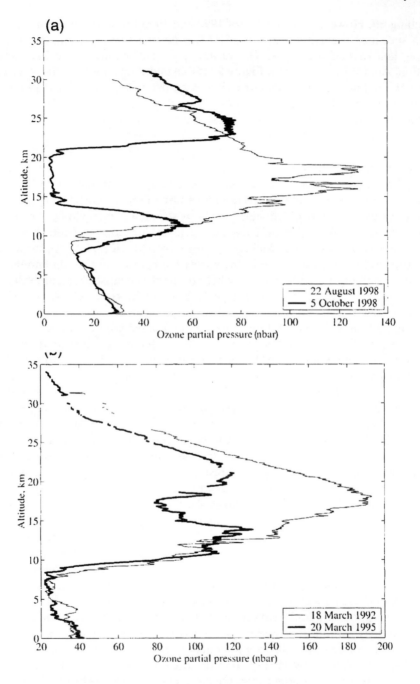

Figure 9.5 (a) *Vertical ozone profiles above McMurdo station for 22 August, 1998 and 5 October 1998. Data courtesy of Terry Deshler, University of Wyoming.* (b) *Vertical ozone profiles above Ny Ålesund (78° N, 11° E) for 18 March 1992 and 20 March 1995. Data courtesy of the Alfred Wegener Institute. All profiles are taken inside the polar vortices*

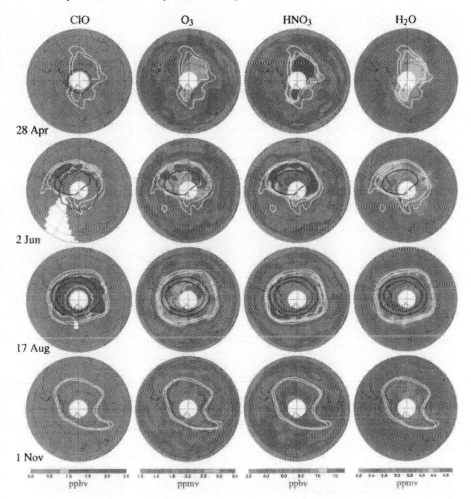

Figure 9.6   *MLS satellite maps of ClO, O_3, HNO_3 and H_2O for selected days during the 1992 southern winter, interpolated onto the 465 K isentropic surface. The maps are polar orthographic projections extending to the equator, with the Greenwich meridian at the top, and the $30°S$ and $60°S$ latitude circles marked. No measurements were obtained poleward of $80°S$, and, for H_2O, equatorward of $30°S$. The edge of the polar vortex is shown by the two white contours. The black contour near S. America on 28 April surrounds air which has come from the polar vortex. Temperatures below which type 1 and type 2 PSCs can form are shown as black contours: the outer white contour is for type 1 PSCs* [Figure courtesy of Michelle Santee (Jet Propulsion Lab, California)]

earlier in the year, resulting in increased radiative cooling, lower stratospheric temperatures, a higher probability of PSC formation later in the season and a delayed break-up of the vortex. When the vortex breaks down, the ozone-depleted air becomes distributed over the hemisphere, leading to ozone reductions at lower latitudes.

9.6 ARCTIC OZONE

The evolution of ozone and related compounds in the Arctic polar vortex during the winter of 1992/93 is shown in Figure 9.7, which can be compared to the evolution in the Antarctic vortex of 1992, shown in Figure 9.6.[20] The initial descent into the vortex is less pronounced on 26 October in the Northern Hemisphere than on 28 April in the Southern Hemisphere, and the northern vortex is less fully developed than its southern counterpart. By 3 December the vortex has developed and descent, bringing down high mixing ratios of ozone, nitric acid and water vapour, has taken place. ClO mixing ratios are only elevated in a small region downwind of a patch of cold air over southern Finland. By 22 February there have been sufficient cold temperatures to activate

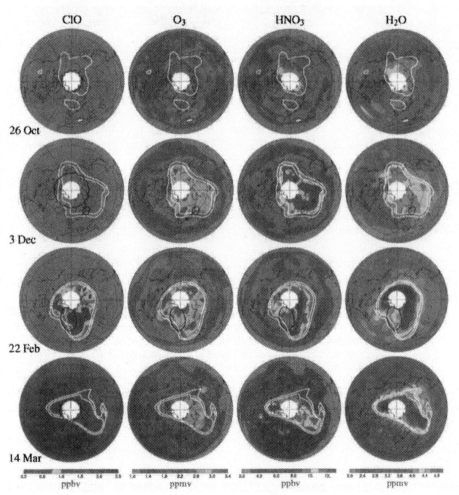

Figure 9.7   *As Figure 9.6, but for selected days during the 1992–1993 northern winter, with the Greenwich meridian at the bottom of each map*
[Figure courtesy of Michelle Santee (Jet Propulsion Lab, California)]

chlorine throughout the vortex, but neither denitrification nor dehydration have taken place. Ozone mixing ratios remain high. By 14 March, ClO mixing ratios have again returned to background levels, nitric acid and water vapour mixing ratios remain high, and no ozone hole has formed. However, that is not to say that chlorine-catalysed ozone destruction has not taken place: ozone mixing ratios in March are lower than those in February. Given the near-continuous descent of ozone-rich air onto this isentropic surface in the northern vortex during winter, this decreased ozone implies chemical loss. Further evidence of chemical loss in the northern polar vortex has been adduced from the careful matching up of ozonesondes that have sampled the same air parcel.[21] A statistically significant correlation is shown between ozone loss and hours of sunlight experienced between samplings.

Figure 9.7 demonstrates that the Arctic has a warmer and weaker vortex than the Antarctic. The greater wave activity in the Northern Hemisphere produces more rapid downward motion in the northern polar vortex relative to its southern counterpart. This downward motion heats the air by compression. Dissipating waves deposit momentum at the vortex edge, thus weakening it. The behaviour of the northern vortex can vary greatly from year to year. For winters which are cold early on, 1991/92 for example, large amounts of ClO can be formed, but it is formed when there is insufficient sunlight to cause severe ozone depletion.[22] Winters that are cold later, 1992/93 for example, produce more severe ozone depletion. The winter of 1994/95 produced severe ozone depletion. Figure 9.5(b) shows results from a pair of ozonesondes, launched from Spitsbergen, into the polar vortex, at about the same time of year in 1992 and 1995. About half the ozone has been destroyed in the 18–20 km region in 1995 relative to 1992. This is a situation that has been likened to that in early spring in the Antarctic. With the present amount of chlorine in the atmosphere, severe ozone depletion is possible in the Arctic in winters that are cold late on.

9.7 MID-LATITUDE OZONE

Analyses of global ozone records, from satellite and ground-based instruments, have shown that ozone loss is not confined to polar regions. Trends for total ozone in the Northern and Southern Hemisphere middle latitudes are significantly negative in all seasons. The trends in the Equatorial region ($20° S–20° N$) are not statistically significant. Measurements of vertical profiles of ozone indicate that at altitudes of 40 km, over the period 1980 to 1996, ozone declined about 7% per decade at northern middle latitudes. Over the same period and region, depletion between 25 and 30 km was about 3% per decade, and about 7% per decade between 15 and 20 km. Integration of the vertical profile trends gives total ozone trends consistent with total ozone measurements. The 'high-low-high' signal in ozone depletion as a function of altitude is due to the effects of gas-phase chemistry at high altitudes and heterogeneous chemistry at low altitudes.

For the first two years after the Mt Pinatubo eruption, anomalously large downward ozone trends were observed in both hemispheres (see also Figure

9.4). Global total ozone values in 1992/93 were 1–2% lower than those expected from the long-term trend. It seems likely that radiative, dynamical, and chemical perturbations, resulting from the injection of huge quantities of aerosol into the stratosphere, were responsible.

The detection of statistically significant downward trends in column ozone in the middle latitudes has provided the spur for extensive measurements of stratospheric chemistry. As well as accurate measurements, models, which couple the transport and the chemistry of long-lived trace species, are required if global ozone depletion is to be quantified. In the middle latitudes, transport of air from both high and low latitudes determines the chemical composition. Transport from the polar vortex into the middle latitudes can produce ozone depletion by exporting ozone-poor vortex air, or by exporting chemically activated ClO-rich air. For example, ozone depletion may be observed as a result of air sinking through the vortex during the period of ozone destruction and then being transported to the middle latitudes.

9.8   GLOBAL MODELLING OF OZONE

The advent of supercomputing has made possible an integrated approach to the modelling of the stratosphere. Models are required ultimately for the prediction of the atmospheric effects of future emissions, so that legislation can be drafted in a timely fashion. Models also perform several, scientifically important, intermediate tasks, and the choice of model is determined by the task that it is required to carry out.

Two-dimensional, in latitude and height, models can be integrated forward for many years, allowing an assessment of likely future ozone destruction to be made. They use the mean circulation, combined with eddy diffusion coefficients,* to transport chemicals. Unfortunately, the 'assessment' models do not resolve polar transport and chemistry sufficiently well to produce the observed chlorine activation and ozone destruction at both wintertime poles. Different assessment models also predict very different lower stratospheric abundances of trace gases. Many of the problems with these models stem from the fact that the eddy diffusion concept does not accurately represent the essentially one-sided isentropic transport occurring at the edge of the polar vortices and the sub-tropical mixing barrier.[2] Fully three-dimensional simulations of the chemistry and dynamics of the stratosphere are also possible (*e.g.* Austin *et al.*[23]). However, such 3D *General Circulation Model* (GCM) integrations are usually limited to a season or a year, for reasons of economy.

For the interpretation of chemical observations, using prescribed meteorological fields to transport chemical tracers in 3D is a very useful alternative to GCM integrations. These *Chemistry and Transport Models* (CTMs) also cannot be integrated forward for as long as the two-dimensional assessment models, but low-resolution multi-year runs are now possible.[23] Alternatively, chemistry

---

* Eddy diffusion coefficients arise from a treatment of atmospheric mixing and stirring which is analogous to the treatment of molecular diffusion in Fick's Laws.

and transport models can be run on a single isentropic surface (making use of the fact that flow across isentropic surfaces is small, with a timescale of about a week). Even without capturing the details of the diabatic flow, single layer models are a useful test-bed for chemical mechanisms and mixing effects. Figure 9.8 shows results from a single layer CTM, run on the 475 K isentropic surface with a horizontal resolution of 5.6 × 5.6°, and which was initiated on 1 January 1992. The results are for 22 February 1993, 53 days into the simulation and a day for which there are measurements of the Northern Hemisphere (Figure 9.7). Although the winds and temperatures of the model are prescribed, the model chemistry is free-running after initialization. Clearly, model and measurements are in good agreement. Notice that on this day the polar vortex was distorted

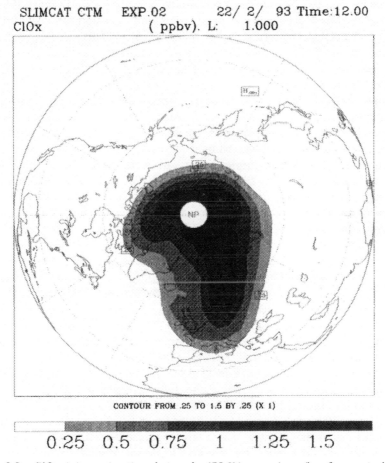

SLIMCAT CTM   EXP.02   22/ 2/ 93 Time:12.00
ClOx   (ppbv). L: 1.000

CONTOUR FROM .25 TO 1.5 BY .25 (X 1)

0.25   0.5   0.75   1   1.25   1.5

Figure 9.8 *ClO mixing ratios (in ppbv) on the 475 K isentropic surface from a single layer CTM, run on the 475 K isentropic surface at 5.6 × 5.6° horizontal resolution, initialized on 1 January 1992 and run out to 22 February 1993. The map is a polar orthographic projection. Cf. Figure 9.7*
[Data courtesy of Martyn Chipperfield (School of the Environment, Leeds University)]

and covered a large part of Europe, bringing ClO-rich air into sunlight, where ozone-destroying catalytic cycles can occur.

9.9 LEGISLATION

On September 16, 1987, a treaty was signed that has been hailed as 'the most significant international environmental agreement in history', 'a monumental achievement' and 'unparalleled as a global effort'.[25] It had been long believed that this particular agreement would be impossible to achieve because the issues were so complex and arcane and the initial positions of the negotiating parties so widely divergent.

The Montreal Protocol on Substances that Deplete the Ozone Layer was adopted in 1987 by 25 countries, and entered into force on 1 January 1989. The Protocol required each party's production and consumption of CFCs -11 ($CFCl_3$), -12 (CF_2Cl_2), -113 ($C_2F_3Cl_3$), -114 ($C_2F_4Cl_2$) and -115 (C_2F_5Cl) first to be frozen at 1986 levels and ultimately to be reduced to 50% of 1986 levels by 1998. Production and consumption of halons 1211 (CF_2BrCl), 1301 (CF_3Br) and 2402 ($C_2F_4Br_2$) were to be restricted to 1986 levels.

Ozone depletion had, from the start, captured the US public imagination. The subject featured prominently in the media and in congressional hearings, and soon began to influence consumer behaviour. In 1978 the US, followed by Canada, Norway and Sweden, banned the use of CFCs as aerosol spray can propellants in non-essential applications.[26] Even before this ban, the US market for spray cans had fallen by nearly two thirds because of public commitment to protection of the environment. In the UK, such a ban was not introduced. Concerns about ozone depletion were defused by ministerial assurances that any terrestrial effects would be inconsequential.[27]

In the light of increasing scientific and public concern over ozone depletion, empirical evidence of global depletion, confirmation that CFCs and other man-made ozone depletion substances were the major factor in creating the Antarctic ozone hole, and generally improved prospects for replacing such substances, significant changes to the Protocol were approved in London (1990) and Copenhagen (1992). In London, the control measures were adjusted to provide for the phase-out of CFC and halon production and consumption by the year 2000. An intermediate cut of 50% of the 1986 level by 1995 was also agreed for CFCs and halons, together with an 85% cut, for CFCs but not halons, by 1997. Additionally, carbon tetrachloride and fully halogenated CFCs were to be phased-out by 2000 and methyl chloroform by 2005.

The 1990 Protocol Amendment introduced the concept of transitional substances, such as HCFCs. These are chemical substitutes for CFCs and other controlled substances, but unlike CFCs, are reactive in the lower atmosphere and so, tonne for tonne, transport less chlorine to the stratosphere. They are necessary in some applications, in the short to medium term, to enable a rapid phase-out of the controlled substances to take place. A non-binding resolution was approved with a view to replacing HCFCs by alternatives that do not deplete ozone by no later than 2040.

Further revisions to the Protocol were agreed in Copenhagen (1992). The phase-out dates were brought forward and controls on several new substances were imposed. For CFCs, carbon tetrachloride and methyl chloroform, the new phase-out date was 1 January 1996. Provision was made to allow production of these substances for 'essential use'. Within the European Union, Member states have adopted even tighter controls. The transitional substances were introduced into the Protocol and controls on HCFCs agreed. These controls set a cap on consumption, which will be reduced stepwise, leading to a phase-out by 2030. The numerical value of the cap is the sum of the quantity of all HCFCs that were produced and used during 1989 plus an amount equal to 3.1% of the calculated CFC consumption during 1989. Contributions to the 3.1% cap from individual HCFCs are adjusted by their ozone depletion potentials.

It is already clear that the Montreal Protocol and its amendments have been adhered to, and have worked.[8] Tropospheric chlorine abundance has peaked and is now beginning to decline. The total inorganic chlorine in the troposphere peaked at about 3.7 ppt in the early 1990s. In the stratosphere the time at which total chlorine peaks is a function of height, since the transport of air to the uppermost stratosphere, *via* the Brewer–Dobson circulation, is slow. The peak in chlorine therefore occurred later in the stratosphere than in the troposphere. At 22 km, for example, the peak was measured in 1998.

9.10 SUMMARY

In the early 1970s, ozone depletion from supersonic transport and CFCs from anthropogenic sources were factors in subsequent decisions in the US to cancel a proposed SST fleet in the early 1970s and impose a ban on aerosol cans in 1978. Nevertheless, long-term depletion of ozone has been observed since the early 1980s in the Antarctic polar vortex, and more recently at mid-latitudes in both hemispheres, with most of the ozone loss occurring in the lower strato-sphere. Intermittently since the spring of 1995, ozone depletion of similar relative magnitude to that observed in the Antarctic spring has been observed in the Arctic.

The present condition of the global ozone layer is much worse than we would wish, but it is easy to forget how close the world came to an environmental problem of much greater proportions. Had the Montreal Protocol and its amendments not been adopted, or if they are not adhered to, the stratosphere in 2050 would look like this:

- About five times more ozone-depleting gases would be in the stratosphere than are currently there. This corresponds to a total mixing ratio about 17 ppb of ozone-depleting substances, having first weighted each sub-stance for its ozone-depletion potential.
- Ozone depletion would be at least 50% over the middle latitudes of the Northern Hemisphere and 70% over the middle latitudes of the Southern Hemisphere. This is about ten times larger than present depletion.
- Surface ultraviolet (UV-B) would be at least double those in an unper-

turbed atmosphere in northern middle latitudes, and four times greater in southern middle latitudes. Currently UV-B levels are up by 5% and 8% in the northern and southern middle latitudes, respectively.

But work remains to be done. The coupling of the ozone layer and global climate is still very poorly quantified, and this makes forecasts of future ozone levels (and climate) uncertain.

Acknowledgements

The contributions of Ian Colbeck and Joe Farman to previous editions of this chapter are very gratefully acknowledged. The task of writing a short description of stratospheric ozone is made immeasurably easier by the existence of the WMO Scientific Assessments: I thank the many scientists who have laboured long and hard on them. I am very grateful also to Martyn Chipperfield, Terry Deshler, Roland Neuber, Michelle Santee, and Jonathan Shanklin for supplying figures and data for figures. And, not least, thanks to Nick Hewitt, Jamie Kettleborough, Adrian Lee, and Roy Harrison for comments and corrections.

9.11 References

1. J. C. Farman, B. G. Gardiner and J. D. Shanklin, 'Large Losses of Total Ozone Reveal Seasonal ClO_x/NO_x Interactions', *Nature*, 1985, **315**, 207–210.
2. M. E. McIntyre, 'Atmospheric Dynamics: Some Fundamentals, With Observational Implications', in '*Proc. Int. School Phys., 'Enrico Fermi' CXV Course, The Use of EOS for Studies of Atmospheric Physics*', eds. J. C. Gille and G. Visconti, North Holland, 1992, pp. 313–386.
3. W. N. Hartley, 'On the Absorption Spectrum of Ozone', *J. Chem. Soc.*, 1881, **39**, 57–60.
4. S. C. Chapman, 'A Theory of Upper Atmospheric Ozone', *Mem. Roy. Met. Soc.*, 1930, **3**, 103–125.
5. A. J. Weinheimer, J. G. Walega, B. A. Ridley, B. L. Gary, D. R. Blake, N. J. Blake, F. S. Rowland, G. W. Sachse, B. E. Anderson and J. E. Collins, 'Meridional Distributions of NO_x, NO_y, and other Species in the Lower Stratosphere and Upper Troposphere During AASE II', *Geophys. Res. Lett.*, 1994, **23**, 2583–2586.
6. C. R. Webster, R. D. May, L. Jaegle, H. Hu, S. P. Sander, M. R. Gunson, G. C. Toon, J. M. Russell, R. M. Stimpfle, J. P. Koplow, R. J. Salawitch and H. A. Michelsen, 'Hydrochloric Acid and the Chlorine Budget of the Lower Stratosphere', *Geophys. Res. Lett.*, 1994, **21**, 2575–2578.
7. M. B. McElroy, R. J. Salawitch, S. C. Wofsy and J. A. Logan, 'Reductions of Antarctic Ozone due to Synergistic Interactions of Chlorine and Bromine', *Nature*, 1986, **321**, 759–762.
8. WMO (World Meteorological Organisation), 'Scientific Assessment of Ozone Depletion: 1998, Global Ozone Research and Monitoring Program Report 44', WMO, Geneva, Switzerland, 1998.
9. R. R. Garcia and S. Solomon, 'A New Numerical Model of the Middle Atmosphere, 2, Ozone and Related Species', *J. Geophys. Res.*, 1994, **99**, 12937–12952.
10. O. B. Toon and N. H. Farlow, 'Particles above the Tropopause', *Annu. Rev. Earth Planetary Sci.*, 1981, **9**, 19–58.

11. WMO (World Meteorological Organisation), 'Scientific Assessment of Ozone Depletion: 1994, Global Ozone Research and Monitoring Program Report 37', WMO, Geneva, Switzerland, 1994.

12. K. S. Carslaw, B. P. Luo, S. L. Clegg, Th. Peter and P. Brimblecombe, 'Stratospheric Aerosol Growth and HNO_3 Gas Phase Depletion from Coupled HNO_3 and Water Uptake by Liquid Particles', *Geophys. Res. Lett.*, 1994, **21**, 871–874.

13. R. A. Cox, A. R. MacKenzie, R. H. Müller, Th. Peter and P. J. Crutzen, 'Activation of Stratospheric Chlorine by Reactions in Liquid Sulphuric Acid', *Geophys. Res. Lett.*, 1994, **21**, 1439–1442.

14. D. R. Hanson, A. R. Ravishankara and S. Solomon, 'Heterogeneous Reactions in Sulfuric Acid Aerosols: a Framework for Model Calculations', *J. Geophys. Res.*, 1994, **99**, 3615–3630.

15. IPCC (1995) Climate Change 1994. Houghton J.T., Filho L. G. M., Bruce J., Lee H., Callander B. A., Haites E., Harris N. and Maskell K. (eds.). Cambridge University Press.

16. H. S. Johnston, 'Reduction of Stratospheric Ozone by Nitrogen Oxide Catalysts from Supersonic Transport Exhaust', *Science*, 1971, **173**, 517–762.

17. IPCC (Intergovernmental Panel on Climate Change), 'Aviation and the Global Atmosphere', eds. J. Penner *et al.*, Cambridge University Press, Cambridge, 1999.

18. M. E. McIntyre, 'The Stratospheric Polar Vortex and Sub-vortex: Fluid Dynamics and Mid-latitude Ozone Loss', *Phil. Trans. Roy. Soc., Ser. A*, 1995, **352**, 227–240.

19. R. L. Jones and A. R. MacKenzie, 'Observational Studies of the Role of Polar Regions in Mid-latitude Ozone Loss', *Geophys. Res. Lett.*, 1995, **22**, 3485–3488.

20. M. L. Santee, W. G. Read, J. W. Waters, L. Froidevaux, G.L. Manney, D. A. Flower, R. F. Jarnot, R. S. Harwood and G. E. Peckham, 'Interhemispheric Differences in Polar Stratospheric HNO_3, H_2O, ClO and O_3', *Science*, 1995, **267**, 849–852.

21. P. van der Gathen, M. Rex, N. R. P. Harris, D. Lucic, B. M. Knudsen, G. O. Braathen, H. Debacker, P. Fabian, H. Fast, M. Gil, E. Kyro, I. S. Mikkelsen, M. Rummukainen, J. Stahelin and C. Varotsos, 'Observational Evidence for Chemical Ozone Depletion over the Arctic in Winter 1991/92', *Nature*, 1995, **375**, 131–134.

22. M. P. Chipperfield, 'A Three Dimensional Model Comparison of PSC Processing During the Arctic Winters of 1991/1992 and 1992/1993', *Ann. Geophys.*, 1994, **12**, 342–354.

23. J. Austin, J. Knight and N. Butchart, 'Three-dimensional Chemical Model Simulations of the Ozone Layer: 1979–2015', *Quart. J. Roy. Meteor. Soc.*, 2000, **126**, 1533–1556.

24. M. P. Chipperfield, 'Multi-annual Simulations with a Three-dimensional Chemical Transport Model', *J. Geophys. Res.*, 1999, **104**, 1781–1805.

25. R. E. Benedick, 'Ozone Diplomacy', Harvard University Press, Cambridge, MA, 1991.

26. S. C. Zehr, 'Accounting for the Ozone Hole: Scientific Representations of an Anomaly and Prior Incorrect Claims in Public Settings', *Sociol. Q.*, 1994, **35**, 603–619.

27. M. Purvis, 'Yesterday in Parliament: British Politicians and Debate over Stratospheric Ozone Depletion, 1970–92', *Environ. Plann. C: Govern. Policy*, 1994, **12**, 361–379.

CHAPTER 10

Atmospheric Dispersal of Pollutants and the Modelling of Air Pollution

M. L. WILLIAMS

10.1 INTRODUCTION

Considerable resources are often devoted to the measurement of air pollutant concentrations in the ambient atmosphere but measurements on their own provide little information on the origin of the pollutants in question, on the dispersal process in the atmosphere and on the impact of new sources or the benefits of controls. There is frequently the need, therefore, for detailed knowledge of the characteristics and quantities of pollutants emitted to the atmosphere and on the atmospheric processes which govern their subsequent dispersal and fate. This knowledge must then be built into an appropriate dispersion model, whether the problem to be addressed is the emission from a single chimney or the emissions from a large multi-source urban/industrial area, or, on a larger scale, from a region or country.

The move within the EC and the UK towards formal air quality limit values, objectives or standards is creating the need for formal air quality management systems and for a more strategic approach to air pollution control. Neither can be accomplished without the use of atmospheric dispersion modelling techniques. A wide variety of techniques are available, ranging from the most simple 'box' model through to numerical solutions of the basic equations of fluid flow, *etc.* For the most chemically reactive pollution, it is also necessary to incorporate the relevant atmospheric chemistry. The spatial and temporal resolution and accuracy of the model output ideally must match the questions being posed. Where emissions vary greatly, both in space and time, or it is necessary to predict the time series of concentrations at specified locations, then the modelling task is extremely difficult, even without the complications of a very chemically reactive pollutant species or substantial topographical effects on the dispersal pattern. On the other hand, if it is not necessary to know when a specified concentration will occur but rather it is the probability of occurrence during a given period (*e.g.* a year) which is of interest, then the modelling task is generally much less demanding. However in general, long-period (*e.g.* annual) average concentrations can be modelled more accurately than can shorter averaging times where the turbulent fluctuations in the atmosphere can result in agreement within factors of 2 to 3 with observed values.

10.2 DISPERSION AND TRANSPORT IN THE ATMOSPHERE

A pollutant plume emitted from a single source is transported in the direction of the mean wind. As it travels it is acted upon by the prevailing level of atmospheric turbulence which causes the plume to grow in size as it entrains the (usually) cleaner surrounding air. There are two main mechanisms for generating atmospheric turbulence. These are mechanical and convective turbulence, and will be discussed in Sections 10.2.1 and 10.2.2 below.

10.2.1 Mechanical Turbulence

This is generated as the air flows over obstacles on the ground such as crops, hedges, trees, buildings and hills. The intensity of such turbulence increases with increasing wind speed and with increasing surface roughness and decreases with height above the ground. If there is only a small heat flux, either to or from the surface, in the atmosphere so that most of the turbulence is mechanically generated the atmosphere is said to be neutral or in a state of neutral stability. In this case the wind speed will vary logarithmically with height z:

$$u(z) = (u_*/k) \ln (z/z_0) \qquad (1)$$

where k is von Karman's constant (~ 0.4), z_0 is the so called surface roughness length (~ 1 m for cities and ~ 0.3 m for 'typical' countryside in the UK), and u_* is the friction velocity and is a measure of the flux of momentum to the surfaces.

10.2.2 Turbulence and Atmospheric Stability

As solar radiation heats the earth's surface, the lower layers of the atmosphere increase in temperature and convection begins, driven by buoyancy forces. The motion of air parcels from the surface is unstable as a parcel in rising finds itself warmer than its surroundings and will continue to rise. Convective circulations are set up in the boundary layer and this form of turbulence is usually associated with large eddies, the effects of which are often visible in 'looping' plumes from stacks. At night when there is no incoming solar radiation and the surface of the earth cools, temperature increases with height, and turbulence tends to be suppressed. During calm clear nights, when surface cooling is rapid and little or no mechanical turbulence is being generated, turbulence may be almost entirely absent.

When considering most dispersion problems it is convenient to classify the possible states of the atmosphere into what are usually referred to as stability

Table 10.1   *Pasquill's stability categories**

Surface wind speed m s^{-1} ($\equiv U_{10}$)	Insolation			Night	
	Strong	*Moderate*	*Slight*	*Thickly overcast or $\geqslant 4/8$ low cloud*	$\geqslant 3/8$ *Cloud*
<2	A	A–B	B	–	G
2–3	A–B	B	C	E	F
3–5	B	B–C	C	D	E
5–6	C	C–D	D	D	D
>6	C	D	D	D	D

Strong insolation corresponds to sunny midday in midsummer in England. Slight insolation in similar conditions in midwinter. Night refers to the period from 1 hour before sunset to 1 hour after dawn. A is the most unstable category and G the most stable. D is referred to as the neutral category and should be used, regardless of wind speed, for overcast conditions during day or night.
*Based on Figure 6.10 in Reference 1.

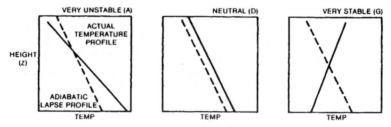

Figure 10.1   *Typical atmospheric temperature profiles and corresponding stabilities*

categories. The typing scheme developed by Smith from the original Pasquill formulation[1,2] is widely used because of its relative simplicity yet dependence on sound physical principles. Stability is classified according to the amount of incoming solar radiation, wind speed and cloud cover. A semi-quantitative guide is given in Table 10.1 and Figure 10.1 shows typical temperature profiles corresponding to the unstable, neutral and stable cases. The adiabatic lapse profile in Figure 10.1 is the vertical temperature gradient for the atmosphere in a stage of adiabatic equilibrium when a parcel of air can rise and expand, or descend and contract, without gain or loss of heat; the temperature of the air parcel is always the same as that of the level surrounding air and the conditions correspond to neutral stability. The numerical value of the adiabatic lapse rate is $\sim 1°C/100$ m. For wind speeds in excess of about 6–8 m s^{-1} mechanical turbulence dominates irrespective of the degree of insolation and neutral stability prevails. Similarly, in areas of high surface roughness like large cities, mechanical turbulence can dominate and the atmosphere can often be considered to be neutral. This can be a useful approximation in modelling studies in cities. Table 10.2 gives typical annual frequencies of occurrence of the different

Table 10.2 *Typical annual frequency of occurrence of stability categories in Great Britain*

Stability category	Frequency of occurrence %
A	0.6
B	6.0
C	17.0
D	60.0
E	7.0
F	8.0
G	1.4

stability categories in Great Britain. For other regions quite different frequencies might apply. In central Continental regions at lower latitudes, for example, the greater incidence of solar radiation would probably result in smaller incidence of neutral conditions and increased frequencies of unstable and stable categories.

Recently, a more quantitative measure of stability, the Monin–Obukhov length, has come into use. This essentially measures the balance between the mechanical and convective contributions to turbulence in a boundary layer, and the Monin–Obukhov length, L, is that height above the ground at which the two contributions are equal. In mathematical terms L is given by

$$L = -u_*^3 / \kappa g F / (\rho c_p T_0)$$

where u_* is the friction velocity, κ is the von Karman constant, F is the surface heat flux, ρ is the air density, c_p is the specific heat of air at constant pressure and T_0 is the surface temperature. Recent developments in boundary layer theory use the Monin–Obukhov length to classify stabilities and calculate dispersion.

10.2.3 Mixing Heights

The stable atmosphere depicted in Figure 10.1 is an example of a ground based temperature inversion, *i.e.* the temperature increases with height unlike the normal decrease. An elevated inversion is often observed where a region of stable air caps an unstable layer below. Pollutants emitted below the inversion can be mixed up to, but not through, the inversion, the height of which is referred to as the mixing height. This term can be used more generally to describe the height of a boundary between two stability regimes. In the case of an emission above an elevated inversion, the pollutant will be prevented from reaching the ground so that for both surface and elevated sources, inversions can have a significant effect on ground level concentrations. The variation of mixing heights throughout the day due to solar heating and atmospheric cooling can have profound effects on ground level concentrations of pollutants. At night the atmosphere is typically stable with a shallow ($\sim$ 1–300 m) layer formed by surface cooling. As the sun rises the surface heating generates convective eddies and the turbulent boundary layer increases in depth, reaching a maximum in the afternoon at a depth of ~ 1000 m. As the solar input decreases and stops, the surface cools and a shallow stable layer begins to form again in the evening. In this idealized day, if emission rates remain constant, concentrations from surface sources will be at a maximum in the periods when the stable layers (with low wind speeds and mixing heights) are present and minimized during the afternoon. Sources emitting above the stable overnight layer will not contribute to ground-level concentrations until the height of the growing convective layer reaches the plume and brings the pollutants to ground level, a process known as fumigation. The patterns of ground-level concentrations from elevated sources can therefore be quite different from those of surface releases. In reality, the diurnal pattern of emissions can, of course, play a significant role.

In assessing air quality impacts, particularly of elevated sources such as power stations, estimates of mixing heights and their frequency of occurrence and variability throughout the day are therefore essential. WSL has used acoustic sounding (SODAR) to determine mixing heights[3] and an example of the diurnal and seasonal variation of mixing heights measured at Stevenage from 1981 to 1983 is given in Figure 10.2. The broad features described above are apparent in this diagram.

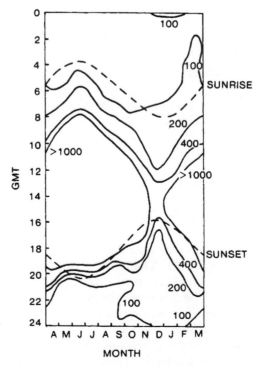

Figure 10.2 *Annual and diurnal variation of mixing height (metres) at Stevenage, 1981-83*

10.2.4 Building and Topographical Effects

Hills or buildings can have significant adverse effects on plume dispersion if their dimensions are large in comparison with the dimensions of the plume or if they significantly deflect or disturb the flow of the wind. Figure 10.3 shows a simplified and idealized representation of the flow over a building. There is a zone on the immediate downwind (or leeward) side of the building which is to some extent isolated from the main flow and within which there is a reversal of the air flow. Further downstream the air flow is highly turbulent. Waste gases escaping through a relatively short chimney attached or adjacent to the building, will be entrained in this characteristic flow pattern and will not disperse according to the conventional Equations 2 and 3 (see Section 10.3.1). Recent wind tunnel studies demonstrate that up-wind buildings can have a significant effect on emissions from a chimney located within a few, say five, building heights; for example, to maintain the same maximum ground level concentration the chimney height required in the presence of one building type studied would be between 1.5 and 2 times the height of the chimney required if the buildings was not present. Further downwind, beyond roughly 10 building heights, the near-field effects of the buildings can be incorporated into dispersion models in a parameterized way as discussed in Section 10.3.1. However, accurate numerical modelling in the near-field to buildings is difficult and most models treat this in an approximate way. Wind tunnel modelling is the preferred solution.

Topographical features such as hills and sides of valleys can have similar effects on dispersion to those described above for buildings. Valleys are also somewhat more prone to problems arising from emission fairly close to the ground. The incidence of low level or ground based temperature inversions can

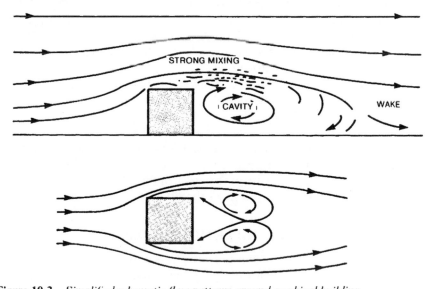

Figure 10.3 *Simplified schematic flow patterns around a cubical building*

be greater, either because solar heating of the ground is somewhat delayed in the morning or because during the night cold air drains down the valley sides (katabatic winds) thus creating a 'pool' of cold air on the valley floor. Any low-level emissions will therefore disperse very slowly and may even accumulate to some extent. The relatively undiluted emissions can also drift along and across the valley thus affecting areas other than the immediate surroundings of the source. Emissions from high chimneys located on the valley floor may not be detected at all on the valley floor while the ground based inversion persists. However, considerable horizontal spreading of the plume aloft can occur and during the morning fumigation period large parts of the valley may experience relatively high concentrations at much the same time.

The effects of hills on the flow of pollutant plumes and the resulting concentrations are complex and will not be discussed here.

10.2.5   Removal Processes – Dry and Wet Deposition

When considering pollution impacts from nearby sources, *e.g.* within say 10 km or so, the various losses of pollutants are generally not important (unless one happens to be interested specifically in such issues as the short range washout of HCl or the deposition of a particularly toxic species in the near field). However, in considering impacts over long ranges and especially on the international scale then consideration of the removal processes is essential.

Dry deposition takes place continously in a turbulent boundary layer as a result of turbulent flux towards the surface. The efficiency of the process is determined by the deposition velocity which in general is a function of the prevailing level of turbulence (high levels of which result in increasing deposition other things being equal) and of the nature of the gas and the surface (for example a reactive gas such as HNO_3 will be deposited more readily than a less reactive species such as NO). The flux to the surface is given by the product of the surface concentration and the deposition velocity, v_g, so that the process is linear. The equivalent first-order rate constant for this process is given by (v_g/H) where H is the depth of the mixing layer through which deposition is taking place. Typical half-lives for this process are ~ 1-2 days for species such as SO_2 and NO_2, but ~ 5 days or more for sulfate aerosols, which is one reason why 'acid rain' is a continental rather than purely a local phenomenon.

Wet deposition is the term given to the removal of gases or particles from the atmosphere in clouds and/or in rain. For species such as SO_2 and aerosols this process is relatively efficient. Typical lengths of dry and wet periods in the UK are such that if transport times are of the order of the dry period duration (~ 70 hours in the UK), wet removal processes should be included in the model.

10.3   MODELLING OF AIR POLLUTION DISPERSION

In Section 10.2 we discussed the underlying physical principles of air pollution dispersion, transport and transformation in the atmosphere. In this Section we summarize the methods used to apply these principles in a quantitative way to model the processes mathematically. The techniques used depend on the distance scales involved in the transport from the source to the receptor or 'target' area. If this distance is small, say of the order of tens or hundreds of metres, then very often buildings and local topography are important and the mathematical description of the ensuing complex flows and turbulence may not be tractable. In such cases (and others such as the dispersion of dense gases or longer distance problems in complex topography) a physical model in a wind tunnel may be the only practicable solution. These complications will be neglected in all that follows will deal with situations of ideal flat terrain.

It is fairly clear that as a plume is transported in the direction of the mean wind, it grows through the effect of atmospheric turbulence producing, very roughly, a cone shaped plume with the apex towards the stack. Now clearly the plume will continue to expand until, in the vertical, it fills the atmospheric boundary layer (~ 1 km deep in neutral conditions). Beyond this point vertical dispersion has no further effect; concentrations are thence reduced only by horizontal dispersion, and by the deposition processes and, if appropriate, by chemical reactions. It can be shown that in neutral conditions this point is reached at downwind distances from a source of very roughly 50–100 km, so for source-receptor distances less than this value, vertical dispersion should be included in a model for an accurate representation of the dispersion. Beyond this region, plumes generally fill the mixing layer and uniformly mixed 'box-models' can be used with some confidence.

We will firstly discuss modelling on scales where vertical dispersion is important, before discussing problems involving longer range transport.

10.3.1 Modelling in the Near Field

In this section we will discuss modelling of pollutant dispersion from 0 to ~ 100 km, using the Gaussian plume approach. This is not to condemn more sophisticated methods, but for most practical applications, the quality of the available emission and meteorological data does not justify the increased resources required to set up and run more complex models. In many cases a sound knowledge of meteorology and aerodynamics can be used to parameterize the Gaussian model to simulate adequately, for example, the effects of buildings on dispersion.

In the Gaussian plume approach the expanding plume has a Gaussian, or Normal, distribution of concentration in the vertical (z) and lateral (y) directions as shown in Figure 10.4. The concentration C (in units of μg m^{-3} for example) at any point (x, y, z) is then given by:

$$C(x, y, z) = \frac{Q}{2\pi\sigma_y\sigma_z U}\exp\left[-\frac{y^2}{2\sigma_y^2}\right]\left\{\exp\left[-\frac{z - H_e)^2}{2\sigma_z^2}\right] + \exp\left[-\frac{(z + H_e)^2}{2\sigma_x^2}\right]\right\} \quad (2)$$

where Q is the pollutant mass emission rate in μg s^{-1}, U is the wind speed, x, y and z are the along wind, crosswind and vertical distances, H_e is the effective

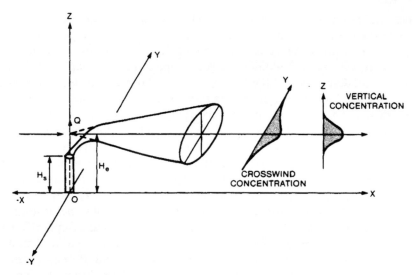

Figure 10.4 *Gaussian plume distribution*

stack height given by the height of the stack plus the plume rise defined below. The parameters σ_y and σ_z measure the extent of plume growth and in the Gaussian formalism are the standard deviations of the horizontal and vertical concentrations respectively in the plume. When $y = z = 0$, this equation reduces to the familiar ground level concentration below the plume centre-line:

$$C(x) = \frac{Q}{\pi\sigma_y\sigma_z U}\exp\left(-\frac{H_e^2}{2\sigma_z^2}\right)$$ (3)

Equation (3) is rather less cumbersome to deal with and several points of interest emerge. Firstly, concentrations are directly proportional to the emission rate, Q, so it is essential that this is known accurately in any practical application. Secondly, unless $H_e = 0$ (*i.e.* unless the source is at ground level) the maximum concentration will occur at a point downwind and this downwind distance will increase with increasing H_e and furthermore the value of C_{max} will decrease with increasing H_e. In fact, C_{max} is roughly proportional to H_e^{-2}. This is the mathematical statement of the so-called 'tall stacks' policy which under-pinned air pollution control in the UK until relatively recently.

The specification of H_e involves calculating the plume rise, which is the height above the point of emission reached by the plume due to its buoyancy (if it is warmer than the surrounding air, as most combustion emissions are) or momentum (plumes may be driven up stacks at relatively high velocities). For most plumes, buoyancy is the dominating force and there have been a large number of studies of methods of determining plume rise. A widely used method however is that due to Briggs,[4] where the plume rise ΔH is given by:

$$\Delta H = 3.3(Q_H)^{1.3}(10H_s)^{2.3}U^{-1} \quad \text{for} \quad Q_H \geqslant 20 \text{ MW}$$

or (4)

$$\Delta H = 20.5(Q_H)^{0.6}(H_s)^{0.4}U^{-1} \quad \text{for} \quad Q_H \leqslant 20 \text{ MW}$$

where Q_H is the sensible heat emission from the stack and U is the wind speed at the stack height (H_s). The dependence on Q_H is not strong and, if measured values of this quantity are not available, an approximation often used is to assume Q_H is equal to one-sixth of the total heat generated in combustion of the fuel. An expression for ΔH due to Moore[5] has been developed for power stations in the UK and is:

$$\Delta H = aQ_H^{1.4}/U \tag{5}$$

where $a = 515$ in unstable and neutral conditions and $a = 230$ in stable atmospheres, with Q_H in MW and U in m s^{-1}. More recent (and more complex) formulations of Moore's and Briggs' formulae have been summarized by the UK Dispersion Modelling Working Group.[6]

The standard deviations of the plume in the vertical and lateral directions, σ_z and σ_y, are extremely important quantities. They are determined by the prevailing atmospheric turbulence in the boundary layer.

Turbulent motions or eddies in the atmosphere vary in size and intensity; the greater their size and/or intensity the more rapid is the plume growth and hence the dilution of the pollutants. Small scale turbulent motions tend to dominate the plume growth close to the point of emission where the plume is still relatively small and the larger scale eddies dominate at greater distances. Furthermore, the small and larger eddies are associated respectively with short and longer time scales. Consequently, σ_y and σ_z increase in value with distance from the source (see Figure 10.4); also they increase with the time or sampling period over which they have been measured. This latter point means that it is essential to state the sampling period to which σ_y and σ_z apply, especially if comparisons are being made between calculated and measured concentration; ideally the two periods should be identical. Most values of σ_y and σ_z to be found in the literature are for sampling periods in the range 3–60 minutes. It is also evident that σ_y and σ_z are dependent on atmospheric stability, being smallest when the atmosphere is most stable (category G) *i.e.* when atmospheric turbulence is least, increasing to their greatest values in highly turbulent very unstable conditions (category A). The underlying surface roughness elements also play a part, σ_y and σ_z increasing with increasing surface roughness so that for a given distance downwind of a chimney σ_y and σ_z will be larger in, for example, an urban area than in an area of open, relatively flat agricultural land.

In general, lateral (horizontal) motion is less constrained than vertical motion with the result that there are larger scale eddies in the horizontal than in the vertical. Fluctuations in wind direction also become important for longer sampling periods. Consequently, σ_y increases more rapidly with increasing sampling or averaging period than does σ_z. This dependence of σ_y on wind direction fluctuation also means that for longer sampling periods, say greater than one hour, σ_y values can increase with increasing atmospheric stability because during low wind speed stable conditions plume meandering can be significant.

Ignoring for the moment the plume meandering component of σ_y the parameters are often conveniently expressed in the form:

$$\sigma_y = \sigma_{yo} + ax^b$$
$$\sigma_z = \sigma_{zo} + cx^d \tag{6}$$

where a, b, c and d are constants dependent on atmospheric stability, x is the downwind distance from the source and σ_{yo}, σ_{zo} are the initial plume spreads generated by, for example, building entrainment. To incorporate plume meander into σ_y, an extra term is added so that:

$$\sigma_y^2 = \sigma_{yt}^2 + 0.0296Tx^2/U \tag{7}$$

where σ_{yt} is given by Equation 6 and T is the averaging time in hours.

A simple expression for σ_z based on Smith's[1] work is:

$$\sigma_z = \sigma_{zo} + 0.9(0.83 - \log_{10}P)x^{0.73} \tag{8}$$

which gives a good representation out to ~ 30 km. Here P is Smith's stability parameter equal to 3.6 for neutral conditions and ranging from 0–1 (stability A) through to 6–7 in stability G.

Values of coefficients specifying σ_{yt}, the so-called microscale σ_y, *i.e.* not including any plume meander effects, are given in Table 10.3. This table also includes typical values of mixing heights in the stability categories A–G. The effect of the mixing height on vertical plume dispersion can be taken into

Table 10.3 *Typical mixing heights and coefficients in $\sigma_y = cx^d$ (x in km) for different stabilities*

	Stability					
	A	B	C	D	E	F/G
Mixing height (m)	1300	900	850	800	400	100
c	213	156	104	68	50.5	34
d	0.894	0.894	0.894	0.894	0.894	0.894

account in the following modification of Equation 2 for ground level concentrations:

$$C(x) = \frac{Q}{\pi \sigma_y \sigma_z U} \exp\left[-\frac{y^2}{2\sigma_y^2}\right]\left\{\exp\left[-\frac{H_e^2}{2\sigma_z^2}\right] + \exp\left[-\frac{(2L - H_e)^2}{2\sigma_z^2}\right]\right\} \quad (9)$$

where L is the mixing height. This equation is not valid for $H_e > L$ when the concentration is zero (*i.e.* the pollutant is emitted above the mixing height).

Where long period averages (*e.g.* annual) are of concern the detailed dependence on σ_y is of much less importance and the pollutant concentrations can be assumed to be uniformly distributed cross-wind within each wind sector. For a 30° sector the first two terms of Equation (9) become:

$$\frac{1.524Q}{U\sigma_z x} \quad (10)$$

The contribution of this wind sector to the overall annual average is then given by Equation (9) modified as in Equation (10) multiplied by the combined frequency of occurrence of that wind sector and stability category.

In Section 10.2.1 we introduced the concept of the logarithmic wind speed profile with height in conditions of neutral stability. In different atmospheric stability conditions the variation will be different, but, in general, the wind speed will increase with height because of the surface drag. As would be expected intuitively this variation is smallest in unstable conditions (since in such boundary layers there is a considerable degree of vertical mixing) and greatest in stable conditions (for the opposite reason). In general one can write:

$$U(z) = U_{10}(z/10)^\alpha \quad (11)$$

where U_{10} is the 10 metre wind speed and α is ~ 0.15 in unstable conditions, ~ 0.2 in neutral and ~ 0.25 in stable conditions.

There are several interesting derivations from the standard Gaussian equation and a particularly useful one is the formula giving the concentrations from a line source (of infinite length) obtained by integrating, for simplicity, Equation (2) over y to yield:

$$C(x) = \sqrt{\frac{2}{\pi}} \frac{Q}{\sigma_z U} \exp\left[-\frac{H_e^2}{2\sigma_z^2}\right] \quad (12)$$

which can be used to estimate the concentration downwind of roads, for example. Here, Q is the mass emission rate per unit length of road ($\mu g\ m^{-1}\ s^{-1}$).

10.3.2 Emission Inventories

We have already seen how important it is to specify the emission rate of a single source in order to model concentrations with confidence. In single stack applications this is often relatively straightforward. However, in multiple source applications such as in the use of an urban air quality model, there can in principle be literally thousands of individual sources. It would be clearly impracticable to attempt to quantify the emission rate of every house, office, shop and car in, say, London so methods have to be devised of making the problem tractable yet retaining as accurate a description of reality as possible.

The usual way of achieving this is to apportion the area to be modelled into a grid and to combine all the numerous small emitters within each grid square (such as individual houses, cars, *etc.*) into so-called 'area sources'. Major sources are usually treated explicitly as individual point sources. The size of the grid square used will usually be determined by the size of the area, or domain, to be modelled and the computing resources available. Typical grid scales are 1 km (or smaller) for urban areas, 20 km for nationwide modelling in a country the size of the UK, and 50–100 km for European or other international scale long range transport.

The specification of emissions is therefore fundamental to modelling and, apart from single source problems, is a difficult task. The usual approach is to collect information on fuel consumption in particular sectors (such as power generation, domestic heating, *etc.*) and multiply this by appropriate emission factors which ideally will have been measured over a range of representative fuels, appliances and combustion conditions.

Very often such data on fuel consumption (or some other measure of industrial commercial activity) are available only on a large scale, *e.g.* at national level, when the area to be modelled is much smaller. The modeller then has to use some means of spatially disaggregating the total domain emissions over the individual grid square of the model. This usually involves the introduction of another level of uncertainty as surrogate statistics have to be employed – for example domestic heating emissions may be assumed to have the same spatial pattern over the model grid as population for which data are often fairly readily available. Other surrogates which can be used are office floor space for emissions from the commercial sector and population for motor vehicle emissions. If the domain is of an appropriate size then questionnaires and other, often labour intensive, techniques can be used to assess the magnitude and spatial pattern of emissions in a particular town or city.[7] Some examples of urban and national emission inventories are given in Figures 10.5 and 10.6. The emissions shown in Figure 10.5 formed part of a study of the London area by the London Research Centre[7] which collected data on fuel use and traffic activity and produced inventories of the emissions of eight pollutants. Figure 10.5 illustrates the emissions of carbon monoxide in the London area (the data are displayed at a resolution of 1 km). Figure 10.6 is a map of total NO_x emissions for the UK for 1997 at a resolution of 1 km and has been derived by NETCEN on the basis of national and regional fuel use and other statistics, and forms part of the National Atmospheric Emissions Inventory.[8]

We have already seen, in Section 10.3.1, how the averaging time inherent in the dispersion model structure and parameters should match that of the

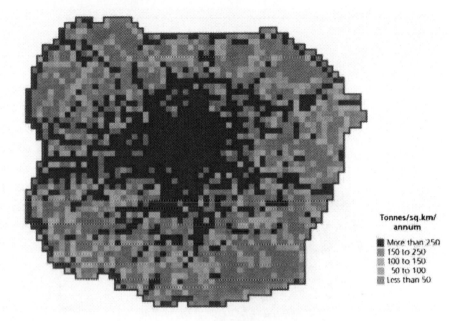

Tonnes/sq.km/
annum

More than 250
150 to 250
100 to 150
50 to 100
Less than 50

Figure 10.5 *Emissions of carbon monoxide in the London area (1 km squares)*

concentrations being modelled. Similarly it is vital that the emissions should be of the appropriate timescale. This is often straightforward in the case of annual averages when annual emissions must be used, but can be more difficult if seasonal or diurnal variations are being modelled. However, various approaches are possible.

Fuel requirements for space heating purposes (residential, commercial and a proportion of industry) depend on the ambient temperature and therefore vary with the season of the year. The Degree–Day principle[9] can be used to calculate these temperature dependent emission rates, $E(T)$, from the annual average emission rate E

$$\begin{aligned}
E(T) &= Eo[0.33 + 0.11(14.5 - T)] \quad &\text{for} \quad T \leqslant 14.5\,^\circ\text{C} \\
E(T) &= 0.33\,Eo \quad &\text{for} \quad T > 14.5\,^\circ\text{C}
\end{aligned} \tag{13}$$

A further factor can be introduced if required to take account of the typical diurnal variation in emissions.

When modelling traffic pollution, diurnal variations in traffic flow are often available so that emissions can be scaled accordingly. One very important feature in dealing with traffic pollution is the variation of emissions with speed. This is particularly important for the pollutants carbon monoxide (CO) and hydrocarbons both of which, being products of incomplete combustion, are formed in the biggest quantities at low speeds. Work at Warren Spring Laboratory using on the road measurements in actual driving conditions[10] has

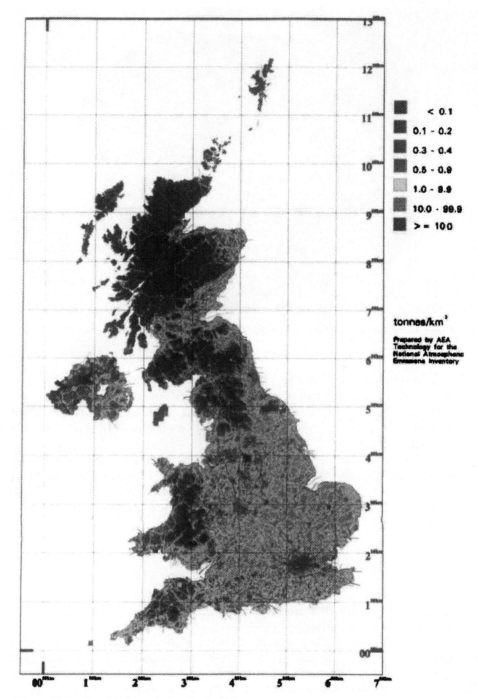

Figure 10.6   *Total NO$_x$ emissions in the UK*

quantified this effect. For example emissions of CO can vary from 25–30 g km^{-1} at 20 kph to 5–10 g km^{-1} at 100 kph, for pre-catalyst cars.

10.3.3 Long Range Transport

With the increasing interest in the problems of acid rain and photochemical ozone in recent years, the modelling of these phenomena has become important. There are several important features which must be considered in modelling long (in this context, greater than 100 km) as opposed to shorter range transport. Firstly the time scales of transport over these distances are such that the removal processes of wet and dry deposition must be incorporated. Secondly large scale meteorological features must be taken into account which involve specifying the movement of an air mass on a synoptic scale. Thirdly chemical reactions will generally be important and must be included. To incorporate all these effects in detail demands immense computing resources, as well as raising questions over how accurately the input data on emissions and meteorology can be known and how well one can describe the detailed physics and chemistry of the processes. In practice, therefore, simplified models have been used. These have generally been of two types, Lagrangian, where a series of air mass trajectories is followed, or Eulerian, where the governing equations are solved for every grid point in the domain at each time step. In many ways Lagrangian models are the simpler in concept and application, as Eulerian models generally require greater complexity of input data and computing requirements, as well as suffering from numerical 'pseudodiffusion' if not appropriately constructed. The Lagrangian models in widest use in the UK take two forms, statistical models and trajectory/box models. The former are typically applied to long period averages of concentration and deposition fields of acidic pollutants such as SO_2 and sulfate, NO_x and nitrate, and use of climatological data or annual frequencies of wind directions, speeds, stability categories, *etc.* Rainfall may be taken into account in several ways, the simplest but not necessarily the most satisfactory being to assume continuous constant rainfall (at an annual average rate). Alternatively the sporadic or stochastic nature of rainfall can be incorporated in a probabilistic way.[11] Trajectory/box Lagrangian models are usually applied to a succession of air mass trajectories arriving at a receptor at relatively short intervals (*e.g.* every six hours in the case of the UN ECE EMEP model[12]). The specification of the trajectories is a fundamental step in using these models and they are usually obtained from the detailed models used in national meteorological services. In practice back-track trajectories up to 96 hours are used; longer timescales would introduce unacceptable errors. As it is, errors in trajectories increase with time back along the path, and particularly in slack pressure areas these can be very large even at relatively short times. These models work by moving the box or air parcel along the trajectory and at each time step the appropriate emissions are introduced from the underlying grid, pollutant is lost by dry deposition at the rate given by $v_g c$ where the deposition velocity v_g is appropriate to the underlying surface type. Wet deposition removes pollutant according to the rainfall field at the particular location of the air parcel, and chemical transformations of reactive species are updated since the previous time step.[12] A good summary of the use of Lagrangian and Eulerian long range transport models of acid rain has been given by Pasquill and Smith.[1]

Much use has been made of Lagrangian box models in modelling photochemical ozone formation in the UK[13] and elsewhere.[14] Because the computational demands of the physical and meteorological aspects of the problems are relatively small, particularly if a one or two layer box model is used, quite complex chemical schemes can be used to describe the chemical processes involved. The model developed by Derwent,[13] for example, uses 339 chemical reactions describing the fate of 40 species. Explicit chemistry is used rather than so-called 'lumped' schemes where for example one hydrocarbon is used as a surrogate for its class (*e.g.* propene could be used to describe all alkenes, *etc*).

In attempting to simulate observed ozone concentrations with Lagrangian models, the correct specification of the air parcel's trajectory is of paramount importance. This may not be easy particularly, as we have noted, when anticyclonic conditions with slack pressure gradients exist. Then, although conditions may be optimal for ozone formation, the uncertainties in the calculated trajectories are often at their greatest. Large differences in the calculated ozone concentration can result, depending on the quantity of precursors (nitrogen oxides and hydrocarbons) picked up along each trajectory.

There have been some recent developments in regional and global modelling which are worth noting. The UNECE/EMEP model, originally developed to address the problem of 'acid rain' in Europe, has been extended to include ozone and more recently particles, as the policy process within the UNECE has expanded its scope. The Lagrangian framework was felt to be inappropriate for this development, and new Eulerian models are being developed for each topic, with the ultimate aim of incorporation into one master model.

Within the UK Meteorological Office, a regional Lagrangian model was developed following the nuclear accident at Chernobyl. This model, NAME, treats the pollutants as particles released into wind, temperature, and rainfall fields from the UK Meteorological Office's numerical weather prediction model. The model gives extremely accurate representations of the timing of pollution events at a given receptor, and is currently being extended to incorporate chemical processes involved in particle and ozone formation.[15] A further development in global modelling at the UK Meteorological Office has been *via* the STOCHEM model.[16] This is also a Lagrangian model but which operates in a global three-dimensional domain, incorporating chemistry as well as physical transport and dispersion. The model is being applied to problems in climate change, acidification and photochemical ozone.

10.3.4 Operational Models

Although much basic research has been, and continues to be, carried out in dispersion modelling, an essential feature of models is their use in practical operational situations. Such uses are often made by non-specialists, and in the last decade or so there have been some significant developments in packaging models to facilitate their use. These developments have been made possible by the rapid expansion in personal and desk-top computing over this period, so that quite major calculations are now possible on PC-based systems without recourse to sophisticated computing facilities. It is worthwhile repeating, however, the crucial importance of high quality input data, particularly on emissions, as the numerical modelling calculations become easier through the use of standard packages.

Even without recourse to computers, some very useful estimates of air quality impacts of single sources can be made by the use of graphical workbooks such as the well-known 'Workbook of Atmospheric Dispersion Estimates' by Turner,[17] and the NRPB report R91 referred to earlier.[2]

While such workbooks are probably most useful for screening calculations, more complex calculations require computer-based models, and the US EPA has produced a set of approved models, based on the Gaussian plume approach, in its Users Network for Applied Modeling of Air Pollution (UNAMAP) system. A variety of models is available on diskette at low cost, covering single and multiple point source, area and line source applications. The models also have options to allow the influence of complicating factors such as building effects and the effects of complex terrain to be taken into account, in a relatively simple and approximate way. To assist the non-specialist in the use of these models, various companies operate training courses or consultancies. Other developments of the basic Gaussian plume model which have been produced by some companies in recent years have involved the addition of user-friendly software to assist in the development and handling of emission inventories and monitoring data for use with the models as part of an integrated package. While the input and output routines are often sophisticated, at heart these systems at present are generally built on the basic Gaussian plume model, although modular upgrading is often possible, to incorporate developments in dispersion theory.

Considerable developments in dispersion theory have been made over the past two decades, and the approximations inherent in the Gaussian plume approach have become more clearly understood. This has led to the development of a 'second generation' of operational models in the UK and in Europe and the USA. In the UK the most widely known of such models is the so-called UK-ADMS (Atmospheric Dispersion Modelling System) funded by a consortium of bodies including regulatory agencies, electricity generators, nuclear agencies and industrial companies. ADMS uses a numerical description of the boundary layer based on the Monin–Obhukov length scale, discussed in section 10.2.2. One practical consequence of the revised boundary layer description is that the standard deviations of the pollutant plume (σ_y and σ_z in the above discussion) can now vary with source height, a feature which accords more closely to observations. Equally, the boundary layer description also facilitates the incorporation of modules to treat building and topography effects in a more coherent manner.

Although these newer models offer the attraction that they incorporate more recent thinking on the properties of the atmospheric boundary layer and its effects on pollutant dispersion, there remains further work to be done in assessing the performance of such models in practical operational applications.

10.3.5 Accuracy of Models

In using models such as those described in Sections 10.3.2 and 10.3.3 for air pollution control purposes, for assessing air quality impacts or for elucidating chemical and transport mechanisms, it is important to assess the accuracy with which the model can reproduce observed concentrations. Detailed model validation exercises can be expensive, involving considerable resources in measurements of pollutants at many locations and timescales, and the appropriate meteorological variables over the domain of the model. In general the more complex the model, the more complex is the validation required. However, the confidence one has in the output of a model depends very much on the questions one is attempting to answer. The prediction of a short-term peak concentration at a specific location and at a specific time will place different demands on a model from answering a question such as whether or not controlling a particular category of source in a region would have a beneficial or an adverse effect.

In general, long-term (*e.g.* annual) average concentrations can be predicted with greater confidence than short-term (hourly or less) averages. Some assessments of the likely accuracy of dispersion models in predicting the concentrations of non-reactive pollutants have been given by Jones[18] for single source situations. In summarizing the work of several authors, he suggests that annual averages from a low level release can be predicted within a factor of about two. At larger distances the factor increases with concentrations at about 100 km being predicted to within a factor of four with high probability. Factor of two accuracy for peak hourly concentrations is also suggested by Jones with the indication that if specification of the time and location of the peak are also required then the accuracy is likely to be worse.

In urban areas, where there are usually numerous sources on all wind directions around receptors, annual average concentrations of relatively inert pollutants, such as SO_2, can generally be predicted to better than a factor of two accuracy on most occasions. A summary of applications of a climatological Gaussian plume model to several urban areas of the UK[19] is shown in Figure 10.7. This diagram shows the frequency distribution of the percentage error of the calculation of annual average SO_2 concentrations. The percentage error here is defined as 100 × (modelled value − observed)/observed. For London, for example, the modelled results were within ± 30% of observed for about 75% of the 20 receptors considered.

In recent years a considerable amount of work has been done on methods for the evaluation of models. A good summary is given by Hanna,[20] and a more recent update of work in the US can be found on the web address, www.dmu.dk/atmosphericenvironment/Harmoni/ASTM_key.htm

In the context of the evaluation of air pollution control strategies it should be noted that the Gaussian plume model and the behaviour of the primary pollutants to which it is usually applied are linear in emission rates so that such

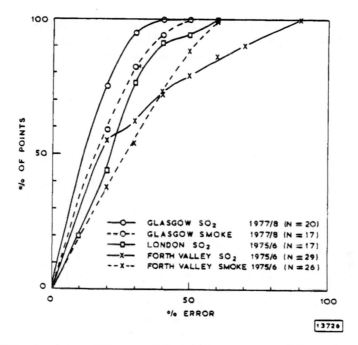

Figure 10.7 *Percentage of Receptor Points within percentages of observed concentrations for annual average SO₂ concentrations*

models will predict concentration reductions proportional to emission reductions which might arise from any postulated control technology. This is very straightforward in single source problems; however, in urban areas a larger number of sources will be present and only one sector (*e.g.* domestic sources under smoke control or motor vehicles under emission regulations) may be subject to controls. The accuracy of prediction of the effects of these controls will then depend on the accuracy with which the proportional contribution of the particular sources is predicted by the model, and this may be difficult to assess.

Turning to the larger scale effects such as acid deposition and photochemical oxidant formation, the question of the linearity, or proportionality, is more important, that is, whether or not for a given reduction in emissions of a particular species (such as SO_2 for example) there is likely to be a proportional reduction in deposition of sulfur. Total sulfur deposition over annual timescales has been shown to be approximately proportional[21] even though in some circumstances the wet deposition component may be non-linear.

The problem of the evaluation of photochemical ozone formation is more complex in that ozone is formed from the atmospheric reactions of nitrogen oxides and many individual hydrocarbon species. The governing reaction schemes are overall non-linear and can be further complicated by the fact that NO_x (and some hydrocarbons) can act as both sources and sinks of ozone, on different scales.

With increasing interest in global problems such as the effects of so-called greenhouse gases on climatic change, and on the effects of man-made pollutants on stratospheric ozone, modelling techniques are now being used to advise pollution control policies over a very wide range of atmospheric problems.

10.4   References

1. F. Pasquill and F. B. Smith, 'Atmospheric Diffusion', Ellis Horwood, Chichester, 1983.
2. 'A Model for Short and Medium Range Dispersion of Radionuclides Released to the Atmosphere', First Report of a UK Working Group on Atmospheric Dispersion, ed. R. H. Clark, NRPB Report R91, HMSO, London, 1979.
3. A. M. Spanton and M. L. Williams, 'A Comparison of the Structure of the Atmospheric Boundary Layers in Central London and a Rural/Suburban Site Using Acoustic Sounding', *Atmos. Environ.*, 1988, **22**, 211–223.
4. G. A. Briggs, 'Plume Rise', US Atomic Energy Commission, Washington, DC, 1969.
5. D. J. Moore, 'A Comparison of the Trajectories of Rising Buoyant Plumes with Theoretical Empirical Models', *Atmos. Environ.*, 1974, **8**, 441–457.
6. 'Models to Allow for the Effects of Coastal Sites, Plume Rise and Buildings on Dispersion of Radionuclides and Guidance on the Value of Deposition Velocity and Washout Coefficients', Fifth Report of a UK Working Group on Atmospheric Dispersion, ed. J. A. Jones, HMSO, London, 1983, NRPB Report 8157.
7. C. Buckingham, L. Clewley, D. Hutchinson, L. Sadler and S. Shah, 'London Atmospheric Emissions Inventory', London Research Centre, 1997.
8. J. W. L. Goodwin *et al.*, 'UK Emissions of Air Pollutants 1970–1997', AEA Technology, National Environmental Technology Centre, Culham, UK, 1999.
9. 'Degree Days', Gas Council, London Technical Handbook No. 101.
10. C. J. Potter and C. A. Savage, 'A Summary of Gaseous Pollutant Emissions from Tuned In-service Gasoline Engined Cars Over a Range of Road Operating Conditions'. Warren Spring Laboratory Report LR 447 (AP), Stevenage, UK, 1983.
11. H. Rodhe and J. Grandell, 'On the Removal Time of Aerosol Particles from the Atmosphere by Precipitation Scavenging', *Tellus XXIV*, 1972, **5**, 442–454.
12. Ø. Hov, A. Eliassen and D. Simpson, 'Calculation of the Distribution of NO_x Compounds in Europe', in 'Tropospheric Ozone, Regional and Global Scale Interactions', ed. I. S. A. Isaksen, Reidel, Dordrecht, 1988.
13. R. G. Derwent and A. M. Hough, 'The Impact of Possible Future Emission Control Regulations on Photochemical Ozone Formation in Europe', HMSO, London, Harwell Report AERE R 12919, 1988.
14. A. Eliassen, Ø. Hov, I. S. A. Isaksen, J. Saltbones and F. Stordal, 'A Lagrangian Long-Range Transport Model with Atmospheric Boundary Layer Chemistry', *J. Appl. Met.*, 1982, **21**, 1645–1661.
15. D. B. Ryall and R. H. Maryon, 'The NAME Dispersion Model: A Scientific Overview', Met O APR Turbulence and Diffusion Note 217b, UK Meteorological Office, 1996.
16. W. J. Collins, D. S. Stevenson, C. E. Johnson and R. G. Derwent, 'Tropospheric Ozone in a Global-scale Three-dimensional Lagrangian Model and its Response to NO_x Emission Controls', *J. Atmos. Chem.*, 1997, **26**, 223–274.
17. D. B. Turner, 'Workbook of Atmospheric Dispersion Estimates', Second Edition, Lewis Publishers, Chelsea, Michigan, 1994.
18. A. Jones, 'What is Required of Dispersion Models and Do They Meet the Requirements?' Paper to 17th NATO/CCMS International Technical Meeting on Air Pollution Modelling and its Applications, Cambridge, 1988.

19. M. L. Williams, 'Models as Tools for Abatement Strategies', in 'Acidification and its Policy Implications', ed. T. Schneider, Elsevier, Amsterdam, 1986.
20. S. R. Hanna, 'Air Quality Model Evaluation', *J. Air Pollution Control Assoc.*, 1988, **38**, 406–412.
21. 'Acid Deposition in the UK 1981–1985', Second Report of the UK Review Group on Acid Rain, Warren Spring Laboratory, Stevenage, UK, 1987.

CHAPTER 11

The Health Effects of Air Pollution

S. WALTERS and J. AYRES

11.1 INTRODUCTION

Worries that air pollution may have significant effects on health have recently been fuelled by publication of new evidence linking low levels of ambient air pollution with small public health effects. However, this is a subject that also attracts a great deal of controversy, with strong views both in favour and against.

This chapter concentrates on health effects of air pollution in the general population at normal ambient levels. It will discuss important factors that need to be considered when reading about health effects of air pollution, enabling readers to appraise critically the extensive literature for themselves, and provide a brief summary of the effects on health of particulates, sulfur dioxide, nitrogen dioxide, ozone and carbon monoxide. It will also address the possible relationship between air pollution and cancer.

11.1.1 Exposure and Target Organ Dose

In order to suffer health effects, an individual must be exposed to a pollutant, and the pollutant must be able to reach those parts of the body that are vulnerable to its effect. The main portal of entry is the respiratory tract which is, however, very effective at dealing with noxious substances before they reach the lower airways and lung tissue. For example the nose and upper airways are excellent at filtering and removing coarse particulate material before it reaches the lower airways and lung tissue, and only a fraction containing the finest particles (under 10 μm) can reach the deep lung. Therefore although an individual may breathe a certain concentration of pollutant into the mouth or nose, this is not necessarily the concentration that reaches the target organ or tissue.

It may be easy to estimate the inhaled concentration of pollutant, but less easy to estimate the dose to which the target organ or tissue is exposed.

11.1.2 Factors Affecting Exposure

Target organ dose may be increased by exposure to greater pollutant concentrations, by exposure for a greater length of time, or by behavioural factors, such as level of exercise and time spent in different micro-environments. For example, the concentration of particulates with an outdoor source is usually greater outdoors than indoors, although for ultrafine particles the difference in concentration may be very small, and the total particulate matter concentration including those of indoor origin may be higher indoors. Conversely, the concentration of nitrogen dioxide may be higher indoors, particularly if gas is used for heating or cooking.[1] Therefore people who work outdoors, or children who play outdoors may be exposed to greater particulate matter dose, whilst those who remain indoors may be exposed instead to higher doses of nitrogen dioxide.

Dose may also be affected by exercise. This both increases the volume of air that is inhaled per minute, and decreases the effectiveness of the nasal filter due to increased mouth breathing. However, faster breathing rates are associated with reduced ultrafine particle deposition due to the reduced time available for sedimentation and diffusive collection.

Dose may also be enhanced by individual sources of exposure, notably cigarette smoking and occupational exposures, and possibly by co-exposure to other factors such as other pollutants or viral infections.

11.1.3 What is a Health Effect?

This is not a simple question. It is easy, for example, to measure small but subtle changes in lung function (for example changes in the amount of air that can be exhaled in one second, or the FEV_1) or bronchial reactivity (how sensitive the lung is to challenge with drugs that make the airways constrict) when an individual is exposed to pollutants in the laboratory. However, these changes are usually transient and fully reversible in the experimental setting, and it may be argued that they are not lasting health effects. It may also be that a small change in bronchial reactivity is of no consequence to a normal individual, but makes a great deal of difference to a person suffering from asthma, who may suffer an attack as a result.

At the other end of the scale, we may observe an increase in deaths or hospital admissions on days following high levels of pollution. However, this increase may represent in the majority of cases an effect on individuals who were already suffering from severe disease, brought forward maybe by just a few weeks or days.

Figure 11.1 represents a pyramid of severity of effects that may be observed after exposure to pollutants. The strata are not meant to be quantitative, since quantification of the effects of air pollution is complex. The further question remains as to where effects cease to be reversible and become lasting health effects.

11.1.4 Time Scales of Exposure–Effect Relationships

Some pollutants may have very rapid effects, for example sulfur dioxide causes constriction of the airways (bronchoconstriction) in sensitive individuals after a

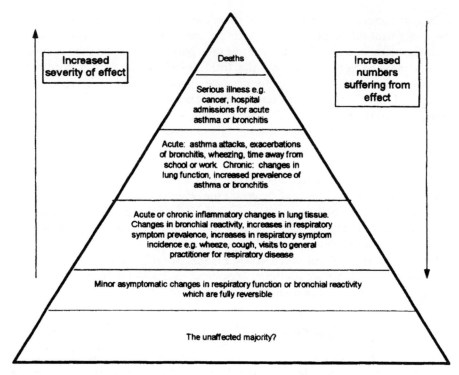

Figure 11.1   *The pyramid of health effects that may be associated with ambient air pollution*

few minutes' exposure.[2] However, some pollutants may only affect individuals after prolonged or repeated exposure, which may take years (Figure 11.2). A good example is that of asbestos and pleural cancer, which occurs 20–30 years after exposure. It is much harder to establish a relationship for conditions with long latencies, such as cancer, since it is difficult to estimate exposure over a long period when both the concentration and the composition of ambient pollution has been changing, and when individuals move their residence, change their jobs, and may change their smoking habits.

Pollutants may also interact with each other[3] or with other environmental factors such as temperature and airborne allergens[4] or even cigarette smoke, to produce a health effect. The exposure required to produce the effect may be either simultaneous or sequential, and the concentration of pollutant contributing to such an effect may vary from episode to episode.

11.1.5   Confounding Factors

In epidemiological studies of the relationship between exposure and disease, a possible cause may appear to be related to the potential effect, but this is only because they are both related to a third *confounding* factor. Confounding may

Short-term health effects (minutes to months)
◊ Inflammatory cells in the lung
◊ Bronchoconstriction
◊ Changes in bronchial reactivity
◊ Minor respiratory symptoms – cough, wheeze, sore throat, sore eyes
◊ Hospital admissions for respiratory and circulatory diseases
◊ Deaths from respiratory and cardiac diseases

Medium-term health effects (months to 10 years)
◊ Increased prevalence of cough, wheeze, asthma, bronchitis
◊ Increased susceptibility to infection
◊ Reduction in lung function
◊ Reduction in lung growth in children
◊ Long-term inflammatory changes in bronchial walls (particularly smoke and SO_2 induced)

Long-term health effects (10 or more years)*
◊ Increased incidence of lung cancer
◊ Increased mortality from cardio-respiratory diseases

Figure 11.2 *Effects that have been postulated for exposure to pollutants in the short, medium and long-term. (*These effects are more debatable but are likely to be real although the size of effect is very small)*

occur either for the disease or the exposure, and may be in the short or long term.

For example, short-term exposure to high levels of particulate matter in winter is associated with still and cold anticyclonic weather conditions. However, cold snaps of this kind are also associated with increased mortality from cardiac and respiratory complaints.[5,6] Therefore an observed association between high levels of particulate matter and increased mortality may be due to confounding effects of temperature, unless this is taken into account in the analysis.

In the longer term, we may observe a high prevalence of asthma and respiratory disease in residents near an industrial site. However, the population living in the zone of high exposure may live in poor social conditions, have high levels of occupational exposure, a high prevalence of cigarette smoking and so on. All these factors are also associated with respiratory disease, and unless they are taken into account the effects of the factory on health may appear exaggerated.

11.2 STUDYING THE HEALTH EFFECTS OF AIR POLLUTION

Bearing in mind the above difficulties in establishing a health effect, and determining exposure to pollutants for individuals and for target organs, the effects of air pollution on health can be studied in a variety of different ways ranging from simple experiments to complex epidemiological studies.

11.2.1 Experimental Studies

Experimental studies are designed to test a hypothesis, and therefore involve a measured exposure in a controlled system designed to test a specific effect. Experiments may be designed either to test the dose at which a toxic response occurs, or to observe more subtle mechanisms of effect. In general, experimental studies involve *in vivo* exposure in animals, or *in vitro* exposure of human or animal tissue or bacterial cultures. Experiments may involve exposure to higher concentrations than in ambient settings to produce a rapid result, or be specifically designed to look at the effect of low level concentrations. The toxicity of many atmospheric pollutants could only be determined in experimental settings. The experimental approach has also been used to determine the mutagenicity of potential carcinogens (cancer-causing agents), and the effects of combinations of pollutants, for example particulates matter coated with weak solutions of acid. They are very valuable in that confounding factors are eliminated by the controlled conditions, but can be misleading if different species have a different susceptibility to a pollutant, or where specific combinations of exposure are needed to produce an effect.

11.2.2 Controlled-chamber Experiments

These consist of experimental exposures carried out under controlled conditions on human volunteers to examine physiological effects in terms of lung function or symptoms. By exposure to different concentration/time combinations it can be possible to establish a dose–response relationship. The volunteers can be normal individuals or those suffering from certain medical conditions such as asthma. Target organ dose may be increased by making subjects exercise. The outcomes measured usually include lung function, bronchial reactivity, markers of inflammation (in washings from the lung or nose), presence of symptoms. If an association between exposure and effect is found, causality may be established because of controlled exposure to a single or known combination of pollutants. They are also useful in establishing the range of inter-subject variability, and whether certain groups of people are more sensitive. However, these experiments can only consider short exposures (hours) and, because they are difficult and expensive to do, small numbers of subjects. People with severe respiratory disease may be unable to participate, and volunteers may not be typical of either the general normal population, or of all people with asthma or respiratory complaints. Nevertheless, these studies have been useful in determining whether effects occur and have been used by standard setters in their deliberations.

11.2.3 Epidemiological Studies

Epidemiological studies estimate the effect on the health of whole populations, rather than individuals in experimental settings. They may deal with either short-term health effects resulting from short-term variations in exposure, or long-term health effects resulting from long-term average exposure.

11.2.3.1 Short-term studies. Short-term epidemiological studies are of two essential types:

(1) *Ecological studies.* These examine the effects of day-to-day changes in air pollution levels on routinely measured health outcomes such as hospital admissions or mortality.
(2) *Panel studies.* In these studies, panels of individual volunteers measure their lung function and record the presence of symptoms every day, which are then related to ambient air pollution levels. These short-term studies generally refer to *incidence* or onset of disease, *e.g.* asthma attacks, people developing a new cough on a particular day and so on.

Epidemiological studies are useful because they reflect real-life exposure conditions. Ecological studies cover the whole population, and therefore no groups are excluded, whilst panel studies can be specifically constructed to look at effects in vulnerable groups, or differences in individual and group sensitivity. Both types of study may be continued over relatively long time periods. However, they do suffer from great problems with confounding due to concurrent fluctuations in other environmental factors (temperature, humidity, pollen count, virus infections), and cyclical fluctuations and trends in hospital admissions and mortality due to other causes. For a measured ambient level of pollution, the exposures of individuals in the population will differ greatly due to behavioural factors. It is also difficult to estimate the relative importance of different pollutants contained in mixtures that tend to rise and fall concurrently with the weather. For this reason statistical analysis needs to be complex, and it is not usually possible to infer causality from these studies. Finally, it is not possible to infer long-term health effects from the presence of short-term changes in response to changes in pollution. For example, we can conclude that asthma is made worse by air pollution exposure, but not that new cases of asthma are caused by air pollution exposure.

11.2.3.2 Long-term Studies. In contrast to short-term studies, those in the long term examine disease *prevalence, i.e.* the proportion of the population suffering from a particular disease or symptom. These studies are of three main types:

(1) *Cross-sectional studies.* In these studies, the prevalence of disease in different communities is compared contemporaneously with the average ambient level of pollution in those communities.
(2) *Cohort studies.* These studies follow up a group of individuals over a

period of time, looking for changes in disease prevalence in relation to changes in average pollution exposure in this group.

(3) *Migrant studies.* These study prevalence of disease in groups whose pollution experience has changed by virtue of migration to an area with a markedly different ambient level.

These studies also reflect real-life exposure conditions. However, they suffer both from the effect of confounding factors (such as effects of cigarette smoking, exercise, occupation, race, socio-economic deprivation) on disease prevalence, and also the problem of estimating exposure over the whole latent period during which disease may develop. Very frequently it is not possible to take account of all such confounding factors in the design or analysis of cross-sectional studies. In addition, current disease patterns may reflect differences in exposure of individuals that were present 30 years ago but no longer exist today. Cohort studies allow more precise control of confounding factors, since these can be determined for all individuals enrolled, but such studies are expensive and time-consuming, and it may be difficult to both trace individuals and estimate their exposure over a long time period. Migrant studies suffer from the problem that migrants often differ fundamentally both from the population they left and the new population that they join, but can be very useful in establishing whether there is, for example, a critical age at which exposure needs to occur.

Finally, there is often a problem with statistical power. In order to be reasonably sure that a finding of no association is true, a large sample size is required, otherwise we may be missing a small but important effect. In this instance, this means comparing disease prevalence in many different sites, which is difficult. Consequently many studies compare only two or three places, making interpretation difficult.

11.2.4 Estimating Public Health Effect

Once a link between pollutant and health effect has been recognized, it is important to establish its potential effect on a population, in order to recommend air quality standards which adequately protect public health. Therefore we need to know the magnitude of the effect, the dose–response relationship, whether there is a threshold for effect, and whether any groups are at particular risk.

These relationships may be different depending on the type of study (*controlled chamber* or *epidemiological*), on the population to whom it refers (studies performed in the United States may not apply to the United Kingdom), on the average ambient levels of pollutants, and on the proportion of the population falling into at-risk groups. For example, studies on effects of nitrogen dioxide appear to be associated with health effects in some ecological studies at concentrations 10 to 100 times lower than are required to produce effects in controlled chamber experiments.[7] Some studies also appear to show no threshold for effect (this has been suggested to be the case for particulate matter, for example), but it is clearly not practicable to require a zero standard for airborne particulates matter because of natural sources of particulates matter such as wind-blown dust, pollen, spores and seawater aerosols, which cannot be controlled. Epidemiological studies are also not often able to distinguish between the pollutant which produces an effect and those which simply co-vary with that pollutant. Finally, there are relatively few studies that provide *direct* comparison between normal individuals and vulnerable groups, or vulnerable groups which may contain a range of susceptibility.

People may be considered to be at risk either because they have greater *exposure* to pollutants (due to age, behaviour, occupation, activity levels), because they are more sensitive to a given level of pollutant (*e.g.* some patients with asthma), or because the consequences of a given effect are greater for some individuals than others (*e.g.* inhaling a respiratory irritant is of greater consequence to somebody with asthma than to a normal individual). At-risk groups might therefore include children and infants, the elderly, people with existing respiratory disease (asthma, chronic obstructive pulmonary disease), people with ischaemic heart disease (angina pectoris), expectant mothers and non-smokers. The special problems of these groups must be taken into account when assessing appropriate air quality standards.

11.3 HEALTH EFFECTS OF PARTICULATE MATTER

Particulate air pollution was one of the first types of pollution demonstrated to have serious health effects, and for which there is greatest evidence of consistent short and long-term health effects at current low ambient levels in the United States and United Kingdom. There is now good evidence to suggest that short-term changes in mortality, hospital admissions, lung function and symptoms are associated with ambient levels of particulate matter, and there is some evidence to suggest that long-term prevalence of symptoms and mortality from respiratory diseases is associated with ambient particulate matter levels.

11.3.1 Experimental Studies

Because of the difficulties in generating particulates in experimental settings that adequately reflect the mix of particulate matter in the ambient air, few experimental studies have been published. Relatively little is therefore known about the mechanisms by which particulate matter produces either acute or chronic health effects, although knowledge in this field is accumulating rapidly.

Animal exposure studies show that toxicity of particles is dependent upon both size and chemical composition, the size broadly determining where it is deposited and the composition broadly determining the type of response. For the ultrafine particles (< 0.1 μm diameter) size also appears to influence toxicity. Human volunteer studies with concentrated ambient particles have shown that markers of inflammation in the lungs tend to increase after exposure, and there is evidence of changes relating to the blood clotting mechanisms.[8] Diesel exhaust particles may also enhance the human response to inhaled allergen.[9]

There is some evidence that sulfuric acid[10] and ammonium bisulfate can induce airway narrowing in subjects with asthma. Whether acidic particles exert a permissive effect by enhancing the effect of other factors is uncertain.

11.3.2   Ecological Epidemiological Studies

Following a severe episode of winter smog lasting four days in London during 1952 over 4000 excess deaths occurred, mainly from respiratory and cardiac conditions and largely in the elderly.[11] During this episode Black Smoke levels rose well over 4500 μg m^{-3}, along with high levels of sulfur dioxide and acid. This led to legislation to control urban pollution in the United Kingdom and elsewhere. In more recent years, concern has focused on the potential health effects of much lower levels of pollutants, particularly inhalable particulate matter (PM$_{10}$). Recent series of analyses from the United States have demonstrated consistent relationships between PM$_{10}$ (or other measures of particulates) and daily mortality, with approximately a 1% rise in all causes mortality for every 10 μg m^{-3} increase in PM$_{10}$,[12-18] although more recent US studies have suggested a rise in mortality closer to 0.5% per 10 μg m^{-3}.[19] The effect seems to be greatest for respiratory and circulatory causes of death, and in the elderly, and for 3–5 day moving average particulate matter levels, rather than hourly peak levels. These relationships appear to have no lower threshold, and to occur at levels well below current US air quality standards for particulate matter. These changes should be considered in the light of an average PM$_{10}$ level of 20–25 μg m^{-3} in the United Kingdom (see Chapter 7).

A large study in several European cities (the APHEA study) showed that similar associations exist between ambient particulate matter pollution and mortality in the European region, although the rise in mortality for every 10 μg m^{-3} is not so great as in North America at around 0.75%.[20]

There is also consistent evidence from the United States and Europe that hospital admissions and emergency room attendances for respiratory complaints and asthma are related to ambient levels of particulate matter, both smoke and PM$_{10}$, again without threshold and well below the ambient air quality standards.[21-23] In Birmingham, UK, hospital admissions for asthma and respiratory disease are significantly associated with levels of both black smoke[24] and PM$_{10}$.[25] A 10 μg m^{-3} rise in PM$_{10}$ was associated with between 1.5 and 5% increase in hospital admissions or attendances in these studies. Where studies have considered other pollutants simultaneously, the association with PM$_{10}$ usually appears to be the strongest, and remains significant in multiple regression analysis.

The Committee on the Medical Effects of Air Pollution attempted to quantify the effects of particulate matter and other pollutants on mortality and hospital admissions. The estimated magnitude of acute health effects were based on meta-analyses by the World Health Organisation, or the APHEA study, as well as relevant local literature, and are summarized in Table 11.1.[26] In the US, estimates of effect are slightly higher for hospital admissions (1–2% per 10 μg m^{-3}).[19]

Table 11.1 *Dose–response coefficients from short-term ecological time series studies for PM$_{10}$, sulfur dioxide and ozone*
(From Committee on the Medical Effects of Air Pollution 'Quantification of the Effects of Air Pollution on Health in the United Kingdom', London: Department of Health (1998))

Pollutant	Health outcome	Change per 10 μg m^{-3} increase in pollutant
PM$_{10}$	Deaths brought forward (all causes)	+0.75% (24 h mean)
	Respiratory hospital admissions	+0.8% (24 h mean)
Sulfur dioxide	Deaths brought forward (all causes)	+0.6% (24 h mean)
	Respiratory hospital admissions	+0.5% (24 h mean)
Ozone	Deaths brought forward (all causes)	+0.6% (8 h mean)
	Respiratory hospital admissions	+0.7% (8 h mean)

Although other pollutants show significant association with deaths and hospital admissions, the estimated effect is greatest for particulate matter.

11.3.3 Epidemiological Panel Studies

Panel studies in the United States and Europe have often but not invariably demonstrated a significant association between ambient levels of particulate matter and lung function, symptom incidence and use of treatment in children.[21,29] Children with symptoms appear in some, but not all, studies to be more sensitive to ambient particulate matter. The overall fall in lung function is small (less than 0.5% reduction in peak flow for every 10 μg m^{-3} rise in particulate matter), and reversible. Recent studies have again suggested that there is no threshold, and that a relationship remains below existing air quality standards.

11.3.4 Long-term Epidemiological Studies

Many early studies did not adequately adjust for confounding factors such as smoking prevalence, and only compared two or three sites. In general, these suggested that the prevalence of respiratory symptoms was higher in polluted sites, but only a few studies have demonstrated a reduction in lung function. An association between particulate matter and prevalence of reduced lung function has been found in at least one large and well-conducted study.[30]

The Harvard Six Cities Study carried out over a long period of time has produced several important findings. They found that, in children, prevalence of cough, bronchitis and chest illness was associated with particulate matter

(PM$_{15}$ and PM$_{2.5}$), although less strongly with gaseous pollutants. There was no association between particulate matter and lung function. Prevalence of cough doubled over the measured range of pollutants (PM$_{2.5}$ 12–37 μg m^{-3}), and the effect on symptom prevalence in children with asthma was greater than for normal children. This, and other studies, revealed that the fine fraction of particulates matter, and sulfate levels, showed strongest association with symptoms.[31,32] More recently, a fourteen year follow-up of a cohort of adults from each of the six cities showed a significant association between ambient particulate matter levels and mortality from cardio-respiratory disorders and lung cancer. This study carefully adjusted for confounding factors, although could take no account of the possible effects of pollution exposure in early life, before subjects were enrolled into the study. The closest association was observed between PM$_{2.5}$ and mortality, followed by sulfate and other particulate matter measures. Mortality was over 30% higher in the most polluted city compared to the least polluted city, although pollution levels were relatively low in all cities.[33]

11.4   HEALTH EFFECTS OF SULFUR DIOXIDE

Sulfur dioxide is a potent bronchoconstrictor at high levels, and patients with asthma are much more sensitive than normal individuals. Because levels of sulfur dioxide and particulate matter co-vary closely it has proved hard to demonstrate effects of sulfur dioxide that are independent from the effects of particulate matter in epidemiological studies. It is likely that sulfur dioxide contributes to respiratory symptoms, reduced lung function and rises in hospital admissions seen during pollution episodes although the effects have been more consistently seen in Europe rather than the USA.

11.4.1   Experimental Studies

Exposure to high levels of sulfur dioxide over a long period produces structural changes in the lung, with thickening of the lung lining, increase in glandular tissue, thickening of the protective mucus layer, and reduction in mucus transport (which clears both mucus and other debris from the lung). Sulfur dioxide may also enhance sensitization to allergens, and allergic response once challenged in animals.[34]

11.4.2   Physiological Studies

Sulfur dioxide produces bronchoconstriction in both normal and asthmatic individuals after exposure for only a few minutes. However, the concentration required to produce an effect in asthmatic individuals is only one tenth of that required for normal individuals.[35] Exercise enhances the effect.[36] Both normal and asthmatic individuals vary in their sensitivity to the effect, so that although mean levels required to produce an effect in asthma patients may still be high in comparison with ambient levels (500–1000 μg m^{-3}, 200–400 ppb), the most sensitive individuals may respond at lower levels (100–300 ppb), or with greater falls in lung function.[37] Studies of the interaction between sulfur dioxide and other pollutants, and the effect of sulfur dioxide on allergen responsiveness have been inconclusive.

11.4.3 Ecological Epidemiological Studies

Early ecological studies of the effect of particulate matter and sulfur dioxide on health did not attempt to separate the effects of particulate matter and sulfur dioxide on mortality, since these co-varied closely. In more recent North American studies, independent effects on mortality have generally been found for particulates matter but not sulfur dioxide. Ambient levels of sulfur dioxide were significantly associated with hospital admissions for respiratory conditions and asthma during the summer in Canada.[38] A careful study of the effects of particulate matter and sulfur dioxide on emergency room visits for chronic obstructive pulmonary disease showed significant associations below current European guide levels in Barcelona, Spain.[19] An association was also found between sulfur dioxide levels and hospital admissions for asthma in Birmingham, UK during the summer, with no association with particulate matter levels.[20]

In a recent large European study of short-term changes in mortality and hospital admissions in relation to ambient air pollution, sulfur dioxide was significantly associated with both health outcomes in some locations.[20,39] The effect was greater in Western than Eastern and Central European cities. Although it is clear that ambient levels of sulfur dioxide may still be associated with health effects below current air quality standards, particularly in Europe, not all studies show this effect.

11.4.4 Epidemiological Panel Studies

As with ecological studies, the effects of sulfur dioxide are difficult to separate from those of particulate matter in most early studies. Panel studies of children following winter episodes of sulfur dioxide and particulate pollution in Europe have shown a reversible 5% fall in respiratory function following the episode, although which pollutant was responsible is not clear.[40] In some panel studies, both normal children and those with chronic respiratory symptoms have shown reductions in lung function with rises in ambient particulate matter and sulfur dioxide pollution, but the evidence for an effect on lung function at ambient levels is weaker for sulfur dioxide than for particulate matter. It is thought unlikely that any effect occurs at concentrations below 200 μg m^{-3} (80 ppb), a level that is rarely exceeded in the United Kingdom or other parts of Western Europe nowadays.

11.4.5   Long-term Epidemiological Studies

Studies considering sulfur dioxide and the prevalence of respiratory disease have also appropriately considered particulate matter and other pollutants. Some early studies did not use sufficiently sophisticated analysis to separate out the effects of individual pollutants. In several early studies the prevalence of respiratory symptoms was considered to be associated with ambient sulfur dioxide and particulate matter levels, at levels between 60 and 140 μg m^{-3} (20–50 ppb) but, in general, association has only been shown with respiratory symptom prevalence and not with lung function changes, at levels higher than current UK ambient levels.[41] These effects could not be distinguished from those of particulate matter, and at current much lower ambient levels, the importance of SO$_2$ in this area must be regarded with caution.

11.5   HEALTH EFFECTS OF NITROGEN DIOXIDE

The rise in nitrogen dioxide emissions has led to concern about its health effects. However, evidence for significant short-term and long-term health effects of nitrogen dioxide is less consistent than for particulate matter and sulfur dioxide.

11.5.1   Experimental Studies

Nitrogen dioxide is an oxidizing agent and can thus theoretically damage lung tissue. At very high doses it acts as a potent initiator of inflammation within the lung, preferentially affecting the small airways, close to the site of gas exchange in the lungs. Animal studies have demonstrated that nitrogen dioxide in high doses can impair ability to fight infection, although all these studies involved nitrogen dioxide exposure to levels far exceeding ambient air, and exposure to high doses of infective agent. One study in humans showed no significant increase in infection rates, although it was higher in nitrogen dioxide exposed individuals.[42] Long-term exposure to high concentrations may produce lung scarring (fibrosis) and emphysema in animals[43] but there is no evidence for this occurring in man.

11.5.2   Physiological Studies

Many controlled human exposure experiments have been carried out, often with conflicting results. This is because exposures took place under different conditions, in subjects with different characteristics, often with different measured end-points. An overview of the various published studies[44] demonstrated that at exposures below 1880 μg m^{-3} (1000 ppb) only 47% of normal individuals showed increases in bronchial reactivity, whereas above this level 79% showed increases. In people with asthma, at exposures under 940 μg m^{-3} (500 ppb) 69% showed changes in bronchial reactivity at rest, and 51% when exercising. The conclusions were that people with asthma were generally more sensitive than normal individuals to nitrogen dioxide, who were unlikely to show significant responses under 1880 μg m^{-3} (1000 ppb) and that exercise modifies the response. The lowest level at which nitrogen dioxide has been *consistently* shown to affect people with asthma is 564 μg m^{-3} (300 ppb),[45,46] which is well above average UK ambient levels (38 μg m^{-3}, 20 ppb) (see Chapter 7).

One study has shown an interaction between nitrogen dioxide and sulfur dioxide,[47] and another more recent study from the UK has shown that low level (762 μg m^{-3}, 400 ppb) nitrogen dioxide exposure enhanced the response to allergen challenge in people with allergic asthma.[48]

11.5.3 Ecological Epidemiological Studies

In most studies of short-term variations in ambient nitrogen dioxide and health, no effect has been found on hospital admissions or mortality, even when these were found for other pollutants in the same study. Studies from Finland have shown an association between nitrogen dioxide and hospital admissions or clinic attendances for asthma, after taking other pollutants into account,[7,49] and a few other studies have shown an association at very low levels. Other studies which have claimed to show associations between nitrogen dioxide and mortality or hospital admissions have not considered whether these were independent of other pollutants. It is therefore likely that the association, if any, between nitrogen dioxide and hospital admissions or mortality is weaker than for other pollutants or that nitrogen dioxide is acting as a surrogate for another pollutant metric whose levels closely match those of nitrogen dioxide.

11.5.4 Epidemiological Panel Studies

Again, the majority of these studies have shown no association between ambient nitrogen dioxide levels and respiratory function. However, some well-designed studies have shown an association between nitrogen dioxide and respiratory symptoms in healthy individuals and people with asthma, the latter studies from Arizona also demonstrating a significant independent effect on lung function in asthmatic individuals.[50] One study from the United Kingdom found an association between nitrogen dioxide levels and lung function in patients with asthma, but particulate matter were was not measured in this study.[51] In general, the association between ambient nitrogen dioxide and effects on panels of individuals appears to be weaker and less consistent than that for other pollutants.

11.5.5 Long-term Epidemiological Studies

Many of the published studies simply compare a polluted with a non-polluted area, and are unable to distinguish between effects of different pollutants. Many published studies have concentrated on indoor exposure to nitrogen dioxide. In general, several cross-sectional studies have demonstrated an association between ambient indoor or outdoor levels of nitrogen dioxide and prevalence of respiratory symptoms, but not lung function, in children.[52,53] The reverse seems true in adults, namely there is an association between nitrogen dioxide exposure and lung function, but not symptoms.[54] One large and well-controlled study from the United States found an independent association between nitrogen dioxide and prevalence of low lung function in young people.[53] It is therefore possible that a long period of nitrogen dioxide is required to affect lung function, these changes becoming manifest only in adults, but the evidence is conflicting.

11.6 HEALTH EFFECTS OF OZONE

There is very good, consistent experimental evidence that ozone has an effect on health, with a consistent dose–response effect on a number of lung function parameters at concentrations close to those seen in ambient air. The evidence from panel studies supports this, although there has been little published evidence of effects on mortality or hospital admissions, particularly in the UK.

11.6.1 Experimental Studies

Ozone is a very powerful oxidizing agent, causing direct cellular damage by damaging the anti-oxidant mechanisms in cells lining the airway walls. It acts preferentially in the small airways and gas-exchange regions of the lung. Prolonged exposure of animals to high doses results in persistent inflammation of the small airways, similar to that induced by cigarette smoking. It may also induce scarring (fibrosis) in the lung. Acute exposure to ozone produces acute inflammation at quite modest levels of exposure (under $2000~\mu g\,m^{-3}$, 1000 ppb). There is also some evidence that prolonged exposure to ozone may impair cellular defences.

11.6.2 Physiological Studies

There have been a large number of these studies carried out, which demonstrate a consistent curvilinear relationship between inhaled ozone concentration and respiratory function at all levels of exercise.[55] Increasing inhaled ozone concentration had a greater effect than either increasing duration of exposure or increasing ventilation (by exercise). Ozone affects a number of lung function parameters, including bronchial responsiveness. Studies have demonstrated effects in exercising individuals at concentrations which frequently occur in ambient air ($160~\mu g\,m^{-3}$, 80 ppb).[56] Individuals can develop tolerance to ozone exposure (diminishing effects with repeated challenge) but it is not known whether this is an adaptive response.[57] Individuals vary widely in susceptibility to inhaled ozone, with some showing great sensitivity. This sensitivity is not confined to those with respiratory disease, but occurs just as frequently in normal individuals.[58] For example people exercising in $240~\mu g\,m^{-3}$ (120 ppb) ozone for 6 hours showed between a 4% and 38% reduction in FEV_1 (forced expiratory volume in one second). Ozone exposure also produces symptoms of cough, breathlessness and chest discomfort.

Exposure to a low level of ozone can also reduce the threshold at which allergic subjects respond to allergen challenge in humans, confirming previous findings in other species, and the effect of NO_2 described earlier.[4]

A review of these physiological studies at the population level suggested that at ambient levels of 200 μg m^{-3} (100 ppb) some sensitive subjects would suffer a 10% decrement in lung function[59] in the presence of exercise.

11.6.3 Ecological Epidemiological Studies

Several studies from North America have shown an association between ambient levels of ozone and hospital admissions for asthma, although not all studies controlled for the effects of temperature and other pollutants.[32,60] Indeed, some have found no association between ozone and hospital admissions for respiratory complaints. More recent studies, with better control of confounding factors, have suggested that between 6 and 24% of the total daily variability in summertime asthma admissions may be due to ozone, alone or in combination with other components of acid summertime haze, but these studies nearly all come from North America.[61] There is no consistent evidence for an acute effect of ambient ozone on mortality. Asthma mortality did not rise following an acute ozone episode in England in 1976.

11.6.4 Epidemiological Panel Studies

The majority of these studies have taken place in children in summer camps in North America,[62-64] although more recently a number of European studies have been published.[65] These show consistent, reproducible and significant reductions in lung function at maximum ambient ozone levels below 500 μg m^{-3} (250 ppb) often below 200 μg m^{-3} (100 ppb) equivalent to a 3% fall in lung function for every 200 μg m^{-3} (100 ppb) rise in ozone, a very small change. A careful analysis in Dutch children showed that there was wide variability in response between children, suggesting that there may be a sensitive sub-group, which is not confined to children with pre-existing respiratory complaints.[66] The characteristics which make these people more sensitive are not known. However, these effects are not always found. Unpublished studies from the UK have not demonstrated adverse responses to ozone in panel studies.

11.6.5 Long-term Epidemiological Studies

There have been few published studies comparing more than two areas, or with good control for confounding factors. Such studies that have attempted to look at the question of whether long-term exposure to ozone increases the prevalence (as opposed to incidence) of asthma have generally been poorly designed. One study in Seventh Day Adventists (who are non-smokers) showed that frequent exceedence of the 200 μg m^{-3} (100 ppb) threshold for ozone was associated with an increased prevalence of asthma in adult males only.[67] The balance of evidence is against the possibility that chronic ozone exposure causes a non-asthmatic individual to develop asthma.

11.7 HEALTH EFFECTS OF CARBON MONOXIDE

It is generally accepted that carbon monoxide exerts its toxic effect by binding very avidly to haemoglobin, thereby reducing the oxygen-carrying capacity of the blood. In very high doses it is fatal due to cerebral and cardiac hypoxia. In lower concentrations it may affect higher cerebral function, heart function, and exercise capacity, all of which are sensitive to lowered blood oxygen content.

It is possible to obtain a direct measure of carbon monoxide exposure in humans by measuring carboxyhaemoglobin levels, normally around 1% of total haemoglobin. Carbon monoxide is present in very high concentrations in cigarette smoke, and cigarettes constitute by far the greatest source of exposure in smokers. This section therefore concentrates on the potential health effects of carbon monoxide in non-smokers exposed to ambient carbon monoxide. There has been relatively little recent research into the potential health effects of low level carbon monoxide exposure.

11.7.1 Experimental Studies

Neurobehavioural effects have been extensively studied in animals. In general these represent reduced ability to carry out complex tasks, although these effects require concentrations greatly in excess of normal ambient exposure. Chronic exposure to low-level carbon monoxide can affect brain structure. Cardiac effects of carbon monoxide include effects on the electro-physiological properties of the heart at quite low levels of carboxyhaemoglobin (5.5%), and may reduce the threshold at which cardiac arrhythmias or arrest can occur.[68]

High levels of carbon monoxide during pregnancy can reduce foetal growth and survival in animals,[69] and carbon monoxide has been shown to preferentially bind to foetal haemoglobin.

11.7.2 Physiological Studies

Neurobehavioural effects in humans suggest that effects on complex task performance (such as driving skills) are unlikely to occur below 5% carboxyhaemoglobin. Many of these studies are not very recent, and do not employ current research design such as double-blind procedures, making results difficult to interpret.

Impairment of exercise performance as measured by maximal oxygen uptake may occur at carboxyhaemoglobin levels between 3–4%, although these levels are still rare in non-smokers.[70] Minor changes in the electrocardiograph may be found in normal individuals but the significance is unclear. Of greater importance are the findings that people with existing ischaemic heart disease who have symptoms of angina pectoris may suffer these symptoms after a lesser degree of exertion following exposure to carbon monoxide. In some good experiments, effects were seen in angina patients at carboxyhaemoglobin levels of 2 to 4%, which may be seen in certain groups of non-smokers with high carbon monoxide exposures.[71,72] In people who have a tendency to irregular heart rhythm, this was enhanced after carbon monoxide exposure (6% COHb), although the clinical significance is unknown.[73]

11.7.3 Ecological Epidemiological Studies

Despite the evidence that patients with angina may be affected by high carbon monoxide levels, there have been relatively few studies of the effects of ambient carbon monoxide. Some studies have found no association between hospital attendances or mortality from ischaemic heart disease and ambient carbon monoxide concentrations. Some studies have found a significant association between carbon monoxide and hospital admissions for myocardial infarction (heart attack), or case-fatality (proportion of cases admitted to hospital who subsequently died) of myocardial infarction once they had reached hospital.[74,75] However, these did not take into account the potential confounding effect of temperature, and cold temperature is a potent cause of coronary artery constriction. Equally, carbon monoxide levels co-vary closely with nitrogen dioxide and some particle metrics so the possibility of these reported effects being at least in part to residual confounding remains.

A recent study found an association between ambient carbon monoxide and hospital admissions for ischaemic heart disease and heart failure. After adjusting for a mass measure of particulate matter, the relationship with heart failure remained significant.[76]

11.7.4 Long-term Epidemiological Studies

The majority of evidence comes from occupational studies looking at the risk of mortality from ischaemic heart disease in groups with high occupational exposure to carbon monoxide, such as bridge and tunnel workers, and drivers.[77,78] Small excess mortality has been reported in these groups after adjusting for smoking prevalence, but it is not clear whether carbon monoxide is causally associated.

Poor foetal outcome is associated with acute episodes of high maternal carbon monoxide exposure,[79] although one case-control study of low birth-weight babies failed to find an association between low birth weight and ambient carbon monoxide concentration in the area of residence of the mother.[80]

11.8   AIR POLLUTION AND CANCER

11.8.1   Problems in Studying Air Pollution and Cancer

Although it is relatively easy to establish whether an airborne chemical has the potential to cause cancer (a *carcinogen*), it is much harder to determine whether it actually does so at ambient concentrations in human populations.

Establishing the potential of a chemical to cause cancer can be done by looking at

(1) Its ability to damage the genetic material in cell cultures or bacterial cultures.
(2) Its ability to cause cancer in animals.
(3) Whether there is a higher incidence of cancer in people who have worked with very high concentrations of the particular chemical.

However, the term 'cancer' represents a variety of different diseases with a complex natural history. Development of cancer may depend on a initial triggering event (*initiation*), followed by exposure to other chemicals which promote the development of tumours or prevent the body from rejecting tumour cells (*promotion*). It may be necessary for these events to occur repeatedly, or in a particular sequence, at particular levels of exposure, and the process from initiation to development of cancer (*latency*) in a human may take decades. Therefore, people developing cancer today may be doing so as a result of exposure to ambient air pollution that no longer exists, and of which there are unlikely to be adequate records.

There is also the problem that inhaled carcinogens tend to be associated either with very common cancers (*e.g.* lung cancer) or very rare cancers (*e.g.* some types of leukaemia). With the former, it is difficult to associate a small effect of low level ambient air pollution from the very great effects of other factors, particularly cigarette smoking. With the latter, it may take a lifetime to accumulate sufficient cases at ambient levels of exposure to have adequate statistical power to detect an effect.

Finally, where a link with human cancer has been established in the occupational setting, this usually involves exposures several orders of magnitude higher than those in ambient air. It is difficult to know if the dose–response curve is valid at much lower exposure levels, or whether there is a threshold for effect. Furthermore the latency may be much longer at lower exposure concentrations, so if epidemiological studies do not take this into account they may have false negative results.

For these reasons, we are still unclear about the potential role of ambient air pollution in the aetiology of cancer in populations. Two examples are explored in this section: the role of benzene in the aetiology of leukaemia, and the role of polycyclic aromatic hydrocarbons in the aetiology of lung cancer.

11.8.2 Airborne Carcinogens

There are many substances which have been shown actually or potentially to cause cancer. These have been classified by the International Agency for Research on Cancer into different categories according to their ability to cause cancer (Figure 11.3).

Group 1 carcinogens present in ambient urban air include benzene. Group 2A carcinogens include benzo[a]pyrene, benzo[a]anthracene and other polycyclic aromatic hydrocarbons (PAH). Group 2B carcinogens include 2-nitrofluorene, 1,6-dinitropyrene and 1-nitropyrene. In general, these carcinogens are present in minute quantities, but there is concern over the potential for low level exposure to cause cancer in human populations although the real effects are likely to be immeasurably small.[81]

11.8.3 Benzene and Leukaemia

Benzene is a group 1 carcinogen, with proven causal association with acute non-lymphocytic leukaemia in humans. The main toxic effects occur on the bone marrow, with toxic exposures producing bone-marrow suppression, and reductions in red cell, white cell and blood platelet production (pancytopenia) which may lead to bone marrow failure (aplastic anaemia).

It is important to adjust for smoking in epidemiological studies because benzene is present in high concentrations in cigarette smoke. Smokers may have up to 10 times the exposure of non-smokers, particularly in rural areas. In the longer term, studies in workers exposed to benzene have clearly demonstrated an excess risk of acute non-lymphocytic leukaemia, but in general this was not detectable in workers exposed to less than 1.5 mg m^{-3} (500 ppb) over a working lifetime,[82] an exposure considerably higher than any achieved by members of the general population. There is evidence of chromosomal abnormalities in

---

Group 1 – proven human carcinogens
Chemicals for which there is sufficient evidence from epidemiological studies to support a causal association between exposure and cancer

Group 2 – probable human carcinogens
Chemicals for which evidence ranges from inadequate to almost sufficient.
    Group 2A: Limited evidence of carcinogenicity in humans and sufficient evidence for carcinogenicity in animals
    Group 2B: Inadequate evidence for carcinogenicity in humans and sufficient evidence for carcinogenicity in animals.

Group 3 – unclassified chemicals
Chemicals which cannot be classified in humans, usually because of inadequate evidence

---

Figure 11.3 *The International Agency for Research on Cancer classification of human carcinogens*

workers exposed to slightly lower levels (0.6 to 40 mg m^{-3}, 200 to 1300 ppb) over a long time period (over 11 years). This contrasts to ambient levels which are usually under 13 μg m^{-3} (4 ppb) in the United Kingdom at the urban roadside and under 3 μg m^{-3} (1 ppb) in rural areas.

Population epidemiological studies are extremely difficult to carry out because of the rarity of this type of leukaemia. Estimates of toxicity at low levels of exposure are therefore made from occupational studies. A combination of estimates of risk from a variety of studies suggest that for a lifetime exposure (70 years) to 1 μg m^{-3} (0.3 ppb) benzene, the excess risk is between three and 30 cases per million population, with the World Health Organisation consensus estimate being an excess risk of around four.[83]

However, acute non-lymphocytic leukaemia is extremely rare. There are only about 6 to 7 cases per million per year in the United Kingdom, or 420–490 per million over a lifetime of 70 years. An additional four cases resulting from ambient levels of 1 μg m^{-3} (0.3 ppb) would be almost impossible to detect in epidemiological studies, and the potential risk at ambient levels of benzene remains difficult to prove at the population level, or indeed in workers with a modest working lifetime exposures. In practice, the risk of leukaemia from ambient benzene exposure is so small as to be unmeasureable.

11.8.4 Polycyclic Aromatic Hydrocarbons and Lung Cancer

Polycyclic aromatic hydrocarbons (PAH) collectively describe a large number of chemicals, many of which, with their metabolites and nitro-derivatives, are known to be animal or human carcinogens. The majority derive from the combustion or organic fuels, including wood, coal, oil, petrol and diesel as products of incomplete combustion. The best studied PAH is benzo[a]pyrene (BaP), which along with others is present in cigarette smoke.

It is known that clearance of BaP is reduced if it is adsorbed onto particles, and that the dose required to produce tumours in animals is reduced if adsorbed onto particles.[84] This may therefore be relevant to the situation in ambient urban air. The fraction of urban particulates containing PAH is known to have carcinogenic effect in animals, and one study demonstrated a dose–response relationship to concentration of BaP in extracts from urban particulates.[85] Human cell mutagenicity has also been demonstrated for urban air particulate extracts. Vehicle exhaust condensates are known to be carcinogenic in animals, and about 40% of this may be attributable to PAH.[86] Carcinogenicity of vehicle exhausts is demonstrable in inhalation experiments and is likely to be mainly due to the particulate fraction and specifically to the 4–7-ring PAH containing fraction.[87] However these experiments establishing carcinogenicity in animals used overwhelming doses of exhaust particulates.

The majority of evidence for carcinogenicity of benzo[a]pyrene comes from occupational studies in coal gasification workers, and coke production. These suggested a dose-dependent risk of lung cancer after some adjustment was made for smoking, but not all studies measured PAH or benzo[a]pyrene exposure directly, but used proxy measures such as where subjects worked in the plant, and the duration of employment. These workers were exposed to levels of benzo[*a*]pyrene over 10 times those which occurred in normal urban air during the 1950s and 1960s.[88,89] More recent studies have detected an excess of lung cancer (for example a relative risk around 1.5 in truck drivers) in workers exposed to high concentrations of vehicle exhausts which could not be attributed to other occupational exposure or smoking.[90] Small excesses have been found in other occupations (relative risks between 1.5 and 7), but some studies have been negative, and none have directly measured levels of PAH to which workers were exposed.

There have been a number of reported epidemiological prevalence studies in the general population demonstrating an excess of lung cancer in urban dwellers over rural dwellers, but many early studies failed to take into account potential confounding factors, particularly cigarette smoking, occupation and socio-economic status. Most, but not all, studies which adjust for smoking prevalence show a raised relative risk of between 1.5 and 2.0.[91-93] However, small differences in the age at which smokers commence regular smoking might produce a relative risk of 1.5 for lung cancer.[94] Urban air pollution has fallen markedly and changed in the sources of particulates and probably the relative content of benzo[*a*]pyrene during the 30 year period during which ambient air pollution might be expected to contribute to current lung cancer rates. Recent studies suggest that only a very small proportion (under 5%) of the excess of lung cancer in urban areas may be attributable to ambient air pollution.[95,96]

It has been estimated that the lifetime risk from exposure to BaP varies from 0.3 to 1.4 deaths per year per 10 000 population per ng m^{-3} BaP. This has been calculated to represent only 3% of lung cancer deaths in Sydney[88] and between 2 and 20% of lung cancer cases in the United States,[97] concurring well with other epidemiological studies suggesting that less than 5% of urban cases of lung cancer may be attributable at least in part to ambient air pollution, of which not all can be attributed to BaP.

11.9 CONCLUSIONS

Although there is clear evidence, at least for some pollutants, of health effects at an individual level, the size of these effects is very small. However, there is increasing evidence that chronic effects may be more substantial and that air pollution remains a significant and measurable threat to health at a population level. For instance a report on the quantification of the health effects of air pollution in the UK has shown that, considering acute effects only, particles contribute to the bringing forward of the date of death of over 8000 individuals per annum. Particles also contribute either to the bringing forward of hospital admissions or may be responsible for admissions which otherwise might not have occurred to a similar extent.[26]

There is a need for more research, particularly into the quantification of both individual and public health effects of air pollution, to guide those who need to control pollution. Research is also needed in the more subtle ways, hitherto unsuspected, in which air pollution may exert effects on health.

11.10   SUGGESTIONS FOR FURTHER READING

1. Department of Health, Reports of the Advisory Group on the Medical Effects of Air Pollution Episodes:
   I   'Ozone', HMSO, London, 1991.
   II 'Sulfur Dioxide, Acid Aerosols and Particulates', HMSO, London, 1993.
   III 'Nitrogen Dioxide', HMSO, London, 1993.
   IV 'Health Effects of Exposure to Mixtures of Air Pollutants', HMSO, London, 1995.
2. Department of Health, Reports of the Committee on the Medical Effects of Air Pollution:
   I   'Asthma and Outdoor Air Pollution', HMSO, London, 1995.
   II 'Non-biological Particles and Health', HMSO, London, 1995.
3. World Health Organisation, 'Air Quality Guidelines for Europe', WHO Regional Publications, European Series No. 23, 1987.
4. D. W. Dockery, C. A. Pope, Xiping Xu, J. D. Spengler *et al.*, 'An Association Between Air Pollution and Mortality in Six US Cities', *New Engl. J. Med.*, 1993, **329**, 1753–1759.
5. Department of the Environment, Reports of the Expert Panel on Air Quality Standards:
   I   'Benzene', HMSO, London, 1994.
   II 'Ozone', HMSO, London, 1994.
   III '1,3-Butadiene', HMSO, London, 1994.
   IV 'Carbon Monoxide', HMSO, London, 1994.
   V   'Particles', HMSO, London, 1995.

11.11   REFERENCES

1. F. E. Speitzer, B. Ferris, Jr, Y. M. Bishop and J. Spengler, 'Respiratory Disease Rates and Pulmonary Function in Children Associated with NO_2 Exposure', *Am. Rev. Respir. Dis.*, 1980, **121**, 3–10.
2. J. R. Balmes, J. M. Fine and D. Sheppard, 'Symptomatic Bronchoconstriction After Short-term Inhalation of Sulfur Dioxide', *Am. Rev. Respir. Dis.*, 1987, **136**, 1117–1121.
3. W. S. Linn, D. A. Shamoo, K. R. Anderson, R.-C. Peng, E. L. Avol and J. D. Hackney, 'Effects of Prolonged, Repeated Exposure to Ozone, Sulfuric Acid and their Combination in Healthy and Asthmatic Volunteers', *Am. J. Respir. Crit. Care Med.*, 1994, **150**, 431–440.
4. N. A. Molfino, S. C. Wright, I. Katz, S. Tarlo, F. Silverman, P. A. McClean, J. P. Szalai, M. Raizenne, A. S. Slutsky and N. Zamel, 'Effect of Low Concentrations of Ozone on Inhaled Allergen Responses in Asthmatic Subjects', *Lancet*, 1991, **338**, 199–203.
5, J. P. Mackenbach, C. W. N. Looman and A. E. Kunst, 'Air Pollution, Lagged Effects of Temperature, and Mortality: The Netherlands 1979–87', *J. Epidemiol. Commun. Health*, 1993, **47**, 121–126.
6. D. B. Frost, A. Auliciems and C. de Freitas, 'Myocardial Infarct Death and Temperature in Auckland, New Zealand', *Int. J. Biometeorol.*, 1992, **36**, 14–17.
7. A. Ponka, 'Asthma and Low Level Air Pollution in Helsinki', *Arch. Environ. Health*, 1991, **46**, 262–270.

8. A. Seaton, A. Soutar, V. Crawford *et al.*, 'Particulate Pollution and the Blood', Thorax, 1999, **54**, 1027–1032.
9. D. Diaz-Sanchez, A. Tsien, J. Fleming and A. Saxon, 'Combined Diesel Exhaust Particulate and Ragweed Allergen Challenge Markedly Enhances Human *in vivo* Nasal Ragweed-specific IgE and Skews Cytokine Production to a T Helper Cell 2-type Pattern', *J. Immunol.*, 1997, **158**, 2406–2413.
10. J. Q. Koenig, W. E. Pierson and M. Horike, 'The Effects of Inhaled Sulfuric Acid on Pulmonary Function in Adolescent Asthmatics', *Am. Rev. Respir. Dis.*, 1983, **128**, 221–225.
11. Ministry of Health, 'Mortality and Morbidity During the London Fog of December 1952', HMSO, London, 1954.
12. J. Schwartz and D. W. Dockery, 'Particulate Air Pollution and Daily Mortality in Steubenville, Ohio', *Am. J. Epidemiol.*, 1992, **135**, 12–19.
13. J. Schwartz, 'Particulate Air Pollution and Daily Mortality in Detroit', *Environ. Res.*, 1991, **56**, 204–213.
14. J. Schwartz and D. W. Dockery, 'Increased Mortality in Philadelphia Associated with Daily Air Pollution Concentrations', *Am. Rev. Respir. Dis.*, 1992, **145**, 600–604.
15. C. A. Pope, J. Schwartz and M. R. Ransom, 'Daily Mortality and PM$_{10}$ Pollution in Utah Valley', *Arch. Environ. Health*, 1992, **47**, 211–217.
16. J. Schwartz, 'Air Pollution and Daily Mortality in Birmingham, Alabama', *Am. J. Epidemiol.*, 1993, **137**, 1136–1147.
17. J. Schwartz, 'Particulate Air Pollution and Daily Mortality in Cincinnati, Ohio', *Environ. Health Perspect.*, 1994, **102**, 186–189.
18. P. L. Kinney, K. Ito and G. D. Thurston, 'A Sensitivity Analysis of Mortality/PM$_{10}$ Associations in Los Angeles', *Inhalation Toxicol.*, 1995, **7**, 59–69.
19. J. Samet, S. L. Zeger, F. Dominici *et al.*, 'The National Morbidity, Mortality and Air Pollution Study: Part II: Morbidity, Mortality and Air Pollution in the United States', *Res. Rep. Health Effects Institute*, 2000, **94**, Part II, 1–84.
20. K. Katsouyanni, G. Toulomi, C. Spix *et al.*, 'Short-term Effects of Ambient Sulphur Dioxide and Particulate Matter on Mortality in 12 European Cities: Results from Time Series Data from the APEA Project', *Br. Med. J.*, 1997, **314**, 1658–1663.
21. C. A. Pope, 'Respiratory Hospital Admissions Associated with PM10 Pollution in Utah, Salt Lake and Cache Valleys', *Arch. Environ. Health*, 1991, **46**, 90–97.
22. J. Schwartz, D. Slater, T. V. Larson, W. E. Pierson and J. Q. Koenig, 'Particulate Air Pollution and Hospital Emergency Room Visits for Asthma in Seattle', *Am. Rev. Respir. Dis.*, 1993, **147**, 826–831.
23. J. Sunyer, M. Aez, C. Murillo, J. Castellsague, F. Martinez and J. M. Anto, 'Air Pollution and Emergency Room Admissions for Chronic Obstructive Pulmonary Disease: a 5-year Study', *Am. J. Epidemiol.*, 1993, **137**, 701–705.
24. S. Walters, R. K. Griffiths and J. G. Ayres, 'Temporal Association Between Hospital Admissions for Asthma in Birmingham and Ambient Levels of Sulphur Dioxide and Smoke', *Thorax*, 1994, **49**, 133–140.
25. J. Wordley, S. Walters and J. G. Ayres, 'Short-term Variations in Particulate Air Pollution and their Association with Hospital Admissions and Mortality in Birmingham', *Thorax*, 1995, **50** (Suppl. 2), A34.
26. Committee on the Medical Effects of Air Pollution, 'Quantification of the Effects of Air Pollution on Health in the UK', The Stationery Office, London, 1998.
27. C. A. Pope and D. W. Dockery, 'Acute Health Effects of PM$_{10}$ Pollution on Symptomatic and Asymptomatic Children', *Am. Rev. Respir. Dis.*, 1992, **145**, 1123–1128.
28. G. Hoek, B. Brunekreef and W. Roemer, 'Acute Effects of Moderately Elevated Wintertime Air Pollution on Respiratory Health of Children', *Am. Rev. Respir. Dis.*, 1992, **142**, A88.

29. W. Roemer, G. Hoek, B. Brunekreef *et al.*, 'Daily Variations in Air Pollution and Respiratory Health in a Multicentre Study: the PEACE Project', *Eur. Respir. J.*, 1998, **12**, 1354–1361.

30. J. Schwartz, 'Lung Function and Chronic Exposure to Air Pollution: A Cross-sectional Analysis of NHANES II', *Environ. Res.*, 1989, **50**, 309–321.

31. J. H. Ware, B. G. Ferris, D. W. Dockery, J. D. Spengler *et al.*, 'Effects of Ambient Sulfur Oxides and Suspended Particles on Respiratory Health of Preadolescent Children', *Am. Rev. Respir. Dis.*, 1986, **133**, 834–842.

32. D. W. Dockery, F. E. Spiezer, D. O. Stram, J. H. Ware *et al.*, 'Effects of Inhalable Particles on Respiratory Health of Children', *Am. Rev. Respir. Dis.*, 1989, **139**, 587–594.

33. D. W. Dockery, C. A. Pope, Xiping Xu, J. D. Spengler *et al.*, 'An Association Between Air Pollution and Mortality in Six US Cities', *New Engl. J. Med.*, 1993, **329**, 1753–1759.

34. F. Riedel, S. Naujukat, J. Ruschoff, S. Petzoldt and C. H. Rieger, 'SO$_2$ Induced Enhancement of Inhalative Allergic Sensitisation: Inhibition by Anti-inflammatory Treatment', *Int. Arch. Allergy Immunol.*, 1992, **98**, 386–391.

35. Department of Health Advisory Group on the Medical Aspects of Air Pollution Episodes: Second Report, 'Sulphur Dioxide, Acid Aerosols and Particulates', HMSO, London, 1992, Ch. 6, pp. 71–100.

36. D. Sheppard, A. Saisho, J. A. Nadel and H. A. Boushey, 'Exercise Increases Sulfur Dioxide Induced Bronchoconstriction in Asthmatic Subjects', *Am. Rev. Respir. Dis.*, 1981, **123**, 486–491.

37. D. Horstman, L. J. Roger, H. Kehrl and M. Hazucha, 'Airway Sensitivity of Asthmatics to Sulfur Dioxide', *Toxicol. Ind. Health*, 1986, **2**, 289–298.

38. D. V. Bates, M. Baker-Anderson and R. Sizto, 'Asthma Attack Periodicity: A Study of Hospital Emergency Visits in Vancouver', *Environ. Res.*, 1990, **51**, 51–70.

39. C. Spix, H. R. Anderson and J. Schwartz, 'Short-term Effects of Air Pollution on Hospital Admissions of Respiratory Diseases in Europe: a Quantitative Summary of APHEA Study Results', *Arch. Environ. Health*, 1998, **53**, 54–64.

40. B. Brunekreef, H. Lumens, G. Hoek *et al.*, 'Pulmonary Function Changes Associated with an Air Pollution Episode in January 1987', *J. Air Pollut. Contr. Assoc.*, 1989, **39**, 1444–1447.

41. World Health Organisation, 'Sulfur Oxides and Suspended Particulate Matter', Environmental Health Criteria No. 8, World Health Organisation, Geneva.

42. S. A. Goings, T. J. Kille, L. R. Sauder *et al.*, 'Effects of Nitrogen Dioxide Exposure on Susceptibility to Influenza A Virus Infection in Healthy Adults', *Am. Rev. Respir. Dis.*, 1989, **139**, 1075–1081.

43. K. Kubota, M. Murakami, S. Tkanaka *et al.*, 'Effects of Long-term Nitrogen Dioxide Exposure on Rat Lung: Morphological Observations', *Environ. Health Perspect.*, 1987, **73**, 157–169.

44. L. Folinsbee, 'Does Nitrogen Dioxide Exposure Increase Airways Responsiveness?', *Toxicol. Ind. Health*, 1992, **8**, 273–283.

45. L. J. Roger, D. H. Horstmann, W. F. McDonnell *et al.*, 'Pulmonary Function, Airway Responsiveness and Respiratory Symptoms in Asthmatics Following Exercise in NO$_2$', *Toxicol. Ind. Health*, 1990, **6**, 155–171.

46. M. A. Bauer, M. J. Utell, P. E. Morrow *et al.*, 'Inhalation of 0.30 ppm Nitrogen Dioxide Potentiates Exercise-induced Bronchospasm in Asthmatics', *Am. Rev. Respir. Dis.*, 1986, **134**, 1203–1208.

47. R. Jorres and H. Magnussen, 'Airways Response of Asthmatics After a 30 minute Exposure at Resting Ventilation to 0.25 ppm NO$_2$ or 0.5 ppm SO$_2$', *Eur. Respir. J.*, 1990, **3**, 132–137.

48. W. S. Tunicliffe, P. S. Burge and J. G. Ayres, 'Effect of Domestic Concentration of Nitrogen Dioxide on Airway Responses to Inhaled Allergen in Asthmatic Patients', *Lancet*, 1994, **344**, 1733–1736.

49. O. V. J. Rossi, V. L. Kinnula, J. Tienari and E. Huhti, 'Association of Severe Asthma Attacks with Weather, Pollen and Air Pollutants', *Thorax*, 1993, **48**, 244–248.
50. M. D. Lebowitz, L. Collins and C. J. Hodberg, 'Time Series Analysis of Respiratory Responses to Indoor and Outdoor Environmental Phenomena', *Environ. Res.*, 1987, **43**, 332–341.
51. B. G. Higgins, H. C. Francis, C. J. Yates, C. J. Warburton, A. M. Fletcher, J. A. Reid, C. A. C. Pickering and A. A. Woodcock, 'Effects of Air Pollution on Symptoms and Peak Expiratory Flow Measurements in Subjects with Obstructive Airways Disease', *Thorax*, 1995, **50**, 149–155.
52. F. E. Speitzer, B. Ferris, Jr, Y. M. Bishop and J. Spengler, 'Respiratory Disease Rates and Pulmonary Function in Children Associated with NO_2 Exposure', *Am. Rev. Respir. Dis.*, 1980, **121**, 3–10.
53. L. M. Neas, D. W. Dockery, J. W. Ware *et al.*, 'Association of Indoor Nitrogen Dioxide with Respiratory Symptoms and Pulmonary Function in Children', *Am. J. Epidemiol.*, 1991, **134**, 204–219.
54. P. Fischer, B. Remijn, B. Brunekreef and K. Biersteker, 'Associations Between Indoor Exposure to NO_2 and Tobacco Smoke and Pulmonary Function Adult Smoking and Non-smoking Women', *Environ. Int.*, 1986, **12**, 11–15.
55. M. J. Hazucha, 'Relationship Between Ozone Exposure and Pulmonary Function Changes', *J. Appl. Physiol.*, 1987, **62**, 1671–1680.
56. W. F. McDonnell, H. R. Kehri, S. Abdul-Saleem *et al.*, 'Respiratory Response of Humans Exposed to Low Levels of Ozone for 6.6 hours', *Arch. Environ. Health*, 1991, **46**, 145–150.
57. S. M. Horvath, J. A. Gliner and L. J. Folinsbee, 'Adaptation to Ozone: Duration of Effect', *Am. Rev. Respir. Dis.*, 1981, **123**, 496–499.
58. D. H. Horstman, L. J. Folinsbee, P. J. Ives, S. Abdul-Saleem and W. F. McDonnell, 'Ozone Concentration and Pulmonary Response Relationships for 6.6 hour Exposures with Five Hours of Moderate Exercise to 0.08, 0.10 and 0.12 ppm', *Am. Rev. Respir. Dis.*, 1990, **142**, 1158–1163.
59. Department of Health Advisory Group on the Medical Aspect of Air Pollution Episodes: First Report, 'Ozone', HMSO, London, 1991.
60. R. P. Cody, Cl. Weisel, G. Birnboaum and P. J. Lioy, 'The Effect of Ozone Associated with Summertime Photochemical Smog on the Frequency of Asthma Visits to Hospital Emergency Rooms', *Environ. Res.*, 1992, **58**, 184–194.
61. G. J. Thurston, K. Ito, P. L. Kinney and M. Lippmann, 'A Multi-year Study of Air Pollution and Respiratory Hospital Admissions in Three New York State Metropolitan Areas: Results for 1988 and 1989 Summers', *J. Expo. Anal. Environ. Epidemiol.*, 1992, **2**, 429–450.
62. M. Lipmann, P. Lioy, G. Leikauf *et al.*, 'Effects of Ozone on the Pulmonary Function of Children', *Adv. Mod. Environ. Toxicol.*, 1983, **5**, 423–446.
63. P. J. Lioy, T. A. Vollmuth and M. Lippmann, 'Persistence of Peak Flow Decrement in Children Following Ozone Exposures Exceeding the National Ambient Air Quality Standard', *J. Air Pollut. Control Assoc.*, 1985, **35**, 1068–1071.
64. D. M. Spektor, M. Lippmann, P. J. Lioy *et al.* 'Effects of Ambient Ozone on Respiratory Function in Active Normal Children', *Am. Rev. Respir. Dis.*, 1988, **137**, 313–320.
65. G. Hoek, P. Fischer, B. Brunekreef *et al.*, 'Acute Effects of Ambient Ozone on Pulmonary Function of Children in the Netherlands', *Am. Rev. Respir. Dis.*, 1993, **147**, 111–117.
66. B. Brunekreef, P. L. Kinney, J. H. Ware, D. Dockery *et al.*, 'Sensitive Subgroups and Normal Variation in Pulmonary Function Response to Air Pollution Episodes', *Environ. Health Perspect.*, 1991, **90**, 189–193.
67. D. E. Abbey, P. K. Mills, F. F. Ptersen and W. I. Beeson, 'Long-term Ambient Concentrations of Total Suspended Particulates and Oxidants Related to Incidence

of Chronic Disease in California USA Seventh Day Adventists', *Environ. Health Perspect.*, 1991, **94**, 43–50.

68. D. A. De Bias, C. M. Banerjee, N. C. Birkhead *et al.*, 'Effects of Carbon Monoxide on Ventricular Fibrillation', *Arch. Environ. Health*, 1976, **31**, 42–46.

69. L. D. Longo, 'The Biological Effects of Carbon Monoxide on the Pregnant Woman, Fetus and Newborn Infant', *Am. J. Obstet. Gynecol.*, 1977, **129**, 69–103.

70. S. M. Horvath, P. B. Raven, T. E. Dahms and D. J. Gray, 'Maximum Aerobic Capacity at Different Levels of Carboxyhaemoglobin', *J. Appl. Physiol.*, 1975, **38**, 300–303.

71. E. Anderson, R. Andelman, J. Strauch *et al.*, 'Effect of Low Level Carbon Monoxide Exposure on Onset and Duration of Angina Pectoris: A Study in Ten Patients with Ischaemic Heart Disease', *Ann. Int. Med.*, 1973, **79**, 46–50.

72. E. N. Allred, E. R. Bleecker, B. R. Chaitman *et al.*, 'Short-term Effects of Carbon Monoxide Exposure on the Exercise Performance of Subjects with Coronary Artery Disease', *New Eng. J. Med.*, 1989, **321**, 1426–1432.

73. A. Hinderliter, K. Adams, C. Price *et al.*, 'Effects of Low-level Carbon Monoxide Exposure on Resting and Exercise-induced Arrhythmias in Patients with Coronary Artery Disease and No Baseline Ectopy', *Arch. Environ. Health*, 1989, **44**, 89–93.

74. S. Cohen, I. M. Deane and J. R. Goldsmith, 'Carbon Monoxide and Survival from Myocardial Infarction', *Arch. Environ. Health*, 1969, **19**, 510–517.

75. A. C. Hexter and J. R. Goldsmith, 'Carbon Monoxide Association of Community Air Pollution and Mortality', *Science*, 1971, **172**, 265–267.

76. J. Schwartz and R. Morris, 'Air Pollution and Hospital Admissions for Cardiovascular Disease in Detroit, Michigan', *Am. J. Epidemiol.*, 1995, **142**, 23–35.

77. G. Paradis, G. Theriault and C. Tremblay, 'Mortality in a Historical Cohort of Bus Drivers', *Int. J. Epidemiol.*, 1989, **18**, 397–402.

78. F. Stein, W. Halperin and R. Horning. 'Heart Disease Mortality Among Bridge and Tunnel Workers Exposed to Carbon Monoxide', *Am. J. Epidemiol.*, 1988, **128**, 1276–1288.

79. C. H. Norman and D. M. Halton, 'Is Carbon Monoxide a Workplace Teratogen? A Review and Evaluation of the Literature', *Ann. Occup. Hyg.*, 1990, **34**, 335–347.

80. B. W. Alderman, A. E. Baron and D. A. Savitz, 'Maternal Exposure to Neighbourhood Carbon Monoxide and Risk of Low Infant Birth Weight', *Public Health Rep.*, 1987, **102**, 410–414.

81. Expert Panel on Air Quality Standards, 'Polycyclic Aromatic Hydrocarbons', The Stationery Office, London, 1999.

82. Department of the Environment, Expert Panel on Air Quality Standards: First Report, 'Benzene', HMSO, London, 1994.

83. World Health Organisation, 'Air Quality Guidelines for Europe', WHO Regional Publications, European Series No 23, WHO, Copenhagen, 1987.

84. A. R. Sellakumar, R. Montesano, U. Saffiotti *et al.*, 'Hamster Respiratory Carcinogenesis Induced by Benzo[*a*]pyrene and Different Levels of Ferric Oxide', *J. Natl. Cancer Inst.*, 1973, **50**, 507–510.

85. F. Pott, R. Tomingas, A. Brockhaus and F. Huth, 'Studies on the Tumorigenic Effect of Extracts and their Fractions of Atmospheric Suspended Particulates in the Subcutaneous Test of the Mouse', *Zbl Bakt I Abt Orig. B*, 1980, **170**, 17–34.

86. J. Misfield, 'The Tumour-producing Effects of Automobile Exhaust Condensate and Diesel Exhaust Condensate', in 'Health Effects of Diesel Engine Emissions', eds. W. E. Pepelko, R. M. Danner and N. A. Clarke, Proceeding of an International Symposium, Vol. 2, USEPA No. EPA-600/9-80-057b, 1980.

87. G. Grimmer, H. Brune, R. Deutsch-Wenzel *et al.*, 'Contribution of Polycyclic Aromatic Hydrocarbons to the Carcinogenic Impact of Gasoline Engine Exhaust Condensates Evaluated by Implantation into the Lungs of Rats', *J. Nat. Cancer Inst.*, 1984, **72**, 733–739.

88. J. W. Lloyd, 'Long Term Mortality Study of Steelworkers: V Respiratory Cancer in Coke Plant Workers', *J. Occup. Med.*, 1971, **13**, 53–68.
89. J. F. Hurley, R. Archibald, McL. Collings *et al.*, 'The Mortality of Coke Workers in Britain', *Am. J. Ind. Med.*, 1983, **4**, 691–704.
90. R. B. Hayes, T. Thomas, D. T. Dilverman *et al.*, 'Lung Cancer in Moto-exhaust Related Occupations', *Am. J. Ind. Med.*, 1989, **16**, 685–695.
91. E. C. Hammond and L. Garfinkel, 'General Air Pollution and Cancer in the United States', *Prev. Med.*, 1980, **9**, 206–211.
92. W. Haenszel, D. B. Loveland and M. G. Sirken, 'Lung Cancer Mortality as Related to Residence and Smoking History: I White Males', *J. Natl. Cancer Inst.*, 1962, 947–1001.
93. W. Haenszel and K. E. Taeuber, 'Lung Cancer Mortality as Related to Residence and Smoking History: II White Females', *J. Natl. Cancer Inst.*, 1964, **32**, 803–838.
94. I. C. T Nisbet *et al.*, 'Review and Evaluation of the Evidence for Cancer Associated with Air Pollution', Clement Associates, Washington D.C., EPA-450/5-83-006, 1983.
95. P. A. Buffler, S. P. Cooper, S. Stinnett *et al.*, 'Air Pollution and Lung Cancer Mortality in Harris County, Texas, 1979–81', *Am. J. Epidemiol.*, 1988, **128**, 683–699.
96. D. J. Freeman and F. C. R. Cattell, 'The Risk of Lung Cancer from Polycyclic Aromatic Hydrocarbons in Sydney Air', *Med. J. Aust.*, 1988, **149**, 612–615.
97. E. Haemisegger, A. Jones, B. Steigerwald and V. Thomson, 'The Air Toxics Problem in the United States: An Analysis of the Cancer Risks for Selected Pollutants', Washington DC: Environmental Protection Agency, No. EPA-450/1-85-001, 1984.

CHAPTER 12

Impacts of Gaseous Pollutants on Crops, Trees and Ecosystems

T. A. MANSFIELD and P. W. LUCAS

12.1 INTRODUCTION

When we look at old photographs of smoke-ridden industrial cities from around the beginning of the 20th century, we are not surprised to discover that plants did not grow well in that sort of environment. The atmospheres of our cities today certainly look much cleaner, but are they any better for the plant kingdom?

The best of the early experiments on air pollution and plant life were those conducted in Leeds by Cohen and Ruston.[1] They found that plants grew three to four times larger on the outskirts of the conurbation than in the foul air of the city centre, and they found a remarkably good correlation between the estimated annual deposition of SO_3 and the stunting of growth (Figure 12.1). Effects of this kind were not confined to the UK, but were later identified throughout Europe and N. America in regions where industrial activity or large urban centres produced localized sources of pollutants from fossil fuel combustion. In 1928 W. W. Pettigrew[2] delivered a lecture which included a graphic description of the problem of growing ornamental plants in Manchester: "While a smoke-laden atmosphere is inimical to both animal and vegetable life, its effects are undoubtedly more apparent if not more deadly in the case of vegetation ...". He noted that in order to maintain Philips Park 'in a presentable condition', it had to be planted up each year with over 7000 trees and shrubs.

In western Europe and N. America we do now consider that the atmospheres of our cities are much cleaner than in the first half of the last century. Air pollution is certainly less visible than it was then, and the exclusion of light by smoke – which was probably an important contributory factor in reducing plant growth – is rarely now a major problem. There are, however, still a lot of concerns about the influence of urban pollution on plants, and about the wider distribution of effects through suburbs and surrounding countryside. In the USA in the 1950s, it became apparent that topographic and climatic factors

296

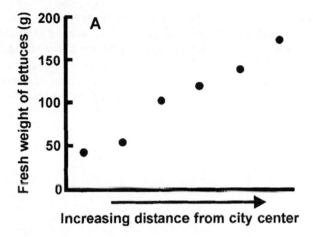

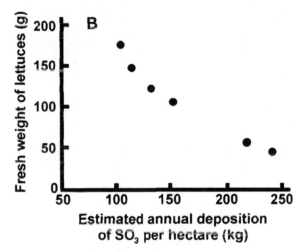

Figure 12.1   A. *Average fresh weights of lettuce plants grown at six different sites in Leeds and its suburbs between July and September 1911.* B. *Estimated annual deposition of SO_3 at the six sites showing an apparently clear, inverse relationship with the growth of the crop*
(Plotted from data in the book by Cohen and Ruston[1])

could combine to produce photochemical smogs, and ozone damage to trees was well documented for areas such as the San Bernardino Forest to the east of the city of Los Angeles.

In the mid-1980s, researchers at Imperial College carried out experiments in London that were similar in concept to those performed by Cohen and Ruston in Leeds in the first decade of the century. They grew peas, clover and barley along a transect extending from central London in a south-westerly direction towards Ascot. There were improvements in plant performance further away from the city centre, and multiple regression analyses showed statistically significant relationships with concentrations of measured air pollutants (NO_2, SO_2 and O_3). Concentrations of NO_2 showed a steeper gradient along the transect than those of SO_2, and in Figure 12.2 we show the data for NO_2 alongside the dry weights of peas (*Pisum sativum* L. cv. Progeta) produced at the different locations. Ashmore *et al.*[3] were correctly cautious in interpreting their findings, pointing out that experiments of this type with multiple environmental variables across sites can never provide conclusive proof of a causal link

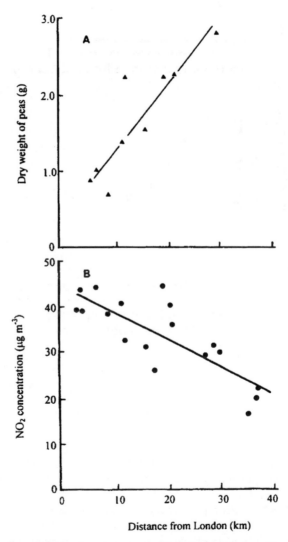

Figure 12.2   A. *Dry weights of garden peas (variety 'Progeta') grown at various distances from London in 1984.* B. *Mean NO_2 concentrations over the same transect during the summer months*
(Compiled from data published by M. R. Ashmore and colleagues[3,4])

between a air pollution and poor plant performance. Nevertheless, the close resemblance between Figures 12.1 and 12.2 is very striking, and it is clear that something in the environment of a modern city remains toxic to plants despite efforts to regulate emissions of smoke. It seems likely, though still unproven, that NO_x may have replaced SO_2 as a major cause of phytotoxicity in developed countries, with contributions from both NO and NO_2. SO_2 does, however, remain a significant problem in many developing countries.[5] Additionally, there may be synergistic effects when there are mixtures of pollutants, and we shall mention some possibilities later.

Apart from changes in the nature of pollutants in urban air, in the last few decades contaminants have become more widely distributed and the possibility that plants in rural areas may be affected has been much studied. In the 1960s and 1970s the research into effects of air pollutants on plants was mainly concerned with crops and grasslands, but during the 1980s the emphasis shifted from herbaceous plants to trees, particularly the economically important conifers grown in Europe and N. America. This shift occurred because of worrying reports of damage to forests in central Europe, Scandinavia and N. America, apparently worsening year by year, from about 1980 onwards.[6–8] Research on trees poses many problems and it was a long time before useful experimental results emerged, but we are now able to identify some effects of individual pollutants, especially ozone, that may contribute to aspects of 'forest decline'. There are however, many facets to this problem and much still needs to be done.

12.2 CURRENT KNOWLEDGE OF THE MECHANISM OF ACTION OF SOME MAJOR AIR POLLUTANTS

12.2.1 Sulfur Dioxide

Sulfur is one of the important mineral nutrients of plants, and it is normally taken into the roots from the soil in the form of sulfate, then transported to the leaves. Enzymes for the reduction of sulfate are located in the leaves, and reduced forms of sulfur are used in the synthesis of amino acids. During the reduction of sulfate ions there is transient formation of sulfite, but there is unlikely to be any significant accumulation of this ion. Gaseous forms of sulfur can also be taken up directly by the leaves. Exchanges of carbon dioxide and water vapour between leaves and the atmosphere are fundamental to many physiological processes such as photosynthesis, respiration and transpiration. These gaseous exchanges are regulated by the opening and closing of stomata, and the uptake of pollutants such as SO_2 is also thought to occur primarily *via* these pores at the surface of the leaf.[9] There is, however, also evidence that some gaseous pollutants may be sorbed by the leaf cuticle which could have important implications in reducing cuticular integrity, leading to enhanced water loss,[10] but there is little likelihood that this sorption of SO_2 is accompanied by its significant permeation into the leaf through the cuticle.[11]

When SO_2 enters the leaves *via* the stomata, it dissolves in a film of water at the surface of the mesophyll cells inside the sub-stomatal cavity, forming sulfite and bisulfite ions. There is little evidence that SO_2 as a dissolved gas persists within the leaf.[12] The aqueous layer of the mesophyll cells is part of the pathway by which water and minerals from the soil are distributed to individual cells of the leaf. Thus sulfate ions will normally be present, but not sulfite or bisulfite in significant concentrations. The proton-pumping activities of the outer cell membrane (the plasmalemma), which are part of normal cellular functioning, alter the pH of the extracellular environment appreciably at different times of the day. Thus the equilibrium between the solution products of SO_2 may change, the ratio between sulfite and bisulfite increasing as the pH increases, but the balance is normally towards bisulfite.

We do not know to what extent the plasmalemma of the cells in a leaf can tolerate the presence of particular concentrations of sulfite and bisulfite, but damage to membranes is known to be one feature of cellular injury by SO_2. The permeability of the plasmalemma is known to be affected by exposure to SO_2 and essential ions such as potassium can then leak out of cells.[13] Nevertheless potassium-leakage rates for leaves are not well correlated with differences in SO_2 sensitivity between cultivars or species, and consequently it is not certain that the plasmalemma is the most susceptible cellular component to SO_2.[14] One reason for this may be the presence of the enzyme sulfite oxidase in the apoplastic (extracellular) water layer, which can act to detoxify sulfite by converting it into sulfate.[15] Particular components of membrane function that are specially susceptible to injury by SO_2 have not been readily identified, and it has been suggested that there are general effects on membrane integrity, which would mean that various membrane-located processes could be affected.[16]

The thylakoid membranes of chloroplasts may be particularly sensitive to the presence of SO_2. Swelling of these structures, which are the locations of the light-harvesting complexes and electron transport components, is seen very soon after the commencement of fumigation with SO_2.[17] Damage of this kind may explain why the process of photosynthesis is inhibited by SO_2.[18,19] Despite this known sensitivity of the photosynthetic apparatus, there is strong evidence that the primary sites of injury within leaves could be located elsewhere, probably in the phloem (see Section 12.4.2.2).

Studies of the effects of SO_2 on plants have often been performed in fumigation chambers subjected to natural illumination and ambient temperatures, and in some of these experiments the growth responses to the pollutant seemed to be affected by climatic changes. A study under controlled conditions with the grass *Phleum pratense* (Timothy) showed that when growth was rapid, in high irradiance and long days, 120 ppb SO_2 (higher than usually found in the most polluted situations nowadays) had no detectable effect on the plants. On the other hand the same concentration applied to plants in low irradiance and short days reduced growth by about 50%.[20] Similar changes in sensitivity were found when rate of growth was reduced by dropping the temperature.

The results of these and other similar experiments have led to the conclusion that there is no critical concentration which can usefully be regarded as the threshold for injury. Different amounts of SO_2 can clearly be tolerated under different climatic conditions, a factor which has not been considered in many of the models so far produced which attempt to evaluate the effects of pollutants on cultivated or natural vegetation. Perhaps our attention should be focused on those effects that occur under the least favourable conditions, particularly if they involve reductions in survival ability under extreme climatic conditions. There is now quite a considerable amount of evidence that plants exposed to SO_2 become more sensitive to frost injury. There is, for example, a substantial reduction in survival of SO_2-polluted ryegrass upon subsequent exposure to sub-zero temperatures.[21] It is not uncommon to see frost injury on pasture grasses during winter in the more polluted parts of the UK. The impact on the productivity of an established sward is, however, probably quite small because recovery can occur in spring and summer. The enhancement by SO_2 of frost sensitivity in woody plants is more likely to be of concern. In the case of *Calluna vulgaris* (Heather) concentrations of SO_2 and NO_2 which enhance growth under favourable conditions can cause a substantial increase in sensitivity to acute frost damage.[22] There is evidence that even dormant deciduous plants may be affected by SO_2 uptake into the shoots in winter, with subsquent death of the terminal buds.[23] This topic is covered in more detail in Section 12.4.2.1.

In some agricultural areas the soil is deficient in sulfur, and it is possible that SO_2 (and also H_2S) in the atmosphere might remedy this deficiency. The total sulfur deposition per unit area may exceed 6 g m^2 per year in parts of the UK, and more than 12 g m^2 per year in central and eastern Europe.[24] Such large inputs exceed the sulfur requirements of most plants. There is evidence that when supplies of sulfate in the soil are inadequate to support growth, then SO_2 in the atmosphere can make up the deficit.[25] When the plants are growing rapidly under favourable conditions, the SO_2 can be regarded as beneficial, but in most industrialized countries in Europe and N. America, SO_2 pollution is greater in winter when plant growth is slowest. To benefit from SO_2 the leaves must be able to use the supply of sulfur as it enters the cells. If metabolism is proceeding too slowly for this to occur then toxic ions such as SO_3^{2-} and HSO_3^- may accumulate with resulting damage to cells and cellular processes. It is thus possible to put mistaken emphasis on the beneficial effects of atmospheric SO_2. Sulfur deficiencies in the soil can be remedied by increases in the sulfate content of fertilizers at small cost to farmers, and the economic advantage of sulfur supplied by SO_2 pollution is therefore very small.

Identification of damage due to SO_2 in the field is difficult because visible symptoms of injury may be secondary, appearing some time after important effects on growth and productivity have occurred. A recent study has, however, suggested that one species, *Amelanchier alnifolia* (Saskatoon serviceberry) displays foliar injury that can be a useful indicator of exposure to SO_2, even though interactions with other pollutants do need to be taken into account. It is suggested that this plant may prove to be a very useful 'bioindicator' for phytotoxic concentrations of SO_2 in the field.[26]

12.2.2 Nitrogenous Pollutants

It has long been recognized that nitrogen, which is a major constituent of proteins and nucleic acids, is essential for plants; and when supplies are deficient, normal growth cannot occur. As a consequence, plants have developed a range of diverse mechanisms to enhance their acquisition of this important nutrient, for example fixation of atmospheric nitrogen by symbiotic bacteria in legumes, associations with mycorrhizal fungi and insectivory. The majority of plants obtain nitrogen through root absorption of the inorganic ions ammonium (NH_4^+) and nitrate (NO_3^-) from the soil solution and have developed pathways of nitrogen assimilation which are now well characterized (Figure 12.3). (Soil processes such as nitrification and mineralization which influence the availability of nitrogen are complex and space does not permit a description of these processes here.) Considerably less is known, however, about how plants that are adapted to low nitrogen inputs, particularly those growing in sensitive ecosystems such as forests or heathlands, react to additional nitrogen deposition from anthropogenic sources, either as gaseous nitrogen compounds (NO, NO_2 and NH_3) or from wet deposition in rain or mist (NO_3^- and NH_4^+). The amount and forms of deposited nitrogen vary depending on the specific region under consideration.[27,28] Inputs of between 10 and 85 kg ha^{-1} yr^{-1}, largely as NH_3 and NO_2, have been suggested for low altitude areas of western Europe, and more than 20 kg ha^{-1} yr^{-1} (predominantly as NO_3^- and NH_4^+ in cloud and rainfall) for some high altitude regions of Britain and Germany. Vitousek[29] estimated that human activities have altered the global biogeochemical cycle of nitrogen to such an extent that 'natural' N fixation has now been overtaken by anthropogenic N fixation. Emissions of nitrogen-containing compounds into the atmosphere have resulted in deposition rates of N that are up to 20 times those under natural conditions.[30] The long-term effects of high inputs of nitrogen to sensitive ecosystems are, however, only partly understood, especially with regard to soil processes, the availability of other nutrients[31] and the possible interaction with environmental stresses such as frost.[32,33] In the southern Pennines in northern England the atmospheric concentrations of SO_2 have declined considerably since the 1950s but there has been little recovery of the *Sphagnum* moss, the decline of which was attributed to pollution from the nearby urban areas. This is thought to be because the *Sphagnum* is now primarily affected by nitrogenous pollutants which have increased in concentrations over the period during which SO_2 has declined.[34,35]

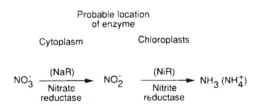

Figure 12.3 *The reduction pathway from nitrate to ammonia which occurs in plants prior to amino acid synthesis*

Over the past ten years there has been a great deal of emphasis on studies of the impacts of NH_3 pollution, amounts of which have risen alarmingly in some countries where farm animal production is very intensive. Duyzer *et al.*[36] estimated that at Speulderbos in the central Netherlands, the average annual flux of NH_3 amounted to about 50 kg N ha^{-1} yr^{-1} by dry deposition alone, which is appreciably higher than the critical load for causing damage to coniferous forests. The precise reasons for the toxicity of NH_3 are not fully understood, but it is likely that the protons released when NH_3 assimilation becomes intensive can cause cellular acidosis.[37] There is also evidence that uptake of excess N into leaves can cause mineral nutrient imbalances in plants.[38,39] The subject of nitrogen deposition and its influence on N-cycling in the soil is a topic of much current interest.[40]

12.2.2.1 Responses to Gaseous NO_x.

Most of the early research was performed with NO_2 rather than with mixtures of NO and NO_2, which would have been more realistic. It was found necessary to fumigate plants with very high concentrations of NO_2 in order to produce visible leaf injury, and up to 1981 it was being suggested that significant damage to trees would only be expected if concentrations around 2 ppm were maintained for 48 hours, or 20 ppm for 1 hour.[41] Concentrations as high as these are rarely encountered in reality, and so it was widely believed that NO_2 (and by implication NO_x) *per se* contributed little to the phytotoxicity of polluted air. This did not mean that there was a lack of concern about NO_x pollution, because its role in the formation of ozone was already established, and evidence was emerging that NO_2 can interact with other air pollutants to enhance their toxicity.[42,43] Little attention was, however, paid to NO as a component of NO_x, indeed an authoritative statement had dismissed the possibility of its being a pollutant bearing any serious threat to plant life.[44] The situation changed quite dramatically in the 1990s, as it emerged that NO has a major biological function that was previously unknown and unsuspected.

Nitric oxide. The aqueous solubility of NO is low, which probably why its rate of uptake into plants is much lower than that of gases such as CO_2, NO_2 and SO_2.[30] Although the NO taken into leaves may form some nitrate and nitrite ions in solution, a significant contribution to a plant's nitrogen budget is unlikely.[45] A very different kind of intervention in cellular function must now be considered, for NO has been identified as an agent of critical importance in signalling transduction pathways. Its role in animal cells and tissues is the subject of a vast literature which was concisely reviewed by Wink and Mitchell.[46] Its wider physiological role in plants is less clear, but it is generally accepted that it plays a part in defence responses, for example against attack by pathogens.[47] An important defence mechanism is believed to depend on a reaction between NO and oxygen species such as superoxide (O_2^-) to form peroxynitrite (ONOO) which then kills the pathogen's cells. An enzyme responsible for production of NO within plant cells, nitric oxide synthase, has been shown to display increased activity when tissues are being attacked by pathogens.

It is not known to what extent the NO in polluted air can penetrate plant tissues and interfere with these cellular processes. Defences which depend on the production of free radicals targeted at invading pathogens must presumably be very accurately located, to avoid damaging vulnerable components of the host's cells. Research in the future will need to give special consideration to the possibility that NO from the atmosphere could disturb the critical defences of plants. These new discoveries about the physiological role of NO may call for reinterpretation of some responses to this pollutant which, although they are clearly documented in the literature, have been given little attention because of any lack of understanding at the mechanistic level. These include:

- Glasshouse crops are exposed to NO_x during CO_2-enrichment when hydrocarbon burners are used as sources of CO_2. The $NO:NO_2$ ratio is around 5:1 and plants display growth inhibition or delayed development which is not seen when they are exposed only to NO_2.[48,49]
- NO appears to have an inhibitory effect on photosynthesis which is, under some circumstances, greater than that of NO_2.[50-53]
- Aphid infestations on crop plants are often enhanced in urban air, and the performance of the aphids is more strongly correlated with the NO content than with other air pollutants. This appears to be the result of a response of the plants to NO rather than as a direct effect of the pollutant on the aphids.[54]
- NO concentrations in the low ppb range have been shown to cause 'stress ethylene' production. Ethylene is an important gaseous hormone in plants (see Section 12.2.3), and its formation under conditions of stress, *e.g.* drought, is well characterized.[55]

The connections between these disparate reports are not clear at present, and we must await further research before we can adequately assess the impacts of NO on plants.

Nitrogen Dioxide. The nitrate and nitrite ions formed when NO_2 enters into solution might be expected to enter normal metabolism *via* the pathway that reduces them to ammonia (Figure 12.3). Nitrate and nitrite are the substrates for two reductase enzymes normally present in leaves, *viz.*, nitrate reductase and nitrite reductase. The usual source of initial substrate is the nitrate (but not nitrite) transported into the leaves from the roots *via* the xylem.

Nitrite ions are known to be toxic to plant tissues and there is metabolic regulation within the reduction pathway to ammonia to prevent them accumulating. The necessary rate control is thought to be provided at the stage of nitrate reduction, the enzyme for which is located inside the cells but outside the chloroplasts[56,57] while nitrite reductase is within the chloroplasts. There must be mechanisms for keeping the toxic NO_2^- ions away from sensitive sites during their movement within the cells, but we know little about these. It is, however, likely that the arrival of NO_2 at the cell surface, leading to NO_2^- ions in the extracellular water, may pose problems because the movement of substantial

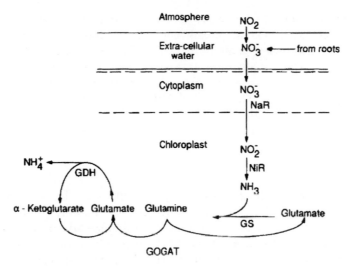

Figure 12.4 *The relationship between the synthesis of amino acids in chloroplasts and the nitrate formed from NO_2 entering the leaf. NaR and NiR are nitrate and nitrite reductase, respectively*

numbers of these ions across the plasma membrane is probably not a normal requirement.

Figure 12.4 shows diagrammatically the locations of nitrate and nitrite reduction, and the incorporation of NH_3 into amino acids in chloroplasts. This is regarded as the normal pattern in the majority of plants but there are exceptions, and these may be important in determining differences in sensitivity to NO_2 between species.

It has been suggested[45,58] that the susceptibility of different plant species to NO_x may depend on the precise location at which they normally carry out nitrate/nitrite reductions. Some woody plants, for example, might be particularly sensitive to NO_x because the reduction pathway is absent or poorly developed in the leaves. These species carry out the reductions of soil-derived nitrates in the roots, and hence nitrogen in an already reduced form is transported to the shoots. Research has so far failed to indicate precisely how such plants are affected by NO_x entering the leaves from the atmosphere.

The exposure of plants to NO_x pollution has been shown to lead to changes in the activities of nitrate and nitrite reductases in leaves,[59] and there are usually marked increases in nitrite reductase, a clear indication that the plant has some ability for metabolic adjustment to detoxify the nitrite ions as they enter the cells. In contrast the activity of nitrate reductase may decline in leaves exposed to NO_x. This is probably because the supply of nitrite ions from the atmosphere reduces the need for the production of nitrite by nitrate reductase. A consequence may be loss of control of the normal rate of the pathway, which could produce serious imbalances in subsequent metabolism. Figure 12.4 shows the routes for the incorporation of ammonium into amino acids and it is these areas of metabolism that may be of much concern in future research.

Additional complexity comes from recent awareness of the ways in which nitrate can act as a cellular signal, activating specific receptors which are connected to complex signalling networks within tissues.[60] Hitherto, the impacts of NO_x have usually been assessed in relation to incoming amounts of N and the capacity of tissues and organs to adjust their metabolism quantitatively and qualitatively. The possibility that NO_3^- ions derived from NO_x might interfere with metabolic control through signalling networks offers new perspectives for mechanistic understanding that we have lacked in the past.

12.2.3 Ozone

Ozone can be highly damaging to both plants and animals, and it is the most thoroughly investigated of a group of phytotoxic compounds that are 'secondary' air pollutants, produced when 'primary' pollutants take part in reactions in the atmosphere. Models have indicated that tropospheric ozone concentrations may increase by up to 1% annually over the next half-century, and these predictions are supported by recent trends[61,62] (see also Chapter 8).

Considerations of the toxicity of O_3 are not complicated by its providing a source of an essential nutrient, as is the case with SO_2 and NO_x. This does simplify the mechanistic evaluation of its impacts to some extent, but nevertheless the precise cause(s) of its toxicity to plants have proved elusive.

At high concentrations (above 100 ppb for most plant species) ozone causes a characteristic brown or white flecking to appear on the foliage, but there is some variation in visible symptoms and sensitivity between species, and even between cultivars of the same species. This variation is particularly wide between different cultivars of the tobacco plant, one of which, the variety Bel-W3, can be visibly injured when hourly mean ozone concentrations rise above 40 ppb. Because of its high sensitivity, Bel-W3 is frequently used as a biomonitor for O_3 to provide a cheap and convenient substitute for chemical monitoring methods, and it has been used in several studies in the UK and elsewhere in Europe.[63]

At lower concentrations, when there is no visible damage to foliage, ozone may reduce the growth rate and physiological activity of plants. Again, however, there is considerable variation from species to species, or between cultivars of one species. There are also marked differences in toxicity in different environmental conditions. Initially this variation caused problems for researchers, but more recently it has also been looked upon as an opportunity to identify those characteristics of a plant, or the physiological states determined by the environment, that lead to enhanced susceptibility. This approach can provide important clues to the mechanisms behind injury.

Cells inside the leaf, especially the photosynthetic mesophyll, have been identified as primary sites of injury. Differences in the susceptibility of these cells do not, however, appear to be the most important cause of variation. It is the ability of a leaf to exclude O_3 from its intercellular spaces that emerges as one critical point of control.

Like other gases in the atmosphere, O_3 gains access into leaves by diffusing through stomatal pores. There has been a great deal of research into the relationship between stomatal conductance and sensitivity to O_3, and there is general agreement that conductance is often a major determining factor. If the stomata are partially closed then some O_3 is excluded from the sensitive sites. Differences in stomatal opening can be inherent (*i.e.* genetically determined) or can result from responses to environmental conditions. Stomata tend to close partially when the atmosphere is dry or when the water content of the soil is low. Differences in O_3 sensitivity are often well correlated with such changes.[64,65] This correlation with stomatal conductance does not appear so clearly when we consider other pollutants such as SO_2 and NO_x. A possible reason for its greater importance in relation to O_3 sensitivity became apparent as a result of the research of Mehlhorn and Wellburn[66] and Mehlhorn, O'Shea and Wellburn.[67] Pea seedlings were fumigated with 50–150 ppb O_3 for seven hours daily for their first three weeks of growth after germination. Even in 150 ppb O_3 (regarded as a high level of pollution) there was no evidence of any injury apart from a slight curling of the leaves. On the other hand, if three-week-old seedlings that had been grown in clean air were fumigated for just one day under the same conditions and in the same concentration of O_3, there was very severe damage to the leaves (Table 12.1).

Subsequent studies showed that the two sets of seedlings differed markedly in the amounts of ethylene they were producing (Table 12.1). Ethylene is normally generated by plants and it is known to be able to regulate various aspects of growth and development. For this reason it is regarded as a gaseous hormone.[68] Increased production of ethylene is associated with many events in plant development such as release from dormancy, stem and root growth and differentiation, abscission of leaves and fruits, flowering and fruit ripening. In particular, ethylene production increases in response to environmental stresses of various kinds (*e.g.* physical wounding, flooding).

Table 12.1   *Ethylene production and leaf injury in pea seedlings*

Treatment	(A) Three-week fumigation (C_2H_4 evolved nmol g^{-1} dry weight h^{-1})	(B) One-day fumigation (C_2H_4 evolved nmol g^{-1} dry weight h^{-1})	Visible leaf injury to plants in (B) (%)
Control	2.5 (0.1)[a]	2.6 (0.2)	None
50 ppbv O_3	1.9 (0.2)[a]	4.7 (0.2)	0–10
100 ppbv O_3	0.3 (0.1)[a]	7.0 (0.4)	10–35
150 ppbv O_3	0.2 (0.1)[a]	5.7 (0.5)	> 50

The data in column (A) are for plants grown for 3 weeks and exposed to O_3 for 7 h daily. Those in column (B) are for plants grown in clean air for 3 weeks, then exposed to O_3 for 7 h for one day only. Results represent means (s.e.m. in parentheses) of at least seven replicates. The differences in effects of the 100 and 150 ppbv treatments were highly significant.
[a] No visible injury.
(Reproduced from data of Mehlhorn and Wellburn[66] with permission from Macmillan Journals Ltd.)

In the experiments of Mehlhorn and Wellburn[66] the rates of ethylene evolution were more than doubled in the three-week-old plants fumigated for seven hours with O_3. The increased production of ethylene after O_3 fumigation was already well known from previous studies, and it was proposed that its production and subsequent reaction with O_3 entering through the stomata was a key factor in determining injury by O_3. They tested this hypothesis by treating plants with an inhibitor of ethylene biosynthesis, aminoethoxyvinylglycine (AVG), prior to exposing them by O_3. This pre-treatment reduced ethylene production by 85% and also almost completely prevented the damage normally caused to the three-week-old plants by the short O_3 treatment.

Ozone can react with biogenic unsaturated hydrocarbons to produce organic peroxides such as hydroxymethyl hydroperoxide.[69] It has been shown that there are free radicals present in leaf tissues, closely correlated with ethylene formation and ozone treatments as postulated.[70,71] Independent evidence also suggested that isoprene, another unsaturated hydrocarbon produced and emitted by leaves, can react with ozone in a similar way and with similar consequences.[72] These findings may explain why high normal stomatal conductance may be of special importance in connection with injury to leaves by O_3, if it is only when the endogenous unsaturated hydrocarbons make contact with the incoming O_3 that the critical formation of toxic peroxides takes place.

The parts of cells most vulnerable to free radicals in the intercellular air are likely to be the 'plasmalemma' or plasma membranes. O_3 fumigation is known to cause damage to membranes so that the retention of solutes is reduced. The double bonds of unsaturated fatty acids in membranes are very vulnerable to attack. These impacts of O_3, and parallel effects on lipid metabolism, have been well covered in a review by Sakaki.[73] Although the plasma membrane is the boundary of what is regarded as the 'living' part of a plant cell, it is surrounded by an aqueous layer known as the apoplast within, and sometimes beyond, the cell wall. In this region there are many substances that have been released across the membrane, including enzymes which can therefore control extracellular reactions. It is now well established that the apoplast provides a very important line of defence against incoming O_3, due to the presence of antioxidants such as ascorbate.[69] There is increasing evidence of a correlation between apoplastic antioxidants and ozone tolerance [74]

12.2.4   Acid Rain and Acid Mist

The presence of SO_2, NO_x and NH_3 in polluted air has a substantial impact on the chemistry of rain and mist. In north-west Europe, the concentrations of ammonium and sulfate can be 1000-fold higher than in areas defined as 'unpolluted', such as the South Island of New Zealand. The wet deposition of acidity has been sufficient to increase the leaching of base cations from most soils in affected areas, but this only increases the acidity within soil types in which the reserves of base cations are low.

There is now good scientific evidence that the pH values of some soils in Scandinavia and central Europe have decreased over the past few decades.[75 77]

Increased base leaching has also been recognized in parts of North America, though here there is little to suggest that soils have yet reached dangerously low pH values.

The inputs of acid (principally sulfuric and nitric) from the atmosphere are not the only way by which soil pH can be decreased. Various biotic processes and humus accumulation can also contribute.[78] There is still some uncertainty about the relative contributions of these different processes, but nevertheless most soil scientists now accept that acidity in rainfall is playing an important part in the changes that are occurring. The direct and indirect effects on plants in the affected soils are much more open to dispute. In Germany in the 1980s the yellowing (chlorosis) of older needles of Norway spruce and silver fir increased markedly on nutrient-poor soils. This phenomenon appears to have been associated with deficiencies in cations, especially magnesium. The problem occurred over large geographical areas, many of which are long distances from the main sources of primary pollutants. It is, however, recognized that acidic deposition can occur far away from such sources. The hypothesis that deterioration of the soil is a major factor in the decline in the health of trees has had much support, but other ideas have also been extensively discussed. In particular, the increased frequency of episodes of O_3 pollution at the damaged sites has attracted attention, and O_3 has been suggested to be either the predominant agent in causing damage, or at least an 'inciting stress'. The injury of membranes caused by O_3 could, it is suggested, be the primary factor in the leakage of nutrients and their subsequent leaching from foliage by acid mist.[79,80] Another suggestion is that the deposition of nitrogen to forests from the increasing atmospheric concentrations of NO_x has caused an imbalance in the major nutrients required for growth[81] (see Section 12.2). Some authors have suggested that 'nitrogen saturation' is beginning to occur in forest ecosystems, but there is not a satisfactory physiological or ecological definition of what 'saturation' means in this context. It is clear, however, that floristic and biodiversity changes can be induced by high rates of acidic and nitrogen deposition.[82]

Fears were being expressed in the early 1980s that 'forest decline' was a developing problem that might lead to widespread destruction of forests, at least in northern Europe and perhaps also in parts of North America. Such forecasts have proved incorrect, and inevitably the strongly expressed opinions of some scientists may now appear to have been discredited. However, there remains an important ongoing debate about the impacts of air pollution on forest health, and complex interactions with other factors are still being actively discussed.[83-85]

Most rainfall chemistry studies rely on the use of rain gauges for the collection of samples. Apart from some other problems[24] simple rain gauges can also underestimate the amount of water from the atmosphere that is captured by vegetation in mist or clouds. This additional input of water from the atmosphere is known as 'occult' precipitation.[86] Not only is the volume of precipitation likely to be underestimated when plants are in cloud, but also the amounts of dissolved materials that are deposited.

Collection of occult deposition separately from rain is difficult and special techniques were developed in the 1980s. On Great Dun Fell in the northern Pennines in England, it has been estimated that the concentration of ions such as H^+, SO_4^{2-} and NO_3^- are two to four times higher in cloudwater than in rain.[86]

There are occasions when cloudwater may simultaneously condense on foliage and evaporate from it, and this can lead to a considerable increase in the concentration of ions in the liquid film or droplets on exposed surfaces.[87] This is thought usually only to occur for short periods, for example when the weather is improving after a dense cloud cover, but in thin cloud when some solar radiation can reach the surface it may be more prolonged. Experimental studies of the effects of simulated rain on leaf surfaces have generally shown that there is little direct damage unless the pH is below 3.0, with the exception of a few cultivars of field crops that appear to be specially sensitive.[88] In some cases the growth of the plants was actually increased, probably because of the fertilizing effects of additional nitrogen and sulfur. Very low pH values are of infrequent occurrence in rain but it seems likely that they could occur much more frequently during occult deposition. For this reason, the effects of cloudwater on forests at high altitude are now being considered as a further contributory factor towards the damage to trees, and there may be impacts on the frost sensitivity of some species (see below).

12.3   INTERACTIONS BETWEEN POLLUTANTS

Air pollutants rarely occur singly, yet much of the literature covers their separate effects rather than their joint action in realistic mixtures. There are two main reasons for this: mechanisms behind injury are necessarily studied in the simplest context, at least in the first instance; and the facilities available to researchers have until recently rarely allowed for the design of experiments involving different pollutants in controlled factorial combinations, especially for long-term studies.

The situation that has emerged so far is complex. The effects of combinations of pollutants are sometimes greater and sometimes less than we would predict from the separate effects, and only rarely are responses simply additive. In the space available here we shall concentrate on specific pollutant combinations whose effects may be specially important.

12.3.1   SO₂ and NO₂

These two primary pollutants occur together in many situations because they are often produced simultaneously by the same sources, for example during the combustion of S-containing fossil fuels. Even though the emissions of SO_2 in many industrialized countries have been reduced dramatically in recent years they are still significant, and in some cases increasing, in developing countries.[5] Furthermore, the possibility of joint action between SO_2 and NO_2 still needs to be considered in all urban areas where there are regular episodes involving both pollutants, or in nearby rural areas where there may be longer, but lower, exposures to both gases. Concentrations of NO_2 in urban situations commonly rise to between 30 and 50 ppb, whereas peaks of SO_2 are nowadays generally lower, ranging from 10 to 30 ppb. Motor vehicle emissions have caused NO_2 concentrations to rise steadily in rural areas over the last three decades, but here there is less likely to be a coincidence of peaks of SO_2 and NO_2.

Short-duration fumigations with SO_2 and NO_2 have often led to severe injury, far greater than that found with the pollutants separately. Tingey et al.[89] were the first to perform a detailed study, and they found surprisingly high toxicity of some mixtures of SO_2 and NO_2 in 4 hour exposures, at concentrations around 10% of those required for inducing visible injury by the individual gases. Since this early work there has been a mixture of confirmatory and contradictory reports, but there have been more than enough observations of synergism between SO_2 and NO_2 to suggest that simultaneous exposure to the two pollutants may sometimes be a cause of unexpected damage in the field.

Studies of long-term effects of SO_2 and NO_2 on grasses and cereals have shown that there are much greater growth reductions during the winter than in spring and summer.[20,90] This suggests that slower rates of metabolism determined by environmental conditions may predispose plants to this form of injury.

The enzyme nitrite reductase is considered to play a critical part in the metabolism of the solution products of NO_2 after its entry into cells. Yoneyama and Sasakawa[91] made use of $^{15}NO_2$ to demonstrate the appearance in cells of nitrite and nitrate ions and to show that after their reduction the ^{15}N was incorporated into amino acids. Analyses by Wellburn and his colleagues[45,92,93] of plastids extracted from leaves of grasses after exposure to SO_2, NO_2 or $SO_2 + NO_2$ revealed important differences in enzyme activities. Fumigation with NO_2 on its own led to a statistically significant increase in the activity of nitrite reductase, but no such increase was found when the NO_2 was accompanied by SO_2. This obviously suggested that SO_2 prevented the induction of greater nitrite reductase activity, as normally occurs in the presence of NO_2. This means that the plants may have been unable to detoxify the solution products of NO_2. Thus there may be a fairly simple biochemical explanation of synergistic effects of SO_2 and NO_2 in damaging plants, and this remains one of the few explanations we have of the mechanistic basis of pollutant interactions at the metabolic level. Petitte and Ormrod,[94,95] who reported strong evidence of $SO_2 \times NO_2$ interactions in potato, found increases in reducing sugars in the leaves within the first day of treatment, and concluded that an early response may be an impairment of dry matter partitioning from shoots to roots.

12.3.2 SO_2 and O_3

Experimental studies of this combination of pollutants have produced a much more confusing picture. The earliest studies involving short-term exposures provided evidence of synergism not unlike that often found with SO_2 and NO_2.[96,97] The types of foliar injury caused by SO_2 and by O_3 are quite different, and it was noted in these and in some later experiments that the enhanced damage in combinations of the two resembled that caused by O_3 rather than by SO_2. This might be explained by the enhanced opening of stomata that can occur in SO_2-polluted leaves (see next section). Thus SO_2 could assist the access of O_3 into the leaf's interior.

In further studies no consistent picture of the combined effects of SO_2 and O_3 emerged. Reviews of the subject[42,43,98,99] have cited evidence for additive, less-than-additive and more-than-additive effects. The only consistency in the data from different sources was that most more-than-additive effects were found in short-term exposures, often involving fairly high concentrations. It seems unlikely that a simple description of the dose–response relationships for effects on plants of these two pollutants, which are commonly co-occurring, can be produced at present. There is too little information on significant interactions between SO_2 and O_3 at realistic concentrations in the field, and on inter-specific variation, and much more research is needed to supply the basic information needed for modelling purposes.[43]

12.3.3 O_3 and CO_2

The impacts of O_3 together with elevated CO_2 concentrations are receiving much attention in current research programmes, since it is predicted that the two pollutants will increase alongside one another in the future. Elevated CO_2 has been found by most authors to reduce the adverse effects of O_3 on the growth of plants, but O_3 does nevertheless reduce the anticipated 'beneficial' effects of additional CO_2 that are based on models derived from its stimulation of photosynthesis.[43] It is well established that increasing concentrations of CO_2 normally cause reductions in stomatal conductance, and this suggests a simple mechanism for any decreased toxicity of O_3, for in elevated CO_2 partial stomatal closure will reduce the access of O_3 to the leaf's interior. The amount of stomatal closure induced by CO_2 enrichment is generally insufficient to prevent an increased intake of CO_2 for photosynthesis, and consequently the impact of O_3 may be further reduced if greater rates of assimilation uphold repair mechanisms.

12.4 POLLUTANT/STRESS INTERACTIONS

In general, experimental fumigations of plants with air pollutants have been conducted under conditions that are favourable for growth. There are good reasons for this: stress factors, biotic and abiotic, are often difficult to maintain and control at a predetermined level, hence their inclusion in experiments that are already technically very difficult has been seen as an unmanageable complication. Yet the evidence from a few well-designed experiments, supported by many observations in the field, suggests that such interactions can play at least a significant role in determining plant responses to pollutants, and that in some cases they may be of overriding importance.

12.4.1   Biotic Factors

One of the earliest known descriptions of the effects of air pollution on insects and plants was made by the diarist John Evelyn, who in 1661 said of the atmosphere of seventeenth-century London: "It kills our bees and flowers abroad, suffering nothing in our gardens to bud, display themselves or ripen." It was a further three hundred years before observational evidence of inter- actions between air pollutants and plant pathogens (insects or fungi) again began to appear in the literature (see Cramer[100] and Flückiger *et al.*[101]) and it is only in the last twenty years that an experimental approach has been adopted to study these interactions.

Most of the evidence from these experiments suggests that exposure to pollutants such as SO_2 and NO_2 can predispose plants to attack by insects, with, in the majority of cases, an increase in insect performance being observed. For example, Warrington[102] produced a dose–response curve showing how precisely the growth rate of aphids can be related to SO_2 concentration (Figure 12.5). There was a linear increase from zero up to about 110 ppb SO_2 after which there was a steep decline. This was thought to be the concentration at which the SO_2 began to be toxic to the aphids. There is little evidence that direct toxicity of pollutants to insects occurs in the field under realistic concentrations and, in general, observed responses appear to be due to an alteration in the quality of the host plant by the pollutants. Aphids, for example, feed specifically

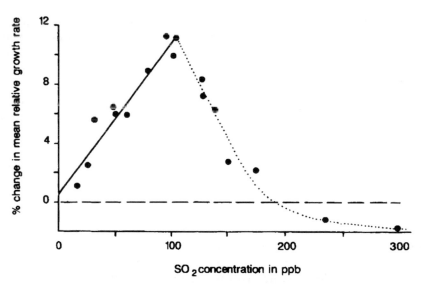

Figure 12.5   *Effects of atmospheric SO_2 concentration on the mean relative growth rate of pea aphids* (Acyrthosiphon pisum). *The aphids fed for four days on garden peas growing in the various SO_2 concentrations, to which the plants had also been exposed for about 20 days prior to the placement of the aphids. Percentage changes are shown compared with aphids on plants in clean air* (From Warrington[102] and used with permission of Applied Science Pub- lishers Ltd)

on the sap of the phloem in plants, which is the main transport pathway for sugars and amino acids. It is possible to raise some species of aphids on artificial diets providing the same nutritional factors as phloem. When such artificially maintained aphids are exposed to low concentrations of pollutants they remain unaffected. Thus there is little doubt that it is changes in the physiology of the plant which are responsible for the differences in aphid activity in polluted air.[103]

The underlying mechanisms responsible for changes in the quality of the host plant have been extensively reviewed,[103] particularly for SO_2 and NO_2 where most studies have shown changes in the nitrogen balance of the host plant, especially amino acid composition. It does, therefore, seem likely that this is the main cause of the increase in the growth rate of aphids, an undesirable effect which may occur when the plants are themselves showing negative growth responses to the pollutants. There are thus likely to be greater final impacts on growth when exposure to pollutants is accompanied by aphid infestation. The precise economic consequences cannot, however, be determined from our present knowledge.

Whilst the impact of pollutants such as SO_2 and NO_2 on insects appears to be fairly clear, this is not the situation for O_3. Evidence from filtration experiments on some crops in rural areas where O_3 is the main pollutant, suggests its presence can significantly improve insect performance. Conversely, experiments using indoor, closed-chamber systems reveal a more complex pattern. For example, chewing insects such as beetles and the Gypsy moth (*Lymantria dispar*) generally show an increase in performance, particularly in food consumption. However, the evidence is far less clear for aphids and both positive and negative effects have been reported (reviewed by Brown[104]). There is also more uncertainty about the mechanisms by which O_3 alters the host plant since no consistent changes have been shown in amino acid composition.

Holopainen and Kossi[105] studied the aphid *Cinara pilicornis* on Norway spruce trees and found that during shoot growth early in the year, O_3 pollution stimulated population development of the aphid, while later on, when the shoots were reaching maturity, O_3 had a negative impact on the performance of the aphids. They thought that this might have been because the O_3 had caused accelerated ageing of the shoots.

The population dynamics of herbivorous insects are determined by a multiplicity of factors, and relatively simple experiments in which a controlled level of pollution is applied to a single insect species on its host plant may provide useful information about changes in that relationship, but may not go far in explaining events in the field. Zvereva and Kozlov[106] have shown that air pollution can have differential effects on herbivorous beetles and their natural enemies, and they suggested that a decrease in predation was a factor in determining increased leaf beetle density near a smelter.

Fungal pathogens may sometimes show increased activity on plants in areas where there are high levels of atmospheric pollution, but caution is required before hasty conclusions are drawn. The low frequency in urban areas of *Rhytisma acerinum*, the fungus responsible for tar spot on the leaves of sycamore (*Acer pseudoplatanus*), and its contrasting abundance in rural areas led to speculation that air pollution may be responsible, particularly SO_2.[107] However, in contrast to *Diplocarpon rosae*, which is responsible for black spot on roses and which has been shown under laboratory conditions to be sensitive to SO_2,[108] there is no direct evidence that SO_2 is the primary cause of the absence of tar spot infection of sycamore in towns and cities. A study by Leith and Fowler[109] in Edinburgh showed that present-day urban concentrations of SO_2 (generally below 20 ppb) are unimportant in determining the distribution of the fungus. The major factor controlling its occurrence appears to be the abundance of overwintered sycamore leaves infested with tar spot, most of which in cities are removed either actively by man or passively by the wind. Clearing of dead leaves in this way would thus remove the source of infection and control the growth of *R. acerinum* populations in urban areas.

12.4.2 Abiotic or Environmental Factors

12.4.2.1 Low Temperatures. The relationship between environmental temperature and individual physiological processes in plants is complex. Different species are adapted to different environments and even within a species individuals from different provenances can display distinctive behaviour. There are two main ways in which plants are normally damaged by low temperatures, *viz.* 'chilling' injury caused by low temperatures above 0 °C and freezing injury. Some plants from warm climates cannot tolerate either chilling or freezing and so their geographical distribution is limited to areas where these stresses do not occur. Plants native to temperate regions can also be severely damaged or killed by chilling or freezing if they are not correctly acclimatized.[110] The process of acclimation is usually called 'hardening' and it is achieved mainly during a period of prior exposure to temperatures slightly above the threshold for injury. In some plants, woody perennials for example, hardening may also require or be enhanced by decreasing photoperiods in the autumn, and it is therefore part of the dormancy cycle.

A full understanding of the mechanisms responsible for acclimation has yet to be established. It is generally agreed, however, that, during the autumn, perennial plants of temperate and Arctic latitudes acclimate to low temperatures by adopting one or a combination of strategies, such as the accumulation of soluble solutes (proline, arginine and simple carbohydrates), changes in cell metabolism away from glycolysis towards the pentose phosphate pathway,[111] or modifications to the cell membranes which involve changes in phospholipids, sterols and glycolipids.[112] The ways in which pollutants may affect these processes are at present uncertain, although it seems likely that alteration of membrane properties caused by pollutants may interfere with the important physiological role of changes in membrane composition during cold hardening. The findings with regard to the likely vulnerability of plasmalemma membranes to free radicals are relevant here (Section 12.3).

Circumstantial evidence that exposure to air pollutants may reduce the frost resistance of woody plants first appeared in the Nordic countries.[113] Keller[23] was amongst the first to demonstrate experimentally that exposure of dormant beech seedlings to SO_2 during the winter led to increased concentrations of sulfur in the leaves when they expanded in spring, and the death of some of the terminal buds which showed symptoms of the type associated with frost injury. Keller[114] also found that Norway spruce suffered frost injury in late winter if exposed to SO_2 from October to April.

Sulfur dioxide can also enhance frost injury in herbaceous plants. Davison and Bailey[21] found that the ability of perennial rye grass to undergo hardening against damage by sub-zero temperatures was reduced by prior exposure to 87 ppb SO_2 for five weeks, and Baker et al.[115] exposed winter wheat to SO_2 in a field-exposure experiment and noticed that a natural frost $-9\,°C$ in mid-January caused increased injury to the fumigated plants. Freer-Smith and Mansfield[116] found that there were small but consistent increases in frost injury on the needles of Sitka spruce cooled to -5 and $-10\,°C$ after exposure to 30 ppb SO_2 and 30 ppb NO_2 followed by a period of hardening in clean air.

Although atmospheric pollution has been implicated as a likely contributing factor to forest declines in Europe and N. America, as yet no clear consensus regarding the primary causes responsible has emerged. The concentrations of SO_2 and NO_x in many locations where there is marked forest decline are very low throughout the year and it is thought unlikely that these pollutants acting alone are responsible. Of the hypotheses put forward as an explanation for tree injury, many that are currently being considered involve ozone as a major factor, either acting alone or in combination with other pollutants or environmental stresses.[117] The involvement of O_3 is based on evidence[118] that, in some of the affected areas, tropospheric concentrations in the summer were found to be in excess of 50 ppb. This is especially true for high elevation sites (>800 m) where, due to a combination of meterological and topographic factors, these sites experience far fewer seasonal and diurnal fluctuations in tropospheric O_3 concentrations than occur at lower elevations.[3] In addition, winter stresses, such as frosts, will occur with a greater frequency at higher altitudes. Controlled fumigations with O_3 alone or in combination with acid mists, either at or in excess of 50 ppb, have failed to reproduce the symptoms of chlorosis or needle loss characteristic of the foliage of damaged trees which suggests that a complex combination of physical and chemical stresses may be involved.

Until the late-1980s there were few studies on the effects of ozone in relation to frost injury, presumably because the conditions necessary for the formation of this pollutant predominantly occur during the summer months (see Chapter 8). However, there is now sufficient evidence from controlled experiments to suggest that exposure to O_3 during the summer months can enhance the susceptibility of conifers to frost injury in the following autumn.[119-121] Although a clear effect of the photo-oxidant was observed in these experiments, the basic mechanism of winter acclimation was not disrupted, because irrespective of treatment or species (Norway spruce, Sitka spruce) the majority of tree seedlings later developed full frost hardiness to temperatures below $-20\,°C$.

Apart from gaseous pollutants, it is now recognized that the frost hardiness of some coniferous tree species growing at high altitude sites can be reduced by exposure to acid mists containing SO_4^{2-}, NH_4^+, NO_3^- and H^+ ions. Such sites frequently receive exposure to mists on 70% of days for up to ten hours per day on average and on 50% of these occasions the mist pH is less than 3.5. Detailed studies by groups in the UK and the USA, summarized by Sheppard,[122] suggest that the uptake and accumulation of high concentrations of SO_4^{2-} ions by younger foliage of red spruce (*Picea rubens*), in which the wax layer of the cuticle is not fully formed, lead to protein denaturation and loss of membrane integrity and consequently a reduction in frost hardiness. It is believed that the presence of NO_3-N can mitigate the toxic effects of SO_4^{2-} but that high concentrations of SO_4^{2-} and H^+ ions can act synergistically and lead to a disruption in the control of cellular pH. In the field these effects may be exacerbated by low concentrations of soil-available Ca^{2+} relative to the availability of Al^{3+} which could further weaken cell membrane integrity. Caution may, however, be necessary in extending these conclusions to other species. In Sitka spruce (*Picea sitchensis*) acid mist had an effect on frost hardiness only on the smallest trees, and only when the trees were already acclimated to $-20\,°C$ and the treatments were applied frequently.[123] In the case of Norway spruce (*Picea abies*) it appears that, when the soil is frozen in winter, acid mist can cause deterioration of the wax layer on the surface of the needles, leading to reduced retention of water, and an enhancement of the damage due to 'frost drought'.[124] Although much of the research on pollution and frost tolerance in perennial plants has been concerned with trees, there is also good evidence that heathland plants may show enhanced frost susceptibility in polluted areas. Foot *et al.*[125] found that heather (*Calluna vulgaris*) showed increased damage in winter after O_3 treatments in summer, and the same species may also be more susceptible to late-winter injury as a result of increased nitrogen deposition in polluted areas.[126]

12.4.2.2 Water Deficits. After temperature, water availability is the main environmental factor governing the distribution of plants. Evolution has provided them with several features and mechanisms for the acquisition of water and to improve the economy of its usage. Some effects of air pollutants on plants have been identified which appear likely to disturb a plant's water economy.

Two pollutants, SO_2 and O_3, and mixtures of SO_2 and NO_2, have often been found to cause a change in the allocation of material between roots and shoots. There is reduced translocation of newly manufactured photosynthates from the leaves to the roots and the eventual effect on the plant can be a considerable increase in the shoot:root dry mass ratio. Noyes[127] fed bean (*Phaseolus vulgaris*) leaves with $^{14}CO_2$ and found that the translocation of ^{14}C-labelled products out of the leaves was inhibited by SO_2 concentrations too small to reduce the rate of photosynthesis. Subsequent studies with several different species have confirmed the high sensitivity of assimilate translocation to SO_2.

Long-distance transport of assimilates in plants involves their movement from sites of photosynthesis (principally the mesophyll cells) into the sieve tubes of the phloem. The final step is known as phloem-loading, and this can take place against a very large concentration gradient, *e.g.* 1:40. The energy required to transport sucrose against such a gradient is almost certainly generated by the consumption of ATP and extrusion of protons, which then return across the membrane 'carrying' sucrose molecules. Presumably the membrane site of this sucrose/proton co-transport may be the point of particular sensitivity of the solution products of SO_2. Minchin and Gould[128] used the very short-lived isotope ^{11}C to show that phloem-loading was almost immediately reduced after exposure of wheat (a C_3 plant) to SO_2. This was not the case, however, with maize, a C_4 plant in which there is a tight ring of cells protecting the phloem from the intercellular air spaces of the leaf. There is no such protective layer in wheat, and consequently SO_2 in the intercellular spaces may have almost uninterrupted access to the phloem tissue. Although it has been suggested that SO_2 might directly inhibit the sucrose-carrier within the plasma membrane, it has not been possible to show that any particular aspect of the sucrose/proton co-transport mechanism is damaged.[16,129]

Like SO_2, O_3 may selectively damage the phloem tissue. Grantz and Farrar[130] found for Pima cotton that although O_3 inhibited assimilation by up to 20%, its effects on phloem translocation were much greater, up to 70%. They used ^{14}C-labelling to study rapid efflux kinetics of recent photoassimilates, and concluded that damage caused by O_3 to plasma membranes of the phloem companion cells would explain their data.

Changes in the balance of growth caused by reduced transport of assimilates to the roots could have important implications for plants growing in drying soil, where the increased penetration of roots is important for the maintenance of water supplies for the shoot. The situation is not simple, however, because there seems to be much variation between species. For example, Taylor and Davies[131] exposed beech seedlings to the ambient atmosphere in southern England during the summer (polluted predominantly by O_3) and found no evidence that the allocation of dry matter between roots and shoots was significantly altered. Nevertheless there were changes in the form of the roots, both root extension and specific root length (root length per unit root dry mass) being significantly greater for beech grown in chambers receiving unfiltered air, which suggests that the root system was longer and also made up of thinner roots. Although these changes in root anatomy did not make the seedlings more susceptible to an artificially imposed drought, the situation could be different for trees growing under natural conditions where the ability of roots to compete for water supplies is more critical. Although the underlying mechanisms are not always clear, and there is much variation between species, sufficient ozone-induced increases in water stress in trees are recorded in the literature for this to be taken seriously as a major threat to forest health.[132-137] There is less information for herbaceous plants and crops, but ozone/drought interactions do appear to be important in some grassland species[138] and in crop species such as tomato.[139]

Effects of pollutants on stomatal behaviour have attracted a lot of attention, since in most plants the strategy for avoiding drought stress involves limiting the rate of water loss through a reduction in transpiration *via* decreases in stomatal conductance. The opening of stomata ideally achieves an acceptable compromise between the plant's need to acquire CO_2 from the atmosphere for photosynthesis and the loss of water by transpiration. Any alteration in stomatal functioning can be serious because of interference with carbon gain or water usage.

There is a considerable amount of evidence in the literature that SO_2, O_3 and mixtures of SO_2 and NO_2 can sometimes cause abnormal stomatal opening.[136,140,141] This can sometimes be attributed to damage to the cells surrounding the stomatal guard cells. The latter are known to be more resilient than the other epidermal cells to adverse treatments.[142] The stomatal pore is opened as result of turgor pressure generated in the guard cells, which push against the resistance offered by the turgor of the surrounding cells. If the surrounding cells lose turgor then greater stomatal aperture occurs and there may not be enough pressure exerted on the guard cells to bring about closure of the pore. Under conditions of severe water stress the stomata are required to close quickly to protect the leaf.

Changes in stomatal function can also be brought about by exposure to ozone. Studies by Heggestad *et al.*[143] found that soil moisture stress enhanced the deleterious effects of O_3 on yields of soybeans so long as the concentrations of O_3 were fairly low (below 80 ppb for seven hours a day). At higher concentrations of O_3 opposite effects were found, *i.e.* there was a reduced impact of water shortage on yield. At higher concentrations, O_3 is known to induce closure of stomata and this may have had the dominant effect by reducing water loss from the leaves.[144]

The complexities surrounding the responses of stomata to O_3 are demonstrated by data from studies on beech.[145] Experiments in which young trees were either subjected to a period of water stress or were watered throughout the experiment showed that in the presence of ozone the stomatal resistance of well-watered trees was increased, whereas in the droughted trees ozone reduced stomatal resistance (*i.e.* the stomata were more open) as the water stress developed (Figure 12.6). These disturbances in stomatal regulation are important in two respects: during periods when water supply is plentiful, an increase in stomatal resistance could reduce the uptake of carbon dioxide for photosynthesis; then, during a drought, which often coincides with conditions favourable for the production of elevated concentrations of O_3 at ground level, it may be difficult for trees to control their water use effectively. Such changes, if they were to occur in the field may, in the long term, ultimately lead to growth reductions and in some cases the death of trees. The method used to determine stomatal behaviour and water loss in these experiments involved detailed measurements on individual leaves of the trees. This makes it difficult to extrapolate from such measurements to determine water loss by the whole tree, due to factors such as leaf age, leaf position in the canopy and interactions between the stomata and the environment. However, there are now methods for

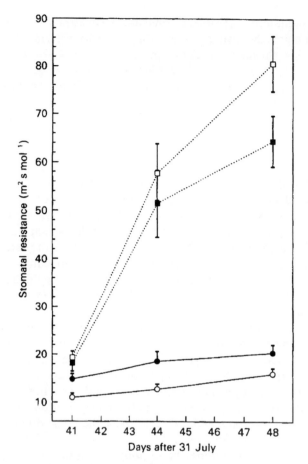

Figure 12.6 *The interaction between ozone and water stress on stomatal resistance in beech. Measurements were made after the commencement of the water stress treatment (day 39). Each point is the mean of data collected at 08.15 and 11.00 hr. The vertical lines represent standard errors of the means (N = 108). Ozone: well-watered (●); droughted (■). Control: well-watered (○); droughted (□).*
(From data of Pearson and Mansfield[145])

measuring whole-tree transpiration and water use, for example the heat-balance technique for measuring sap flow.[146] This procedure was adopted by Wiltshire et al.[147] to study the effects of O_3 exposure on water use by ash trees (*Fraxinus excelsior*). Both daily and seasonal changes in transpiration were monitored and they showed a clear trend for the ozone-exposed trees to use more water per day than the controls during the early part of the growing season (June–July). Later in the season, the reverse was the case. It seems likely that these changes in water use were caused by increases in stomatal resistance during the later stages of the experiment, and thus O_3 needs to be considered as a causal factor in the decline of ash which has been observed in some parts of England.[148]

12.5 AIR POLLUTION AND FOREST HEALTH

Debates about the role of air pollution in causing deterioration in the health of forest trees have been ongoing since the 1980s, and many new papers are published each year. At first sight these may appear to present manifold conflicting views, but close examination of the data shows that some deterioration caused by air pollution in one or more physiological functions is usually reported, and many of the discrepancies are at the quantitative level, or might be attributed to differences between species. In a few of the studies there have been indications that relatively small effects detected at the physiological level in the first year of treatment are detectable in the next year as impacts on growth.[149,150] Most of the evidence available to us has come from short-term exposures of young trees, and it is obvious that this brings severe limitations to our understanding of long-term effects on mature trees in forests. A priority of future work must be to learn more of carry-over effects which may be of cumulative importance on trees in the field.

12.6 CRITICAL LOADS AND CRITICAL LEVELS

One of the major objectives of research of the kind we have discussed is to provide a basis for air quality control standards for the protection of crops, forests and both natural and managed ecosystems. The United Nations Economic Commission for Europe (UNECE) has adopted this definition of a *critical load* for ecosystems: "a quantitative estimate of exposure to one or more pollutants below which significant harmful effects on sensitive elements of the environment do not occur according to present knowledge". The term critical load refers specifically to the deposition of pollutants or of their derivatives. Threshold gaseous concentration exposures are termed *critical levels* and are defined as "the concentrations in the atmosphere above which direct adverse effects on receptors such as plants, ecosystems or materials, may occur according to present knowledge". Both critical loads and critical levels are based on the precautionary principle with the aim of protecting the most sensitive elements of any ecosystem from long-term damage. Both critical loads and levels are often represented on maps, allowing various physical, chemical, ecological, geological and hydrological factors which may affect deposition or the sensitivity of organisms to be taken into account. Information on pollutant distribution and deposition loads then enables critical load or level exceedance maps to be produced.

The concept of critical loads could be applied to any toxic factor affecting any component of the biosphere, but it is mainly used in relation to the total deposition of sulfur, nitrogen or hydrogen ions, and the primary aim is to prevent acidification or eutrophication of sensitive terrestrial or aquatic ecosystems. Information and critical load values for different types of vegetation, soils and freshwaters have been provided by Hornung and Skeffington.[151] The concept of critical loads often leads to pressure from administrators for scientists to stretch the conclusions from their data beyond what they consider to be reasonable limits, and the validity of the procedure is frequently questioned. A recent review of the concept by Malcolm Cresser is strongly recommended.[152] Despite the criticisms and qualifications, the concept is now extensively used in Europe,[153] and increasingly throughout the world. For example, critical loads for acidic deposition have been completed for China, taking soil types and other factors into account, and it has been concluded that a quarter of the land in the south-east of the country is at risk of acidification.[154]

In contrast to critical loads, critical levels refer to the direct effects of atmospheric pollutants on specific components of above-ground vegetation or on ecosystems. The establishment of critical levels for different types of vegetation, *e.g.* lichens and bryophytes, crop species, forest trees and natural vegetation can only be estimated from experimental exposure experiments, most of which involve some form of chamber. Different plant species, and even cultivars or genotypes of the same species, may differ in their sensitivity to pollutants and in the type of effect which might be considered adverse. It is also well established that pollution resistance can evolve over very short time periods in some species, whereas others may remain vulnerable.[155,156] Defining what constitutes an adverse effect can be very important. For example, a reduction in yield is of primary concern for arable crops, whereas both yield and conservation value are important for trees, and the maintenance of biodiversity is of major importance for semi-natural ecosystems. Assigning critical levels to a wide range of different species, representative of different categories of vegetation in different locations and at different times of year, is an unenviable task.

12.7 REFERENCES

1. J. B. Cohen and A. G. Ruston, 'Smoke: A Study of Town Air', Edward Arnold, London, 1912.
2. W. W. Pettigrew, 'The Influence of Air Pollution on Vegetation', Lecture to Smoke Abatement League of Great Britain, Harrogate, 1928.
3. M. R. Ashmore, J. N. B. Bell and A. Mimmack, *Environ. Pollut.*, 1988, **53**, 99.
4. M. R. Ashmore and C. Dalpra, *London Environ. Bull.*, 1985, **3**, 4.
5. M. Yunus and M. Iqbal (eds.), 'Plant Response to Air Pollution', John Wiley, Chichester, 1996.
6. D. C. Adriano and M. Havas (eds.), 'Acidic Precipitation – Volume 1: Case Studies', Springer-Verlag, New York, 1989.
7. C. Eager and M. B. Adams (eds.), 'Ecology and Decline of Red Spruce in the Eastern United States', Springer-Verlag, New York, 1992.
8. D. L. Godbold and A. Hüttermann, 'Effects of Acid Rain on Forest Processes', Wiley-Liss, New York, 1994.
9. T. A. Mansfield and P. H. Freer-Smith, in 'Gaseous Air Pollutants and Plant Metabolism', eds. M. J. Koziol and F. R. Whatley, Butterworth, London, 1984, p. 131.
10. K. J. Lendzian, *Aspects Appl. Biol.*, 1988, **17**, 97.
11. K. Lendzian and G. Kerstiens, *Rev. Environ. Contam. Toxicol.*, 1991, **121**, 65.
12. V. Black and M. H. Unsworth, *J. Exp. Bot.*, 1980, **31**, 667.
13. E. Nieboer, D. H. S. Richardson, K. J. Puckett and F. D. Tomassini, in 'Effects of Air Pollutants on Plants', ed. T. A. Mansfield, Cambridge University Press, Cambridge, 1976, p. 61.

14. D. T. Tingey and D. M. Olszyk, in 'Sulfur Dioxide and Vegetation', eds. W. E. Winner, H. A. Mooney and R. A. Goldstein, Stanford University Press, Stanford, CA, 1985, p. 178.
15. H. Rennenberg and A. Polle, in 'Plant Responses to the Gaseous Environment', eds. R. G. Alscher and A. R. Wellburn, Chapman and Hall, London, 1994, p. 165.
16. C. E. Russell, J. Pitman, N. M. Darrell, L. E. Williams and J. L. Hall, *Plant, Cell Environ.*, 1999, **22**, 221.
17. A. R. Wellburn, O. Majernik and F. A. M. Wellburn, *Environ. Pollut.*, 1972, **3**, 37.
18. A. Ranieri, A. Castagna, G. Lorenzini and G. F. Soldatini, *Environ. Exp. Bot.*, 1997, **37**, 125.
19. A. Ranieri, F. Pieruccetti, A. Panicucci, A. Castagna, G. Lorenzini and G. F. Soldatini, *Plant Physiol. Biochem.*, 1999, **37**, 919.
20. T. Jones and T. A. Mansfield, *Environ. Pollut. A*, 1982, **27**, 57.
21. A. W. Davison and I. F. Bailey, *Nature*, 1982, **297**, 400.
22. S. J. M. Caporn, T. W. Ashenden and J. A. Lee, *Environ. Exp. Bot.*, 2000, **43**, 111.
23. T. Keller, *Environ. Pollut.*, 1978, **16**, 243.
24. 'Acid Rain', Report no 14, Watt Committee on Energy, London, 1984.
25. S. P. McGrath and F. J. Zhao, *Soil Use Manage.*, 1995, **11**, 110.
26. S. V. Krupa and A. H. Legge, *Environ. Pollut.*, 1999, **106**, 449.
27. F. T. Last, J. N. Cape and D. Fowler, *Span*, 1986, **29**, 2.
28. D. Fowler, in 'Air Pollution Effects on Biodiversity', eds. J. R. Barker and D. T. Tingey, Van Nostrand Reinhold, New York, 1992, p. 31.
29. P. M. Vitousek, 1994, *Ecology*, **75**, 1861.
30. I. Stulen, M. Perez-Soba, L. J. De Kok and L. Van der Ferden, *New Phytol.*, 1998, **139**, 61.
31. G. G. Gebauer and E.-D. Schulze, *Aspects Appl. Biol.*, 1988, **17**, 123.
32. B. Nihlgard, *Ambio*, 1985, **14**, 2.
33. A. J. Friedland, G. J. Hawley and R. A. Gregory, *Plant Soil*, 1988, **105**, 189.
34. M. C. Press, S. J. Woodin and J. A. Lee, *New Phytol.*, 1986, **103**, 45.
35. R. Baxter, M. J. Emes and J. A. Lee, *New Phytol.*, 1992, **120**, 265.
36. J. H. Duyzer, H. L. M. Verhagen, J. H. Westrate and F. C. Bosveld, *Environ. Pollut.*, 1992, **75**, 3.
37. J. Pearson and G. R. Stewart, *New Phytol.*, 1993, **125**, 283.
38. J. B. Aber, K. J. Nadelhoffer, P. Steudler and J. M. Melillo, *BioScience*, 1989, **39**, 378.
39. E.-D. Schulze, *Science*, 1989, **244**, 776.
40. K. W. T. Goulding, N. J. Bailey, N. J. Bradbury, P. Hargreaves, M. Howe, D. V. Murphy, P. R. Poulton and T. W. Willison, 1998, *New Phytol.*, **139**, 49.
41. W. H. Smith, 'Air Pollution and Forests', Springer-Verlag, New York, 1981.
42. T. A. Mansfield and D. C. McCune, in 'Assessment of Crop Loss from Air Pollutants', eds. W. W. Heck, O. C. Taylor and D. T. Tingey, Elsevier Applied Science, London, 1988, p. 317.
43. J. D. Barnes and A. R. Wellburn, in 'Responses of Plant Metabolism to Global Change', eds. L. J. De Kok and I. Stulen, Backhuys Publishers, Leiden, 1998, p. 147.
44. 'Nitrogen Oxides', National Academy of Sciences, Washington D.C., 1977, pp. 147–158.
45. A. R. Wellburn, *New Phytol.*, 1990, **115**, 395.
46. D. A. Wink and J. B. Mitchell, *Free Radical Biol. Med.*, 1998, **25**, 434.
47. J. Durner and D. F. Klessig, *Curr. Opin. Plant Biol.*, 1999, **2**, 369.
48. R. M. Law and T. A. Mansfield, in 'Effects of Gaseous Pollutants in Agriculture and Horticulture', eds. M. H. Unsworth and D. P. Ormrod, Butterworths Scientific Press, London, 1982, p. 93.
49. L. M. Mortensen, *New Phytol.*, 1985, **101**, 103.
50. H. Saxe and O. V. Christensen, *Environ. Pollut. A*, 1985, **38**, 159.

51. H. Saxe, *New Phytol.*, 1986, **103**, 185.
52. S. J. M. Caporn, T. A. Mansfield and D. W. Hand, *New Phytol.*, 1991, **118**, 309.
53. S. J. M. Caporn, D. W. Hand, T. A. Mansfield and A. R. Wellburn, *New Phytol.*, 1994, **126**, 45.
54. G. Houlden, S. McNeill, A. Craske and J. N. B. Bell, *Environ. Pollut.*, 1991, **72**, 45.
55. S. Nussbaum, M. Geissmann and J. Fuhrer, *Water, Air, Soil Pollut.*, 1995, **85**, 1449.
56. R. M. Wallsgrove, P. J. Lea and B. J. Miflin, *Plant Physiol.*, 1979, **63**, 323.
57. P. J. Lea, J. Wolfenden and A. R. Wellburn, in 'Plant Responses to the Gaseous Environment', eds. R. G. Alscher and A. R. Wellburn, Chapman and Hall, London, 1994, p. 279.
58. R. G. Amundson and D. C. Maclean, in 'Air Pollution by Nitrogen Oxides', eds. T. Schneider and L. Grant, Elsevier, Amsterdam, 1982, p. 501.
59. A. R. Wellburn, J. Wilson and P. H. Aldridge, *Environ. Pollut.*, 1980, **22**, 219.
60. C. MacKintosh, *New Phytol.*, 1998, **139**, 153.
61. A. M. Thompson, *Science*, 1992, **256**, 1157.
62. W. L. Chameides and E. B. Cowling, 'The State of the Southern Oxidants Study: Policy-relevant Findings in Ozone Pollution Research 1988–94', North Carolina State University, 1995.
63. M. R. Ashmore, J. N. B. Bell and C. L. Reily, *Nature*, 1987, **327**, 417.
64. R. L. Heath, in 'Plant Responses to the Gaseous Environment', eds. R. G. Alscher and A. R. Wellburn, Chapman and Hall, London, 1994, p. 121.
65. T. A. Mansfield, *Environ. Pollut.*, 1998, **101**, 1.
66. H. Mehlhorn and A. R. Wellburn, *Nature*, 1987, **327**, 417.
67. H. Mehlhorn, A. R. O'Shea and A. R. Wellburn, *J. Exp. Bot.*, 1991, **42**, 17.
68. P. J. Davies, 'Plant Hormones: Physiology, Biochemistry and Molecular Biology', Kluwer, Dordrecht, 1995.
69. A. Polle, in 'Responses of Plant Metabolism to Global Change', eds. L. J. De Kok and I. Stulen, Backhuys Publishers, Leiden, 1998, p. 95.
70. H. Mehlhorn, B. J. Tabner and A. R. Wellburn, *Physiol. Plant.*, 1990, **79**, 377.
71. V. C. Runeckles and M. Vaartnou, *Plant Cell Environ.*, 1997, **20**, 306.
72. C. N. Hewitt, G. L. Kok and R. Fall, *Nature*, 1990, **344**, 56.
73. T. Sakaki, in 'Responses of Plant Metabolism to Global Change', eds. L. J. De Kok and I. Stulen, Backhuys Publishers, Leiden, 1998, p. 117.
74. T. Lyons, J. H. Ollerenshaw and J. D. Barnes, *New Phytol.*, 1999, **141**, 253.
75. E. Matzner and B. Ulrich, in 'Effects of Atmospheric Pollutants on Forests, Wetlands and Agricultural Ecosystems', eds. T. C. Hutchinson and K. M. Meema, Springer-Verlag, Berlin, 1987, p. 25.
76. G. Abrahamsen, in 'Effects of Atmospheric Pollutants on Forests, Wetlands and Agricultural Ecosystems', eds. T. C. Hutchinson and K. M. Meema, Springer-Verlag, Berlin, 1987, p. 321.
77. J. J. Colls and M. H. Unsworth, in 'Air Pollution Effects on Biodiversity', eds. J. R. Barker and D. T. Tingey, Van Nostrand Reinhold, New York, 1992, p. 93.
78. S. I. Nilsson, H. G. Miller and J. D. Miller, *Oikos*, 1982, **39**, 40.
79. R. A. Skeffington and T. M. Roberts, *Oecologia*, 1985, **65**, 201.
80. T. M. Roberts, R. A. Skeffington and L. W. Blank, *Forestry*, 1989, **62**, 179.
81. H. Rennenberg, K. Kreutzer, H. Papen and P. Weber, *New Phytol.*, 1998, **139**, 71.
82. T. V. Armentano and J. P. Bennett, in 'Air Pollution Effects on Biodiversity', eds. J. R. Barker and D. T. Tingey, Van Nostrand Reinhold, New York, 1992, p. 159.
83. P. H. Freer-Smith, in 'Plant Response to Air Pollution', eds. M. Yunus and M. Iqbal, John Wiley, Chichester, 1996, p. 437.
84. T. Izuta, *J. Plant Res.*, 1998, **111**, 471.
85. R. G. Sayre and T. J. Fahey, *Can. J. Forest Res.*, 1999, **29**, 487.
86. M. H. Unsworth and D. Fowler, in 'Air Pollution and Ecosystems', ed. P. Mathy, Reidel, Dordrecht, 1988, p. 68.
87. M. H. Unsworth, *Nature*, 1984, **312**, 262.

88. L. S. Evans, K. F. Lewis, E. A. Cunningham and M. J. Patti, *New Phytol.*, 1982, **91**, 429.
89. D. T. Tingey, R. A. Reinert, J. A. Dunning and W. W. Heck, *Phytopathology*, 1971, **61**, 1506.
90. P. C. Pande and T. A. Mansfield, *Environ. Pollut. A*, 1985, **39**, 281.
91. M. T. Yoneyama and H. Sasakawa, *Plant Cell Physiol.*, 1979, **20**, 263.
92. A. R. Wellburn, in 'Effects of Gaseous Air Pollutants in Agriculture and Horticulture', eds. M. H. Unsworth and D. P. Ormrod, Butterworth, London, 1982, p. 169.
93. A. R. Wellburn, in 'Gaseous Air Pollutants and Plant Metabolism', eds. M. J. Koziol and F. R. Whatley, Butterworth, London, 1984, p. 203.
94. J. M. Petitte and D. P. Ormrod, *Am. Potato J.*, 1988, **65**, 517.
95. J. M. Petitte and D. P. Ormrod, *J. Am. Soc. Hort. Sci.*, 1992, **117**, 146.
96. R. L. Engle and W. H. Gabelman, *Proc. Am. Soc. Hort. Sci.*, 1966, **89**, 423.
97. H. A. Menser and H. E. Heggestad, *Science*, 1966, **153**, 424.
98. R. Kohut, in 'Sulfur Dioxide and Vegetation', eds. W. E. Winner, H. A. Mooney and R. Goldstein, Stanford University Press, Stanford, CA, 1985.
99. W. E. Winner, in 'Air Pollution and Plant Metabolism', eds. S. Schulte-Hostede, N. M. Darrall, L. W. Blank and A. R. Wellburn, Elsevier Applied Science, London, 1987, p. 317.
100. H. H. Cramer, *Forstw. Cbl.*, 1951, **70**, 42.
101. W. Flückiger, S. Braun and M. Bolsinger, in 'Air Pollution and Plant Metabolism', eds. S. Schulte-Hostede, N. M. Darrall, L. W. Blank and A. R. Wellburn, Elsevier Applied Science, London, 1987, p. 366.
102. S. Warrington, *Environ. Pollut.*, 1987, **43**, 155.
103. J. B. Whittaker and S. Warrington, in 'Pests, Pathogens and Plant Communities', eds. J. R. Burdon and S. R. Leather, Blackwell Scientific Publications, Oxford, 1990, p. 97.
104. V. C. Brown, in 'Insects in a Changing Environment', eds. R. Harrington and N. E. Stork, Academic Press, London, 1995, p. 219.
105. J. K. Holopainen and S. Kossi, *Entomol. Exp. Appl.*, 1998, **87**, 109.
106. E. L. Zvereva and M. V. Kozlov, *J. Appl. Ecol.*, 2000, **37**, 298.
107. G. N. Greenhalgh and R. J. Bevan, *Trans. Br. Mycol. Soc.*, 1978, **71**, 491.
108. M. Fehrmann, A. Von Tiedemann, L. W. Blank, B. Glashagen and T. Eisenman, in 'How are the Effects of Air Pollution on Agricultural Crops Influenced by Interactions with Other Limiting Factors?' report of Working Party III, Concerted Action on 'Effects of Air Pollution on Terrestrial and Aquatic Ecosystems', CEC, Brussels, 1986, p. 98.
109. I. D. Leith and D. Fowler, *New Phytol.*, 1987, **108**, 175.
110. J. Levitt, 'Responses of Plants to Environmental Stresses, Vol. 1, Chilling, Freezing and High Temperature Stresses', Academic Press, New York, 1980.
111. C. J. Andrews and M. K. Pomeroy, *Plant Physiol.*, 1979, **64**, 120.
112. B. Yoshida and M. Uemura, *Plant Physiol.*, 1984, **75**, 31.
113. S. Huttunen, in 'Air Pollution and Plant Life', ed. M. Treshow, John Wiley, Chichester, 1984, p. 321.
114. T. Keller, *Gartenbauwissenschaft*, 1981, **46**, 170.
115. C. K. Baker, M. H. Unsworth and P. Greenwood, *Nature*, 1982, **299**, 149.
116. P. H. Freer-Smith and T. A. Mansfield, *New Phytol.*, 1987, **106**, 237.
117. S. B. McLaughlin, *J. Air Pollut. Control Assoc.*, 1985, **35**, 512.
118. B. Prinz, G. H. Krause and H. Stratman, 'Forest Damage in the Federal Republic of Germany', Land Institute of Pollution Control of the Land North-Rhine Westphalia, Wessen, West Germany, 1982.
119. P. W. Lucas, D. A. Cottam, L. J. Sheppard and B. J. Francis, *New Phytol.*, 1988, **108**, 495.
120. K. A. Brown, T. M. Roberts and L. W. Blank, *New Phytol.*, 1987, **105**, 149.

121. J. D. Barnes and A. W. Davison, *New Phytol.*, 1988, **108**, 159.
122. L. J. Sheppard, *New Phytol.*, 1994, **127**, 69.
123. L. J. Sheppard, A. Crossley, F. J. Harvey, D. Wilson and J. N. Cape, *Environ. Pollut.*, 1997, **98**, 175.
124. A. Esch and K. Mengel, *Environ. Exp. Bot.*, 1998, **39**, 57.
125. J. P. Foot, S. J. M. Caporn, J. A. Lee and T. W. Ashenden, *New Phytol.*, 1997, **135**, 369.
126. J. A. Carroll, S. J. M. Caporn, L. Cawley, D. J. Read and J. A. Lee, *New Phytol.*, 1999, **141**, 423.
127. R. D. Noyes, *Physiol. Plant. Pathol.*, 1980, **16**, 73.
128. P. E. H. Minchin and R. Gould, *Plant Sci.*, 1986, **43**, 179.
129. L. Maurousset, R. Lemoine, O. Gallet, S. Delrot and J. L. Bennemain, *Biochim. Biophys. Acta*, 1992, **1105**, 230.
130. D. A. Grantz and J. F. Farrar, *J. Exp. Bot.*, 1999, **50**, 1253.
131. G. Taylor and W. J. Davies, *New Phytol.*, 1990, **116**, 457.
132. F. A. M. Wellburn and A. R. Wellburn, *J. Exp. Bot.*, 1994, **45**, 607.
133. S. B. McLaughlin and D. J. Downing, *Can. J. Forest Res.*, 1996, **26**, 670.
134. M. Dixon, D. Le Thiec and J. P. Garrec, *Environ. Exp. Bot.*, 1998, **40**, 77.
135. E. Paakkonen, M. S. Gunthardt-Goerg and T. Holopainen, *Ann. Bot.*, 1998, **82**, 49.
136. U. Maier-Maercker, *Tree Physiol.*, 1999, **19**, 71.
137. W. A. Retzlaff, M. A. Arthur, N. E, Grulke, D. A. Weinstein and B. Gollands, *Tree Physiol.*, 2000, **20**, 195.
138. P. Bungener, S. Nussbaum, A. Grub and J. Fuhrer, *New Phytol.*, 1999, **142**, 283.
139. I. A. Hassan, J. Bender and H. J. Weigel, *Gartenbauwissenschaft*, 1999, **64**, 152.
140. T. A. Mansfield and P. H. Freer-Smith, 'Gaseous Air Pollutants and Plant Metabolism', eds. M. J. Koziol and F. R. Whatley, Butterworth, London, 1984, p. 131.
141. V. J. Black, in 'Sulfur Dioxide and Vegetation', eds. W. E. Winner, H. A. Mooney and R. A. Goldstein, Stanford University Press, Stanford, CA, 1985, p. 96.
142. G. R. Squire and T. A. Mansfield, *New Phytol.*, 1972, **71**, 1033.
143. H. E. Heggestad, T. J. Gish, E. H. Lee, J. H. Bennett and L. W. Douglas, *Phytopathology*, 1985, **75**, 472.
144. N. M. Darrall, *Plant Cell Environ.*, 1989, **12**, 1.
145. M. Pearson and T. A. Mansfield, *New Phytol.*, 1993, **123**, 351.
146. J. M. Baker and C. H. M. Van Bavel, *Plant Cell Environ.*, 1987, **10**, 777.
147. J. J. J. Wiltshire, M. H. Unsworth and C. J. Wright, *New Phytol.*, 1994, **127**, 349.
148. S. K. Hull and J. N. Gibbs, 'Forestry Commission Bulletin 74', HMSO, London, 1991.
149. M. Pearson and T. A. Mansfield, *New Phytol.*, 1994, **126**, 511.
150. E. Oksanen and A. Saleem, *Plant Cell, Environ.*, 1999, **22**, 1401.
151. M. Hornung and R. A. Skeffington (eds.), 'Critical Loads: Concepts and Applications', ITE Symposium No. 28, Proceedings of a Conference held on 12–14 February in Grange-over-Sands, Cumbria, HMSO, London, 1993.
152. M. S. Cresser, *Sci. Total Environ.*, 2000, **249**, 51.
153. R. A. Skeffington, *Env. Sci. Technol.*, 1999, **33**, 245A.
154. L. Duan, S. D. Xie, Z. P. Zhou and J. M. Hao, *Water, Air, Soil Pollut.*, 2000, **118**, 35.
155. A. W. Davison and J. D. Barnes, *New Phytol.*, 1998, **139**, 135.
156. J. D. Barnes, J. Bender, T. Lyons and A. Borland, *J. Exp. Bot.*, 1999, **50**, 1423.

CHAPTER 13

Control of Pollutant Emissions From Road Traffic

C. HOLMAN

13.1 INTRODUCTION

There are many ways of reducing emissions from traffic such as improving standards of maintenance, reducing the volume of traffic and smoothing traffic flow. These are all important elements in an emissions control strategy. However, this chapter focuses on the role of vehicle and fuel technologies. Specifically, it discusses the potential improvements from engine design, after-treatment systems (catalysts and traps) and fuel quality to reduce future emissions.

Emissions from motor vehicles have been reduced significantly over the last 30 years. Emissions of the gaseous regulated pollutants – nitrogen oxides (NO_x), carbon monoxide (CO) and hydrocarbons (HC)* – per vehicle kilometre are now only a few percent of their unregulated levels. These improvements have largely been in response to progressively more stringent emission limits for new vehicles in the European Union (EU), North America and Japan. There have also been significant, but smaller, reductions in emissions of particulate matter (PM) per vehicle kilometre. Further limits are to be introduced for new vehicles sold in these markets over the next few years. It is likely that the main focus of future legislation for cars, buses and commercial vehicles will be on PM.

Whilst emissions per vehicle kilometre have declined in the last three decades since emission controls were first introduced, in the 1970s and 1980s the rapid increase in vehicle mileage offset the progress made in introducing cleaner technology, and total emissions increased. In the 1990s, however, total emissions have begun to decline and are predicted to decline still further. This is shown for the European Union (EU) in Figures 13.1, 13.2 and 13.3 for NO_x, PM_{10} (PM_{10} refers to particulate matter that is less than 10 μm in diameter) and carbon dioxide (CO_2) respectively. These figures show the total estimated EU emissions as well as that for road transport. For PM_{10} the road transport

---

*The road vehicle emissions legislation regulates hydrocarbons (HC) emissions. In this context hydrocarbons are assumed to be equivalent to volatile organic compounds (VOCs).

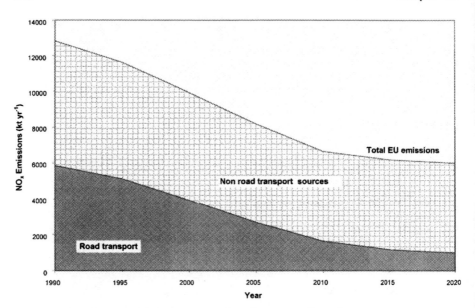

Figure 13.1 *EU NO$_x$ emissions: total and road transport 1990–2020*
(Source: 2nd European Auto Oil Programme, SENCO 1999[1])

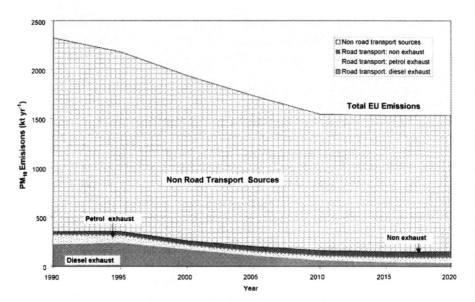

Figure 13.2 *EU PM$_{10}$ emissions: total and road transport 1990–2020*
(Source: 2nd European Auto Oil Programme, SENCO 1999[1])

emissions have been further split between diesel exhaust, petrol exhaust and non-exhaust emissions.

Road transport emissions of NO$_x$ are predicted to be about three-quarters of their 1995 level by 2000, one-half by 2005, and one-third by 2010. For PM$_{10}$ the

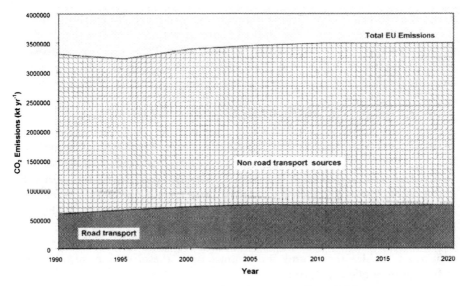

Figure 13.3 *EU CO₂ emissions: total and road transport 1990–2020*
(Source: 2nd European Auto Oil Programme, SENCO 1999[1])

emission reduction is lower, and is predicted to be 45% of the 1995 level by 2010.[1] Despite this decline it is thought that further reductions in emissions will be required in some cities to meet the EU air quality objectives. The expectation, based on current policies, is that emissions will continue to decline until 2015 to 2020. If no further measures are taken it is expected that there will then be a slow increase in emissions as the growth in road traffic offsets the reductions in emissions from individual vehicles.

For CO_2 the picture is quite different. Road transport emissions in the EU are forecast to increase slightly such that by the year 2010 emissions are forecast to be 12% greater than in 1995. This estimate takes account of the impact of the voluntary agreements between motor manufacturers and the European Commission (EC) to reduce fleet average new car CO_2 emissions by 25% over the period 1996 to 2008. The EU is committed within the United Nations Framework Convention on Climate Change to *stabilize* emissions of carbon dioxide (CO_2) at 1990 levels by 2000. Under the December 1997 Kyoto Conference, as parties to this convention the EU accepted a target to *reduce* its greenhouse gas emissions by 8% by 2008–12 compared to 1990 levels.

Vehicles are used for many years before they are scrapped (typically over 10 years for a passenger car, longer for other vehicles in the United Kingdom). It therefore takes some years for the benefits of cleaner/more fuel-efficient new vehicles to have maximum impact on the environment. To encourage a more rapid turnover of the fleet, and the purchase of cleaner vehicles ahead of their mandatory introduction, fiscal incentives have been used in many EU countries.

Controls have also been introduced on fuel quality to complement the investment in new vehicle technologies. The introduction of unleaded petrol allowed the use of catalysts for petrol cars, and limits on the sulfur content of diesel allowed heavy-duty engine manufacturers to meet PM emission limits in the 1990s. Further controls on sulfur in both petrol and diesel were introduced in the year 2000, with lower limits scheduled for 2005. Further controls are probable as more sulfur sensitive catalyst technology is used to reduce emissions. Fiscal incentives have also been used to encourage the use of cleaner fuels.

Higher quality fuels enable the use of cleaner vehicle technology, but also have an important role in reducing emissions on their own account. For example, cleaner petrol and diesel products introduced onto the UK market in the late 1990s (*e.g.* Tesco's low benzene petrol and City diesel) can reduce urban traffic emissions by 5 and 15 per cent for nitrogen oxides and fine particulate matter respectively compared to conventional products.[2] These cleaner fuels can be used most effectively in cities where pollution levels are higher, and particularly where vehicle fleets (buses, taxis, delivery vans) make a large contribution to urban pollution. Other advantages of cleaner diesel are that it eliminates visible smoke and the characteristic diesel odour, resulting in a reduction in public complaints.

13.2 ROAD TRANSPORT EMISSIONS

In the EU in 1995 road transport contributed an estimated 44% of total NO_x emissions (see Figure 13.1). There was, however, significant variation between EU Member States. The road transport contribution varied between 29% and 62% of the total NO_x emissions, depending on the strength of other sources. Road transport is also an important source of CO (60%), and non-methane volatile organic compounds (37%) in the EU. Relatively little PM (16%) and CO_2 (20%) comes from road transport (see Figures 13.2 and 13.3).

At the urban scale road transport, in general, makes a much greater contribution to total emissions than the national/EU figures suggest, particularly for NO_x and PM. This is illustrated in Table 13.1. In the centre of these conurbations there are likely to be pollution 'hot-spots' where the road traffic

Table 13.1   *Contribution of road traffic to 1995 emissions in UK nationally and selected conurbations*

	Percent			
Area	*NO_x*	*NMVOC*	*PM_{10}*	*CO_2*
Greater London	75	53	78	29
Greater Manchester	63	19	31	21
West Midlands	85	46	56	43
Merseyside	42	28	8	18
Bristol	61	60	36	26
Glasgow	76	56	73	29
UK	53	37	25	20

(Sources: London Research Centre[3] [7] and Department of the Environment.[8])

contribution is even higher, particularly in city centres close to major roads. Those cities with a relatively low contribution from road transport have large industrial sources.

It should be noted that all emission estimates are uncertain. According to the UK National Environmental Technology Centre emission estimates for past years are accurate within ± 30 per cent for nitrogen oxides and ± 20–25 per cent for black smoke (no estimates of accuracy are given for PM_{10}). Future forecasts are even more uncertain as they have to rely on forecasts of activity in different economic sectors such as the National Road Traffic Forecast, which are also subject to uncertainty.[9] In estimating how much further emissions need to be reduced to meet air quality objectives these uncertainties need to be taken into account.

13.3 FUEL CHOICE AND EMISSIONS

13.3.1 Introduction

The two main automotive fuels currently used worldwide are petrol and diesel. These fuels are likely to continue to dominate in the foreseeable future.

The role of alternative fuels in reducing emissions from traffic has received considerable attention in recent years. Government and environmental organizations have published many reports comparing the relative merits of the different fuels.[10] These have drawn on information from studies in Europe and elsewhere. In addition, many fleet operators have undertaken trials with alternative fuels. The UK Government has completed an evaluation of some of the first trials in terms of their emissions, costs and operational performance.[11] From all these studies a general consensus emerges. There may be a role for the gaseous fuels in the short term for some types of vehicle. Biofuels, hydrogen, and electricity, however, are unlikely to be widely used in the next ten years for a variety of reasons depending on the fuel. These include being economically uncompetitive, the unavailability of technology suitable of meeting customer requirements in terms of performance, safety, and durability, and uncertain security of fuel supply. In some countries these fuels have been made to be economically more attractive by the use of fiscal incentives.

When comparing emissions from vehicles using different fuels it is important to compare similar vehicles. Comparisons should be between vehicles of similar function and performance. Emissions from a petrol car cannot realistically be compared to those of a 32 tonne lorry. Nor should comparisons be made between a high performance car operating on one fuel, and an economy model on another. Similarly comparisons must be made with similar generations of vehicle. Essentially all petrol cars sold in Europe are now fitted with a three-way catalyst (which reduces emissions of CO, HC and NO_x), and most diesel cars are fitted with an oxidation catalyst (which reduces emissions of CO, HC and PM). In addition comparisons must be made for vehicles tested using the same test conditions, as some vehicles have lower emissions under cruising conditions while others have advantages under the stop–start driving conditions of urban areas.

13.3.2 Petrol and Diesel

Due to the different physical characteristics of petrol and diesel these fuels are burnt in different types of internal combustion engines. Petrol requires a spark to ignite the fuel, and thus is used in spark ignition engines.

Compressing the fuel ignites diesel, and therefore the engines using this fuel are known as compression ignition engines. There are two types of compression ignition – direct and indirect injection, commonly known as DI and IDI respectively. In a DI engine the fuel is injected directly into the combustion chamber, whereas in an IDI engine the fuel is first mixed with air in a pre-chamber, and the mixture is injected into the combustion chamber.

Modern diesel engine technology and the use of turbochargers, which boost the power from an engine, have to a large extent closed the gap in performance between petrol and diesel engines. This has resulted in the popularity of diesel cars and vans increasing in recent years in many European countries. Small vehicles with diesel engines are, however, a European phenomenon, with few models on the Japanese and US markets. It is thought that the light commercial vehicle market will become almost exclusively diesel in the coming years.

Cars and vans can have either DI or IDI diesel engines, whereas only DI engines are used in buses and lorries due to their superior fuel efficiency. Another advantage of DI engines is their great durability. However, most of the diesel cars and vans until the late 1990s used IDI engines because vehicles with DI engines could not meet the emission limits for NO_x, PM and noise. However, recent improvements to the technology has resulted in the range of DI cars on the market widening and it is anticipated that DI will replace IDI in the light duty market over the coming years.

Each of these engine types – spark ignition, and DI and IDI compression ignition – have different emission characteristics. Diesel cars are more fuel-efficient, and have lower CO_2 emissions despite the higher carbon content of the fuel. They also have lower CO and HC emissions compared to similar modern petrol vehicles. On the other hand emissions of NO_x and PM are greater. Emissions of CO_2 are lowest, but the NO_x emissions are greatest from DI compression ignition engines. DI engines may be 15% more fuel-efficient than IDI engines, and up to 40% more than petrol engines.

The types of HC emitted from petrol and diesel engines vary considerably, reflecting the differences in the fuel. Typically petrol cars have greater emissions of low molecular weight HC (*e.g.* benzene) whilst diesel has greater emissions of high molecular weight compounds (*e.g.* polycyclic aromatic compounds).

13.3.3 Alternative Fuels

The main alternatives to petrol and diesel available on the European market are natural gas (compressed or liquefied), liquefied petroleum gas (LPG), and electric vehicles. In general there has been little support for the use of alcohol fuels (*i.e.* ethanol and methanol) in Europe, although these fuels have been used in other parts of the world. In general alcohol is used mixed with petrol. There is a long history of ethanol use as an automotive fuel in Brazil and parts of North America. However, there are a number of technical problems with its use such as its miscibility with water, and emissions of some important photochemical ozone forming HC can increase. Its widespread use is only likely where governments wish to encourage its use to support local agriculture, for example, to find a market for locally produced sugar as in Brazil. There are also problems with the use of methanol as an automotive fuel; for example, it burns with an invisible flame, making it potentially very dangerous in accidents.

The main emission advantage of gaseous fuels is their significantly lower PM emissions compared to diesel vehicles. They can also have lower NO_x emissions, but not always.

Both spark and compression ignition engines can be used to run on natural gas (NG). Vehicles with spark ignition engines can also be used as dual-fuel vehicles, operating on gas or petrol. Where the engine is not optimized for gaseous fuels, for example in conversions or dual-fuel vehicles, the vehicle is likely to give worst emissions than a dedicated vehicle. Where a catalyst is fitted it might not be optimized for NG operation. Gaseous fuel technology is rapidly improving, as is emission control for conventionally fuelled vehicles. The following comparison is largely based on the results of a UK sponsored review of the performance of alternatively fuelled vans and buses undertaken in the late 1990s.[11] Some heavy-duty engine manufacturers have stated that, despite the current benefits of gaseous fuelled vehicles, meeting stringent emission standards (*i.e.* Euro V standards in the year 2008) will be easier with diesel engines than gaseous engines. Therefore NG/LPG vehicles may fill a niche market for a relatively short period of time.

Today's dual-fuelled vehicles have lower emissions of CO_2, CO, and PM emissions when operating on compressed natural gas (CNG) compared with petrol. However, HC emissions can be greater or less than when using petrol, depending on the driving conditions. Natural gas mainly consists of methane, and this is a significant component of the HC emissions from CNG vehicles. Methane is an important greenhouse gas. Under CNG operation the NO_x emission is the same or less than with petrol operation.

Dedicated CNG vehicles offer greater benefits than dual-fuelled vehicles. Compared to petrol vehicles emissions of all the regulated pollutants are either lower or similar. Compared to diesel vehicles emissions of CO, NO_x and PM are generally lower with CNG vehicles. In the case of particulate matter, emissions are lower by 80 to 90% for smaller vehicles such as vans, and 50 to 80% for large vehicles such as buses. Total HC (mainly methane) emissions tend to be higher with CNG under all driving conditions. The most recent dedicated CNG vehicles can have hydrocarbon emissions similar to those from diesel vehicles except under congested driving conditions.

CNG buses show an increase in carbon dioxide emissions compared to similar diesel buses. This increase is greater than the expected loss of efficiency resulting from switching a compression ignition to a spark ignition engine and may be due to poor matching of the engine to the bus operating cycle.

The most recent LPG vehicles have lower CO and HC emissions than petrol vehicles. Typically the NO_x emissions are lower than those from petrol cars and vans. Compared to diesel vehicles LPG vehicles have about 80% lower PM emissions. Some LPG vehicles have much higher carbon monoxide and hydrocarbon emissions than diesel, but for those LPG vehicles fitted with catalysts the emissions are lower than comparable diesel vehicles. NO_x emissions from LPG vans are lower under all driving conditions, but from buses can be higher under some driving cycles compared to equivalent diesel vehicles.

The CO_2 emissions from LPG vans are lower than the equivalent diesel vehicles except under heavily congested driving conditions. For buses CO_2 emissions are greater when using LPG than with diesel, possibly for the same reason as for the CNG buses.

There are a number of electric vehicles on the market. These vehicles have zero emissions at the point of use, but their life-cycle emissions depend on the fuel used at the power station. They are best suited to urban operation because of their relatively limited range. Widespread use of these vehicles will depend upon the development of more advanced batteries, which can extend their range economically, and the development of a re-charging infrastructure.

Possibly more promising in the short term is the use of hybrid electric vehicles, the first of which came onto the UK market in the year 2000. This vehicle combines a petrol engine and electric motor. Energy-saving features include automatic engine shutdown when the vehicle is stopped and regenerative braking that converts kinetic energy into electricity to charge the battery. Hybrid electric vehicles have been on the Japanese market since 1997. These vehicles are claimed to have half the fuel consumption and significantly lower emissions compared to conventional vehicles (under the relatively undemanding Japanese test cycle).

In the longer term hydrogen and fuel cells may become important alternatives to petrol and diesel. The widespread use of hydrogen is likely to depend on the development of a clean and economical production method for the fuel. Hydrogen vehicles have very low emissions of the regulated pollutants during use. Currently more than 90% of the hydrogen produced comes from fossil fuels. For hydrogen to have low life-cycle emissions clean production using the electrolysis of water by electricity generated by renewable energy sources is likely to be needed.

Fuel cells generate electricity on-board the vehicle, typically from hydrogen and atmospheric oxygen. Fuel cells need hydrogen stored either as a metal hydride, or produced on-board from liquid hydrocarbons. Petrol could be an early fuel for fuel cell vehicles as the infrastructure is already in place, allowing more rapid market penetration. The reaction is:

$$2H_2 + O_2 \longrightarrow 2H_2O$$

The Ballard fuel cell, which is being developed for automotive applications with Daimler Benz and Ford Motor Company, uses two electrodes separated by a polymer membrane electrolyte (proton exchange membrane). The electrodes are coated on one side with a thin platinum catalyst layer. Hydrogen dissociates into free electrons and protons in the presence of the platinum catalyst at the anode. The electrons are conducted in the form of useable electric current through the external circuit. The protons migrate through the membrane electrolyte to the cathode. At the cathode the combination of oxygen from the air, electrons from the external circuit and protons produces water and heat. Individual fuel cells produce about 0.6 V and are combined into a fuel cell stack to provide the amount of electrical power required.

Methanol can also be used as a fuel in fuel cells but it produces CO_2. The basic chemical reaction is:

$$2CH_3OH + 3O_2 \longrightarrow 4H_2O + 2CO_2$$

Several motor manufacturers have announced that the first fuel cell vehicles will be on the market by around 2005. However, it is likely that it will be many years before they achieve a significant market penetration due to the high cost of these systems.

13.3.4 Summary

Tables 13.2 and 13.3 give the relative emissions of light duty vans and heavy-duty buses vehicles respectively, when operating with petrol, diesel, LPG and

Table 13.3 *Relative emissions of buses using different fuels*

Fuel (vehicle type)	CO_2	CO	HC	NO_x	PM
Diesel	*	**	*	**	***
Ultra low sulfur diesel (ULSD)	**	**	*	**	**
CNG (converted vehicle)	***	***	***	***	*
CNG (dedicated, non-catalyst vehicle)	***	***	***	**	*
CNG (dedicated, catalyst vehicle)	***	*	*	*	*
LPG (converted, non-catalyst vehicle)	***	***	**	***	*
LPG (converted, catalyst vehicle)	***	*	*	***	*
LPG (dedicated, catalyst vehicle)	***	***	*	**	*

Key: * Lowest emissions; ***highest emissions
(Source: Adapted from Cleaner Fuels Forum, 'Cleaner Air: The Role of Cleaner Fuels', National Society for Clean Air, Brighton, 1998.[2])

Table 13.2 *Relative emissions of light duty vans using different fuels*

Fuel (vehicle type)	CO_2	CO	HC	NO_x	PM
Petrol	***	***	***	*	*
Diesel	**	*	*	**	***
Ultra low sulfur diesel (ULSD)	*	*	*	**	**
LPG	***	**	**	*	*
CNG (bi-fuel vehicle)	*	**	***	*	*
CNG (dedicated vehicle)	**	*	**	*	*

Key: * Lowest emissions; ***highest emissions.
(Source: Adapted from Cleaner Fuels Forum, 'Cleaner Air: The Role of Cleaner Fuels', National Society for Clean Air, Brighton, 1998.[2])

CNG, using different technologies. Ultra low sulfur diesel (ULSD) has a lower sulfur content and density than conventional diesel, and meets the fuel quality requirements to be introduced throughout the EU in 2005.

13.4 CONTROLLING EMISSIONS

13.4.1 Introduction

If complete combustion of the fuel were possible vehicle exhaust would contain only carbon dioxide and water vapour. However, as a result of a number of factors, including the short time available for combustion in the engine, the poor mixing of the fuel and air, and the high temperature of combustion, vehicle exhaust also contains CO, HC, PM and NO_x.

Emissions from motor vehicles depend on a large number of factors. The main ones are the vehicle technology, including the fuel used, and the mileage driven. Other factors such as fuel quality, vehicle maintenance, driver behaviour, and the rate of scrappage of older vehicles can also be important.

For each vehicle there is a compromise during its design between optimizing for fuel consumption, emissions, and performance. For a luxury vehicle, fuel consumption has tended to be a lower priority for the manufacturer than performance, whilst for a small compact car the emphasis has been more on fuel consumption. Since agreements between the motor manufacturers and regulators to reduce new car CO_2 emissions made in the late 1990s there has been renewed interest in fuel consumption amongst car manufacturers selling products in Europe and Japan.

In the early phases of emission control the emphasis was on improving the precision and timing of the injection of fuel into the combustion chamber and the design of the combustion chamber itself to improve combustion and minimize the deposit of fuel on the walls of the chamber, which leads to exhaust HC emissions.

Increasingly, as emission limits have become more stringent, new techniques for reducing emissions have emerged. Both petrol and diesel vehicles depend on the use of 'after-treatment', that is the removal of the pollutants in the exhaust rather than controlling its formation in the engine. Now more than 275 million of the world's 500 million cars and over 85% of all new cars produced worldwide are equipped with after-treatment systems.[12] However, improvements to the engine-out emissions (*i.e.* before the catalyst) continue to be made. This is an evolutionary process of gradually improving the combustion process and fuel management system.

For petrol engines one of the most important factors influencing emissions is the ratio of air to fuel in the engine. Figure 13.4 shows that there is no ideal ratio at which all the main emissions are low and the engine power is at an acceptable level. Indeed where CO and HC emissions are at their lowest, the NO_x emission is at its maximum. The best compromise is found in the lean burn region. Lower emissions, however, can be achieved using a three-way catalyst, which, as

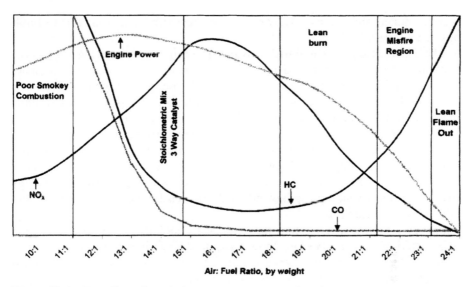

Figure 13.4 *The effect of air : fuel ratio on emissions and engine power*

explained below, needs a richer fuel and air mixture. The main advantage of lean burn engines is that they are more fuel-efficient than those running richer.

13.4.2 Reducing Carbon Dioxide Emissions

Motor manufacturers selling in the European market are now under increasing pressure to improve fuel consumption of new cars. In 1998 the European motor manufacturers association (ACEA) agreed with the European Commission (EC) that average new car CO_2 emissions would be 140 g km^{-1} by 2008, a 25% reduction from 1996 levels. This commitment is having a major impact on the design of future cars.

There is no add-on technology, such as a catalyst, that can be fitted to reduce CO_2 emissions from motor vehicles. A number of factors affect the fuel consumption and hence CO_2 emissions from passenger cars. These include the weight of the vehicle, its aerodynamic resistance and the efficiency of the engine and transmission. New materials, particularly plastics, have been increasingly used which have successfully reduced vehicle weight. At the same time the aerodynamic resistance has been reduced. Much of the emphasis today is on improving the engine efficiency. However, as the motor manufactures agreement relates to average new car CO_2 emissions, measures will also be needed to ensure that the current trend towards larger and more powerful passenger cars is halted.

There are trade-offs between the technology that might be applied to reduce fuel consumption and that required to reduce the regulated emissions. For example, if improved fuel consumption were the only concern it is likely that lean burn engines would be in widespread use now. However, because of greater concern during the last two decades over the regulated emissions this technology is not generally used.

There have also been small fuel consumption penalties to meeting most of the emission standards introduced over the last ten years. For example, vehicles meeting the Euro III limits, introduced in 2000, use on average 2 to 3% more fuel over the legislated test cycle than those meeting the Euro II emission limits that were introduced in 1996. This penalty may be greater in the real world.[13]

Conventional petrol engines requiring stoichiometric air to fuel mixtures are likely to become increasingly incompatible with the need to *significantly* improve fuel consumption of new vehicles to meet the ACEA/EC CO_2 target. Recognizing the need to improve fuel efficiency whilst meeting increasingly stringent emission standards has led to the search for new petrol engine and after-treatment technologies.

Probably the most important new engine technology is the direct injection spark ignition engine (also known as a gasoline direct injection (GDI) engine), produced by several Japanese manufacturers, but also being developed by most European car companies. Over the European test cycle vehicles with these engines may have up to 20% better fuel economy than those with a conventional stoichiometric petrol engine.[14]

Exploitation of GDI technology will depend on the development of new NO_x after-treatment systems. It is thought likely that to meet the current European emission limits (Euro III) vehicles will require both a three-way catalyst and NO_x after-treatment to control all regulated emissions when running stoichiometrically and NO_x emissions when it is running lean. The greater the time spent running stoichiometrically the lower the fuel economy benefits of the technology. GDI engines need to periodically run stoichiometrically to allow the regeneration of the NO_x after-treatment device (see Section 13.1.5.6).

Mitsubishi Motors brought the first GDI engine car to the EU market in 1998. This met Euro II emission standards. Control of the NO_x emissions was mainly through exhaust gas recalculation (EGR). EGR uses a proportion of the exhaust gases to replace some of the air input into the engine. This reduces the amount of oxygen in the combustion chamber, reducing the formation of NO_x. This process needs to be carefully controlled to avoid excess CO and PM formation.

To meet Euro III (2000) and Euro IV (2005) emission limits the engine-out NO_x emissions from a GDI engine will need to be reduced by 80–90%. Currently the most feasible way to achieve this level of NO_x removal is through the use of NO_x adsorbers. This technology is discussed later in this chapter.

There is a range of other options for improving fuel economy of petrol cars including improved transmission design, variable valve timing, improved ignition, and cylinder deactivation. These technologies have a smaller fuel saving potential compared to GDI but may offer some emissions benefits. Some of these technologies are already used on advanced vehicles. Other approaches, such as downsizing are also possible. This is unlikely to be acceptable to the consumer if performance is lost. To avoid performance losses variable compression ratio technology is being developed but is not yet commercially available. Another promising technology, brought to the

European market in 2000, is the hybrid car that combines an internal combustion engine with an electric motor. Other low CO_2 options such as fuel cells and hydrogen produced using renewable energy are unlikely to be economically feasible in the short term.

13.4.3 Reducing Regulated Emissions

The following discussion of engine improvements has been divided into petrol (spark ignition) and diesel (compression ignition). This is followed by a discussion of after-treatment technologies. It should be noted that in the development of advanced petrol and diesel engines there is a convergence between the technologies, thus there is also likely to be convergence in the emissions characteristics in the longer term.

Emission control for petrol cars is further advanced than that for diesel cars, which in turn is further advanced than that for heavy-duty diesel engines. This is illustrated by about 65% of new petrol car models sold in 1997 meeting Euro III (2000) limits and about 15% meeting Euro IV (2005) limits. On the other hand less than 10% of diesel car models met the Euro III standards (with pre-2000 diesel quality) in 1997. None achieved the Euro IV limits. In spring 1999 there were no heavy-duty engines on the market that met Euro III (year 2000) emission limits.[15]

13.4.4 Petrol

Over many years vehicle emissions have been reduced through progressive improvements in engine design, fuel systems and ignition control. Significant reductions in exhaust emissions have been made through the use of electronic fuel injection, EGR and control of the air to fuel ratio.

The greatest advance in controlling emissions from road vehicles was the introduction of the closed loop (or controlled) three-way catalyst for petrol vehicles. Essentially all new petrol vehicles sold in the world's major markets are now fitted with these catalysts. This technology can remove more than 75% of emissions of CO, HCs and NO_x and is described below in the section on after-treatment devices.

Evaporative emissions are controlled through the use of a carbon-filled canister.

From Euro III (2000) new petrol vehicles have been fitted with on-board diagnostic systems (OBD), which monitor the performance of a number of components of the exhaust, and evaporative emission control systems. When the system detects a fault a light, warning the driver, is lit. As engine management systems are developed an increasing number of components can be monitored, and in the distant future on-board systems capable of directly measuring emissions may be developed.

In the future, whilst vehicle manufacturers will continue to reduce engine-out emissions, most of the effort will be on improving catalyst efficiency, particularly on reducing the time taken to warm up the catalyst (see Section 13.5 below).

13.4.5   Diesel

European emissions legislation has essentially treated light duty vehicles (cars and vans) differently from heavy duty vehicles (such as buses and lorries). There have been huge refinements to diesel engines over recent years, both those used in light duty and heavy duty vehicles.

With the introduction of electronic systems the diesel engine is fast becoming as well controlled as the petrol engine. Like petrol engines much of the focus of emission control has been on improving the precision of injection of the fuel into the combustion chamber. Most pre-1996 diesel engines relied on mechanical fuel injection systems, while most Euro II (1996) diesel vehicles have electronic fuel injection systems. Today all new diesel engines will be electronically controlled to a greater or lesser extent. These systems allow more precise control of the fuel metering and timing, as well as the introduction of OBD, which has been mandated from 2003 for all cars and small vans, and 2005 for larger vans. Similar systems are likely to become mandatory in Europe within the next decade for heavy duty diesel vehicles.

One of the key technologies for reducing emissions from diesel engines has been to increase the pressure at which the fuel is injected into the engine. The higher the pressure the more efficient the combustion and, in general, the lower the emissions. Future developments are likely to include faster response fuel injection and varying the amount of fuel injected. These technologies will allow the combustion process to be more closely linked to the vehicle needs, depending on the operating conditions.

EGR systems have been used on diesel cars and vans since the early 1990s for NO_x emission reduction and improved performance (see Section 13.1.4.2). EGR is also likely to be introduced to heavy duty vehicles, but, due to the high mileage of these vehicles (*ca.* 1 000 000 km), there are additional problems of wear to valves and other components that will need to be resolved. Many manufacturers anticipate having to use advanced EGR systems, such as cooled EGR, where the exhaust gas is cooled prior to input into the combustion chamber, to further reduce NO_x emissions. This occurs because the rate of NO_x formation is dependent on the combustion temperature; lowering the temperature of the engine input gases lowers the NO_x formation.

Improvements to existing technologies such as turbochargers and inter-cooling may also be introduced. Turbochargers essentially boost the power from an engine by using its exhaust gas to turn a turbine to compress the engine's intake air. This effectively increases the amount of oxygen available in the combustion chamber. It is possible to reduce emissions by matching the level of turbocharging to the vehicle operating conditions and by cooling the air input into the engine using electronic control of the inter-cooler. Recent years have also seen improvements in engine design to reduce oil burning. Developments in this area will continue by improving material and tolerances for cylinder liners and pistons.

Nearly all Euro II diesel light duty vehicles have oxidation catalysts fitted to reduce the PM emissions. These devices reduce the organic compounds adsorbed onto the surface of the particles, but do not affect the carbon core of the particles. To meet Euro III limits virtually all light duty diesel vehicles will have oxidation catalysts. Currently these catalysts are not as durable as the gasoline three-way catalyst and particles can temporarily de-activate the active sites under some operating conditions.

To meet Euro IV limits it is possible that some, particularly the larger light duty vehicles, will need particulate traps and de-NO_x systems (see Section 13.5 on Exhaust After-treatment). The first light duty vehicle with a particulate trap came onto the European market early in 2000. This technology may be standard by 2005.

Oxidation catalysts are in less common use for heavy duty vehicles. Many heavy duty diesel engines can meet current emission limits without using after-treatment.

For diesel engines there is a trade-off between techniques to reduce PM and those to reduce NO_x emissions. The use of after-treatment devices allows the engine design to be optimized to minimize one emission, and the other to be reduced through after-treatment. The Euro IV (2005) standards and beyond are severe. As well as significant advances in basic engine design and control, exhaust after-treatment (including de-NO_x and/or particulate traps) may be required. The Euro V (2008) NO_x limits will require further reduction of NO_x probably using de-NO_x technology.

13.5 EXHAUST AFTER-TREATMENT

13.5.1 Introduction

A range of after-treatment systems has been developed to reduce emissions from motor vehicles. These include three-way catalysts, diesel oxidation catalysts and particulate traps. New technologies are being developed to remove NO_x from lean burn engines. Recent developments with these technologies are described below.

13.5.2 Three-way Catalysts

Virtually all petrol cars in the EU have been fitted with controlled three-way catalysts since the beginning of 1993. In other parts of the world this technology has a much longer history. Three-way catalysts were widely used from 1981 and 1978 in the United States and Japan respectively, when new emission legislation for cars was introduced. Today the EU has caught-up with these other countries and is now among the world leaders in emission control.

A three-way catalyst simultaneously removes CO, HC and NO_x from a vehicle's exhaust through chemical processes on the catalyst surface. The reactions are:

Oxidation reactions

$$2CO + O_2 \longrightarrow 2CO_2$$
$$HC + O_2 \longrightarrow CO_2 + H_2O$$

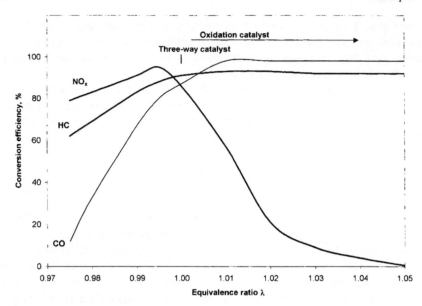

Figure 13.5   *The effect on catalyst efficiency of moving away from the stoichiometric air/ fuel ratio*

Reduction reactions

$$2CO + 2NO \longrightarrow 2CO_2 + N_2$$
$$HC + NO \longrightarrow CO_2 + H_2O + N_2$$

For the effective removal of all three pollutants the air to fuel ratio needs to be close to the stoichiometric ratio (*i.e.* 14.7 : 1). Cars fitted with these catalysts require an oxygen sensor to monitor the exhaust gas composition and electronically manage the fuel system to control the air to fuel ratio. The effect on catalyst efficiency of moving away from the stoichiometric air to fuel ratio (*i.e.* the equivalence ratio $\lambda = 1$) is shown in Figure 13.5.

This technology cannot be used in the oxygen-rich exhaust of a lean burn petrol or diesel engine. For these engines CO and HC emissions can be reduced using a simple oxidation catalyst. This type of catalyst also oxidizes the soluble organic fraction of the PM in the exhaust of diesel engines, reducing the total PM emission. Special oxidation catalysts have been developed to operate efficiently in the low exhaust temperatures of diesel vehicles. These catalysts will need to be improved to meet future emission limits.

Possibly the greatest challenge facing the automotive industry today is the development of an efficient system of removing NO_x emissions from lean burn engines. New technologies capable of doing this are beginning to enter the market. This is discussed below.

Most of the emissions from petrol cars fitted with three-way catalysts occur at the beginning of journeys while the catalyst is warming up. This is known as the light-off time. In recent years considerable effort has gone into developing fast light-off catalysts by decreasing the exhaust temperature required for the catalyst to operate. Changes to the thermal capacity of substrates and type and composition of the precious metal catalyst have brought light-off times down from one to two minutes to less than 20 seconds. Modern catalysts can reduce the catalyst light-off temperature from around 330 °C to less than 280 °C. The more palladium used, the lower the light-off temperature.

A complementary approach to reducing the catalyst light-off temperature is to heat the catalyst by moving it closer to the engine, heating it (electrically or by burning a very small amount of fuel under carefully controlled conditions) or by heating the exhaust up-stream of the catalyst (using exhaust gas ignition). The development of more thermally durable catalysts with increased stability at high temperature has allowed these developments. Thermally durable catalysts also have a longer life particularly during demanding driving when there are high exhaust temperatures.

The technology of the substrates on which the active catalyst is supported has also been improved. Thirty years ago substrates had a density of about 30 cells cm^{-2} and a wall thickness of 0.3 mm. By the end of the 1970s the cell density had increased up to 60 cells cm^2 and wall thickness had been reduced by 50% to 0.15 mm. In the late 1970s substrates derived from ultra thin foils of corrosion resistant steels came onto the market. In the beginning the foils could be made from material only 0.05 mm thick allowing very high cell densities to be achieved. Complex internal structures can be developed. Today's catalysts have a wall thickness of 0.03 mm and cell densities of 120 cells cm^{-2}.

These developments have allowed a larger catalyst surface area to be incorporated into a given converter volume. This allows better conversion efficiency and durability. The thin walls reduce thermal capacity and avoid the penalty of increased pressure losses. Alternatively the same performance can be incorporated into a smaller converter volume, making the catalyst easier to fit close to the engine, the smaller size being important for compact cars.

13.5.3 Diesel Oxidation Catalyst

A more recent development has been the diesel oxidation catalyst, which has to be specially formulated to operate efficiency at the lower exhaust temperatures of diesel vehicles, particularly of light duty diesel vehicles. These can lower PM by up to 50%, by destroying the organic fraction, as well as making significant reductions in CO and HC (including the characteristic diesel odour). However, the number of particles is unchanged.

13.5.4 HC Adsorber

Another new area of development is the HC adsorber. This incorporates special materials, such as zeolites, into or ahead of the three-way catalyst in petrol cars. HC emissions are collected when the exhaust temperatures are too low for effective catalyst operation. The HCs are then desorbed at higher temperatures when the catalyst has lit-off. This technology has the potential to reduce HC to less than half the levels emitted from a three-way catalytic converter on its own.

13.5.5 Particulate Traps

It is increasingly likely that particulate traps will be needed to meet future emission limits for both heavy duty and larger light duty diesel vehicles in the EU. Already over 5000 trucks and buses have been retrofitted with them. Due to the relatively high cost of retrofitting this has generally occurred only where fiscal incentives have been provided or where clean vehicle zones, entry to which is restricted to certain specified 'clean' vehicles, have been introduced. One European car manufacturer already supplies one of its models fitted with a particulate trap.

Ceramic wall-flow filters or traps remove well over 90% by weight of the total PM contained in diesel exhaust. Current limit values are set on the mass of PM, in grams per kilometre for cars and vans, and grams per kilowatt-hours for larger engines used in buses and lorries. However, recent work on particles produced by diesel engines concentrates on the number and size of particulates, which are thought to be more critical indicators of health impact. Some studies have shown that traps may increase emissions of ultra-fine particles under certain engine conditions whilst others have shown a decrease. It appears that the number of these extremely small particles measured may be a product of the experimental conditions and therefore further work is required to understand whether in the real world the number of particles emitted increase or not.[16] This issue is discussed again in the last section of this chapter.

Since the trap becomes plugged with PM in a short time, it is necessary to 'regenerate' the filtration properties of the filter by burning off the collected PM on a regular basis. In the presence of oxygen regeneration occurs when the exhaust temperature exceeds 550 °C. However, diesel exhaust temperatures are considerably lower. Finding durable, efficient and affordable regeneration systems, with small fuel consumption penalties, has proved to be the most challenging aspect in the development of particulate traps over the last twenty years.

The most successful methods to achieve regeneration include:

- Electrical heating of the trap, either on or off the vehicle.
- Incorporating a catalytic coating on the filter to lower the temperature at which particulate burns so that it is in the diesel exhaust temperature range.
- Using very small quantities of fuel-borne catalyst, such as ceria
- Incorporating an oxidation catalyst upstream of the filter that, as well as operating as a conventional oxidation catalyst, also increases the ratio of NO_2 to NO in the exhaust. Trapped particulates burn off at normal exhaust temperatures using the powerful oxidative properties of NO_2.

These technologies generally require low sulfur fuel (see discussion in Section 13.6.2).

13.5.6 De-NO$_x$ Technologies

With the development of GDI engines and increased use of diesel engines, controlling NO$_x$ emissions in lean combustion is the greatest challenge for automotive catalyst producers. The conventional three-way technology used with petrol engines needs a richer environment to reduce NO$_x$, so a radical new approach is required.

The most promising technologies for removing NO$_x$ in a lean environment are selective catalytic reduction (SCR) and the NO$_x$ adsorber.

SCR continuously removes NO$_x$ by creating a rich 'microclimate' where NO$_x$ can be reduced to nitrogen, while the overall exhaust remains lean. Several systems are under development and they can be categorized according to the reducing agent used. Hydrocarbon SCR (HC-SCR) can be further divided into active and passive systems. Passive systems use hydrocarbons in the fuel as the reducing agent, whilst active systems use additional fuel which is added upstream of the catalyst. The other reducing agent being investigated for use in SCR is ammonia (NH$_3$-SCR).

The NO$_x$ reduction achievable with a passive HC-SCR system is up to about 15%. With an active system this increases to about 30%.[13] However, the active system gives rise to an undesirable increase in HC, CO and PM emissions and fuel consumption. In some HC-SCR systems there may also be an increase in the proportion of NO$_x$ emitted as N$_2$O, a potent greenhouse gas. A large number of factors influence the performance of HC-SCR systems. These include the concentration of HC and NO$_x$ in the exhaust, the exhaust temperature, the type and loading of the catalyst material, fuel sulfur content and the concentration of certain HC in the exhaust.

NH$_3$-SCR systems are more efficient at removing NO$_x$. In light duty vehicles it can be as high as 80%, but can also be as low as 20% when the exhaust temperatures are low. The conversion efficiencies for heavy duty vehicles are generally more consistent, typically 60–85% during transient operations and up to 99% under steady state conditions. However, there are practical difficulties in using NH$_3$-SCR systems in vehicles. These systems generate NH$_3$ from a urea solution, which needs to be carried on-board and added to the exhaust gases using a sophisticated dosing and monitoring system. This in turn will require a distribution system. To minimize NH$_3$ emissions an oxidation catalyst will be needed downstream of the SCR catalyst.

NO$_x$ adsorbers (NO$_x$ traps) are probably a more promising technology as they are less constrained by operational temperatures. They store NO$_x$ under lean engine operations. A brief return to stoichiometric or rich operation, for one or two seconds, is enough to desorb the stored NO$_x$ and provide the conditions for a conventional three-way catalyst mounted downstream to destroy the NO$_x$. Prototype systems have demonstrated over 80% NO$_x$ removal with new systems. To maintain good conversion efficiency extremely low fuel sulfur levels are required. This technology is currently being used in Japan, although not in Europe where higher fuel sulfur levels inhibit its use.

The main problem with NO_x adsorbers is their sensitivity to the effects of fuel (and lubricant) sulfur. The NO_x removal efficiency can decline from over 90% to about 50% after 100 h use with 30 ppm sulfur fuel. The maximum permitted in the year 2005 in the EU will be 50 ppm. The sensitivity of these systems to sulfur has led to the EC reviewing whether 50 ppm is an appropriate limit or not.

The sulfur adsorbed on the NO_x trap as sulfate can be removed. Higher temperatures and richer conditions are needed than those to remove nitrates. Thus the engine needs to run rich more often, losing some of the fuel economy benefits of lean burn technology. With sulfate removal the NO_x removal efficiency does not decline as rapidly.

13.6   FUEL QUALITY

13.6.1   Introduction

Many studies have showed that several fuel parameters, for example, distillation characteristics and the aromatics content, have an effect on emissions. As a result fuel quality standards in Europe, and elsewhere, have been extended to include a wider range of parameters than in earlier legislation. These new European standards came into effect in 2000 with further requirements from 2005.

In general, the effect of fuel quality changes on emissions is relatively small, compared to improvements achievable by improvements to engine and after-treatment technologies. However, for certain pollutants (*e.g.* lead, sulfur and benzene) their emission is directly related to their content in the fuel. For benzene the aromatic content also affects the benzene emission as it is formed in the combustion process from other aromatic compounds. Reducing the content of these substances in the fuel can have a large impact on emissions.

There are also certain substances in fuels that affect the performance of emission control systems. Lead has been added to petrol since the 1920s when it was discovered that certain lead compounds are a cheap way of increasing its octane rating. Now, however, it is banned in many countries as it poisons automotive catalysts, and has been shown to have public heath impacts, particularly on children.

In recent years there has been a growing body of evidence on the effects of sulfur on the performance of emissions after-treatment devices. Sulfur is naturally present in crude oil. The level of sulfur in petrol and diesel depends on the sulfur content of the crude oil feedstock, the refinery production processes and the extent to which sulfur is removed. The impact of sulfur on emissions is discussed in the next section.

Many different types of additives are used in automotive fuels and lubricants. For example, diesel may contain an ignition improver, cold flow improver, anti-corrosion additive, detergent and anti-foam compounds. Most diesel with less than 500 ppm sulfur also requires a lubricity additive to stop excessive fuel injection pump wear and, in some cases, pump failure. Whilst most additives have been shown to be beneficial, others have proved to have an adverse effect on the environment.

Methylcyclopentadienyl manganese tricarbonyl (MMT), for example, has been marketed as an octane-enhancing fuel additive for petrol and a combustion improver for diesel. European motor manufacturers believe manganese additives adversely affect emission control systems and cause the engine to misfire. The manufacturer of MMT has refuted these claims, and the US Environmental Protection Agency has allowed its use in petrol.

Combustion generated deposits build up over time in engines impairing driveability and emission control systems. Detergents and other additives have been developed to reduce deposits in carburettors, fuel injectors, inlet valves, and combustion chambers. These were first used in the US in the 1950s and in California their use in petrol has been mandated. Additives are not however, required in Europe, but are nevertheless widely used.

13.6.2 Impact of Sulfur on Emissions

In many parts of the world there has been a rapid reduction in fuel sulfur levels over the last few years as new regulations and fiscal incentives have been introduced.

California has the world's most stringent requirement on sulfur in petrol, requiring the average to be no more than 30 ppm. By 2003, new requirements will half this to 15 ppm. In the European Union the maximum permitted is currently 150 ppm, which will be reduced to 50 ppm in the year 2005.

The European Union has the most stringent restriction on the sulfur content of diesel. Between 1989 and 2005 the maximum permitted sulfur content in diesel has been reduced from 3000 to 50 ppm. The reduction in sulfur in diesel is shown in Table 13.4.

Table 13.4 *EU legislation limiting the sulfur content of diesel*

Directive	Date of implementation	Applicability	Maximum limit
EEC/85/716	1st Jan 1989	All gas oil except used by shipping or for further processing	0.3% (3000 ppm)
93/12/EEC	1st Oct 1994	All gas oils including diesel	0.2% (2000 ppm)
	1st Oct 1995	Progress availability of diesel fuel with 0.5% sulfur (to allow implementation of the HDV directive)	
	1st Oct 1996	Diesel fuel	0.05% (500 ppm)
98/70/EC	1st Jan 2000	Diesel fuel	0.035% (350 ppm)
	1st Jan 2005	Diesel fuel	0.005% (50 ppm)

Test programmes undertaken over the last decade in Europe, US and Japan have shown the effect of fuel sulfur on emissions from three-way catalyst petrol vehicles. These studies have shown that sulfur in petrol increases exhaust emissions. This is because the sulfur in the fuel is absorbed onto the catalyst surface blocking its active sites. Sulfur also interferes with the management of the oxygen on the catalyst surface, which must be precisely controlled to maximize the reduction of NO_x emissions. The magnitude of the sulfur impact on emissions depends on a number of factors including catalyst formulation, catalyst location, emission control design, vehicle calibration and the sulfur level of the fuel.

The latest generation of emission control systems have been shown to be more sensitive to the impact of sulfur than earlier generations. For example, for NO_x emissions, the latest generation of vehicles are roughly ten times more sensitive to the impact of sulfur than early three-way catalyst vehicles. The sensitivity of individual vehicles to sulfur varies dramatically. However, as emission limits decline and demands for improved fuel economy increase, it is likely that the less sensitive technologies will become obsolete.

For diesel vehicles the use of oxidation catalysts is increasing. However, as these engines operate lean little sulfide is produced to strongly bind to the active catalyst sites. Instead sulfate is formed. This can result in a net increase in PM emissions.

Diesel emission control technologies, particularly particulate filters, generally require less than 50 ppm sulfur. Even those that can run on fuel with higher sulfur levels perform better on low sulfur fuel.

The US Diesel Emission Control – Sulfur Effects (DECSE) Program has shown that with 3 ppm sulfur diesel the traps tested removed over 95% of the PM over a transient test cycle, but with 30 ppm sulfur fuel the removal efficiency declined to just under 75%. With 150 ppm sulfur fuel there is virtually no PM removal.[17]

13.6.3 Methyl Tertiary-Butyl Ether (MTBE)

Methyl tertiary-butyl ether (MTBE) has been a controversial component of petrol since it was first used in oxygenated/reformulated fuels in the United States as a result of the Clean Air Act Amendments of 1990. Evidence from both the United States and Europe shows that it reduces CO and, to a lesser extent, HC emissions from catalyst equipped cars. For NO_x there is a greater variability of response with an increase in emissions in some studies and a decrease in others. MTBE is also used as a high octane component of petrol, allowing the use of cheap low octane components in the blending of the final fuel.

MTBE has traditionally been thought to have a low toxicity, especially when compared to other components of petrol. The main concern regarding its use is that it is extremely soluble in water and has low taste and odour thresholds.

A Health Effects Institute (HEI) review, undertaken in 1995, found that most people do not experience unusual symptoms in response to short-term exposure to MTBE, although some individuals may experience acute symptoms. There is some evidence of neurotoxicity at high levels of exposure in rats, and it is a possible carcinogen. However, the study concluded that adding oxygenates to petrol is unlikely to substantially increase the health risks associated with automotive fuels, hence the potential health risks were not sufficient to warrant an immediate reduction in oxygenate use.[18]

Since then MTBE has been banned in California and consideration is being given to banning it throughout the US. In November 1998 the US Environmental Protection Agency set up a 'Blue Ribbon Panel' of experts to investigate the concerns raised by the discovery of MTBE in drinking water. The Panel broadly agreed that the use of MTBE should be reduced substantially and that Congress should act to provide clear federal and state authority to regulate or eliminate the use of MTBE and other petrol additives that threaten drinking water supplies.

In areas using reformulated gasoline 5% to 10% of the drinking water supplies have detectable amounts of MTBE. The great majority are below levels of public health concern. Detection at lower levels has raised consumer taste and odour concern and has caused water suppliers to stop using some water suppliers and to incur costs of treatment and remediation. The major source of the groundwater contamination appears to be releases from underground storage tanks.

13.7 EFFECT OF TECHNOLOGY ON PM EMISSIONS

Only the mass of PM emitted from diesel motor vehicles is currently controlled under legislation worldwide. There are neither specifications for specific sizes of particles nor any limits on the number of particles emitted. In addition, there is no specific legislation to control the emissions of PM from petrol vehicles.

Different vehicle technologies give rise to different masses of PM emission. That from a diesel light duty vehicle is perhaps one to two orders of magnitude greater than from an equivalent petrol vehicle. In general, diesel engines produce a much greater number of particles than petrol engines. However, there may be little difference under high speed/load driving conditions. This is due to an increase in the petrol emissions rather than a decrease in the diesel emissions. Recent three-way catalyst cars have the lowest PM emissions. The new GDI technology is thought to have PM emissions more similar to a diesel than a conventional petrol engine.[19]

In terms of mass more than 85% of light duty diesel PM emissions are < 1 μm. This corresponds to 99% by number. There is some evidence that petrol vehicles emit a higher proportion of smaller particles than diesel vehicles. That is, more than 99% of the number of particles are < 1 μm.

There has been some concern that as the mass of PM emitted from diesel vehicles declines in response to increasingly more stringent legislation that the number of very small particles may increase. Since particle mass is approximately proportional to diameter cubed, bigger particles are more significant to total mass than smaller ones. However a recent test programme run by the European Automobile Manufacturers Association suggests that measures taken to reduce PM mass emissions from light duty diesel vehicles also reduce PM number emissions, and that there is little difference in the size distribution of PM emitted from 'conventional' and 'advanced' diesel vehicles.[20]

The number of particles measured in vehicle exhaust is very sensitive to the dilution technique used. Significant gas into particle conversion takes place as the exhaust dilutes and cools. It is thought that about 90% of the number and 30% of the mass is formed during dilution. New particles are formed by nucleation while pre-existing particles grow by adsorption or condensation. Nucleation and adsorption are competing processes, and are dependent on the dilution rate. There is evidence that high dilution ratio and high relative humidity both favour the formation of nanoparticles.[21]

The majority of work on the measurement of vehicle particle size and number has been carried out using a standard dilution tunnel (*i.e.* as in the legislated emission test). On the road dilution of a vehicle exhaust can reach 1000 to 1 in 1 to 2 s; that is considerably more rapid that in a dilution tunnel. This has raised questions as to how adequately the size/number of particles measured in the dilution tunnel relate to the real world.

13.8 REFERENCES

1. SENCO, 'Contract to Provide Liaison on Air Quality and Related Modelling for the Auto Oil II Programme', Final Report to Directorate General XI of the European Commission, Part III: 'AOPII Emissions Base Case', July 1999, Bristol.
2. Cleaner Fuels Forum, 'Cleaner Air: The Role of Cleaner Fuels', National Society for Clean Air, Brighton, 1998.
3. London Research Centre, 'West Midlands Atmospheric Emissions Inventory', London, 1996.
4. London Research Centre, 'Greater Manchester Atmospheric Emissions Inventory', London, 1997.
5. London Research Centre, 'Greater London Atmospheric Emissions Inventory', London, 1998.
6. London Research Centre, 'Atmospheric Emissions Inventories for Four Urban Areas – Merseyside, Bristol, Southampton and Portsmouth and Swansea and Port Talbot', London, 1997.
7. London Research Centre, 'Glasgow, Middlesborough and West Yorkshire – Atmospheric Emissions Inventories', London, 1998.
8. Department of the Environment, 'Digest of Environmental Statistics', No. 18, HMSO, London, 1996.
9. A. G. Salway, H. S. Eggleston, J. W. L. Goodwin and T. P. Murrells, 'UK Emissions of Air Pollutants 1970–1995', National Atmospheric Emissions Inventory, AEA Technology, Culham, 1997.
10. For example, M. L. Poulton, 'Alternative Fuels for Road Vehicles', Computational Mechanics, Publications, Southampton, 1994; ETSU, 'Alternative Road Transport Fuels – a Preliminary Life-cycle Study for the UK', Vols. 1 and 2, Harwell, 1996; Friends of the Earth, 'Fuelling the Debate: An Appraisal of the Role of Advanced Technologies in Reducing Vehicle Emissions', London, 1996.
11. ETSU, 'Comparative Field Trials of Alternative Road Transport Fuels', Overview Report and Main Report, Harwell, 1997.
12. R. A. Searles, 'Catalyst Technologies – Challenges and Opportunities', Paper presented to 'Health Effects of Vehicle Emissions', held London 23rd and 24th February 2000, Energy Logistics International, 2000.

13. FEV Motorentechnik, 'Influence of the Sulphur Content in Fuel on the Fuel Consumption and Pollutant Emissions of Vehicles with Gasoline and Diesel Engines', Report Commissioned by the German Automobile Industry Association (Verband der Automobilindustrie, VDA), Aachen, 1999.
14. Ford Motor Company, 'Review of the Impact of Fuel Sulphur on Advanced Aftertreatment Systems', 2000, Basildon.
15. J. S. McArrager, R. F. Becker, P. J. Bennett, G. Claus. J. Graham, G. Lang, C. J. van Leeuwen, D. Richeard, F. Schuermann and P. Heinze, 'Fuel Quality, Vehicle Technology, and their Interactions', Report No. 99/55, CONCAWE, Brussels, 1999.
16. I. S. Abdul-Khalek, D. B. Kittleson and F. Brear, 'Diesel Trap Performance: Particle Size Measurments and Trends', SAE 982599, 1998.
17. Diesel Emission Control – Sulfur Effects (DECSE) Program, 'Phase 1 Interim Data Report No 3: Diesel Fuel Sulfur Effects on Particulate Matter Emissions', 1999 and 'Program Phase 1 Interim Data Report No. 4: Diesel Particulate Filters', Final Report, 2000.
18. Health Effects Institute, 'The Potential Health Effects of Oxygenates Added to Gasoline: A Review of the Current Literature', Cambridge, 1996.
19. D. E. Hall and C. J. Dickens, 'Measurement of the Number and Size Distribution of Particles emitted from a Gasoline Direct Injection Vehicle', SAE, 1999-01-3530, 1999.
20. Association des Constructeurs Européens d'Automobiles (ACEA), 'ACEA Report on Small Particle Emissions from Passenger Cars', Brussels, 1999.
21. J. Ping Shi and R. M. Harrison, 'Investigation of Ultrafine Particle Formation during Diesel Exhaust Dilution', *Environ. Sci. Technol.*, **33**, 3730–3736.

CHAPTER 14

Soil Pollution and Land Contamination

B. J. ALLOWAY

14.1 INTRODUCTION

Soil is an essential component of terrestrial ecosystems because the growth of plants and biogeochemical cycling of nutrients depend upon it. Of the total area of the world's land mass (13.07×10^9 ha), only 11.3% is cultivated for crops; permanent grazing occupies 24.6%, forest and woodland 34.1% and 'other land' including urban/industry and roads, accounts for 31%.[1] From a resource perspective, soil is vitally important for the production of food and fibre crops and timber and it is therefore essential that the total productive capacity of the world's soils is not impaired. Pollution, along with other types of degradation, such as erosion, and the continuing spread of urbanization poses a threat to the sustainability of soil resources. Soil pollution can also be a hazard to human health when potentially toxic substances move through the food chain or reach groundwater used for drinking water supplies.

In comparison with air and water, soil is more variable and complex in composition and it functions as a sink for pollutants, a filter which retards the passage of chemicals to the groundwater, and a bioreactor in which many organic pollutants can be decomposed. As a consequence of its occurrence at the interface between the land and the atmosphere, soil is the recipient of a diverse range of polluting chemicals transported in the atmosphere. Further inputs of pollutants to the soil occur as a result of agricultural and waste disposal practises but, in general, the most severe pollution usually results from industrial and urban uses of land.

It is generally accepted that most of the soil in technologically advanced regions of the world is polluted (or contaminated), at least to a slight extent.[2] However, in many cases the relatively small amounts of pollutants involved may not have a significant effect on either soil fertility or animal and human health. More severe 'chemical pollution' which poses a greater hazard has been estimated by a Global Assessment of Soil Degradation ('GLASOD') to affect a total of 21.8×10^6 ha of land in Europe, Asia, Africa and Central America.[3] Realistic estimates of areas affected by soil pollution are difficult owing to unreliable official figures and inadequate data for many parts of the world.

352

Industrially contaminated land tends to contain higher concentrations and a greater possible range of pollutants than other sources of pollution. It has recently been estimated that there are between 50 000 and 100 000 contaminated sites in the United Kingdom which are estimated to occupy around 300 000 ha. In the USA, 25 000 contaminated sites have been identified; 6000 sites are being cleaned-up in the Netherlands, there are known to be at least 3115 sites in Denmark and 40 000 suspect areas have been identified on 5000–6000 sites in the former western part of Germany.[4] Land contamination as a result of warfare, military training and the manufacture and storage of explosive materials, such as TNT (trinitrotoluene), have resulted in very large areas of many countries, such as Germany, being contaminated with very persistent organic and inorganic chemicals.

Old industrial sites are generally characterized by being very heterogeneous, both with regard to the distribution of pollutants and also to the properties of the soil materials that control the behaviour of these chemicals. In contrast, atmospherically deposited pollutants tend to have more even distribution with gradual changes in concentrations, which tend to decrease with distance from the source. The upper horizons of the soil are contaminated to the greatest extent by atmospheric deposition.

All contamination/pollution situations comprise the following components: (i) a source of pollutant, (ii) the pollutant itself, (iii) a transport mechanism by which the pollutant is dispersed, and (iv) the receptor where the transport phase terminates. Transport can be by moving air or water, by gravity movement downslope, or by direct conveyance and placement such as the spreading of waste materials on land. Although this is a very simple conceptual model which does not take account of variations in time and quantity, it does provide a useful basis from which to consider the pollution of soils and other environmental media.

14.2 SOIL POLLUTANTS AND THEIR SOURCES

14.2.1 Heavy Metals

Heavy metals have a density of greater than 6 g cm^{-3} (but some authors use a value of >5 g cm^{-3}) and an atomic number greater than 20, and occur naturally in rocks and soils but concentrations are frequently elevated as a result of pollution.[5,6] The term 'heavy metal' is imprecise but is widely used although others such as 'toxic metals', 'potentially toxic elements' and 'trace metals' are possible alternatives. Heavy metals belong to the group of elements described geochemically as 'trace elements' because they collectively comprise $< 1\%$ of the rocks in the earth's crust. All trace elements are toxic to living organisms at excessive concentrations, but some are essential for the normal healthy growth and reproduction by either plants or animals at low but critical concentrations. These elements are referred to as 'essential trace elements' or 'micronutrients' and deficiencies can lead to disease and even death of the plant or animal. The essential trace elements include: Co (for bacteria and animals),

Cr (animals), Cu (plants and animals), Mn (plants and animals), Mo (plants), Ni (plants), Se (animals) and Zn (plants and animals). In addition, B (plants), Cl (plants), Fe (plants and animals), I (animals) and Si (plants and animals – probably) are also essential trace elements but are not dense enough to be classed as heavy metals.

Other elements, including: Ag, As, Ba, Cd, Hg, Tl, Pb, Sb, have no known essential function and, like the essential trace elements, cause toxicity above a certain tolerance level. The most important heavy metals with regard to potential hazards and occurrence in contaminated soils are: As, Cd, Cu, Cr, Hg, Pb and Zn.[6]

14.2.1.1 Sources of Heavy Metals. (a) Metalliferous Mining. This is an important source of contamination by a wide range of metals, especially As, Cd, Cu, Ni, Pb and Zn, because ore bodies generally include a range of minerals containing both economically exploitable metals (in ore minerals) and uneconomic elements (in gangue minerals). Most mine sites are contaminated with several metals and accompanying elements (*e.g.* sulfur). Wind-blown tailings (finely ground particles of ore and country rock) and ions in solution from the weathering of ore minerals in heaps of tailings tend to be the major sources of pollution from abandoned metalliferous mine sites.

(b) Metal Smelting. This is the process of producing metals from mined ores and so can be a source of many different metals. These pollutants are mainly transported in air and can be in the form of fine particles of ore, aerosol-sized particles of oxides (especially important in the case of the more volatile elements, such as As, Cd, Pb, and Tl) and gases (SO_2). In some cases, pollution is directly traceable in soil < 40 km downwind of smelters.

(c) Metallurgical Industries. Pollution can include aerosol particles from the thermal processing of metals and solid wastes, effluents from the treatment of metals with acids and solutions of metal salts used in electroplating.

(d) Other Metal-using Industries. These can be a source of metals in gaseous/particulate emissions to the atmosphere, effluents to drains and solid wastes. These include: the electronics industry where metals are used in semiconductors, contacts, circuits, solders and batteries; plating (Cd, Ni, Pb, Hg, Se, and Sb); pigments and paints (Pb, Cr, As, Sb, Se, Mo, Cd, Co, Ba, and Zn); the plastics industry (polymer stabilizers such as Cd, Zn, Sn, and Pb); and the chemical industry which uses metals as catalysts and electrodes including Hg, Pt, Ru, Mo, Ni, Sm, Sb, Pd, and Os.

(e) Waste Disposal. Municipal solid waste, special wastes and hazardous wastes from many sources can contain many different metals.

(f) Corrosion of Metals in Use. Corrosion and chemical transformation of metals used in structures, *e.g.* Cu and Pb on roofs and in pipes, Cr, Ni and Co in stainless steel, Cd and Zn in rust preventative coatings on steel, Cu and Zn in brass fittings, and Cr and Pb from the deterioration of painted surfaces.

(g) Agriculture. This mainly includes As, Cu, and Zn which are (or have been) added to pig and poultry feeds, Cd and U contaminants in some

phosphatic fertilizers, and metal-based pesticides (historic and current) such as As, Cu, Mn, Pb and Zn.

(h) Forestry and Timber Industries. Wood preservatives containing As, Cr and Cu have been widely used for many years and have caused contamination of soils and waters in the vicinity of timber yards. Several organic chemicals, including tar derivatives (creosote) and pentachlorophenol, are also used as wood preservatives.

(i) Fossil Fuel Combustion. Trace elements present in coals and oils include Cd, Zn, As, Sb, Se, Ba, Cu, Mn and V and these can be present in the ash or gaseous/particulate emissions from combustion. In addition, various metals are added to fuels and lubricants to improve their properties (Se, Te, Pb, Mo and Li).

(j) Sports and Leisure Activities. Game and clay pigeon shooting involves the use of pellets containing Pb, Sb and As but alternatives to these metals such as steel, Mo and Bi are also being introduced.

14.2.2 Hydrocarbon Pollutants

Hydrocarbon pollutants from petroleum mainly comprise a range of saturated alkanes from methane (CH_4), ethane (C_2H_6) and propane (C_3H_8) through straight and branched chains to $C_{76}H_{154}$. Aromatic hydrocarbons and organic components containing nitrogen and sulfur can also be important constituents of some petroleum deposits. The hydrocarbons derived from coal and petroleum tend to form the main group of organic macropollutants in soils.

A commonly encountered group of aromatic molecule hydrocarbon pollutants are the BTEX compounds (benzene, toluene, ethyl benzene and xylene) which commonly occur in plumes in the groundwater beneath a wide range of industrial sites. Organic solvents are used widely in industry and can be important soil pollutants at industrial sites. These can include butane and *n*-hexane, benzene, toluene and organochlorine compounds such as vinyl chloride, chloroform, carbon tetrachloride and trichloroethane. Apart from any toxicity hazard associated with the ingestion or inhalation of hydrocarbons, there is also a high risk from fires and explosions.

On the basis of their behaviour in soils and groundwaters, hydrophobic organic liquid contaminants, such as solvents, are often grouped as non-aqueous phase liquids (NAPLs) and depending on their density they are subdivided into light non-aqueous phase liquids (LNAPLs) and dense non-aqueous phase liquids (DNAPLs).

14.2.2.1 Sources of Hydrocarbon Pollutants. (a) Fuel Storage and Distribution. Leaking underground storage tanks, spillages at distribution depots and from road accidents can lead to pollution of soils and aquifers with petrol and diesel fuels. It is possible that around 30% of filling stations in the United Kingdom may be causing some subsurface pollution though leakages from underground storage tanks. In view of the very large volumes of petroleum fuels used, this source must account for a high proportion of soil pollution by hydrocarbons. However, despite their ubiquitous occurrence, hydrocarbons are more readily degraded in soils and pose less of a toxicological risk than organomicropollutants such as PAHs, PCBs, dioxins and many pesticide derivatives. However, Pb-containing petrol will continue to pose a long-term Pb contamination hazard.

   (b) Disposal of Used Lubricating Oils. In addition to hydrocarbons and powdered metal, used lubricating oils contain PAHs. Some do-it-yourself car mechanics sometimes dispose of used motor oils onto garden soils, and land around garages, farm yards and scrap yards can be polluted with this material.

   (c) Leakage of Solvents from Industrial Sites. Hydrocarbon solvents are used widely in industry and spillages/leaks into soils frequently occur. In addition to specialist manufacturers and distributors of solvents, an important source of pollution by these compounds are manufacturers of semiconductors and other electronic components which use solvents for degreasing. Organic solvents are DNAPLs which, although having a very low water solubility, can cause serious groundwater pollution problems. Pollution of groundwater (in aquifers) necessitates remedial actions, such as 'pump and treat' processes, including air stripping to maximize the volatilization of these compounds.

   (d) Coal Stores. Coal is a solid form of hydrocarbon and the main hazard associated with it is the risk of fires. Coal mines, and the sites of former coal storage facilities in factories, railway depots and similar places are likely to contain significant amounts of coal mixed into the soil which could constitute a combustion hazard. Some coals can contain significant amounts of iron pyrites (FeS_2) which undergoes oxidation when exposed to the air resulting in the formation of iron hydroxides and sulfate anions. These anions have a major acidifying effect on soils which can result in any heavy metal contaminants becoming more mobile and bioavailable. If the acidification is very pronounced, decomposition of clay minerals occurs at around pH 4 which results in the formation of free Al^{3+} and $Al(OH)_2^+$ ions which are toxic to most plants.

14.2.3   Toxic Organic Micropollutants (TOMPs) (also called Persistent Organic Pollutants (POPs)) (see also Chapter 17)

The most commonly encountered toxic organic micropollutants include: polycyclic aromatic hydrocarbons (PAHs), polyheterocyclic hydrocarbons (PHHs), polychlorinated biphenyls (PCBs), polychlorinated dibenzodioxins (PCDDs), polychlorinated dibenzofurans (PCDFs) and pesticide residues and metabolites. Many of these organic pollutants are discussed in more detail elsewhere in this book.

   Pesticides.   These comprise a very large range of different types of organic molecules which are used with the intention of destroying pests of various types, including: insects, mites, nematodes, weeds, and fungal pathogens. The types of compounds used as pesticides include (examples are given in parenthesis):

Insecticides: Organochlorines (DDT, BHC)
Organophosphates (Malathion, Parathion)
Carbamates (Aldicarb)

Herbicides: Phenoxyacetic acids (2,4-D, 2,4,5-T)
Toluidines (Trifluralin)
Triazines (Atrazine, Simazine)
Phenyl ureas (Fenuron, Isoproturon)
Bipyridyls (Diquat, Paraquat)
Glycines (Glyphosate)
Phenoxypropionates (Mecoprop)
Translocated carbamates (Barban, Asulam)
Hydroxyl nitriles (Ioxynil, Bromoxydynil)

Fungicides:
Non-systemic
Inorganic and heavy metal compounds (Cu in Bordeaux Mixture)
Dithiocarbamates (Maneb, Zineb)
Phthalimides (Captan, Captafol)
Systemic
Antibiotics (Cycloheximide, Blasticidin-S)
Benzimidazoles (Carbendaim, Benomyl, Thiabendazole)
Pyrimidines (Ethirimol, Triforine)

As a consequence of the variety of compounds involved, there are major differences in their behaviour in soil and toxicity to plants, animals, soil organisms and humans. Many pesticides break down into toxic derivatives and may cause phytotoxicity problems in sensitive crops. The most serious problems associated with pesticide pollution of soils are the contamination of surface and groundwaters and entry into the food chain through crops or livestock.

Typical rates of pesticide application in agriculture are 0.2–5.0 kg ha^{-1} but frequently higher rates of some pesticides may be used for non-agricultural purposes, such as weed clearance on rail tracks and urban paths.[7] In the United Kingdom, the total tonnage of pesticide active ingredient used decreased by 20% between 1980 and 1990 but the area treated with these substances increased by 9%.[8] Generally, less than 10% of the pesticide reaches its intended target; the remainder may reside in the soil, some will be volatilized and some will be leached through soils to groundwater or *via* under drainage to water courses. Most pesticides have water solubilities greater than 10 mg L^{-1} and are therefore highly prone to leaching through soils. The half-lives of many pesticide compounds in fertile soils range from 10 days to 10 years and so, in many cases, there is sufficient time for some leaching to occur. Atrazine with a half-life of 50–100 days gives rise to widespread groundwater contamination. The concentrations of soil-acting pesticides in the soil solution are thousands of times greater than the EC guideline concentration for potable waters (0.1 μg L^{-1} per compound, total concentration 0.5 μg L^{-1}) and so there is a strong probability of groundwater contamination above EC limits.[7] (See also Chapter 1.)

14.2.4   Other Industrial Chemicals (see also Chapter 1)

It is estimated that between 60 000 and 90 000 chemicals are in current commercial use and thousands of new compounds are being brought into use (and dispersed in the environment) each year. Although not all of these constitute potential toxicity hazards, many will cause pollution of soils as a result of leakage during storage, from use in the environment, or from their disposal either directly, or from wastes containing them. Apart from industrial uses, a wide range of chemicals are used in domestic products and so their use is over a much larger geographical area and their disposal is less controlled than most industrial chemicals (which are subject to strict regulations).

The total world production of hazardous and special wastes was 338×10^6 t in 1990.[9] Although the Red, Black and Grey Lists of hazardous chemicals contain a large number of priority substances which can pollute soils, only a few examples can be given in Table 14.1.

14.2.5   Nutrient-rich Wastes

(a) Sewage Sludges (also called Biosolids) (see also Chapter 5).   These are the residues from the treatment of wastewater and large quantities are produced worldwide (6.3×10^6 t in the original 12 countries of the European Union in 1990 and 5.4×10^6 t dry solids in the USA). This sludge has usually been disposed of onto agricultural land (43% of total in UK and 22% in the USA), into the sea (30% in the UK), landfilled or incinerated. In the European Union,

Table 14.1   *Priority hazardous chemicals (based on the UK Red List and EC Lists I and II)*

Mercury and its compounds	Dichlorvos
Cadmium and its compounds	1,2-Dichloroethane
γ-Hexachlorocylohexane (Lindane)	Trichlorobenzene
DDT	Simazine
Pentachlorophenol and its compounds	Organotin compounds
Hexachlorobenzene	Cyanide, Fluorides
Hexachlorobutadiene	Trifluralin
Fenitrothion	Azinphos-methyl
Aldrin, Dieldrin	Organophosphorus compounds
Endrin	Endosulfan
Carbon tetrachloride	Atrazine
Polychlorinated biphenyls	
Persistent mineral oils and hydrocarbons of petroleum	

disposal at sea was banned from December 1998 and many other countries have discontinued oceanic disposal. As a consequence of this, the other disposal routes are being used to a much greater degree and disposal to land is increasing. Sewage sludge is a valuable source of plant nutrients (especially N and P) and a useful source of organic matter which has beneficial effects on soil aggregate stability. However, its value is somewhat diminished by its content of potentially harmful substances which include heavy metals, especially Cd, Cu, Ni, Pb and Zn, and organic pollutants. The most important POPs in sewage sludges include: (a) halogenated aromatics, *e.g.* polychlorinated biphenyls (PCBs), polychlorinated terphenyls (PCTs), polychlorinated naphthalenes (PCNs) and polychlorinated benzenes, polychlorinated dibenzodioxins (PCDDs), (b) halogenated aliphatics, (c) polycyclic aromatic hydrocarbons (PAHs), (d) aromatic amines and nitrosamines, (e) phthalate esters, and (f) pesticides.[10] Sewage sludges can also contain some pathogenic organisms which were not destroyed during the sewage treatment. Following concerns about the transmission of these pathogens to humans through food crops, an additional sterilization stage is being introduced into sewage treatment to overcome this problem.

(b) Livestock Manures. These contain large amounts of N, P and K and are valuable sources of these nutrients for crops but they can also contain residues of food additives which can include As, Cu and Zn, veterinary medicines fed to pigs and poultry and some pathogens.

14.2.6 Radionuclides (see also Chapter 18)

Nuclear accidents like those at Windscale (UK) in 1957 and Chernobyl (Ukraine) in 1986 resulted in many different radioactive substances being dispersed into the environment. The greatest long-term pollution problem is considered to be caused by ^{137}Cs which has a half-life of 30 years and behaves in a manner similar to K in soils and ecosystems. Atmospheric testing of nuclear weapons dispersed large amounts of ^{90}Sr which has a half-life of 29 years and behaves similarly to Ca in biological systems and poses a hazard to humans because it is stored in the skeleton.

14.2.7 Pathogenic Organisms

Soils can be contaminated with pathogenic organisms (bacteria, viruses, parasitic worm eggs) from various sources, including the burial of the dead bodies of animals and humans, manures and sewage sludges. The soil can act as a reservoir of these pathogens which can reach groundwater, can infect livestock and humans through soil particles consumed directly by children or on unwashed hands or attached to herbage and vegetables. The latter is the main reason for the introduction of the sterilization of sewage sludges which are applied to agricultural land.

14.3   TRANSPORT MECHANISMS CONVEYING POLLUTANTS TO SOILS

Pollutants reach soils by four main pathways:

- atmospheric deposition of particulates (washout or dry deposition) (see Chapter 7),
- sorption of gases (*e.g.* volatile organic compounds) from the atmosphere,
- fluvial transport and deposition/sorption from flood waters,
- placement (agricultural amendments, dumping, injection, surface spreading *etc.*).

Fluvial transport is important in land subject to flooding. This has been an important pollution pathway in the United Kingdom in areas of metalliferous mining in the nineteenth century. Before pollution controls were introduced in 1876, Pb–Zn mines discharged waters from ore dressing operations directly into streams and rivers. This led to the alluvial soils in most flood plains of rivers draining mining areas being severely contaminated with Pb, Zn and other metals.[11] Soils on the flood plains of many major rivers in the world which drain industrial and urbanized areas have been significantly contaminated with a diverse range of substances through flooding.

Volatilization involves a substance changing from a liquid to a gas and this is a very important mechanism by which many organic compounds become dispersed in the atmosphere and, conversely, sorbed from the atmosphere onto soils or plants.

Placement of pollutants can occur in many ways; the most obvious being the spreading of wastes, such as sewage sludges or metal-rich manures from pigs or poultry. Phosphatic fertilizers can also contain significant concentrations of Cd and U contaminants and have been at least partially responsible for the significantly elevated concentrations of these elements in agricultural soils in many parts of the world. The spraying of pesticides onto crops and soils is a good example of placement although they may be further dispersed in the environment afterwards. Industrially contaminated land with its associated demolition of old buildings and manufacturing plant, the redevelopment of sites, leakages of stored chemicals, accumulations of wastes (as well as fires and even explosions) provides other examples of the placement of contaminants onto soils.

14.4   THE NATURE AND PROPERTIES OF SOILS RELATED TO THE BEHAVIOUR OF POLLUTANTS

14.4.1   The Nature of Soils

Soil is the geochemically and biochemically complex material which forms at the interface between the atmosphere and the earth's crust and is highly heterogeneous in composition and spatial distribution. Soil comprises a mixture of mineral and organic solids, permeated by voids containing aqueous and gaseous components and a microbial biomass. Soils are usually differentiated vertically into a series of distinctive layers, called 'horizons', which differ both morphologically and chemically from the layers above and below them. These horizons collectively form the soil profile (or pedon) which is the unit of classification of soils. Soil formation results from interactions between the weathered geological material on which the soil has formed, and climatic conditions, vegetation cover, landscape position and the time over which the soil has been forming. Soil formation is a dynamic process and major changes in any of the environmental factors (such as climate, drainage or vegetation) will result in changes in the nature of the soil horizons. As a result of the wide range of rock types and environmental conditions around the globe, soils differ markedly in physical, chemical and biological characteristics. Nevertheless, there are several properties which most soils have in common which relate to the behaviour of pollutants.

All soils contain humus which is highly polymerized organic material synthesized by microorganisms from the decomposition products of dead plant material. The organic matter contents of most soils lie in the range 0.1–10% but peaty soils can contain more than 70% organic matter. Soils in hot, dry climates tend to contain much lower amounts of organic matter than soils in humid and cooler regions.

Soils contain varying amounts of different primary and secondary minerals. Primary minerals occur in unweathered fragments of igneous rock either from the parent material, or erratic stones deposited by ice or water, and their weathering provides plant nutrients and gives soils distinctive colour and chemical properties. Secondary minerals have been synthesized from the products of weathered rocks and can include hydrated oxides of Fe, Al and Mn (Fe oxides give soils their characteristic brown colour) and clay minerals which are thin layered forms of aluminium silicates. These secondary minerals (clays and precipitates of Fe oxides) and humus together form the colloidal fraction which provides soils with significant sorptive properties and is very important in determining the fate of pollutants (and plant nutrients) in soils. Other secondary minerals, such as calcium carbonate, can be precipitated from groundwater in semi-arid and arid climates, but also occur as part of the parent material of soils which have developed on limestones in humid climates. In general, as soils mature their content of primary minerals gradually decreases until they contain only secondary minerals. Tropical soils tend to contain mainly Fe and Al oxides and kaolinite (clay mineral).

It is important to stress that the soils found at many heavily contaminated sites can differ markedly from natural (pedologically derived) soils. Very often, the use of rubble for hard core to act as foundations for buildings results in the pedological constituents being heavily diluted by other material, such as highly alkaline concrete, mortar, wastes and product residues. These extraneous materials will have a major effect on the behaviour of pollutants at these sites both chemically, such as adsorption/desorption, and physically, especially with regard to porosity and the movement of chemicals downwards towards the groundwater. A generalized comparison of contaminated agricultural and industrial soils is given in Section 14.4.3 of this chapter.

14.4.2   Chemical and Physical Properties of Soils Affecting the Behaviour of Pollutants

Space does not allow a detailed consideration of the chemistry related to the behaviour of pollutants in soils. (Readers are referred to texts by McBride,[12] Sposito[13] and White[14] and others for a more detailed coverage.) However, the main considerations are the factors which control the sorption and desorption of ionic and uncharged compounds in soils. Most heavy metals exist in the soil predominantly as cations but important elements such as: As, B, Mo, V, Sb and Se occur as anions. Although some pesticides are ionic, most organic contaminants are uncharged and tend to be hydrophobic.

Sorption of pollutants can be by several mechanisms, including:

(a) Non-specific Cation and Anion Adsorption (also referred to as cation or anion exchange).   Cations are adsorbed onto negatively charged surfaces on the soil colloidal fraction. This comprises the aluminosilicate clay minerals, hydrous oxides of Fe and Mn, and humic organic material. Anions are adsorbed on positively charged sites on the colloidal fraction which are mainly contributed by hydrous oxides of Fe. These Fe oxides have a variable charge which is dependent on the pH of the soil. The point of zero charge (PZC) on the pure oxides is around pH 8.0 but when present in the soil it is around pH 7.0. Below this PZC (pH) value these oxides are positively charged and adsorb anions, but above the PZC they are negatively charged and adsorb cations. Soil organic matter also has a pH dependent surface charge but this tends to be predominantly negatively charged above pH 2.5. The negative charges result from the deprotonation of carboxy and phenolic groups on the surfaces of humic polymers. Clay minerals possess permanent negative charges on their surfaces due to charge imbalances where isomorphous substitution of a major constituent has occurred in the crystal lattice of the mineral during its formation substitution (*e.g.* Al^{3+} replacing Si^{4+}, or Mg^{2+} replacing Al^{3+}).

The soil pH is the most important single physico-chemical parameter controlling the sorption–desorption of ions in soils. The normal range of pH in soils throughout the world is 4–8.5 owing to the buffering by Al at the lower end and by $CaCO_3$ at the upper end. In general, soils in humid regions, which are subject to leaching of bases, tend to have a pH range of 5–7 (although organic upland soils may have values of less the 4.0). Soils in arid regions tend to have pH values of 7–9 owing to the accumulation of $CaCO_3$ and other salts in the predominantly evaporating moisture regime.

Cation exchange involves a higher concentration of cations being held in the zone of attraction of the negative charges on the soil colloid surfaces. In general, it is found that the Cation Exchange Capacity (CEC) of a soil increases with a rise in pH, at least up to pH 7.0. These cations are in a dynamic state of flux dependent on the nature of the charged surface, the nature of the ion (its valency and hydrated size) and its concentration and the concentrations of other ions in the soil solution. There is a general order of replacement whereby it is found that those ions which are most strongly attracted replace other cations in the zone of attraction. This order tends to vary with the adsorbent surface, but for the clay mineral illite the order of increasing selectivity was given by Bittel and Miller:[15]

$$Mg > Cd > Ca > Zn > Cu > Pb.$$

Anions are retained on positively charged surface sites and these tend to arise as a result of pH values below the PZC of hydrous oxides of Fe, Mn and Al, and by 'ligand exchange' where a surface complex forms between an anion and a metal, usually Fe or Al in a hydrous oxide or a clay mineral. The sorptive capacity of a soil (*i.e.* the Cation or Anion Exchange Capacity) is expressed in units of centimoles of charge per kg of soil ($cmol_c\ kg^{-1}$).

(*b*) *Specific Adsorption.* This occurs where metals such as Cd, Cu, Ni and Zn form complex ions (MOH^+) on surfaces that contain hydroxyl groups, especially hydrous oxides of Fe, Mn and Al. These complex ions do not undergo cation exchange but can be displaced by strong acids or complexing agents. Specific adsorption is strongly pH dependent and is responsible for the retention of much larger amounts of metals than cation exchange. The general order of increasing strength of specific adsorption of heavy metals was given by Brummer:[16]

$$Cd > Ni > Co > Zn \gg Cu > Pb > Hg.$$

(*c*) *Organic Complexation of Metals.* This occurs when the solid state humic material binds metals into a ring type structure, most commonly a chelate. Humic compounds with hydroxy, phenoxy, and carboxy reactive groups can form coordination complexes with metals. The stability constants of chelates vary for different elements and different ligands. In general, the stability constants of humic complexes tend to decrease in the order: $Cu > Fe = Al > Mn = Co > Zn$. Organic ligands can render many metals, especially Cu and Pb, relatively immobile. However, low molecular weight complexes of metals, not necessarily of humic origin, tend to be soluble and can prevent metals from being adsorbed onto soil surfaces and thus render them more mobile and possibly more available for uptake by plant roots.

(*d*) *Sorption of Organic Contaminants on Humic Material.* This is the main mechanism by which non-polar, hydrophobic organic molecules are bound in soils. This may be by physical means or by chemical bonding.

(*e*) *Chemisorption of Elements.* This occurs when the element is incorporated into the structure of the compound. The most common example of this is when metals, such as Cd, replace Ca in the mineral structure of calcite ($CaCO_3$).

(*f*) *Co-precipitation of Elements.* This is the simultaneous precipitation of a chemical agent in conjunction with other elements. The elements typically found co-precipitated with secondary minerals in the soil include:[17]

Fe oxides: V, Mn, Cu, Zn, Mo;
Mn oxides: Fe, Co, Ni, Zn, Pb;
Calcite: V, Mn, Fe, Co, Cd;
Clay minerals: V, Ni, Co, Cr, Cu, Pb, Ti, Mn, Fe.

(g) Precipitation. This occurs when the concentrations of metal and accompanying ions exceed the solubility product of compounds such as $CdCO_3$, CdS and $Pb_5(PO_4)_3Cl$.

Those contaminants which are sorbed tend to be held against leaching and are less readily available for uptake by crops than those remaining in unbound forms within the soil. Volatilization of organic molecules and methylated forms of certain inorganic elements (As, Hg and Se) is also important. Some of the volatile compounds lost to the atmosphere may be decomposed by UV light (photolysis) or can also be sorbed onto the waxy cuticle of plant leaves and possibly enter the food chain.

The relative balance of reduction and oxidation (redox status) of a polluted soil plays an important role in the behaviour of some pollutants. Firstly, it will determine whether there will be an appreciable concentration of Fe and Mn oxides present. These are especially important for the sorption of As, Mo, and Cd. Secondly, some elements such as Cd, which readily form insoluble sulfide precipitates (CdS) under strongly reducing conditions, will be very immobile in waterlogged soils. However, if these soils become aerated due to drainage and drying out, the sulfide will oxidize to form sulfuric acid and so the liberated Cd^{2+} ions will be highly mobile and available for uptake. This occurs in contaminated paddy soils used for growing rice.

Organic pollutants bearing electrostatic charges will also be adsorbed onto oppositely charged sites on the soil colloids (*e.g.* the herbicides paraquat and diquat are strongly cationic). However, many organic pollutants are non-polar and uncharged and are normally bound to the soil organic matter by physical mechanisms, such as hydrophobic bonding. In soils which have received sewage sludge, the sludge material acts both as a source of several organic and inorganic pollutants and also as the major adsorbent for them.

The sorptive properties of a soil for both inorganic and organic pollutants can be described mathematically; in most cases sorption is found to fit either the Langmuir or the Freundlich adsorption isotherm equations but space does not permit its coverage here.

14.4.3   Comparison Between Soils of Rural and Industrial Sites

Many contaminated soils tend to occur on derelict or active industrial sites and these may differ greatly from a normal (pedological) soil in both constituents and physico-chemical properties. A summary of the differences which may be encountered between soils in rural and industrial sites in Northern European and other temperate climatic zones is given in Table 14.2.

14.4.4   Degradation of Organic Pollutants in Soils

Organic pollutants can be degraded in soils and, possibly, aquifers by either (a) non-biological mechanisms, including: hydrolysis, oxidation/reduction, photo-decomposition and volatilization, or (b) microbial decomposition ('biodegradation').

Table 14.2 *Comparison of typical soils from rural and industrial sites with regard to factors affecting the behaviour of pollutants*

Parameter	Rural soils	Industrial sites
pH	5–8	2–13
Organic matter (%)	1–10	< 1
Clays and oxides	Abundant	Variable
Plant nutrients	Abundant	Low
Concs of Cl^-, SO_4^{2-}	Low	High
Concrete/brick rubble	Absent	High
Heavy metals	< 0.05–0.1(< K_{sp})	< 1% (> K_{sp})
Organic pollutants	Low	Often high
Spatial heterogeneity	Low–moderate	Very high
Toxicity hazards	Food chain	Inhalation
	Soil ingestion	Soil ingestion
	Ecotoxicity	Ecotoxicity
		Potable ground and surface waters

When a pollutant chemical comes into contact with microbial colonies in biofilms lining the voids within the soil matrix, various extracellular enzymes will be secreted. These may partially degrade the chemical which is then absorbed into the microbial cell where intracellular enzymes may catalyse further decomposition reactions which bring about the release of energy and nutrients. Several species of bacteria and different enzymes may be involved at the same time and bring about a sequence of degradation steps producing increasingly simpler compounds which are either used in 'anabolic' (cell building) or 'catabolic' (energy releasing) processes.[18]

The greatest energy yield is obtained when the catabolic decomposition of an organic substrate by microorganisms occurs in the presence of free oxygen, the next in order of decreasing energy yield is the reduction of Fe and Mn oxides, then the reduction of NO_3^-, followed by SO_4^{2-} and finally the reduction of CO_2 to CH_4.

Although most organic molecules are biodegradable, the rate at which this takes place can vary greatly. The susceptibility of an organic molecule to biodegradation depends largely on its structure. Compounds which are most resistant to degradation tend to have halogen atoms in their structure, especially a large number of halogens, or which are highly branched, have a low solubility in water or an atomic charge difference.[18] Straight chain aliphatic compounds are easily degraded, but unsaturated aliphatics are less degradable than saturated forms. Simple aromatic compounds are usually degradable by several ring-cleavage mechanisms, but the presence of halogens stabilizes the ring and makes the compound less readily degradable. In general, unless there are specifically adapted microorganisms present, there will be a tendency for the more readily degradable molecules to be catabolized first. Adaptation can occur in a population of microorganisms as a result of a selection pressure. Individual organisms with the ability to degrade or tolerate a toxic compound may arise as a result of mutations and gene transfer and these individuals will have a competitive advantage if they can utilize an abundant supply of an organic pollutant as a source of energy and nutrients.[18] There is a time lag while adaptation occurs before there are sufficient microorganisms present which have the ability to degrade the pollutant; this tends to give a two-phase curve for concentration in the soil against time. The first phase is normally a steep decrease in concentration with time due to physical processes such as volatilization. The second phase is slower but goes on until the concentration approaches close to zero when the microorganisms have become adapted. 'Co-metabolism' can also occur and this is the term applied to the degradation of the pollutant (secondary substrate) by enzymes secreted by microorganisms to degrade another substance (the primary substrate).

Biodegradation requires appropriate conditions for the growth of the microorganisms which include adequate moisture, a temperature between 10 and 45 °C, a pH which is preferably in the range 6–8, and a supply of macronutrients (N and P). The redox conditions will depend on the types of microorganisms and pollutants involved. The anaerobic process of reductive dehalogenation can bring about the degradation of some halogenated pollutants such as perchloroethylene under reducing conditions. Some pollutant chemicals are highly toxic to soil microorganisms and so the inability to degrade may be linked to the lack of tolerance to the toxicity.[18]

The organochlorine pesticides have been used for more than 50 years and are regarded as the most persistent of all groups of pesticides. The order for decreasing persistence is: DDT > dieldrin > lindane (BHC) > heptachlor > aldrin with half-lives of 11 years for DDT down to 5 years for aldrin.[19] The most persistent organic pollutants of all in soils are probably the more highly chlorinated PCBs and dioxins (PCDDs). However, the persistence of a chemical in any soil is determined by the over-all balance between its adsorption onto soil colloids (usually organic matter), the extent of volatilization, uptake into or binding on, plant roots, and transformation/biodegradation processes. This balance depends on the nature of the pollutant, its concentration, the soil organic matter content, the soil pH and redox status, the available moisture, general soil fertility (level of nutrient supply, activity of soil microorganisms) and the time taken by microorganisms to adapt to the new pollutant substrate.

Bioremediation of soils contaminated by organic chemicals exploits the processes of biodegradation which would occur naturally but often at a much slower rate. The natural degradation processes are promoted by optimizing conditions for the soil microorganisms with regard to factors such as pH, aeration, moisture content and nutrient supply.

14.5 THE CONSEQUENCES OF SOIL POLLUTION

Soil pollution can give rise to toxicological problems in humans, livestock, crops, ecosystems and also in damage to structures and services. This can happen directly through contact with the soil itself, or indirectly through soil pollution causing groundwaters to become contaminated and these giving rise to toxicological and structural problems. As a consequence this pollution can restrict the uses to which land is put and there is therefore a need to be able predict problems by the use of soil quality standards and guidelines. These are based on the analysis of soils for the inorganic and organic contaminants considered to constitute the greatest hazards.

14.5.1 Soil Analyses and Their Interpretation

14.5.1.1 Methods of Soil Chemical Analysis. Analysis of soils for contaminants involves the collection of representative samples from suspected polluted sites and local controls. These samples are subsequently prepared for analysis, usually by drying and grinding followed by sieving. Analytical procedures involve either determination of the total concentration of the pollutant or a partial extraction procedure which can be correlated against a critical concentration. Space does not permit a detailed description of analytical procedures commonly used, but a brief summary is given in Table 14.3.

Table 14.3 *Summary of analytical procedures for soil pollutants*

Pollutant	Analytical procedure
Heavy metals (in conc. acid digests or partial extractants)	Flame atomic absorption spectrophotometry (FAAS), inductively coupled plasma atomic emission spectrophotometry (ICP-AES) or ICP-mass spectrometry (ICP-MS)
As, Bi, Hg, Sb, Se, Sn, and Te (in acid digests or partial extractants)	Hydride generation atomic absorption spectrophotmetry (HGAAS) using sodium borohydride in NaOH
Borate (water soluble)	ICP-AES (using quartz or plastic apparatus instead of Pyrex)
Organic pollutants (dissolved in appropriate organic solvents)	Gas chromatography with flame ionization detector (GC-FID) or with electron capture detector (GC-ECD) or GC combined with mass spectrometry (GC-MS)
Cyanides	Colourimetrically – using pyridine pyrazalone (blue) reaction
Sulfates (water soluble)	Elution through exchange column followed by titration with NaOH (or ion chromatography)
Sulfates (total)	Dissolution in HCl, precipitation of Al and Fe followed by gravimetric determination using $BaCl_2$
Chlorides (total – in HNO_3)	Volhard's Method – back titration with ammonium thiocyanate after initial precipitation with $AgNO_3$ (or ion chromatography)

Adapted from Alloway.[20]

Table 14.4 *The Netherlands guideline values for selected pollutants in soils and sediments*

| | Critical values for soils and sediments $(\mu g\ g^{-1})$ | | | | |
| | 1986 values | | | 1994 values | |
Substance	A	B	C	Target Value	Intervention Value
Metals					
Arsenic	20	30	50	29	55
Cadmium	1	5	20	0.8	12
Copper	50	100	500	36	190
Mercury	0.5	2	10	0.3	10
Nickel	50	100	500	35	210
Lead	50	150	600	85	530
Zinc	200	500	3000	140	720
Inorganic pollutants					
Cyanides (free)	1	10	100	1	20
Cyanides (complex) (pH < 5)	5	50	500	5	650
Cyanides (complex) (pH > 5)	5	50			
Sulfur	2	20	200		
Polycyclic aromatics					
PAHs (total of 10)	1	20	200	1	40
Naphthalene	0.1	5	50	15	
Anthracene	0.1	10	100	50	
Benzo[a]pyrene	0.05	1	10	25	
Chlorinated hydrocarbons					
CH (total)	0.05	1	10		
PCBs (total of 7)	0.05	1	10	0.02	1
Chlorophenols (total)	0.01	1	10		10
Aromatic compounds					
Aromatics (total)	0.1	7	70		
Benzene	0.01	0.5	5	0.05	1
Toluene	0.05	3	30	0.05	130
Phenols	0.02	1	10	0.05	1
Other organic compounds					
Pyridine	0.1	5	60	0.01	1
Gasoil	20	100	800		
Mineral oil	100	1000	5000	50	500

1986 Values: A = reference value, B = test requirement, C = intervention value (clean-up!).[21]
1994 Values: Target Value is the concentration which ought to be aimed for in the longer-term; Intervention Value = action to clean-up (Netherlands Ministry of Housing, Spatial Planning and Environment, 1994).[22]

14.5.1.2 Soil Quality Standards for the Interpretation of Soil Analytical Data. Having obtained concentrations of pollutants in soils, the interpretation of the data will be dependent upon national or international guideline concentrations or legal limits. These critical concentrations vary between countries and states. Perhaps the most widely known are the values used in the Netherlands ('Dutch Values') but these are relatively conservative being based largely on ecotoxicological data (which tend to have lower critical concentrations than human toxicological values) and were originally based on the principle of 'multifunctionality of use' whereby soil should be maintained, or cleaned up to a standard which will allow it to be used for any purpose including growing food crops. This is not universally accepted, and other countries, such as the United Kingdom, have used the principle of 'fitness for use'. This means that land to be used for domestic gardens where food crops will be grown should have much lower concentrations of hazardous chemicals than land to be built on for non-residential purposes. However, owing to the problems of cost and time necessary to clean up soils, a more 'fitness for use' approach has also been adopted in the Netherlands. Values are given in Table 14.4 for selected pollutants. This is not the complete list and the values are given for a standard soil containing 10% organic matter and 20% clay. The A, B, and C values, originally introduced in 1986, have been superseded by Target Values and Intervention Values and the latter have been revised (usually downwards).

In the United Kingdom, the guideline values used are based on the ICRCL provisional document of 1987.[23] In due course, these will be replaced by a different set of values obtained from a quantitative risk assessment model (CLEA – Contaminated Land Exposure Assessment) based on exposure pathways. The United Kingdom ICRCL values are based on fitness for purpose and have maximum total concentrations of elements in domestic gardens of: 3 μg g^{-1} Cd, 130 μg g^{-1} Cu and 500 μg g^{-1} Pb. Space does not permit the various guideline values to be given in detail and readers wishing to find out more about this are recommended to consult the most recent values for their own countries.

14.5.2 Hazards Associated with Soil Pollutants

A wide range of possible harmful effects can be associated with polluted soils and the more important types of possible hazard are summarized in Table 14.5.

14.5.3 Remediation of Contaminated Soils

Where the concentrations of pollutants exceed statutory or guideline quality standards there is a need to make a careful assessment of the risks at the particular site and in many cases it will be necessary to carry out some form of clean-up or remediation. In many cases, this will involve excavating the most polluted soil and either disposing of it safely in a licensed landfill (often referred to as 'dig and dump'), or cleaning it up off-site (*ex situ* remediation). However, with the high cost of haulage and of landfilling special wastes, there is increasing

Table 14.5   *Examples of the hazards caused by soil pollutants*

Hazard	Pollutants
(1) Direct ingestion of contaminated soil (by children, animals, gardeners and on unwashed vegetables)	As, Cd, Pb, CN^-, Cr^{6+}, Hg, coal tars (PAHs), PCBs, dioxins, phenols, pathogens
(2) Inhalation of dusts and volatile compounds from contaminated soil	Organic solvents, radon, methyl forms of Hg, metal-rich particles, asbestos
(3) Uptake by plants and movement through the food chain to livestock and/or humans	As, Cd, ^{137}Cs, Hg, Pb, ^{90}Sr, Tl, PAHs, various pesticides
(4) Phytotoxicity	SO_4^{2-}, Cu, Ni, Zn, Cr, B, and CH_4
(5) Toxicity to soil biota (especially microbial biomass)	Cd, Cu, Ni, Zn
(6) Deterioration of building materials and services	SO_4^{2-}, SO_3^{2-}, Cl^-, tars, phenols, mineral oils, organic solvents
(7) Fires and explosions	CH_4, S, coal and coke dust, residues of explosives (*e.g.* TNT), tar, rubber, plastics, high calorific value wastes (*e.g.* old landfills)
(8) Contact of people with contaminants during demolition or excavation of sites	Tars (PAHs), phenols, asbestos, radionuclides, PCBs, TCDDs, pathogens
(9) Contamination of surface and ground waters compounds	CN^-, SO_4^{2-}, metal salts, organic (LNAPLs and DNAPLs) surfactants, farm wastes, pesticides

Adapted from Becket and Sims,[24] ICRCL[23] and Alloway.[20]

use made of *in situ* remediation methods. The main types of remediation methods in use are briefly outlined in Table 14.6.

In the case of organic liquid pollutants, these methods include: (a) Containment – where the contamination is retained by physical barriers (such as trenches filled with impermeable clay, such as bentonite) or hydraulic barriers based on maintaining fluid pressure differentials by the extraction or injection of water to confine the contaminant plume in aquifers (permeable underground strata); (b) 'Pump and Treat' technique applied to contaminated groundwaters which are pumped up to the surface from specially drilled wells and then treated to remove the contaminants and the cleaned water is then usually returned to the aquifer. In the case of volatile organic compounds, such as chlorinated solvents, this can include bubbling air through the extracted water to enhance the volatilization of the organic pollutants (called 'air sparging'). When water is injected into contaminated soil to dissolve soluble contaminants and then extracted, this technique is referred to as '*in situ* soil washing' which can be successfully applied to the removal of a wide range of contaminants (not all necessarily water soluble), including salts such as sulfates. *In situ* enhanced recovery techniques include soil vapour extraction by soil venting, where air is

Table 14.6 *Techniques available for the remediation of contaminated land*

Method	Contamination problem
Civil engineering based methods	
(1) Excavation and safe disposal (usually to landfill)	'Hot spot' contamination at concentrations too high for safe *in situ* treatment, especially inorganic chemicals such as heavy metals
(2) Physical containment – using in-ground barriers and covers	Organic NAPLs, landfill leachates
(3) Hydraulic controls	Supporting containment and/or for treatment of contaminated surface or groundwaters
Process-based remediation	
(1) Thermal treatment – to remove, stabilize or destroy contaminants	Incineration of soils contaminated with POPs such as PAHs, PCBs and PCDDs
(2) Physical treatment to separate contaminants from soils or different fractions of contaminated media (usually *ex situ*)	Soil washing to remove sulfates, PAHs metals and other pollutants adsorbed on silt and clay particles
(3) Chemical treatment – reactions to remove, destroy or modify contaminants (*in situ*, or *ex situ*)	Dechlorination of PCB contaminated soils, acid leaching of metal-contaminated soils
(4) Biological treatment – using microorganisms to remove, destroy or modify contaminants (*in situ* or *ex situ*)	Organic contaminants, including hydrocarbons, pesticides, solvents, tars, PAHs
(5) Stabilization/solidification – where chemicals are rendered less available to receptors (*in situ* or *ex situ*)	Heavy metals, persistent organic pollutants

Adapted from Harris and Herbert.[25]

pumped into soil which has sufficient macropores to provide adequate air permeability and collected from extraction wells from which the volatile compounds can be either incinerated on site or adsorbed onto activated carbon.

Soil pollutants adsorbed to the soil solids are generally more difficult to treat. In the case of organic compounds, bioremediation, which involves optimizing the conditions for the microbial decomposition of the chemicals, is increasingly being employed. For heavy metals, chemical treatment involving the use of extractants, such as acids or chelating agents, is possible but not widely used. Excavation and disposal to a licensed landfill is the most common approach. A relatively recent development is phytoremediation, which uses plants with a high capacity to accumulate metals to deplete the plant-available fraction of metals in the soil. However, although several hyperaccumulator plant species have been identified and show considerable potential, most do not produce sufficient biomass to effectively remove significant amounts of metals from contaminated soil. Nevertheless, it is likely that this method will be developed for use on soils with low to moderate levels of pollution but severely polluted soils will probably still need to be removed.

Apart from the application of the various possible remediation methods, it is often found that some organic pollutants undergo natural processes of adsorption and degradation, and this is referred to as 'natural attenuation'. There is an increasing body of evidence to show that these natural processes account for a great deal of diminution of risk from soil pollution at many types of sites. However, it does need careful monitoring to ensure that improvements are taking place over an appropriate period of time. The range of available remediation techniques is exemplified by Table 14.6.

14.6   CASE STUDIES OF CONTAMINATED INDUSTRIAL LAND

14.6.1   Former Gasworks Sites

By the end of the nineteenth century most towns and cities in North West Europe, North America and many other parts of the world had their own gasworks to produce coal gas for heating and lighting in industrial and domestic properties. When the total number of sites, including large municipal works and small units in some industrial premises are considered, there are probably many thousand sites polluted by gas manufacture throughout the world. In the United Kingdom, there are between 2000 and 5000 gas manufacturing sites and in the Netherlands 234 former gasworks sites have been identified as being in need of major remediation. The production of the gas was based on the heating of coal in a non-oxidizing atmosphere. In addition to the main products of methane and carbon monoxide, this process also produced tars, phenols, cyanides and various other impurities which had to be removed from the gas and have consequently accumulated at gas works sites. The contamination problems associated with gasworks include contamination of soils and both surface and goundwaters by tarry wastes and phenols, and cyanides. Atmospheric pollution by volatilization of naphthalene, benzenes, acenaphthalene and cycloalkenes, thioprene, pyridine, and hydrogen sulfide from contaminated soils at gasworks sites is also reported. In addition to the chemical pollution associated with gas manufacture, additional environmental pollution can also arise from asbestos insulation and Pb from old paintwork.[26]

In the UK, the National Millennium Exhibition was housed in a dome structure on the Greenwich Peninsula in the River Thames in East End of London. The site had been occupied by one of the largest gasworks in Europe (125 ha) for about 90 years until the 1970s and had become heavily contaminated with the usual materials and waste products from gas production, plus those of ancillary works which included chemicals, tar distillation, and a benzol plant. The remediation works required before the Millenium Dome could be constructed and opened to the public involved: the excavation and safe disposal of $> 200\,000$ m^3 of contaminated soil material, soil vapour extraction over

4.3 ha, the washing of $> 30\,000$ m^3 of sand and gravel, and an on-site water treatment facility (Griffin[27] and Barry[28]).

14.6.2 Sites Contaminated with Solvents

Chlorinated hydrocarbon solvents such as trichloroethene, 1,1,1-trichloro-ethane, tetrachloromethane and tetrachloroethene are widely used in industry and are frequently found as soil pollutants at industrial sites. In most cases, pollution occurred as a result of spillages and this has resulted in the chlorinated hydrocarbons being found in the following four types of sites: (i) as isolated droplets within the pollution plume in the groundwater (aquifer) or trapped in pools in low permeability material, (ii) dissolved in porewaters, (iii) as vapour in the unsaturated zone (above the aquifer), and (iv) sorbed onto solid phase soil organic matter.[29] Soil gas monitoring has revealed that some derelict industrial sites with relatively low concentrations in soil have given rise to marked pollution of groundwater in boreholes below the sites at depths down to 187 m. The concentrations of chlorinated hydrocarbon solvents in soil gas at derelict industrial sites were found to vary by seven orders of magnitude up to a maximum value of 2000 μg L^{-1} trichloromethane.[29] Volatile compounds, such as chlorinated solvents, can be removed from groundwaters by pump and treat methods in which air is bubbled through the water to enhance the volatilization of the chemicals which are then trapped and destroyed.

A fire at a solvent recovery works in the village of Carrbrook, in Cheshire, England, in 1981 caused severe contamination of soils in the local area with benzene, PCBs, and dioxins.[30] Concentrations of up to 304 mg kg^{-1} total solvents, 1160 mg kg^{-1} PCBs and 168 μg kg^{-1} dioxin (expressed as TEQ = toxic equivalents of 2,3,7,8-tetrachlorodibenzodioxin TCDD) were found in soils at the site. Domestic pets kept in houses and gardens near to the site were found to show abnormal health effects. Worst affected were guinea pigs which developed a terminal wasting disorder.[30] It is generally recognized that guinea pigs have the lowest tolerance of all animals tested to dioxins. Concentrations of 100–1000 ng kg^{-1} TCDD are considered typical for industrially contaminated sites but the levels at the solvent recovery works site were much higher than the maximum in this range. Perhaps the worst case of dioxin pollution occurred at Times Beach in Missouri where waste oil contaminated with distillation residues from organochlorine production had been used to reduce dust problems on dry soils. This contaminated oil gave rise to concentrations of up to 33 mg kg^{-1} TEQ in the soils and caused the death of horses, cats, dogs, chickens and birds exposed to the soil. Children playing in the area developed chloracne, a characteristic symptom of exposure to dioxins.[31]

14.6.3 Lead and Arsenic Pollution in the Town of Mundelstrup in Denmark

Severe Pb and As contamination was found in housing in the town of Mundelstrup Stationsby west of Aarhus in Denmark in 1987. The pollution had arisen from the disposal of fill from a former fertilizer factory which had used metal-rich pyrite ore for the manufacture of sulfuric acid. Concentrations of up to $67\,562$ mg kg^{-1} Pb and 5481 mg kg^{-1} As were found in the soil of gardens of the houses affected. A comprehensive survey was carried out involving the analysis of approximately 1000 samples. The contaminated area covered 6700 m^2 and varied in depth between 0.5 and 8 m. The worst affected area was at the site of the original factory. It was decided to remove all soil from around the houses which exceeded the Danish soil quality criteria of 40 mg kg^{-1} Pb and 20 mg kg^{-1} As. A borrow-pit was dug to supply clean soil to replace the $50\,000$ m^3 of soil excavated from around the houses with the contaminated gardens. This same pit was used as a special landfill to receive the contaminated soil after appropriate engineering with layers of lime to prevent leaching and the whole landfill was covered by a new motorway roundabout. Only 30 houses were affected and these had their garden soil completely replaced and professional landscape gardeners created new gardens. This is an example of the ideal clean-up of an urban area. It was carried out at great cost (32×10^6 DKK) and this may not be possible in other countries with a greater legacy of historically polluted urban sites.[32] It is interesting to note that the Danish soil quality standard of 40 mg Pb kg^{-1} is extremely conservative. If this value was to be used in more highly industrialized countries, such as the United Kingdom, almost all urban soils would be classed as having excessive concentrations of lead.

14.6.4 Pollution from a Pb–Zn Smelter at Zlatna, Romania

Acid precipitation from a Pb–Zn smelter at Zlatna in Transylvania, Romania has led to obvious toxic effects on vegetation and damage to buildings within an area of at least $50\,000$ ha. The smelter lies in a valley and the fumes are trapped within the valley. The precipitation of SO$_2$ fumes and metal aerosol particles has killed many plant species and this has led to massive soil erosion and associated deterioration of the soil structure due to the loss of the surface organic matter. A layer of colluvium up to 46 cm thick has developed on the top of the soil profiles at the base of the valley side slopes. The metals deposited from the smelter fumes are rendered relatively highly mobile in the acid soils but SO$_2$ is the most destructive pollutant. The most urgently required pollution control measures are scrubbers to reduce the SO$_2$ concentrations in the smelter emissions but this would be prohibitively expensive.

14.6.5 Cadmium Pollution in the Village of Shipham in England and the Jinzu Valley in Toyama Province, Japan

Zinc and Pb mining was carried out around the village of Shipham in Somerset, England during the nineteenth century and the land has been left highly contaminated with these metals and Cd which was present in considerable concentrations in the Zn ore. Garden soils used for the growing of vegetables were found to contain up to 360 μg g^{-1} Cd, 37 200 μg g^{-1} Zn, and 6540 μg g^{-1} Pb and the mean Cd concentration in almost 1000 samples of garden vegetables was 0.25 μg g^{-1} which was nearly 17 times the national average of 0.015 μg g^{-1} Cd (in the dry matter). The highest Cd concentrations occurred in leafy vegetables such as spinach, lettuce and cabbage which contained up to 60 times more Cd than the same species grown locally in uncontaminated soils. Although the Cd concentrations in the vegetables were relatively high, no adverse health effects were found in the population and this was ascribed to the low percentage of home-grown vegetables in the diet, a generally varied diet and a public water supply which conformed to national and international standards for Cd.[33]

In Japan, soils used for growing paddy rice in the Jinzu Valley had also been contaminated with Cd and other metals from mining operations upstream but, in contrast to Shipham, many people had suffered ill-health effects. Two hundred elderly women who had given birth to more than one baby had been disabled by a Cd-induced skeletal disorder know as 'itai-itai' disease and 65 women had died of this condition. The disease occurred during and immediately after the Second World War when diets were more deficient in Ca and protein than normal. Average Cd concentrations in rice (0.7 μg g^{-1} DM) in the contaminated soils were ten times greater than in local controls and the maximum Cd content in rice grown in the Jinzu Valley was 3.4 μg g^{-1} DM. In the paddy soils the Cd is present as insoluble CdS during the flooded period but when the paddy fields are drained in readiness for harvest the sulfide oxidizes to release Cd^{2+} ions and sulfuric acid which increases the bioavailability of the Cd.[34] In the Shipham soils, a high content of CaCO$_3$ from mining waste helps to reduce the bioavailability of the Cd which was present at total concentrations more than a hundred times higher than in the Jinzu paddy soils.

14.7 CONCLUSIONS

Soils and associated surface and groundwaters can become polluted by a wide range of chemicals arising from many human activities and in every case the bioavailability of the pollutant will depend on: (a) the combined effects of the type and concentration of the polluting substance; (b) the composition of the soil or site matrix (in the case of heavily developed industrial sites), especially its organic matter, sand, clay and free carbonate contents; (c) the soil physico-chemical conditions (pH and redox status); (d) the genotype of the plants growing on the soil (species and varieties vary greatly in their capacity to accumulate pollutants); and (e) the climate (soil moisture status, temperature). Although these interactions can be relatively complex, there is enough information becoming available to allow predictive and risk assessment models to be developed. However, much more is currently known about the behaviour of inorganic pollutants, such as heavy metals which are easier (and less expensive) to analyse for, than many of the organic pollutants. However, analytical techniques for persistent organic pollutants are becoming more rapid and less expensive and will thus provide more data for use in predictive models. In order to be able to monitor both naturally attenuated and remediated sites effectively, a new generation of sensors and techniques to measure concentrations of selected organic pollutants and their degradation products on-line in the field (such as in wells) is being developed. From the food safety aspect, it is widely considered that the priority pollutants of soils and crops are dioxins and PCBs among the organic pollutants and As and Cd out of the inorganic pollutants.[35] However, in addition to these elements and substances, the more ubiquitous PAHs and BTEX compounds (benzene, toluene, ethyl benzene and xylene) and heavy metals such as Hg and Pb will continue to pose major problems in polluted soils and/or waters which will need to be attended to.

14.8 REFERENCES

1. World Resources Institute, 'World Resources 1994–95: A Guide to the Global Environment', Oxford University Press, New York, 1994.
2. K. C. Jones, *Environ. Pollut.*, 1991, **69**, 311–325.
3. L. R. Oldeman, R. T. A. Hakkeling and W. G. Sombroek, 'World Map of the Status of Human Induced Soil Degradation', CIP-Gegevens Koningklijke Bibliothek, Den Haag, 1991.
4. E. M. Bridges, *Soil Use Manage.*, *1991*, **7**, 151–158.
5. S. M. Ross, 'Toxic Metals in Soil Plant Systems', John Wiley and Sons, Chichester, 1994.
6. B. J. Alloway (ed.), 'Heavy Metals in Soils', 2nd edn, Blackie Academic and Professional, Chapman and Hall, London, 1995.
7. S. S. Foster, P. J. Chilton and M. E. Stuart, *J. Inst. Water Environ. Manage.*, 1991, 186–193.
8. A. Brown, 'The UK Environment', Government Statistical Service, HMSO, London, 1992.
9. Organization for Economic Co-operation and Development (OECD), 'The State of the Environment', OECD, Paris, 1991.
10. D. Sauerbeck, in 'Scientific Basis for Soil Protection in Europe', eds. H. Barth and P. L'Hermite, Elsevier, Amsterdam, 1987.
11. G. Merrington and B. J. Alloway, *Appl. Geochem.*, 1994, **9**, 677–687.
12. M. B. McBride, 'Environmental Chemistry of Soils', Oxford University Press, New York, 1994.
13. G. Sposito, 'The Chemistry of Soils', Oxford University Press, New York, 1989.
14. R. E. White, 'Principles and Practice of Soil Science: The Soil as a Natural Resource', 3rd edn., Blackwell Science, Oxford, 1997.
15. J. E. Bittel and R. J. Miller, *J. Environ. Qual.*, 1974, **3**, 243–244.
16. G. W. Brummer, 'The Importance of Chemical Speciation in Environmental Processes', Springer Verlag, Berlin, 1986.
17. G. Sposito, in 'Applied Environmental Geochemistry', ed. I. Thornton, Academic Press, London, 1983.
18. M. D. LaGega, P. L. Buckingham and J. C. Evans, 'Hazardous Waste Management', McGraw-Hill, New York, 1994.
19. F. A. M. de Haan, in 'Scientific Basis for Soil Protection in Europe', eds. H. Barth and P. L'Hermite, Elsevier, Amsterdam, 1987, p. 181.
20. B. J. Alloway, in 'Understanding Our Environment', ed. R. M. Harrison, Royal Society of Chemistry, Cambridge, 1992.
21. J. E. T. Moen, J. P. Cornet and C. W. A. Evers, in 'Contaminated Soils', eds. J. W. Assink and W. J. van den Brink, Martinus Nijhoff, Dordrecht, 1986.
22. Netherlands Ministry of Housing, Spatial Planning and Environment, 'Intervention Values and Target Values – Soil Quality Standards', Directorate-General for Environmental Protection, Department of Soil Protection (625) Rijnstraat 8, The Hague, The Netherlands, 1994.

23. Interdepartmental Committee on the Redevelopment of Contaminated Land, 'Guidance on the Assessment and Redevelopment of Contaminated Land', Guidance Note 59/83, Department of the Environment, London, 1987.
24. M. J. Beckett and D. L. Sims, in 'Contaminated Soil', eds. J. W. Assink and W. J. van den Brink, Martinus Nijhoff, Dordrecht, 1986.
25. M. Harris and S. Herbert, 'Contaminated Land', Institution of Civil Engineers, Thomas Telford Publishers, London, 1994.
26. A. O. Thomas and J. N. Lester, *Sci. Total Environ.*, 1994, **152**, 239–260.
27. M. M. H. Griffin, *Land Contam. Reclam.*, 1999, **7**(1), 20–26.
28. D. L. Barry, *Land Contam. Reclam.*, 1999, **7**(3), 177–190.
29. P. R. Eastwood, D. N. Lerner, P. K. Bishop and M. W. Burston, *J. Inst. Water Environ. Manage.*, 1991, 163–171.
30. T. Craig and R. Grzonka, *Land Contam. Reclam.*, 1994, **2**, 19–25.
31. C. D. Carter, R. D. Kimbrough, J. A. Liddle, R. E. Chine, M. M. Zack, W. F. Barthel, R. E. Koehler and P. E. Philips, *Science*, 1975, **188**, 738–740.
32. P. Clement, N. J. Olsen and P. Madsen, *Land Contam. Reclam.*, 1995, **3**, 39–46.
33. H. Morgan and D. L. Sims, *Sci. Total Environ.*, 1988, **75**, 135–143.
34. T. Asami, in 'Changing Metal Cycles and Human Health', ed. J. Nriagu, Springer Verlag, Berlin, 1984.
35. Ministry of Agriculture, Fisheries and Food, 'Surveillance and Applied R & D on Food, Requirements Document, 1996–97', MAFF, London, 1995.

CHAPTER 15

Solid Waste Management

G. EDULJEE and D. ARTHUR

15.1 INTRODUCTION

Waste is not a unique material in terms of its constituents: the main distinguishing feature relative to the products from which it derives is its perceived lack of value. Dictionary definitions of waste include the descriptions 'useless' or 'valueless'. In the UK, the legal definition of waste as given in Section 75 of the Environmental Protection Act 1990 includes phrases such as "scrap material or other unwanted surplus substance" and "any substance which requires to be disposed of as being broken, worn out, contaminated or otherwise spoiled".

Up to the 1970s, the perception of waste as unwanted, 'useless' material with no intrinsic value shaped society's approach to waste management. The ultimate disposal of waste was the overriding priority. Waste generators (domestic, commercial and industrial) sought disposal at the lowest cost (overwhelmingly in landfills) and had little or no incentive to 'manage' waste as opposed to merely dispose of it.

However, the past two decades have witnessed a sea change in attitudes to waste. Public sensitivity towards waste disposal outlets has put enormous pressure on operators and regulators alike; the siting of new facilities has at best proceeded after lengthy delays and in the teeth of intense opposition, and at worst become all but impossible. Waste generators and policy makers have perforce turned their attention to upstream waste-related activities in an effort to minimise waste production and hence make more efficient use of ultimate disposal capacity.

Another key driver has been the concept of sustainable development,[1] defined as "development that meets the needs of the present without compromising the ability of future generations to meet their own needs". The slogan more or less encapsulates the current thinking: extracting the maximum value and benefit from products and services, using the minimum of energy and rejecting the minimum of waste materials or emissions to the environment. Arising out of this is the concept of sustainable waste management.[2,3] In essence, waste is given value. Viewed against the principle of the conservation and nurturing of natural resources, the production of waste is in itself seen as a manifestation of the

378

inefficient management of the earth's raw materials. Commencing with waste reduction during the production of goods and the provision of services, sustainable waste management calls for the recovery and reuse of materials so as to conserve raw materials, the use of waste as a source of energy in order to conserve non-renewable natural resources, and finally for the safe disposal of unavoidable waste. A new policy approach has been developed to instil a sense of common responsibility towards waste, and to create overall strategies for waste management that do more than pay mere lip service to the concept of sustainability.

The aim of this chapter is to illustrate how this approach has been applied to the management of municipal solid waste (MSW). For the purpose of this chapter, we define MSW as solid waste collected from households, commercial and industrial premises. In many ways MSW presents the greatest challenge to waste management. Waste arising from domestic, commercial and industrial premises is extremely heterogeneous in nature, and as a result components of potential value (such as glass, metals or biodegradable) cannot be beneficially used without significant effort being put into segregation and sorting schemes. Even with such systems, it is very difficult to achieve products of value due to cross-contamination and fragmentation of the constituents. In many countries, the poor standards of MSW disposal also leads to a significant threat to public health.

In keeping with the importance governments now place on sustainable development, this chapter emphasizes the preventative aspects and the resource potential of waste management as opposed to the impact of MSW disposal *per se*: other publications have examined this latter aspect of waste management in some detail.[4–6] The chapter commences with an introduction to the so-called waste management hierarchy and the need to integrate various strands of policy into a coherent waste management strategy. Next, technical options for waste prevention and recycling are presented, followed by a discussion of the policy options that can be applied to encourage or to mandate waste prevention and reuse. Bulk waste reduction technologies, in particular incineration and composting, are then presented. Finally, some of the issues concerning the development of an integrated waste management strategy are discussed.

15.2   AN INTEGRATED APPROACH TO WASTE MANAGEMENT

15.2.1   The Waste Management Hierarchy

Options for waste management are often arranged in a hierarchical manner to reflect their desirability.[2,3] The first priority is waste avoidance, that is not producing the waste in the first place. If the waste must be produced, then the quantities should be minimized. Once that has been achieved, the next priority is to maximize recovery, reuse and recycling of suitable waste materials. Taken together, these three options are often called waste prevention, although strictly speaking only the first two are prevention whereas the third is already an end-of-pipe solution. Once the possibilities for waste prevention have been exhausted,

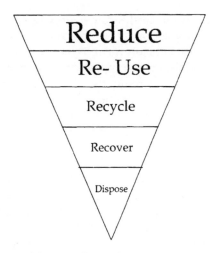

Figure 15.1   *Components of the waste hierarchy*

the next priority is to reduce the volume of residual wastes being passed on for final disposal, extracting resources in the form of products and/or energy in the process. The waste management hierarchy is illustrated in Figure 15.1.

To move from the current situation in which the majority of wastes are dealt with by final disposal or other end-of-pipe solutions to one where waste prevention becomes dominant, it is necessary for governments to provide some support through policy measures. Policies which encourage waste avoidance and waste minimization are to be preferred over those which focus purely on further encouraging present recycling, recovery and reuse.

The waste management hierarchy is discussed further in Section 15.6.

15.2.2   An Integrated Approach

If waste management is to change significantly, the behaviour of individuals and groups in society will have to change. Three groups in society are key to this process: government, industry and commerce, and individuals. A fourth group, lobby groups and NGOs, is effective as a conduit of ideas and energy. Policies need to be designed to change behaviour of all these groups in order to reduce or even reverse the growth in waste generation which accompanies increasing wealth.

Effective policies generally operated on a two pronged approach. Where possible, waste reduction policies should be implemented on a voluntary basis, by all the above groups. In order to kick-start some initiatives, government can provide an incentive or 'carrot' in the form of financial or other support. Combined, these form the first prong. For the second prong, governments implement policies to tackle the underlying causes relating to waste generation and positively discourage it. Some such measures need to be implemented alongside those from the first prong. Others may only be required if measures from the first prong, for whatever reason, prove to be ineffective. This second prong of policies can be viewed as the disincentives, sanctions or 'sticks' which complement the 'carrots' above.

In addition, it is essential that policies build on rather than undermine existing strengths. Where there are already efforts to reduce waste generation or recover materials for recycling and reuse on a voluntary basis or using current market mechanisms, policies must aim to reinforce these efforts, and not be so invasive as to undermine such existing initiatives.

Finally, even with all these policies and measures in place and operational, some wastes will still be generated and require disposal. The provision of bulk waste reduction technologies will help to reduce the volume and weight of these remaining residues prior to final disposal.

An integrated strategy for waste reduction requires a combination of all the types of measure detailed above. The following sections address these measures, along with technical options for waste prevention and segregation.

15.3 TECHNICAL OPTIONS FOR WASTE PREVENTION AND RECYCLING

15.3.1 Opportunities for Waste Avoidance and Minimization

Waste avoidance and minimization are the most desirable options in the waste management hierarchy as discussed in Section 15.2.1 above. This section explores the ways in which waste avoidance and minimization have been or may be practised to reduce the amounts of domestic, commercial and industrial wastes that arise. Table 15.1 lists some options which may be appropriate and these are discussed in greater detail under each sector. Some of these waste avoidance and minimization options may be applicable to all waste generators

Table 15.1 *Ways to achieve waste avoidance and minimization*

Waste management options	Opportunities for avoidance and minimization
Product design change	• Product design with less waste • Increase product life
Package change	• Product in bulk or concentrate form • Reusable or recyclable pack
Materials change	• Substitution of less toxic materials • Use of reusable or recyclable materials
Technological change	• Improved/more efficient equipment • Cleaner technology
Management practices	• Good housekeeping • Proper operating procedures and regular maintenance • Inventory control • Training and clear instructions • Waste segregation

while some would be applicable only to industry (or even to specific industry sectors).

15.3.1.1 Domestic Sector. There are many ways in which individuals can avoid or minimize the amount of waste they put out for disposal. As consumers, they may select product types, packaging types and material types that would lead to the generation of less waste. For example, waste can be reduced by buying in bulk, utilizing reusable shopping bags, buying reusable and more durable products and by buying equipment that generates less waste (for example electrical equipment which runs off mains power or uses rechargeable batteries to reduce the quantity of primary cells used). Thoughtless household practices can also lead to waste generation. In a Dutch study on consumer habits relating to the purchase of milk and bread[7] it was found that about 15% of bread was wasted because it was stored too long in the home, resulting in about 70 000 tonnes of bread being rejected as waste and hence landfilled. The accompanying wasted packaging amounted to some 6000 tonnes of plastic and paper.

A major factor is to change the public's perception of waste and how to deal with waste materials. Education and communication programmes therefore need to be built into any approach aimed at the domestic sector before significant take-up of the options offered could be achieved.[8,9] Inducements and incentives can be deployed to support information dissemination and communication, encouraging communities to change their perception of waste.[10] However, lack of attention to practical obstacles such as the provision of recycling bags or adequate collection facilities can negate the drive towards waste avoidance, even where there is a general motivation to avoid, minimize and recycle waste.[11,12]

Some waste avoidance and minimization measures taken within the domestic sector would not be possible without the supporting actions of both industry and commerce. The manufacturing, packaging and labelling of products and the provision of some services that would lead to the generation of less waste are necessary to enable householders to have some choice. This type of responsibility relating to producers is examined in later sections.

15.3.1.2 Commercial and Industrial Sectors. As manufacturers, companies may produce longer life products, requiring less maintenance. Manufacturing processes could use less toxic substitutes and recycling materials and may be improved in terms of reduction in material wastage. Proper operating procedures and regular maintenance would also reduce wastage as well as reducing other emissions. Such measures are likely to be industry specific, *i.e.* their application and potential to reduce wastes will vary industry by industry. For example, in the plastics industry, internal recycling of segregated clean plastics may contribute 10% to 15% of total consumption of plastic materials. In metal products fabrication industry, improved maintenance of cutting machines has been identified as a way of preventing contamination of scrap metal waste which may then be recycled. In the wearing apparel industry, use of a wider roll of cloth and re-design of garments may significantly reduce cutting wastes. The experience of McDonald's in the US shows that, among other initiatives, a simple trimming back of one edge in their paper napkins reduced the amount of paper used by 21%.[13] Improving management and control of processes can lead to significant waste reduction, which in turn can lead to considerable savings in costs due to less wastage and more efficient procedures. For example, between 1986 and 1990, consumers in the UK consumed 20% more drinks, using as a result 16% more energy and generating 4% more waste. In the same period, industry has used 7% less energy and generated 14% less solid waste per litre of drink produced.[14]

It should be noted that waste avoidance and minimization measures are already adopted by many individuals and companies, partly for environmental concern and partly for financial concern. Experience from running cooperative waste minimization programmes around the world shows that less wastes mean higher profit. However, there are barriers and constraints identified in these programmes and Government policies and measures to encourage the adoption of these methods would be required, as discussed in Section 15.4.

15.3.2 Collection and Sorting

A prerequisite for the cost-effective reuse and/or recycling of potentially valuable components in MSW is that these materials be separated out from the bulk waste. A number of options are available, as discussed below.

15.3.2.1 Source Separation. Collections of waste materials at source are termed 'kerbside collection systems'. This method involves the householder putting out recyclable materials for collection separate from the normal refuse. These recyclable materials may either be mixed, all materials being placed into one container for future sorting either by the collector at a reclamation facility, or separated into individual materials.

In the mixed at source scheme the householder places all recyclable materials into one container which is emptied into the collection vehicle and taken to a central facility for sorting. This central facility is usually referred to as a Materials Reclamation Facility (MRF). In MRFs, co-mingled materials are re-separated, stored and perhaps some initial processing is carried out, prior to selling on to a manufacturer or into the secondary materials markets. Two processing lines are usually set up; one for the separation of co-mingled containers (aluminium cans, steel cans, plastic bottles and glass bottles) and the other for co-mingled paper (newspaper, domestic paperboard, white paper). The process separates these and prepares them for onward sale to recycling companies. The initial processing of waste at a MRF (for example, washing, baling, *etc.*) can add value to the separated materials.

Source separated materials which reach the MRF should already be fairly clean and uncontaminated. Hence, a MRF should not give rise to significant air or water pollution so long as the civil works of the facilities are properly designed. Proper enclosure and management of MRFs would reduce the impact of noise and odour.

To a large extent, the economics of source separation and MRFs are determined by the price for recovered materials, which can fluctuate widely. The relative quantities of recyclable collected and sorted are also key, because of different prices and different sorting regimes for different materials.

In the separation at source scheme, the householder is either required to place recyclable materials into one container for sorting by the collector at the kerbside when the materials are collected, or the recyclables are placed in separate containers. An example of the former is the 'Blue Box' scheme in Sheffield, where in the first year of operation some 277 tonnes of paper, 78 tonnes of glass, 39 tonnes of cans and 27 tonnes of plastic was collected from 3300 properties, resulting in a reduction of 17% in the quantity of materials entering the domestic waste scheme. The scheme has not greatly expanded since its inception. Multi-container collection has been in operation in Leeds, Bury and Milton Keynes. The Milton Keynes plant currently processes approximately one-half to one-fifth of its potential capacity.[15,16] Sorting from a co-mingled source can also be done at the kerbside using a specially adapted vehicle.

15.3.2.2 Bring Systems. Bring systems are very widely used in many parts of the world, and are typically employed for the recovery of glass or paper. In the UK, the recycling centres at Civic Amenity sites and recycling facilities, such as bottle banks and paper banks outside supermarkets and in town centres, are examples of 'bring systems'. At most Civic Amenity sites, recycling centres have been set up to permit the deposit of recyclable materials in separate facilities. Typically, separate facilities are used for glass, metals, aluminium and steel cans, paper, cardboard, oils and textiles. Occasionally containers for hardcore, wood, car and domestic batteries, *etc.* are also provided. Refrigerators containing CFCs can also be recycled.

The overall diversion rate of materials from the waste stream destined for direct landfill varies widely, but is generally higher for kerbside collection, presumably because this involves less effort on the part of the householder.

15.4 POLICY OPTIONS TO MAKE WASTE PREVENTION AND RECYCLING WORK IN PRACTICE

15.4.1 Introduction

For recycling to become more effective, it is necessary to introduce measures to 'level the playing field' so that secondary materials can compete more fairly with virgin stock, and to provide direct support to increase the size of local markets. These measures can be *via* voluntary participation, *via* positive encouragement, generally in the form of some sort of government support (*i.e.* the carrots referred to in Section 15.2.2), or *via* persuasion or mandatory measures, enacted by government to more forcefully encourage the adoption of waste reduction measures (*i.e.* the sticks discussed in Section 15.2.2). The central objective of the application of these policy measures is to encourage waste avoidance, minimization, reuse and recycling. In general, they each seek to change behaviour by making waste avoidance, minimization, reuse and recycling more attractive options than disposal. Most successful schemes for waste prevention and recycling work by combining the 'carrot' of financial incentives and other measures to encourage positive behaviour with the 'stick' of either financial penalties or legal requirements to discourage negative behaviour. Policies and other measures may target either those who provide the goods (*e.g.* manufacturers and importers) or those who use and dispose of the goods (*e.g.* householders), or both.

Five types of policy options that have been considered or are emerging as key policy drivers are discussed in this section:

- Producer responsibility
- Eco-labelling
- Charges and economic incentives
- Persuasion measures
- Integrated product policy

To illustrate the mix of measures that can be adopted, Table 15.2 provides examples of how five waste management authorities have approached the problem of diverting waste from landfill, and raising recycling and recovery rates.

15.4.2 Producer Responsibility

15.4.2.1 Principles. Waste avoidance, waste minimization and the separation of waste materials at source to facilitate recycling all require the active participation of the waste producers and waste generators, including householders and commercial/industrial companies. The concept of 'producer responsibility' is one which has become the norm rather than the exception around the world for the management of such wastes. The concept is that the manufacturer or importer of the products giving rise to the waste should take responsibility for those wastes. These groups are thereby encouraged to consider the implications of disposal of their product and are given an incentive to investigate methods of reducing, reusing or recycling their wastes. The producer either levies a charge on the product to finance the cost of recovery and collection of materials, or uniquely in the UK waste packaging legislation, system costs are financed by reprocessing of waste material (see below). The concept is illustrated in Figure 15.2.

This concept has been implemented in a number of countries, often on a voluntary basis, whereby industry negotiates agreed targets for waste prevention and recycling with government, and is then left to implement these in the most cost-effective manner. All voluntary schemes worldwide have been negotiated between industry and government on the understanding that if a satisfactory agreement is not reached or if agreed targets are not met then a mandatory scheme will be introduced. However, within the European Union (EU) the introduction of legislation such as the Packaging and Packaging Waste

Table 15.2 Different approaches designed to raise the level of recycling of household waste

Criterion	Germany (Wiesbaden)	Netherlands (Arnhem)	California (Santa Monica)	Australia (Canberra)	Ontario (Region of Peel)
Principal recycling drivers	• Landfill disposal costs • Organic waste landfill ban • Consumer waste charges • Strong public relations	• Landfill ban • High cost of incineration	Volume-based billing system	Community awareness and endorsement	• Public awareness • Easy accessibility
Recycling targets	• Recycling of 80% of collected recyclable materials • Bio-waste – a maximum of 120 kg per capita • Bulky waste – 50% of generation to be collected	• At national level, 60% for reuse and recycling • Arnhem has set a target of 60% + 10%	Reduce amount of MSW by 50% by 2000, from a baseline of 1990	No waste by 2010	50% reduction of waste disposal on 1987 levels
Reached target?	Yes. Recyclate production of 210 kg per capita	No. 37% recycling achieved	Yes, but not sustained. 38% recycling achieved	No. 57% recycling achieved	No. Reduction of 29% of waste to disposal
Legal/regulatory	• Banning of organic landfill after 2005, phasing out of landfilling household waste by 2020 • Eco-labelling • Packaging Ordnance	• Landfill ban on household waste from 1995 • Producer responsibility for various waste streams	• Reduction of solid waste by 50% by 2000 • California Beverage, Container Recycling and Litter Reduction Act • State law on purchase of recyclates	None	Municipalities with more than 5000 residence are required to implement the Blue Box programme, parks and garden waste collection and municipal waste recycling depots
Markets for end products	• Compost: agricultural markets • Cork: insulating materials • Wood: chipboard and thermal treatment • Paper and glass: retailers	Markets are available for paper and compost. Paper price volatility is addressed by giving recovered product free	• Bottles/cans/paper: sold to Pacific rim • Yard/green waste: landfill cover, mulch • Tyres/wood: fuel for cement kilns and boilers	• Paper/compost: used by government • Aggregates/compost: construction firms and civil contractors	Large market in Toronto
Public awareness	• Quarterly newsletters • Collection diary for DSD yellow bags • Literature on recycling practice	Waste calendars	• Quarterly city newsletter • Cable channel providing information on services • Customers service line	• Literature dissemination • Media advertising	• Customer services line • Door to door visits • Literature dissemination

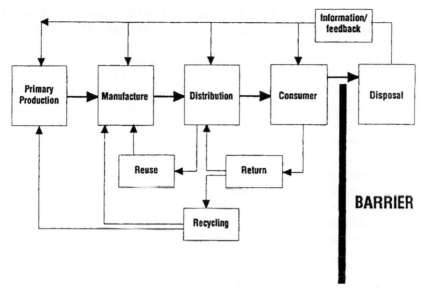

Figure 15.2 *The concept of producer responsibility*

Directive has generally resulted in voluntary quotas being replaced by national targets (see below). The concept has been applied to a range of products, including batteries, waste paper, end of life vehicles, *etc*. Two examples relating to waste packaging and end of life electrical and electronic equipment are described below.

15.4.2.2 Waste Packaging. One of the first and most studied producer responsibility schemes is that which applies to waste packaging. The best known producer responsibility scheme is the Duales System Deutschland (DSD) or 'Green Dot' packaging waste system, introduced in Germany through the 1991 Packaging Ordnance.[17,18] The Ordnance obliged distributors and producers to take back packaging from consumers. Quotas were set for collection, sorting and recycling of different packaging materials. An amendment to the Ordnance in 1998 transposed into German legislation the EU-wide Packaging and Packaging Waste Directive 94/62/EC, which came into force on 31 December 1994.[19]

The German system has come under criticism because the collected materials flooded markets in other countries, distorting local markets. In 1998 the European Commission served the German government with a Letter of Formal Notice calling upon Germany to justify the quotas contained within the Ordnance, and expressing concern over mismatches between German legislation and the EU Directive. In response to the German reply in 1999, the Directorate General Environment recommended to the Commission that the case be dropped. Efforts have continued in 2000 to seek a compromise and to reconcile differences in definitions.[19]

In 2000, Landbell GmbH obtained accreditation to compete with DSD AG for packaging waste in a number of Landers, through their 'Blue Bag' and 'Blue Sack' schemes. A 'Green Arrow' scheme proposed by another waste management firm as competition to DSD was disallowed by the Courts.[20]

In the UK, the Producer Responsibility Obligations (Packaging Waste) Regulations came into force in 1997 (subsequently amended in 1999), formalizing an earlier voluntary initiative and bringing into UK legislation the EU-wide Packaging and Packaging Waste Directive. The Regulations established a shared responsibility for each participant in the packaging chain, from supply to waste creation, as well as an incentive to reduce the cost of compliance by reducing the amount of packaging handled. Obligated businesses could implement their obligations independently, or join one of fourteen registered compliance schemes which certified compliance on behalf of its members and met their obligations in the aggregate. These compliance certificates are called Packaging Waste Recovery Notes (PRNs). Surplus PRNs can be traded, the intention being that the resale value of the certificates would provide reprocessors with an incentive to increase capacity.

Directive 94/62/EC also charged the European Committee for Standardization (CEN) with developing standards as a means of companies demonstrating that all packaging on the EU market complies with the essential requirements of the Directive. In 2000 five CEN standards were adopted by 19 European countries: the so-called umbrella standard or guidance document (EN 13427), the prevention standard (EN 13428), the reuse standard (EN 13429), the recycling standard (EN 13430), the energy recovery standard (EN 13431) and the composting standard (EN 13432).

15.4.2.3   The Proposed WEEE Directive.   The concept of producer responsibility is being extended to other product types. For example, the EU is proposing to introduce a Directive on Waste Electrical and Electronic Equipment (WEEE). The proposed Directive has three aims: to prevent waste from end-of-life electrical and electronic equipment; to set out measures for reuse, recycling and other forms of recovery; and to minimize the risks and impacts associated with the production and the treatment and disposal of end-of-life electrical and electronic equipment. The proposed Directive sets targets for reuse, recycling and recovery and seeks to ensure that products are designed to make recycling easier by removing hazardous substances, such as heavy metals, from the waste stream. When implemented, it will result in high recovery targets for electrical and electronic goods and strict take-back obligations for product manufacturers.

A key aspect to the proposed Directive is that it places the onus for the collection, recycling and reuse of equipment firmly on the electrical and electronic equipment (EEE) producers. The Directive will affect all players within the electronics industry including producers, importers and distributors. The draft Directive is an example of the increasing tendency of legislation to address the totality of the waste management cycle rather than with one or other specific aspects, and includes the following:

- Minimizing the use of dangerous substances, preparations and certain types of plastics. Certain substances will have to be phased out completely by 1 January 2004.
- Reducing the number of types of plastic in equipment and setting a minimum level for generic identification and marking of plastic products.
- Promoting design for recycling/dismantling (with the Commission developing European standards).
- Promoting the use of recycled materials in new products. Producers will be responsible for the development of collection procedures associated with end-of-life goods. The recycling targets range from 70–90% of end-of-life goods (by weight), depending on the product category.

Some countries have anticipated the extension of producer responsibility to WEEE goods. For example, since January 1999 a Decree on the disposal of so-called 'white' goods (such as kitchen equipment) and 'brown' goods (such as hi-fi equipment) has been in force in The Netherlands. The Decree requires manufacturers and importers to set up collection and recycling systems for electrical and electronic goods and appliances used in private households and in industry, from the point at which they are discarded. The scheme is financed through funds raised by means of a disposal levy, charged to consumers, and is implemented through The Netherlands Association for Disposal of Metalectro Products (NVMP), a body created by manufacturers and importers to manage the collection and recycling system. Austria, Germany, Spain and the UK are also considering or have launched WEEE collection and recycling initiatives.

15.4.3 Eco-labelling

Eco-labelling schemes for products are designed to encourage consumers to purchase environmentally friendly (including low waste) products. Labels may help to facilitate recycling for materials which would otherwise be difficult to segregate (*e.g.* different types of plastics). Eco-labels seek to give a product a rating for environmental impact and assist consumers in making environmentally based purchasing decisions and to educate consumers about environmental issues. Eco-labelling helps to support and reinforce other policy measures. For example, if a product charge were introduced, an eco-label could briefly explain the purpose of the product charge. The earliest eco-labelling schemes date from the late 1970s – for example, the Blue Angel scheme in Germany, introduced in 1977. Eco-labelling schemes can be found in most regions, ranging from the Green Seal scheme in the USA, to the Nordic Swan scheme in the Nordic countries, the ECOMARK scheme in India, the Green Label scheme in China, the Green Mark Program in Taiwan, and the Eco-Mark scheme in Japan. The Forest Stewardship Council (FSC) scheme, launched in 1993, applies internationally. The FSC evaluates, accredits and monitors forest products and their sources. Currently, 60 forests covering 3 million hectares in 16 countries have been certified, including nine sites or organizations in the UK.

Within the EU, Council Regulation EEC 880/92 set up an Eco-Label Award Scheme in 1992, with the objective of establishing an easily recognizable label for all Member States. A major revision was proposed in 1996, introducing a graduated label, and an independent European Eco-label Organization (EEO). The new scheme was formally recognized in 2000. Further revisions to the scheme are being discussed, including a provision to allow national eco-labelling schemes to exist alongside the EU scheme, a move away from the use a full life cycle analysis approach to criteria development to the more flexible approach of life-cycle considerations, and extension of the scheme from manufacturers to retailers and service providers.

In recent years, there has been general criticism over the level of uncontrolled, dubious and confusing environmental information supplied to consumers on products. The Standards community have therefore been investigating different types of environmental claim. Under the umbrella of the International Standards Organization (ISO), national experts have developed standards related to both self-declared environmental claims and the formal 'eco-labels'. A terminology has developed related to the various types of environmental claim:

- Type I claims are based on criteria set by a third party. The awarding body may be either a governmental organization or a private non-commercial entity.
- Type II claims are based on self-declaration by manufacturers.
- Type III claims are based on product information on environmental aspects/impacts, but without comparing or weighting against other products.

The emerging policy/standards framework is broader than labelling, covering environmental claims made on product or packaging labels, through product literature, advertising, or other applications. International standard ISO 14021 covers Type II claims, ISO 14024 covers Type I claims, while ISO 14025 covers Type III claims.

15.4.4 Charges and Economic Incentives

Four types of economic incentive are examined in this chapter:

- Grants and subsidies
- Recycling credits
- Preferential purchase policies
- Taxes and tradeable permits

15.4.4.1 Grants and Subsidies. Grants in the form of low interest loans or one-off cash payments and/or tax allowances for investments may be used to encourage innovative projects on waste avoidance, waste minimization, collection and processing of recyclable material. This type of support is compatible with funding schemes for demonstration projects (such as the UK's Recycling

Programme operated by the Department of Trade and Industry) introduced at the inception stage before the innovation becomes a commercially viable option in the free market system. Grants and subsidies tend to be aimed at supporting the initial capital outlay for the introduction of new technologies or for 'kick starting' other waste reduction initiatives. Normally, grants would not exceed 50% of the total capital required, though in some cases up to 80% of the project costs can be funded.

Several countries use grants or subsidy systems to encourage the development of recycling schemes or initiatives. Subsidy systems can be used to encourage both supply (encouragement of collection schemes) and demand (support for processing facilities and innovative project development) for materials for recycling. Examples include the subsidized introduction of home composting units in Canada, grants towards the establishment of separate collection programmes for household recyclables in Luxembourg, and grants to cover the capital cost of thermal recycling facilities for waste plastic in Japan. Ireland has a system of grants which cover up to 50% of the cost of approved recycling developments. Several countries also provide assistance to industry to adopt cleaner, low-waste, technologies.

Taiwan allocated over $400 million to finance low interest loans for pollution abatement investments by private sector companies. The incentive is the interest rate of 2% below prime. Similar schemes have been set up in Germany, Japan, Norway, Spain and USA.

15.4.4.2 Recycling Credits. A recycling credit can be defined as a payment to those who divert materials from final disposal for recycling. The recycling credit would reflect the saving in reduced collection, transfer and disposal costs. The aim of recycling credits is to encourage the recovery of materials in situations where the economics of doing so would be marginal. Typically, these situations include operations where the amounts of materials recovered are small compared with the effort required and where the value of the recovered material is sufficiently low to make the recovery process marginal in financial terms. Recycling credits are not intended to support recovery activities that are already economically viable or those that are completely untenable, in that, for example, there is no market for the recovered material.

Recycling credits are also aimed at the initial recovery end of the recycling process. Their role is to increase the amount of material removed from the waste stream in marginal situations, so that, later in the process, they are in sufficient quantity or are of sufficient quality to make the remainder of the process viable. They are not intended to support bulking up operations or, indeed, recycling industries.

Recycling credit schemes have been introduced in the UK, Canada and Australia. In the UK, a scheme was introduced under Section 52 of the Environmental Protection Act, and implemented by Waste Disposal Authorities (WDAs) paying a credit to Waste Collection Authorities (WCAs) operating recycling schemes, and also to third parties collecting waste for recycling. WCAs in turn can make a discretionary payment to recyclers of a collection credit commensurate with the saving in household waste collection costs. In 1992 the value of the credit was set at half the long-run marginal cost of disposal, but in April 1994 the disposal credit was raised to its full marginal value.

15.4.4.3 Preferential Purchase Policies. A concern with many recycling initiatives relates to identification of reliable markets for the recovered materials. Many recycling initiatives in the past have failed because either markets for the recovered materials could not be found or because they collapsed shortly after the scheme was initiated. Importance must therefore be given to stimulating the demand side of the recycling process to ensure that sufficient markets exist to absorb the recovered materials. Preferential purchasing schemes is one of the options for achieving this: under a voluntary participation approach, such schemes will relate only to the private sector.

The need for a preferential purchasing policy implies that the products targeted cost more than other products. If the products receiving the preference could operate under normal market conditions, the preference would not be needed. For example, if recycled paper is more expensive than paper produced from virgin pulp because there is greater demand for the latter, then, until the demand for recycled paper increases, the small market for recycled paper means that the prices are higher (due to dis-economies of scale). The objective is to encourage greater purchases which will stimulate price reductions and, in the long run, encourage further purchases of recycled paper.

Other examples of demand side initiatives include guarantees of a certain minimum price for waste paper, as practised by local authorities in The Netherlands, and the subsidizing of energy from the incineration of waste under the Non Fossil Fuel Obligation (NFFO) scheme in the UK.

15.4.4.4 Taxes and Tradeable Permits. These economic measures are used both to discourage specific waste management practices, and to influence environmentally beneficial alternatives. For example, Denmark, Germany and France have introduced taxes on materials destined for landfill, ranging from £2–8 tonne^{-1} in France, to £20 tonne^{-1} in Denmark and £10–40 tonne^{-1} in Germany. The taxes are intended to raise landfilling costs to better reflect the environmental impact of this form of waste disposal, while also discouraging producers from producing waste. For similar reasons, the UK introduced a two-tier landfill tax in 1996, of £2 tonne^{-1} for inactive or inert waste and £7 tonne^{-1} for all other waste. In 1998 the latter band was raised from £7 tonne^{-1} to £10 tonne^{-1}, with indications of a further rise by £1 per year till 2004. The charge is borne by waste producers and, to offset their costs, revenues from the Landfill Tax are used to reduce the level of employers' National Insurance Contribution. The tax is collected by Customs and Excise through the landfill operators. In parallel with this charge is a Landfill Tax Credit Scheme, in which landfill operators can claim credit against their landfill tax payments if they make a voluntary contribution of up to 20% of their annual landfill tax bill to an approved Environmental Body. Up to 90% of the contribution can be reclaimed. The contribution to the Environmental Body is intended to finance environmental improvement schemes approved by the Environmental Trust

Scheme Regulatory Body Limited (ENTRUST). By 2000, the estimated total amount raised by the tax was approximately £500 million, with 20% of this (£100 million) available through tax credits for environmental projects. In 1999 the Landfill Tax Regulations were amended to permit spending of tax credits on a wider range of projects, while also tightening the controls on the funding of environmental bodies.

The concept of a tradeable permit was introduced in Section 15.4.2.2, in relation to the PRN system associated with the packaging waste regulations in the UK. This market-based approach was intended to incentivize reprocessors to provide more facilities, and raise the collection, recovery and recycling rate of waste packaging. Under the scheme, about £100 million has been provided to reprocessors; however, insufficient funds from PRN sales appear to have been directed into investment in recycling projects. Alternative schemes designed to increase the collection and recycling rate have been proposed by the waste industry, but any intervention which attempts to influence the free market is likely to result in undesirable distortions similar to those encountered in the German DSD system (see Section 15.4.2.2). The UK is also considering tradeable permits for newspapers and magazines, in order to fund the implementation of mandatory recycling targets for newsprint.

15.4.5 Persuasion Measures

Several types of persuasion measures are available, either based wholly on economic incentives, or linking economic measures with mandatory 'command' policies.

15.4.5.1 Take Back Requirements. In the 'take back' system, householders are required to return the waste to the retailers to enter a parallel, private waste collection/recycling/disposal system rather than the public system (*e.g.* the Austrian scheme); or it could be a compulsory deposit–refund system (*e.g.* as in Taiwan); or it could involve a charge levied on either raw materials used or on products sold.

15.4.5.2 Product Charges and Product Taxes. Product charges are levied at the point of consumption on products that are harmful to the environment. The charges can be set at such a level as to achieve the desired reduction in usage, or to incorporate some or all of the costs of recycling or disposing of the product. The former is preferable.

Italy and Austria have introduced product charges on disposable carrier bags, in an effort to reduce waste and encourage consumers to use durable bags. Immediately after the introduction of the charge in Italy in 1988, the consumption of plastic bags declined by 5–10%. The long term reduction is estimated at between 20–30%. These reductions have been achieved through a combination of responses, namely: use of durable bags; reuse of plastic bags; substitution to paper bags; and, to a lesser extent, use of the biodegradable corn-based bag. It is understood that where consumers purchase plastic bags, they reuse them several times on subsequent shopping trips.

The use of product taxes is growing in many European countries as a means of raising the price of disposal or non-recyclable goods relative to less environmentally demanding alternatives. Taxes are commonly used as an incentive to set up deposit refund schemes, as is the case in Norway and Finland, where tax exemptions are allowed if a suitable return rate is achieved. Other countries considering product taxes as a way of influencing consumer behaviour and providing subsidies for recycled materials include Denmark (beverage containers and tableware), The Netherlands (PET soft drinks bottles), Belgium and Switzerland.

15.4.5.3   Refunds.   Deposit refund schemes are another policy instrument for waste reduction. For example, there have been several initiatives in Scandinavia, while in Germany and The Netherlands deposit schemes for plastic beverage containers were introduced. There are also examples of deposit refund schemes used for a number of other products including car bodies (Sweden), Norway), batteries (Denmark, The Netherlands, the US) and disposable cameras (Japan). In Korea an industrial deposit refund scheme has been set up whereby manufacturers are required to pay a deposit to the government which is refunded if, after customer use, the company collects and treats the product itself.

15.4.5.4   Compulsory Collection and Recycling.   Legislation to force local authorities to collect and recycle materials is becoming a widely used means of achieving a reduction in the quantity of MSW which is sent for disposal. Measures of this type proliferated in Europe during the early 1990s and usually focus on particular components of the MSW stream. For example, legislation enacted in The Netherlands and Austria obliges municipal authorities to set up source-separated organics collection programmes to collect and compost household organic waste. Similar legislation has been proposed in Denmark, Germany and Luxembourg.

Compulsory collection and recycling programmes usually have to achieve certain waste reduction or recycling targets (for example 50% diversion from landfill by 2000 in Ontario, or 100% diversion from landfill by 2010 in Canberra; see Table 15.2). The programmes are often funded by state or country-wide waste disposal levies.

15.4.6   Integrated Product Policy (IPP)

Integrated Product Policy (IPP) represents an important paradigm shift in policy development within the EU and its Member States. The concept of IPP differs from traditional policy approaches (which are fundamentally process focused, being source, substance and media specific) in that it is based on a life cycle perspective, and seeks to avoid shifting environmental problems from one stage of the life cycle to another. It further avoids the risk of conflicting life cycle legislation. IPP has been defined as "public policy which explicitly aims to modify and improve the environmental performance of product systems", and in the context of waste management, integrates different, but related, aspects of product consumption:[21]

- preventing and managing wastes generated by the consumption of products;
- introducing more environmentally sound products;
- creating markets for environmentally sound products;
- transmitting information up and down the product chain;
- allocating responsibility for managing the environmental burden of product systems.

Thus, many of the initiatives discussed individually above, such as waste minimization, waste prevention, chemicals policy, eco-labelling, and standardization, and producer responsibility can be integrated under the umbrella of IPP.

In 2000 the European Commission published a Green Paper on IPP, which set out the challenges to its successful up-take within the Community. Such challenges include:

- a successful transition from the traditional policy approach to IPP;
- the balance between successful global trade and environmental protection;
- constructive co-operation between all stakeholders;
- the provision of tools necessary for the integration and communication of the IPP concept.

These challenges are dependent on, amongst other things, the transfer of sound product information between the various members of the product chain, and particularly from producers and retailers to consumers. This, in turn, necessitates the need for a consolidated approach to environmental labels at a Community level, an issue that has been discussed in Section 15.4.3 above.

15.5 BULK WASTE REDUCTION TECHNOLOGIES AND FINAL DISPOSAL

Whatever success is achieved in reducing waste arisings and in separating materials for recycling, some waste will always remain. To achieve high waste reduction rates in terms of landfill demand, a technology component is required. Some options are listed in Table 15.3.

Other than the physical size and weight reduction technologies such as baling and separation, of the options listed in Table 15.3, waste fired power generation and composting are perhaps the most widely used. These two options are discussed in this section, along with landfilling, which remains the most common waste disposal option.

Table 15.3   *A list of bulk waste reduction technologies*

Options	Technologies
Size reduction technologies	• Baling • Pulverization/shredding • Homogenization/wet pulping
Weight reduction technologies	• Separation • Materials recycling facilities • Waste derived fuels
Waste to energy generation	• Mass burn incineration • Fluidized bed incineration • Combustion of prepared waste derived fuels
Other combustion technologies	• Aggregate/block production • Cement kiln firing • Wood burning power/CHP stations • Tyre burning power/CHP stations • Gasification • Pyrolysis
Biological systems	• Composting • Vermiculture • Hydrolysis • Anaerobic digestion

15.5.1   Mass Burn Incineration

Mass burn waste incineration has been practised as a waste management and volume reduction technique since the 1890s. It is an extremely effective bulk waste reduction technology, typically reducing waste volume by 90% and mass by around 70%. In terms of waste processing, mass burn incineration is a relatively simple option, with unsorted waste being fed into a furnace and, by burning, reduced to one-tenth of its original volume. Typically the only materials removed from the waste stream prior to burning are large bulky objects such as refrigerators and mattresses, or potentially hazardous materials such as gas bottles.

The combustion gases then typically pass through a boiler system to recover energy. The most flexible means of recovering energy from the hot gases is to produce steam for direct use (at lowered temperature) or for electricity generation. To generate electricity, superheated steam is passed from the boiler system through a turbine generator. Depending on the nature of the turbine, the resulting temperature of the steam can still be sufficient for its subsequent use in district or process heating. Around 550 kW h of electricity can typically be produced from a tonne of waste input, compared to 2300 kW h of electricity from industrial coal. Where both steam and electricity are produced by waste burning, the thermal efficiency of such schemes is between two and three times that of electricity-only production, although such co-generation schemes requires specialized turbine technology.[6]

The gases are then cleaned prior to discharge to atmosphere through a tall stack, in order that the discharge conforms to the strict emission limits laid down by national governments. Gas cleaning strategies aim to remove the following components:

- Acid gases such as nitrogen and sulfur oxides, and hydrogen chloride. A variety of wet and dry scrubbing systems can be applied, but in essence the gas cleaning method involves neutralization of the acidity by dosing with an alkaline reagent such as lime. Chemical methods of controlling nitrogen oxide emissions (so called De-NOX systems) have also been developed.
- Particulate matter, often associated with trace metals and semivolatile organic micropollutants. Fabric filters are generally used for this purpose. The particulate matter is retained in the ESP or the filter, and is periodically removed from the system as ash which requires land disposal.
- Organic micropollutants such as dioxins. Control of these emissions is achieved by control of combustion conditions, as well as by dosing in the gas cleaning train with activated carbon or other adsorbents. A fabric filter located further downstream removes the adsorbent material along with the reaction products of acid gas with the neutralizing agent. Simultaneous reduction of emissions of metals such as mercury and cadmium is also achieved.

In Europe, approximately 40 million tonnes a year of waste was incinerated in 1998 in a total of 295 waste to energy plants. This is expected to rise to approximately 62 million tonnes by 2006, with an incineration stock of 474 plants.[22] Given the intense public opposition to such facilities, this projection is likely to be overoptimistic: a further 34 plants are currently under construction in Europe.

The impact of prior materials separation and recycling on the calorific value of waste sent for incineration has been studied. UK research has indicated that the calorific value will fall if large quantities of only paper and plastics are recovered. However, if putrescibles are also separated from the waste stream, there is little effect on the overall calorific value.[6]

15.5.2 Other Thermal Processes

Gasification and pyrolysis are thermal treatment techniques that are being developed as alternatives to mass burn incineration. Unlike conventional combustion, pyrolysis involves heating the waste in the complete absence of air. The waste decomposes to produce a char-like solid, liquid hydrocarbons and a hydrocarbon-rich gas.[6] In contrast, combustion of the waste in excess of air in a conventional incinerator will produce a highly oxidized and degraded gas with very little residual hydrocarbons, and an inert solid residue in the form of ash.

The manner in which pyrolysis works offers some advantages over conventional incineration. Firstly, because no air is fed into the combustor, far less waste gases are produced and therefore the gas cleaning system can be smaller and hence less costly. Secondly, the waste itself must be pre-prepared to make it homogeneous and to remove bulky materials. This results in a smaller volume of waste being treated, and hence, overall, a smaller plant. Thirdly, in theory the solid, liquid and gaseous streams can be further processed into useful products (oil, soil conditioner, *etc.*) and hence there should be less material to discard to landfill. While these merits are attractive, the technology is still relatively unproven insofar as commercial size and applicability to MSW are concerned. Further, there is little practical experience in the refining and reuse of the products, especially in relation to liquid effluent.

Gasification operates along similar lines to pyrolysis, except that oxygen is fed in sub-stoichiometric quantities into the combustor. The waste is partially oxidized at high temperature to again produce a hydrocarbon-rich gas, and an effluent.[6] The potential benefits and disbenefits of gasification units are similar to those for pyrolysis.

15.5.3 Composting

Composting is essentially the controlled aerobic decomposition of putrescible material. Of the various methods of aeration, windrowing (mechanical or manual turning of the material) and forced aeration of the static pile are the most common methods used. However, closed systems in which the composting process is regulated in rotating drums or packed towers are increasingly being applied in order to reduce environmental nuisance from dust and odours. During composting, putrescible material is progressively broken down by microorganisms in a series of distinct stages. In the mesophilic stage, microorganisms begin to actively break down the organic material, the temperature of the composting material rising to around 50 °C in about two days. During the second, or thermophilic stage, temperatures begin to rise so that only the most temperature resistant microorganisms survive. As the microorganism population reduces, the composting material cools and anaerobic conditions may develop unless sufficient air is introduced. In the third stage, the material continues to cool and microorganisms begin to compete for the remaining organic material, in turn leading to breakdown of cellulose and lignin in the waste. During the final, maturation stage, levels of microbial activity continue to fall as the remaining organic material is broken down, and the microorganisms die off as their food sources deplete.[23]

Overall, maintaining the correct balance of oxygen, and therefore temperature, is crucial for the successful degradation of wastes. If the process becomes anaerobic, odour problems can be severe and microbial activity can cease. Temperature and moisture content can be monitored by electronic sensors. Depending on the type of organics and moisture levels in the material to be composted, volume and weight reduction levels for that waste fraction are typically in the range of 40–60% and 40–50% respectively.

Despite widespread use as a waste management strategy, in the past many composting schemes have been unsuccessful. Maintaining a consistently high quality and as such a marketable compost product, and producing it in an efficient and environmentally acceptable manner, were the key problems facing early schemes. However, composting is currently enjoying a resurgence of interest, and prospects for composting schemes look promising. The key to this change of fortune has been a switch to the composting of uncontaminated source separated organic wastes rather than of mixed MSW. Operational experience in the US and elsewhere with mixed MSW feedstock has indicated that the end product is often of poor quality and has limited end use. Recent composting schemes have therefore increasingly focused on processing source segregated kitchen and garden wastes, 'green' wastes from parks and gardens, and food wastes from the food processing industry and large commercial generators.

15.5.4 Landfilling

Landfills are the final destination for the residues from incineration and other treatment and processing options, as well as for the primary waste stream – in the UK about 90% of domestic waste and 85% of commercial waste is dispatched directly to landfill.[2,3] It is therefore a critical element in a waste management strategy since, despite the best attempts at waste minimization, recycling and recovery, its use is unavoidable. The challenge is to design and manage the process of landfilling in a sustainable manner so as not to leave a long term potential for environmental damage.

The process of landfilling consists of the following steps:[6]

- Preparation: waste may be processed prior to landfilling, for example by shredding, baling or compaction, which often takes place as part of the transfer system or in a MRF rather than at the landfill site.
- Waste placement: waste is deposited in phases or cells by tipping from the delivery vehicles, followed by spreading over the area by on-site plant. The deposited waste is often compacted by specialist plant or by the movements of the bulldozers to increase the density of the waste, to reduce subsidence and to preserve valuable void space. At regular intervals, typically daily, the waste is covered with an inert material to protect it from scavenging birds and animals, and to reduce odours and airborne dispersal.
- Landfill completion: after the void space within the site is filled, the landfill is capped with a low permeability layer to minimize the ingress of rainwater, and covered with soil to return the area to the surrounding landscape.

In practice, landfill restoration is a continuing process throughout the lifetime of the site: void space is typically utilized in a phased manner, with progressive contouring, capping and restoration.

Within the landfill, the constituents of the waste undergo biodegradation and stabilization. The infiltration of rainfall and surface waters into the waste mass, coupled with the biochemical and physical breakdown, produces a leachate which contains soluble components of the waste.

The breakdown products also include so-called 'landfill gas', which can be harnessed for its energy content. The maximum gas volume which is generated from the decomposition of organic matter is in the region of 350–400 m^3 per tonne of MSW, amounting to an average of about 6 m^3 per tonne per year. In reality, gas generation peaks in the early years and then gradually decreases as the waste ages, diminishing to non-commercial recovery levels after 30–40 years. The gas typically consists of 50–70% methane and 30–50% carbon dioxide with traces of nitrogen, hydrogen, oxygen, hydrogen sulfide and a range of trace organic compounds. Landfill gas utilization schemes can capture 50–60% of the gas released by the site: its combustion can potentially generate about 350–370 kW h of energy per tonne of MSW.

15.5.5   Environmental Considerations

15.5.5.1   Waste Incineration.   All bulk waste reduction technologies have the potential to produce gaseous, liquid and solid contaminants. For waste incineration, the key environmental issues are:

- Dust and odour from waste handling and storage.
- Ash management (grate ash and flyash from the gas cleaning system).
- Atmospheric emissions from the stack.

The potential effects of dust and odour from waste handling can be controlled by the following measures:

- Location of the waste unloading area within an enclosed building.
- Extraction of air from above the storage bunker for use as combustion air in the incinerator, where any odorous compounds and dust entrained in the air will be destroyed. This also results in a slight negative pressure within the building which draws air inwards, thus minimizing the escape of dust or odours.
- The use of dust suppression waste sprays at the waste tipping bays.

The first two of these measures are standard in all modern incinerator plants, and are not identifiable as separate environmental mitigation within a scheme.

Grate ash and flyash are produced at a rate of 20–30% and 3–4% of the waste input respectively.[6] Grate ash is essentially inert; flyash has relatively high levels of heavy metals and is classed as a hazardous waste in some countries. Due to its inert characteristics, the disposal of grate ash is not generally considered a significant pollution risk and it is often used for beneficial purposes such as road building, in construction materials, for intermediate cover at landfill sites, *etc.* Prior to disposal, the ash is normally scavenged for ferrous metals by magnetic separation: up to 10% of the ash may be recoverable ferrous material.[6]

The cleaning of combustion gases prior to their release to atmosphere has been discussed in Section 15.5.1. Additional information of the environmental impact of atmospheric releases can be obtained from other references.[24,25]

15.5.5.2 Composting. The key environmental issues of the composting process are as follows:[6,26]

- Fugitive emissions of litter and dust. These emissions may arise from wind dispersal of the waste feedstock, especially paper, during preparation and from compost piles.
- Odour, arising from trace organics produced in anaerobic conditions are permitted to occur in the composting process.
- Leachate generation, since the requirement for moisture in the composting process may lead to excess water and run-off from the composting pile, which may be contaminated by materials present in the waste.

Dust, litter and odour from waste material preparation can be mitigated by enclosure of the operation, both with respect to individual items of equipment (shredders, *etc.*) and by locating the process within a building with appropriate ventilation systems. Dust, litter and odour from the composting process can also be mitigated by enclosure of the activity, or by in-vessel composting. However, since these measures are relatively expensive due to the large volume of material which needs to be processed, windrow composting is most commonly applied. This is normally conducted out of doors, although Dutch barn buildings are sometimes used (*i.e.* roofed buildings with open sides). Regular wetting of the windrows and proper management of the composting process assist in minimization of dust, odour and litter.

The requirement to keep the compost pile wet increases the potential for leachate generation from excess water and run-off. The control of this leachate is important for the protection of water resources. Typical design features include organization of the windrows or compost piles to retain water inflow and avoid run-off, or the use of a concrete plinth equipped with a controlled drainage system to collect run-off.[27]

15.5.5.3 Landfilling. Of the potential releases and environmental effects of landfilling, the following have raised most concern:[28-30]

- Leachate: leachate arises from the moisture contained in the deposited waste, from the infiltration of water into the site, and from the biodegradation process itself. Escape of leachate, for example due to engineering failures of landfill caps, covers and liners, has been linked to contamination of water resources.
- Landfill gas: landfill gas, a mixture of methane and carbon dioxide, can cause damage to vegetation, and is also an explosion hazard. Methane and carbon dioxide are also greenhouse gases.
- Trace organics: a variety of trace organic compounds can be entrained with landfill gas, for example vinyl chloride, benzene, toluene, alkanes,

organosulfur compounds, *etc*. While many of these compounds are potentially toxic, their concentration in offsite air is generally too low to pose a threat to public health. Odour nuisance is potentially a more common problem.

- Litter, vermin, noise, *etc*. The nuisance aspects of landfills and their operation are potentially the most intrusive in terms of disturbance and disruption to the amenities enjoyed by the surrounding population. Careful consideration is given to operational work plans, traffic movements on and off site, the fitting of screens to reduce the visual intrusion and dispersal of litter, daily cover of the waste, *etc*. to mitigate against the possibility of nuisance.

Landfilling provides a method of reclaiming existing excavations as well as the development of new landforms. Hence, it can be used to return unproductive land to beneficial use. However, after landfilling operations are complete and the site has been capped, the *in situ* processes of biodegradation continue for a significant length of time, measured in decades. Hence, the generation of leachate and landfill gas also continues, as does the potential for offsite migration of these releases. Post closure management of the landfill is therefore a key consideration: typically this takes the form of the installation of a landfill gas utilization/control scheme, provision for leachate collection and treatment, and regular monitoring of releases from the site.

15.6 INTEGRATED WASTE MANAGEMENT STRATEGIES

15.6.1 Revisiting the Waste Management Hierarchy

In order to achieve an integrated approach to waste management, the following need to be included in any strategy:

- Policy and other measures to encourage the avoidance and minimization of waste.
- Policy and other measures to encourage the recovery, recycling and reuse of materials that would otherwise enter the waste stream.
- Adoption of bulk waste reduction technologies which will effectively reduce the volume of materials remaining in the waste stream after the above measures have been put into place, thus minimizing the amount of materials requiring final landfill.

While there are options within each of these categories, a strategy which omits the inclusion of measures from any one complete category is likely to achieve much lower levels of overall waste reduction than one which uses a more integrated approach. In conceptual terms, what an integrated strategy seeks to achieve is to convert a linear transference of materials through the producer-user-disposer chain into a series of circular and as far as possible, closed systems where beneficial aspects of the waste (materials, energy, *etc.*) are drawn out at each stage.

However, the phrase 'waste management hierarchy' may suggest that, after waste prevention and minimization, the remaining options invariably represent more environmentally acceptable solutions and, conversely, options such as landfilling invariably represent less environmentally acceptably and less sustainable solutions. This view could lead to anomalous and unbalanced waste management strategies which take insufficient account of local situations and economic sustainability.[20-22] For example in a predominantly rural area in which MSW is generated in sparsely populated and widely separated conurbations, long transport hauls to a central composting, or waste to energy incineration, facility may not represent an environmentally or economically preferable option relative to local landfilling if the cost, energy and emissions relating to transportation are taken into account. Reusable containers need to withstand repeated cleaning and handling, and therefore require more material and energy in their manufacture than do single-trip containers.[31] The overall energy consumption of a bring system which involves the householder in a 2 km car ride to a Civic Amenity site to deliver recyclables is over double that of a basic system involving non-segregated kerbside collection followed by landfilling.[32]

In order to optimize the overall waste management strategy, both with respect to the environment and to economics, it is necessary to address the entire strategy in a holistic sense. The final section of this chapter discusses the concept of life cycle assessment (LCA) and its role in the optimization of waste management strategies.

15.6.2 Principles and Stages of Life Cycle Assessment (LCA) (see also Chapter 16)

Life cycle assessment (LCA) is an environmental management technique in which the inputs and outputs of an activity are systematically identified and quantified from 'cradle to grave'; that is, from the extraction of raw materials from the environment to their eventual assimilation back into the environment. The inputs and outputs include raw material and energy consumption, emissions to air and water and the production of solid waste. These environmental flows are then assessed in terms of their potential to contribute to specific environmental impacts which might include global warming, acidification, ozone depletion, eutrophication and toxic effects. By taking the comprehensive life cycle approach, one can identify the environmental advantages and disadvantages of alternatives with the assurance that all their upstream and downstream consequences have been taken into account. Comparisons made between competing systems can highlight counter-productive initiatives, which might at first sight appear to be worthwhile, but in reality merely shift an environmental burden to a different part of the waste management chain.

The process of life cycle assessment is generally regarded as consisting of four distinct activities, as shown in Figure 15.3;[33]

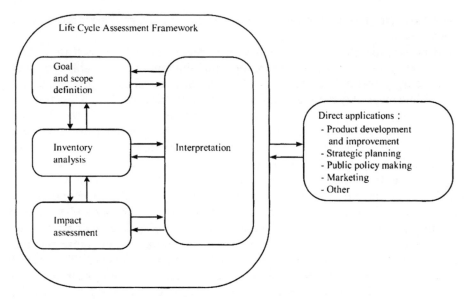

Figure 15.3   *Framework and stages of Life Cycle Assessment*

- *Goal Definition and Scoping* which defines the purpose and scope of the study and sets out the framework in which it will carried out.
- *Life Cycle Inventory Analysis* which is the quantification of environmental burdens throughout the life cycle of the product, process or activity.
- *Life Cycle Impact Analysis (LCIA)* which involves assessing the effects of the burdens identified in the inventory, and is often divided into:
     classification: grouping burdens into impact categories;
     characterization: quantifying the significance of burdens in each category;
     valuation: assessing the relative importance of the categories.
- *Life Cycle Interpretation* in which the results are assessed and applied.

The life cycle process is iterative, with preliminary results prompting a reassessment of goals and fresh data collection.

Typically a waste management system would be described in terms of the management of a quantity of waste of a given composition, which allows the comparison of alternative systems which might perform this service in very different ways (*e.g.* systems based on recycling, incineration or landfill). A schematic diagram of the elements of a life cycle inventory of a generic system is shown in Figure 15.4.

The system is defined as part of the Goal Definition and Scoping Phase of LCA, but in waste management studies might include, for example, the collection, transport, treatment and disposal of wastes, together with the recovery of materials and energy and there replacement for virgin materials and fuels.

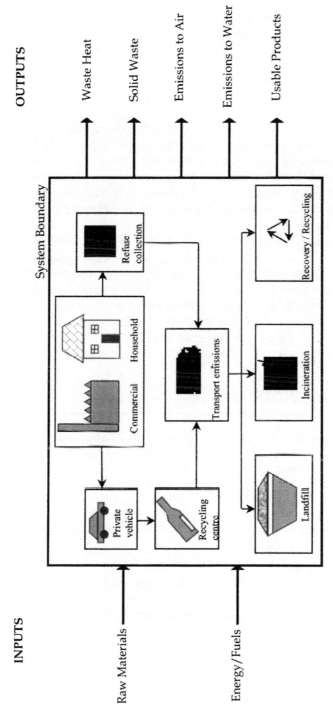

Figure 15.4. *Environmental effects of waste management systems*

15.6.3   Waste Management Strategy for Gloucestershire

An example of the application of LCA in aiding the decision-making process is the study of alternative waste management options conducted by the County of Gloucestershire in the UK.[34] The base case and alternative waste management options are described in Table 15.4.

Environmental burdens to air and water were calculated for the base case and for each of the options. For the baseline scenario, the largest emissions to air were of carbon dioxide and methane, the former primarily from the degradation of organic material in the landfill (with smaller contributions from collection, transport and sorting activities) while the latter derived solely from landfill. Smaller amounts of acid gases and heavy metals were released from the composting and landfilling stages. Discharges to water comprised BOD, COD, organics and heavy metals, primarily as leachate from landfilling.

Carbon dioxide and methane emissions are compared in Figure 15.5, together with net energy savings.

Methane emissions decreased with decreasing dependence on landfilling, while carbon dioxide emissions increased with increasing dependence on incineration, the lowest emissions being for the non-thermal options 1 and 3 and the highest emissions for option 2, without energy recovery. Net energy savings were highest for the incineration options with energy recovery. Overall, options 4 and 5 were regarded as the most suitable scenarios, achieving a landfill diversion rate of 50–51% relative to 6% for the baseline, and a 21–22% recovery rate for materials. Option 4 had the lowest energy consumption and

Table 15.4   *Waste management scenarios for Gloucestershire*[34]

Option	Description
Base case	• 92% of total waste to directly landfill • No thermal treatment, no energy recovery • Composting of 18% of organic waste fraction • 19% of household waste recycled
Option 1	Composting of 84% of organic waste (equal to 17% of the household waste stream); remainder to landfill; maximum utilization of landfill gas
Option 2	50% of total waste stream incinerated, remainder (including incineration residues) to landfill. No energy recovery
Option 3	Recycling of household waste through kerbside sort systems; remainder to landfill
Option 4	100% of unsorted waste sent to materials recycling facility, with overall recovery rate of 30%. Unsorted residues combusted in a gasification plant with energy recovery; residues to landfill
Option 5	Front end collection of recyclables through collection banks and recycling centres sent to materials recycling facility, with an overall recovery rate of 30%. Unsorted residues combusted in a gasification plant with energy recovery. Residues to landfill

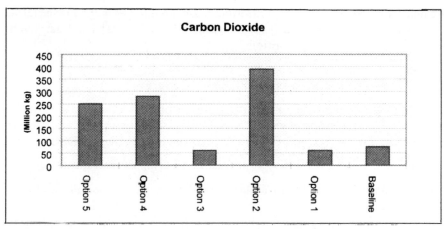

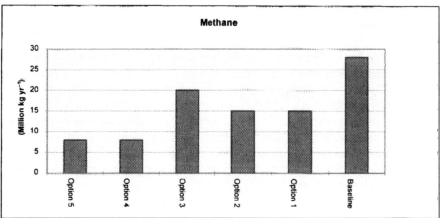

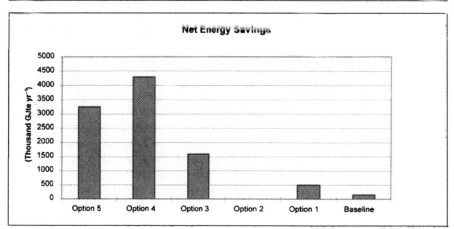

Figure 15.5 *Comparison of environmental performance of alternative waste management options for Gloucestershire*[34]

the highest net energy savings. Offsetting these advantages is the higher release of carbon dioxide, generated from the gasification of waste. The least favourable option was regarded as option 2.

15.6.4   Recycling and Recovery of Plastics in Germany

LCA can also be applied to specific components of the waste stream. For example, a German study assessed the following plastics recycling and recovery methods in the context of the Duales System Deutschland (DSD) system discussed in Section 15.4.2.2:[18,35]

- Mechanical recycling (production of bottles, cable conduits and waste sacks);
- feedstock recycling (gasification, thermolysis, utilization in blast furnaces, hydrogenation);
- energy recovery *via* combustion.

The recovery methods included collection close to the consumers, transport, sorting, preparation, and recovery through to production of useful end products. The sorting stage yielded a range of plastics: film, containers, mixed plastics and residues. Each recovery and recycling process was offset against an equivalent process for plastics manufacture from primary resources, which allowed comparisons of resource consumption and environmental pollution from different recovery methods producing different products. The lifecycle inventory took into consideration primary resources, energy requirements, emissions and waste disposal, while the impact categories assessed were global warming potential, acidification, eutrophication, resource depletion, energy use, and waste generation.

The study was able to deliver a detailed analysis of the relative merits of different operating systems for the DSD scheme. For example, the additional collection, sorting and transport required for mechanical sorting of plastics, while incurring a high cost, did not pose a sufficiently significant environmental impact to rule out this process, since compensatory gains were made in terms of reduced consumption of resources, and reduced emissions in comparison to feedstock recycling and energy recovery. The critical factor was the degree of substitution of mechanically recycled regrind product for highly refined primary material. A substitution rate of 1 kg of regrind for 1 kg of primary granules provided an improvement in the recovery operation for consumption of energy and primary resources, global warming potential, acidification and eutrophication. However, if the substitution rate was significantly less, then mechanical recycling resulted in a net deterioration in environmental performance relative to the best feedstock recycling or energy recovery processes.

This raised the issue of the expenditure required within the waste management system to achieve the critical levels of substitution associated with break-even environmental performance. On the basis that plastics from DSD collections could not be completely recovered by mechanical recycling, a balanced recovery strategy recommended, combining mechanical and feedstock recycling with energy recovery of the waste plastics that could not be recovered.

15.7 CONCLUSIONS

Policy development in the arena of waste management is moving away from piecemeal approach towards a systems approach, in which push–pull effects arising from different parts of the life cycle of a product are addressed within a single, integrated policy framework. Following on from policy development, the trend towards formulating sustainable waste management strategies is now firmly established. Strategies are increasingly being developed as a combination of policy and other measures, with certain core policies driving the strategy forward. These now typically include mandatory producer responsibility combined with contract-bound voluntary agreements and incentive schemes.

Optimization of waste management plans in terms of their operational effectiveness has moved from purely technical and capacity considerations of individual unit processes to an assessment of the entire plan from a life cycle perspective, with particular regard to the overall environmental impact of the scheme. The use of formal tools, such as LCA in waste management planning, enables comparisons and optimization to be performed in a structured, transparent manner, open to examination and audit. This is particularly important if greater public participation is to be encouraged in seeking consensual solutions to the problem of waste management. The preparation of guidance documents, backed up by expert systems and software packages, will accelerate the wider application of LCA, while ensuring a consistent approach to the decision-making process.

15.8 REFERENCES

1. World Commission on Environment and Development, 'Our Common Future', Oxford University Press, Oxford, 1987.
2. Department of the Environment and the Welsh Office, 'A Waste Strategy for England and Wales', London, 1995.
3. Department of the Environment, Transport and the Regions (DETR), 'A Way with Waste. A Draft Waste Strategy for England and Wales', DETR, London, 1999.
4. R. E. Hester and R. M. Harrison (eds.), 'Issues in Environmental Science and Technology No. 2 – Waste Incineration and the Environment', The Royal Society of Chemistry, Cambridge, 1994.
5. R. E. Hester and R. M. Harrison (eds.), 'Issues in Environmental Science and Technology No. 3 – Waste Treatment and Disposal', The Royal Society of Chemistry, Cambridge, 1995.
6. P. T. Williams, 'Waste Treatment and Disposal', John Wiley & Sons, Chichester, 1998.
7. M. Kooijman, *Environ. Manage.*, 1993, **17**, 575-586.
8. F. L. Margai, *Environ. Behav.*, 1997, **29**, 769–792.
9. W. J. Bryce, R. Day and T. J. Olney, *J. Consumer Affairs*, 1997, **31**, 27–52.
10. H. E. Jacobs and J. S. Bailey, *J. Environ. Systems*, 1982, **12**, 141-152.

11. D. Birley and J. Hummel, 'Householder Participation in Recycling Schemes: Experiences from ERRA Programmes', European Recovery and Recycling Association, Brussels, 1996.
12. P. Tucker, *Environ. Waste Manage.*, 1999, **2**, 55–63.
13. Biffa Waste Services, 'Waste: A Game of Snakes and Ladders?', Biffa Waste Services, High Wycombe, 1995.
14. I. Boustead, 'Resource Use and Liquid Food Packaging', Incpen, London, 1993.
15. S. Leadbeater, *J. Waste Manage. Resource Recov.*, 1994, **1**, 13–18.
16. Environmental Resources Management, Oxford. Unpublished database of international waste management practice.
17. European Commission DGXI.E.3, 'European Packaging Waste Management Systems', Brussels, December 2000.
18. B. K. Fishbein, 'Germany, Garbage and the Green Dot – Challenging the Throwaway Society', INFORM Inc, New York, 1994.
19. Perchards, 'Packaging Legislation in Europe', Perchards, St. Albans, July 2000.
20. Perchards, 'Packaging Legislation in Europe', Perchards, St. Albans, 7 February 2001.
21. European Commission, 'Workshop on Integrated Product Policy', Directorate-General XI, Brussels, 8 December, 1998.
22. 'The European Market for Waste to Energy Plants', Frost & Sullivan, London, 1999.
23. L. F. Diaz, G. M. Savage, L. L. Eggerth and C. G. Golueke, 'Composting and Recycling of Municipal Solid Wastes', Lewis Publishers, Boca Raton, FL, 1993.
24. G. H. Eduljee, in 'Issues in Environmental Science and Technology No. 2 – Waste Incineration and the Environment', eds. R. E. Hester and R. M. Harrison, The Royal Society of Chemistry, Cambridge, 1994, pp. 71–94.
25. P. T. Williams, in 'Issues in Environmental Science and Technology No. 2 – Waste Incineration and the Environment', eds. R. E. Hester and R. M. Harrison, The Royal Society of Chemistry, Cambridge, 1994, pp. 27–52.
26. P. R. Bardos, 'Survey of Composting in the UK and Europe', Paper No. W93036, Warren Spring Laboratory, Stevenage, 1993.
27. P. Stenbro-Olsen and P. Collier, *J. Waste Manage. Resource Recov.*, 1994, **1**, 113–118.
28. K. Westlake, in 'Issues in Environmental Science and Technology No. 3 – Waste Treatment and Disposal', eds. R. E. Hester and R. M. Harrison, The Royal Society of Chemistry, Cambridge, 1995, pp. 43–67.
29. D. J. Lisk, *Sci. Total Environ.*, 1991, **100**, 415–468.
30. G. H. Eduljee, in 'Issues in Environmental Science and Technology No. 9 – Risk Assessment and Risk Management', eds. R. E. Hester and R. M. Harrison, The Royal Society of Chemistry, Cambridge, 1998, pp. 113–135.
31. J. Bickerstaffe, *J. Waste Manage. Resource Recov.*, 1994, **1**, 91–96.
32. F. McDougall, P. R. White, M. Franke and P. Hindle, 'Integrated Solid Waste Management – A Lifecycle Inventory', 2nd edn., Blackwell Science, Oxford, 2000.
33. Society for Environmental Toxicology and Chemistry (SETAC), 'A Technical Framework for Life-Cycle Assessment, and Guidelines for Life-Cycle Assessment: A Code of Practice', SETAC-Europe, Brussels, 1993.
34. J. Powell, N. Sherwood and T. Robson, *Environ. Waste Manage.*, 1998, **1**, 221–233.
35. Association of Plastics Manufacturers in Europe (APME), 'Life Cycle Analysis of Recycling and Recovery of Household Plastics Waste Packaging', APME, Brussels, 1996.

CHAPTER 16

Clean Technology and Industrial Ecology

R. CLIFT

16.1 INTRODUCTION: CHANGING PARADIGMS

A major development in reducing pollution and, at the same time, reducing resource consumption has been the realization that pollution is to be avoided rather than merely abated. In the USA, this realization has led to the drive for *Pollution Prevention*, which is defined as[1] "source reduction and other practices that reduce or eliminate the creation of pollutants through:

(i) increased efficiency in the use of raw materials, energy, water, or other resources; or
(ii) protection of natural resources by conservation."

The US Pollution Prevention Act (1990)[2] defines *Source Reduction* as "any practice which:

(i) reduces the amount of any hazardous substance, pollutant, or contaminant entering any waste stream or otherwise released into the environment prior to recycling, treatment or disposal; and
(ii) reduces the hazards to public health and the environment associated with the release of such substances, pollutants or contaminants."

The US National Commission on the Environment[3] subsequently made it clear that pollution prevention should be interpreted to include recycling and re-use of material which would otherwise be released as waste.

Internationally, the equivalent movement is *Cleaner Production*, defined by the United Nations Environment Programme (UNEP) as[4] "a conceptual and procedural approach to production that demands that all phases of the life cycle of a product or of a process should be addressed with the objective of prevention or minimization of short- and long-term risks to human health and to the environment".

411

Articulated this way, Cleaner Production concentrates on reducing polluting emissions, with resource use and material efficiency included only by implication. However, Cleaner Production introduces something which is not made explicit in Pollution Prevention, that it is necessary to consider the entire *life cycle*; *i.e.* the complete supply chain of materials and energy leading to the product or process.

Going beyond Cleaner Production, *Clean Technology** has been defined as[5] "a means of providing a human benefit which, overall, uses less resources and causes less environmental damage than alternative means with which it is economically competitive".

'Overall' means 'over the whole life cycle', as in the UNEP definition of Cleaner Production. However, the definition of Clean Technology includes two elements which go beyond Clean Production (and possibly Pollution Prevention, although this is not clear from the definitions quoted above). The first is that Clean Technology concentrates not on products or processes, but on the benefit which the product or process provides. This focus on meeting human needs rather than providing material artefacts links Clean Technology explicitly to Sustainable Development.[6] The second new element is the idea of economic efficiency; *i.e.* efficient use of resources which could be deployed elsewhere. Almost any product, process or service can be 'cleaned up' if enough resources are devoted to it; Clean Technology implies improvement through redesigning the product or process or rethinking the way the service is delivered. This underlines the distinction between *Clean Technology* and *Clean-up Technology* (or 'End-of-pipe Technology') which lies behind Pollution Prevention and is explored further below.

This chapter explores the themes which emerge from the new paradigms of Pollution Prevention, Cleaner Production and Clean Technology: Life Cycle Thinking, Clean Technology itself and the systematic use and re-use of materials and energy which has become known as Industrial Ecology. Life Cycle approaches are set in the broader context of the emerging field of Environmental System Analysis. The general idea and practice of Clean Technology are examined, with examples but not in terms of specific technologies or processes, because Clean Technology is a way of thinking rather than a set of hard technologies.[7] Recovery, re-use and recycling of materials are reviewed in the context of Industrial Ecology using the framework provided by Life Cycle Assessment.

16.2 ENVIRONMENTAL SYSTEM ANALYSIS

16.2.1 Economy and Environment

Clean Technology is concerned with obtaining maximum human benefit from the material and energy flows through the human economy, summarized in the

---

*It has sometimes been argued that Clean Technology should more correctly be termed Cleaner Technology, on the basis that no technology can be absolutely clean. However, given that the definition given here is already comparative, there seems little point using the longer word.

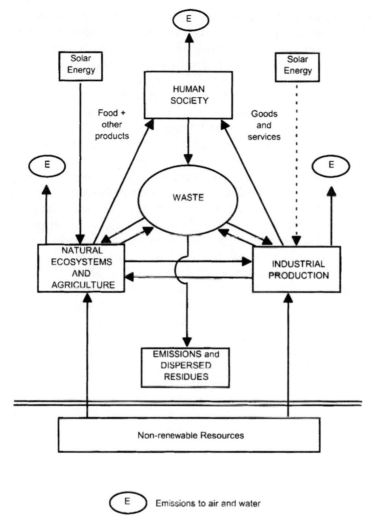

Figure 16.1 *Resource flows in the human economy*[6,8]

most general terms in Figure 16.1. Human society is maintained by food and other products obtained by agriculture and from natural ecosystems, and by goods and services provided by industrial production. The energy flows needed to bring about the material transformations derive in part from renewable sources (ultimately from incident solar energy), and in part from non-renewable fossil sources. From the industrial revolution, the process of industrial development has involved a shift from renewable to non-renewable energy sources.[9] This applies even to agricultural production, where the use of fossil energy is large, mainly for production of synthetic fertilizers.[10] A shift back to carbon-neutral energy sources, primarily to mitigate global environmental change due to accumulation of carbon dioxide in the atmosphere, is an inevitable conclusion from applying the Precautionary Principle[11] in the interests of Sustainable Development. Reduced use of fossil fuels also requires systematic increase in the efficiency of energy use, for example using the low-grade heat rejected from electrical generating stations for purposes such as space heating.[11] This is an example of the approach of Industrial Ecology, explored below.

Materials used in the human economy derive from agricultural production (including silviculture) which can in principle be sustainable, and also from non-sustainable exploitation of biotic and abiotic resources. In much the same way as for energy, material resource efficiency depends on the systematic use and re-use of materials, so that they remain within the broad human economy shown in Figure 16.1. This is the principle recognized by, amongst others, the US National Commission on the Environment in interpreting recycling and re-use as components of pollution prevention (see above). Therefore Figure 16.1 shows *waste* as part of the human economy. As long as they are not dispersed, wastes can in principle be re-used: obvious instances are the re-working of old 'mine tailings' and the quarrying of old landfills.[6] Waste from one sector may be used in a completely different sector; examples are the use of agricultural wastes in structural composites and the use of 'Thomas-meal' – waste from some steel-making processes -- as a fertilizer.[12] These are further examples of Industrial Ecology. Wastes are lost from the human economy to the environment when they are dispersed.

Systematic study of the interactions between the human economy and the environment (used here in the sense of 'that which surrounds the economic system'*) is the subject of the emerging field of *Environmental System Analysis*. The general approach is summarized in Figure 16.2. *Environmental interventions* pass between the environment and the part of the economic system which delivers a human benefit. Following the definition of Clean Technology and the definition of 'waste' implicit in Figure 16.1, material products remain within the economic system. Environmental System Analysis includes analysing the relationship between the service and the environmental interventions needed to provide it, along with the significance of the interventions. Different forms of Environmental System Analysis differ in the way in which the system boundary between the economic activity and its environment is drawn.

16.2.2 Life Cycle Assessment

According to the International Organization for Standardization[14] (ISO), an environmental *Life Cycle Assessment* (LCA) studies the environmental aspects and potential impacts of a product or process or service throughout its life, from raw material acquisition through production, use and disposal. The importance

---

*The word 'environment' has acquired a range and depth of meanings, including the one used here: 'physical surroundings and conditions'. This is the original meaning. The word entered the English language *via* the thermodynamic literature as a translation or transliteration of the French *environnement* in the work of Sadi Carnot. 'Environment' is about 20 years younger than another word which entered the language *via* thermodynamics: 'Energy', which appears to have been first used by Young in a lecture to the Royal Society in 1805.

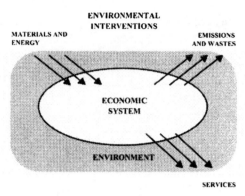

Figure 16.2 *Environmental System Analysis*[15]

of LCA in Clean Technology and Cleaner Production is that it provides a structured approach to defining and evaluating the total environmental load associated with providing a service.[15]

In the terms of Figure 16.2, the economic system is that within boundary 2 in Figure 16.3: all the operations along the material and energy supply chains, including transport between them, to obtain a complete 'cradle to grave' assessment of environmental interventions.

Where the product is not controlled following sale to the user, the analysis may be on a 'cradle to gate' basis, including the supply chain only up to the point of sale. In either case, analysis of the life cycle (boundary 2 in Figure 16.3) rather than just the immediate production process (boundary 1 in Figure 16.3) avoids the risk of apparently improving the process to prevent pollution but actually transferring it to some other point in the material or energy supply chain.[16] This has implications for regulating environmental performance which are discussed below. It should be noted that the focus of attention is not the material product which is the output of the activities indicated as 'processing' in Figure 16.3, but the use to which it is put; *i.e.* as in the definition of Clean Technology above, the benefit or service rather than the product.

While the complete system 2 must be examined to provide a full environmental assessment, this type of analysis necessarily means that the location of some of the environmental interventions must be undefined. This is most obviously the case for materials and energy produced in the background economic system and bought in through a homogeneous market so that the plants providing them cannot be identified.[17] Therefore LCA has developed primarily as a way of comparing technological routes; *i.e.* the classic application of LCA is to decisions like A in Figure 16.4, an application which includes the generic assessment of clean technologies. LCA should therefore be seen as complementary to other tools such as Environmental Impact Assessment (EIA) which have been developed to aid in selecting sites for a preselected technology; *i.e.* decisions like B in Figure 16.4. However, decisions which involve selecting a technology for a pre-identified site (C in Figure 16.4) or, in the most difficult

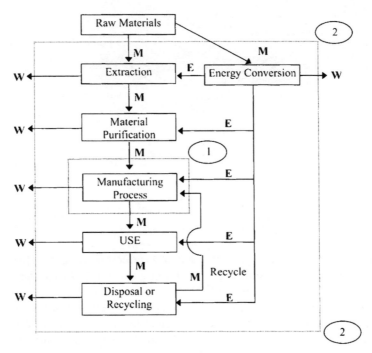

Figure 16.3 *System boundaries:[15] 1 – process or plant; 2 – life cycle assessment; M – material flow; E – energy flow; W – wastes and emissions*

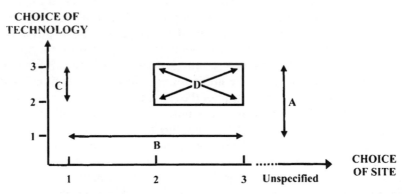

Figure 16.4 *Application of Life Cycle Assessment (LCA) and site-specific approaches such as Environmental Impact Assessment.[19] A – Established application of LCA; B – Established application of EIA; C,D – Decisions requiring both LCA and EIA*

case (D in Figure 16.4), selecting both technology and site from amongst several options, require both LCA and a site-specific tool such as EIA.[18]

The methodology of LCA has developed so that a study passes through four phases, shown in Figure 16.5.[14,20 24]

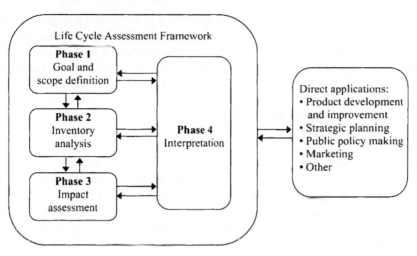

Figure 16.5 *Phases in a Life Cycle Assessment (after ISO 14040[14])*

16.2.2.1 Goal and Scope Definition. It is essential at the outset to define why the LCA is to be carried out, and what decision is to be informed by the results. This involves defining the system boundary, whether 'cradle to grave' or 'cradle to gate', to ensure that no relevant processes are omitted. Ways of defining the system boundary to facilitate or simplify the analysis are discussed below in the context of waste recycling.

When LCA is used to compare alternative ways of providing a service, the basis for comparison is the *Functional Unit*, defined as a number of 'units of service'. To take an obvious example, in comparing alternative packaging materials or systems the basis for comparison would be a common number of units of specified volume of a beverage delivered to the consumer, rather than a common mass of packaging material.

16.2.2.2 Inventory Analysis. The next phase identifies and quantifies the materials and emissions crossing boundary 2 in Figure 16.3, *i.e.* the environmental interventions in this embodiment of Figure 16.2. The complete set of interventions, in the form of quantified exchanges with their chemical compositions, constitutes the *Inventory Table*.

In process engineering terms, Inventory Analysis amounts to compiling material and energy balances over the processes and operations making up the Life Cycle. However, Life Cycle Inventory Analysis differs from conventional process analysis in two respects. It is necessary to include trace emissions of substances with large actual or potential impacts even when the material flows are trivial. Second, many of the processes within the system boundary will be shared with other supply chains, so that it is necessary to find a rational basis for allocating the environmental interventions between the supply chains. This problem, termed *Allocation*, can be contentious, especially when the processes in question provide more than one function or product, *e.g.* references 17 and 25.

16.2.2.3   Impact Assessment.   The Inventory Table comprises a great body of detailed numerical information which rarely displays clearly the most important environmental impacts. Life Cycle Impact Assessment (LCIA) aims to evaluate and clarify the magnitude and significance of the potential environmental impacts of the system within boundary 2 of Figure 16.3. LCIA and Interpretation include several elements, some of which are held by ISO to be mandatory.[14] The approach is illustrated by Figure 16.6.

The first mandatory element is selection of a manageable number of categories defining resource usage and environmental *Impact Categories,* together with indicators for the categories and models to quantify the contribution of the different interventions to these categories. Table 16.1 lists the set of environmental impact categories most commonly used in LCIA.

The next mandatory element, *Classification,* identifies which of the interventions in the Inventory Table contribute to which of the classified impacts (see Figure 16.6). Thus, for example, carbon dioxide emissions only contribute to global warming, while chlorofluorocarbons (CFCs) and hydrochlorofluorocarbons contribute both to global warming and to depletion of stratospheric ozone. *Characterization* is the ensuing mandatory step, in which the contribution to each impact category is quantified by expressing the effect of each intervention in terms of an equivalent quantity of a reference species. Thus Greenhouse Warming Potential (GWP), for example, is assessed by expressing

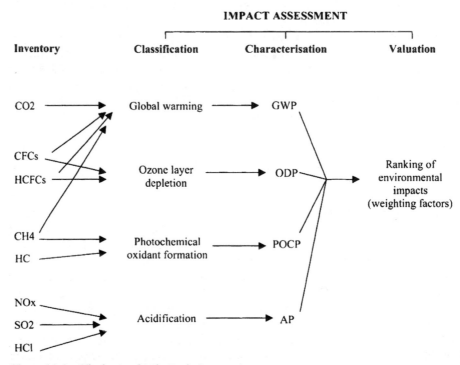

Figure 16.6   *The basis of Life Cycle Impact Assessment*

Table 16.1 *Common set of environmental impact categories used in LCIA*[20,23,24,26]

Abiotic Depletion Potential:	Extraction of non-renewable raw materials such as ores
Energy Depletion Potential:	Extraction of non-renewable energy carriers; can be included in Abiotic Depletion Potential
Global Warming Potential:	Contribution to atmospheric absorption of infrared radiation leading to increase in global temperature
Ozone Depletion Potential:	Contribution to depletion of stratospheric ozone, leading to increase in ultraviolet radiation reaching earth's surface
Human Toxicity:	Contribution to human health problems through exposure to toxic substances *via* air, water or soil (especially through the food chain)
Aquatic/Terrestrial Ecotoxicity:	Contribution to health problems in flora and fauna caused by exposure to toxic substances
Acidification Potential:	Contribution to acid deposition onto soil and into water
Photochemical Oxidant Creation Potential:	Contribution to formation of tropospheric ozone within photochemical smog (see also Chapter 8)
Nutrification Potential:	Contribution to reduction of oxygen concentration in water(or soil) through providing nutrients which increase production of biomass

the effect of each emission contributing to this impact category in terms of an equivalent mass of carbon dioxide, and summing these to obtain the total contribution to GWP per Functional Unit.

This approach is in principle straightforward for those impacts which are truly global – *i.e.* greenhouse warming and stratospheric ozone depletion – so that the nature and extent of the impact are independent of the location of the relevant emissions. It is more problematic for the other impact categories in Table 16.1, which are site-dependent. It was noted above that it is inevitable in a full LCA that the locations of some parts of the life cycle cannot be defined. Therefore the usual practice in Life Cycle Impact Assessment is to estimate potential global rather than actual localized impacts.[26] Thus, for example, Acidification Potential (AP) is assessed as the stoichiometric production of hydrogen ions by complete reaction of the emission, referred to sulfur dioxide as the reference species to give a value for AP in SO_2 equivalents. For some impacts, notably acidification, approaches have been proposed for inclusion of site-dependent impacts[27,28] but this is not yet established practice. On the positive side, the approach of comparing emissions on a common globalized basis lends itself logically to measuring and reporting the environmental performance of multinational companies with global operations.[29]

16.2.2.4   Interpretation.   What is done with the estimated impacts differs between different studies and different practitioners.[14,21,26] *Normalization* relates the impacts to reference values, such as the total contribution to each impact category by all the economic activity in the relevant country. Normalization helps to show the significance of the different impacts. Other forms of normalization can be used, for example to compare different products or different parts of the supply chain.[30]

Other elements, summarized in Figure 16.6, are more contentious. *Grouping* entails sorting and possibly ranking the (normalized) impacts in order of importance. *Weighting* aims to attach to each Impact Category a factor indicating its importance, opening the possibility of aggregating the different impacts to a single score. Weighting is sometimes expressed in terms of monetary 'damage costs', in which case the process may be called *Valuation*; this amounts to a form of Cost–Benefit Analysis.

The usefulness of Weighting and Valuation are much debated; critics regard such a process of aggregation as conceptually and theoretically ill-founded,[31,32] misleading[32,33] and unnecessary.[32,34] The view taken here is that LCA is to be seen as a tool for structuring and presenting information in a way that is intelligible and accessible to those making decisions;[18] *i.e.* it is one decision-support tool amongst others, not an algorithm which generates the 'right answer'. The kind of gross aggregation represented by Weighting and Valuation obscures information. Therefore aggregation beyond normalized impacts is to be avoided. Instead, it is preferable to recognize that environmental improvement usually requires balancing performance in non-commensurable impact categories; multi-objective optimization can then be used to inform decisions.[34-37] However, it must be recognized that this has certain implications in identifying the Best Available Technology (BAT) for any product, process or service.[38]

16.2.3   Legislation and Regulation (see also Chapters 20 and 21)

Figure 16.3 also serves to illustrate the way in which UK and European legislation has been developing. Prior to the Environmental Protection Act 1990 (EPA90), emissions from industrial processes in the UK were subject to single medium regulations covering releases to air, water or land. In 1976, the Royal Commission on Environmental Pollution[39] had pointed out that this approach is not the most effective for minimizing the total impact of pollution on the environment: control of one form of pollution can transfer impacts to another medium. EPA90 introduced the system of *Integrated Pollution Control* (IPC), which takes account of the combined emissions from a process. However, while the regulated emissions were integrated across all receiving media, IPC considered only emissions from the process itself, *i.e.* the system within boundary 1 in Figure 16.3.

Subsequent legislative developments, particularly in the EU, have generally extended the regulatory system boundary.[40] The Directive concerning *Integrated Pollution Prevention and Control* (IPPC) came into force in 1996, and has recently been implemented in the UK through the Pollution Prevention and Control Act. The IPPC Directive requires that processes be assessed on a wider basis than under EPA90, including the control of whole installations rather than individual processes and covering a broader range of environmental aspects. IPPC requires consideration of energy efficiency, raw material use, off-site waste disposal and site restoration. Moreover, IPPC favours reduction of pollution at source, as distinct from end-of-pipe abatement. Thus IPPC goes beyond the focus of IPC on emissions from the process (boundary 1 of Figure 16.3) to consider the principal environmental impacts integrated over the whole life cycle (boundary 2). In turn, this implies that Life Cycle Assessment should be used as the basis for establishing Best Available Technology guidelines.[38]

The development from IPC to IPPC is part of a progressive widening in the scope of industrial pollution control, illustrated schematically in Figure 16.7. Related developments, covering products rather than processes, include take-back legislation and product policy; these are discussed further in a later section of this chapter. Again, the trend is towards reducing pollution over the whole life cycle and requiring material resources to be contained within the human economy in the way illustrated by Figure 16.1. These developments are all driving towards adoption of the Pollution Prevention, Cleaner Production and Clean Technology approaches.

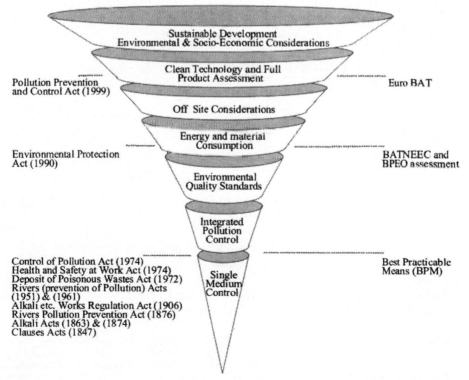

Figure 16.7 *The widening scope of industrial pollution control*[38]

16.3   CLEAN TECHNOLOGY

16.3.1   Clean-up, End-of-Pipe and Clean Technologies

16.3.1.1   Distinctions.   Following the definitions given in Section 16.1 of this chapter, the distinction between Clean Technology and Clean-up Technology can be illustrated by Figure 16.8. Each of the curves in Figure 16.1 represents the trade-off between the economic cost of providing a specified benefit or service and the associated environmental impact. The economic cost represents the conventional cost of providing the service – the concept of Clean Technology can be developed without invoking the neo-classical economic concept* of 'external costs'.[15] The environmental impact is estimated over the whole set of economic activities needed to provide the surface, established by Life Cycle Assessment as outlined above. Representation of environmental impact by a single parameter is purely for explanatory purposes; the arguments against reducing non-commensurable impacts to a single index were adumbrated in Section 16.2.2.

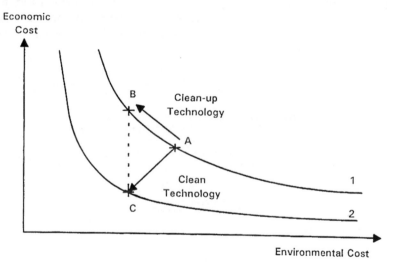

Figure 16.8   *Clean Technology and Clean-up Technology*[5]: *1 – Established technology for providing a service; 2 – Alternative 'cleaner' technology*

In fact, the curves are multi-dimensional surfaces; see Azapagic and Clift.[34–37] They form *Pareto surfaces* (also known as *decision frontiers* or *non-inferior surfaces*) which represent the set of design or operating conditions such that it is impossible to improve one objective describing economic or environmental performance without worsening at least one other objective.

---

*Conventional economics has difficulties with the whole concept of Clean Technology, on the argument that if it were possible to introduce a technology which complies with existing regulations and meets demand for a service more cheaply, then it would necessarily be implemented and would displace older technology. In reality, as distinct from economic theory, there are barriers to the adoption of Clean Technology which have been the subject of several careful studies.[e.g. 41,42]

Current standard practice might correspond to technology 1 (Figure 16.8) operated at the conditions corresponding to point A. It is possible in principle to reduce environmental impact by adding some End-of-pipe abatement device (bearing in mind that the End-of-pipe approach may introduce a new environmental problem to replace the emission being contained). This Clean-up approach by definition involves adding to the process, or treating or reprocessing the product which provides the benefit or service. Thus it involves moving around the curve to point B shown in Figure 16.8, increasing the economic cost to reduce the environmental impact. Following the definition given in Section 16.1, introducing Clean Technology involves shifting to a new way of providing the service which has less environmental impact and is also less expensive. The Clean Technology is represented by curve 2 in Figure 16.8, so that introduction of a Clean Technology corresponds to shifting from point A to point C.

16.3.1.2 An Example: Power Generation. To give a specific example, the service in question might be provision of electrical power distributed to industrial and domestic consumers *via* a grid. In the UK, particularly in previous decades, the dominant technology for power generation has been combustion of coal, usually in the form of a pulverized fuel. In the terms of Table 16.1, the dominant environmental impacts from such a process are in the categories of Global Warming (resulting from the carbon dioxide generated) and Acidification (primarily from sulfur and nitrogen oxides). Acidification impacts can be reduced by installing Flue Gas Desulfurization (FGD) as an End-of-pipe measure. However, this involves additional capital and operating costs, the typical feature of a Clean-up Technology. It also reduces the efficiency of the process, so that Acidification is reduced at the expense of increased Global Warming. Furthermore, it introduces a new set of environmental impacts associated with mining limestone for use in the FGD plant, and possibly disposing of the gypsum produced if this cannot be used, for example in construction materials.

The Clean Technology approach is to look for alternative ways to provide the service. If it is essential to use coal as the fuel, for example because of political priorities, then alternative means of converting the coal must be sought. Rather than a combustion process, the coal might be gasified in an Integrated Gasification Combined Cycle plant. This type of process typically has a higher conversion efficiency than pulverized fuel combustion, so that the Global Warming impact is reduced. As a further step, the fuel might be changed from coal to natural gas. This reduces Global Warming emissions even further and also reduces acid gas emissions, while the operating and capital costs can also be reduced. Therefore switching from coal to gas represents a Clean Technology shift (always provided that this is considered politically to be an appropriate use for gas). The technology is improved even further if the waste heat from the generating station is put to beneficial use (see Section 16.1); this is the approach known as Combined Heat and Power (CHP).

Further reduction in Greenhouse Warming might in principle be achieved by capturing the carbon dioxide from generating stations and 'sequestering' it in

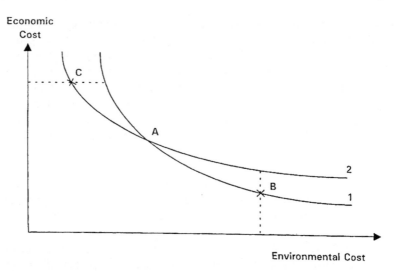

Figure 16.9 *Environmental regulation as a driver for introduction of Clean Technology[5]*
(see text)

rock strata rather than emitting it to the atmosphere.[11] This represents a significant extra cost: it is a Clean-up approach. The Clean Technology approach looks for energy sources which are closer to carbon-neutral, such as biomass as a fuel or wind and wave as direct energy sources. However, no generating technology is free of environmental impacts;[11] switching to renewable sources introduces new impacts, some of which – such as visual intrusion – go beyond the list in Table 16.1.

16.3.1.3 Regulation as a Driver for Clean Technology.   It is rarely the case that one technology is 'cleaner' than any other over the whole range of possible operation. Figure 16.9 shows schematically the case where two technologies are available, each optimal over a particular range of costs and environmental impacts. When expectations of environmental impact are relatively low – point B for example – technology 1 will be preferred. However, as standards of environmental performance rise, conditions move to the left. At point A, technology 2 becomes competitive. For yet higher environmental standards, like point C in Figure 16.9, technology 2 is the Clean Technology. The shift from fossil to renewable energy sources is an example of this kind of change.*

This is a particularly straightforward case, because one or two impact categories dominate. In general, however, given that curves like Figure 16.9 are really multi-dimensional surfaces, one technology rarely shows the best performance in all categories. Definition of Best Available Technology, with its possible implications for forcing the adoption and development of new

---

*Imposing a sufficiently high levy on carbon dioxide emissions would have a similar effect. However, the development here is written in terms of environmental regulations, to emphasize the point that Clean Technology can be understood without invoking economic instruments.

technologies, more usually requires trade-off between different categories, although this can be done on a case-by-case basis without universal weighting or aggregation of impacts.[37,38]

16.3.1.4 Improvement Threshholds. The need to consider the full material and energy life cycles in assessing improvements in performance has already been noted. Figure 16.10 illustrates this for a specific case: reduction of human health impacts from acid gas and dust emissions from a glass furnace.[38]

The End-of-pipe technology considered is 'dry scrubbing', achieved by injecting dry lime into the furnace flue gas, followed by a reaction tower and an electrostatic precipitator to remove the reaction products and furnace particulates. Because of the increased pressure drop, this system requires a fan for the flue gases. The fan requires additional energy consumption which is only weakly dependent on the degree of gas cleaning achieved. Therefore this clean-up system actually causes increased environmental impact overall; the overall Improvement Threshold is only reached if there is sufficient abatement of the emissions from the glass process itself to offset the emissions from energy conversion (see Figure 16.10).

The Clean Technology considered is a process involving chemical reduction of the NO_x through injection of hydrocarbon fuel immediately prior to the furnace regenerator. Use of additional fuel represents an overall increase in environmental impact, but this is much less significant than for the End-of-pipe approach. Therefore the Improvement Threshold is much lower: almost any abatement of emissions from the plant now represents an overall environmental improvement.

Figure 16.10 shows that, for this case, the overall (*i.e.* life cycle based) impacts favour the Clean Technology whatever the degree of abatement from the plant itself. It also shows schematically how attempting to achieve too much abatement can reach an upper Improvement Threshold. At very high levels of

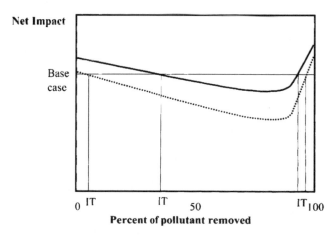

Figure 16.10   *Potential net impacts for two technologies:*[38] *(—) End-of-pipe; (- - -) Clean Technology; (IT) Improvement Thresholds*

pollutant removal the increase in material and energy use may outweigh the abatement. This kind of upper Improvement Threshold has also been observed for other processes.[36,43] At this point, in general dependent on the thermodynamic constraints on the process, the net impact can increase sharply (see Figure 16.10). Abatement approaching or going beyond the upper Improvement Threshold will require a completely new Clean Technology.

16.3.1.5   Process Inefficiencies and 'Profligate Environmentalism'.   Although this illustration of the need for careful consideration of regulatory limits and complete life cycles refers to a specific process, the qualitative features are more general. Therefore the relationship between cost and environmental impact is not so simple as Figure 16.8 suggests: the curve can actually take the form shown by Figure 16.11. Around point B, investment in process or equipment improvements generally improves environmental performance. However, the cost cannot reduce indefinitely even if environmental impact is not abated: at a point like C, problems arising for example from poor maintenance and material losses cause the cost of providing the product or service to rise. This phenomenon has been noted in Third World industries, particularly in mining.[44 46]

At the other extreme, around point A, increasing expenditure can lead to increases in environmental impact. The upper Improvement Threshold for the glass furnace in Section 16.3.1.4 illustrates how this can arise for a processing operation. The phenomenon is perhaps more common in waste management,[47] where it arises from the common assumption that material recovery and recycling necessarily involve less environmental impact than other forms of waste management. In this context, insistence on recycling even when it increases both economic cost and environmental impact has been called 'profligate environmentalism'.[48] The application of Life Cycle Assessment to

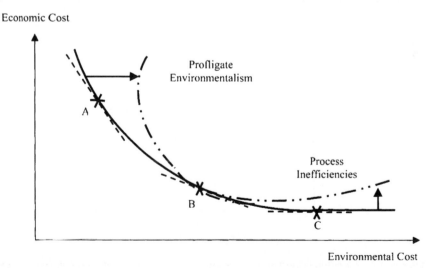

Figure 16.11   *Process inefficiencies and 'profligate environmentalism'*[5]

detect and avert 'profligate environmentalism' is considered in a later section of this chapter.

16.3.2 Cleaner Production

It is useful to distinguish between three levels of change in producing any particular product by a cleaner route:[49]

(1) Minimize waste arisings, effluent and energy consumption within the process or operating unit. This approach – source waste reduction – essentially remains within system boundary 1 in Figure 16.3.
(2) Modify the technology to use the same materials but more efficiently and with less energy. In effect, the system boundary in Figure 16.3 is extended to include the process along with upstream material preparation and energy supply.
(3) Redesign the process completely and change input materials. This implies changing the entire cradle-to-gate supply chain.

This classification is not sharp, but it provides a useful framework for discussing Cleaner Production. Substitution of the product by a different way of delivering its function is a more radical step, and is discussed in Section 16.3.3 below.

16.3.2.1 Waste Reduction. Because waste reduction at source can both improve environmental performance and save money, it usually constitutes a Clean Technology change in the sense of Figure 16.8. The financial risks are usually small, so that the likely return can in principle be estimated.[49] Therefore the barriers to embarking on a waste reduction programme should be low. However, there is little evidence that many companies do estimate the financial benefits and then embark on such a programme.[41,42] Nevertheless for many companies, systematic waste reduction is the first step in implementing Clean Technology. It can lead on to the other levels of change, particularly when an appropriate systematic approach is used.[50-52] For example, a waste audit can prioritize waste streams for further attention according to their environmental impacts, treatment costs and safety aspects rather than merely their quantity.[52] Losses during loading and unloading and, particularly for batch operations, from cleaning operations can be particularly significant, and sometimes relatively straightforward to contain.[52]

16.3.2.2 Technology and Raw Material Changes. If experience with the modified technology or something resembling it is available, then it may be possible to estimate the financial implications of a level 2 change.[49] Furthermore, the uncertainties are likely to be greater than for waste reduction, so that the barriers may be greater. However, this is not always the case; for example, the kind of NO_x suppression system introduced as a Clean Technology in Section 16.3.1.4 is relatively inexpensive and risk-free. Particularly if life cycle considerations are introduced at the stage of process selection and design, the overall environmental improvements can be substantial and the risk small.[35,36,53-55] Examples include matching abatement systems to the concentration of the relevant pollutant, finding the best balance between on-site energy generation and energy imports[35-37] and selection of the best solvents and reagents[56] and process configuration.[57] More conventional approaches such as integration of heat[58] and mass[59,60] exchange networks, and use of catalysts with improved selectivity and re-usability[52] generally apply to process improvements at this level.

16.3.2.3   Redesigning the Life Cycle.   Level 3 changes usually involve substantial financial uncertainty and require significant capital investment. Both of these aspects represent significant barriers to this level of Cleaner Production,[41] particularly where there has been relatively recent investment in updating existing plant.[42] Even so, some informative examples have been reported.

Replacement of mercury cells by diaphragm cells in the electrolysis of brine amounts to a redesign of the life cycle, with major implications in reducing emissions of toxins, most notably mercury; in many industrialized countries, electrolysis of brine has represented the largest use of mercury and the largest source of mercury emissions.[9] In some countries, of which Sweden is a leading example, firm dates have been set for the complete elimination of mercury – an example of legislation designed to force the adoption of a Cleaner Technology.*

An interesting example of systematic design of a product to cover the entire life cycle, including recovery and closed-loop recycling, is provided by a polyurethane material used primarily for cushioning in domestic and institutional furniture.[61] No single stage in the life cycle dominated the environmental impacts; in fact, the different stages contributed to different impact categories. However, raw material manufacture proved to be highly significant; for example, it contributed nearly 80% of the total energy consumption. Therefore waste reduction and modest technology change could not provide a significant Clean Technology shift; hence a level 3 approach was essential. By consulting stakeholders along the supply chain, the key life cycle design criteria were identified. Given the importance of feedstock production, the key improvement was chemical recycling; *i.e.* breaking the polymer down into basic chemicals: monomers or hydrocarbon feedstocks. By maintaining high purity and quality and avoiding contamination of the used product, it proved possible to recycle the polyurethane using a split-phase glycolysis process carried out in standard equipment to produce separate phases containing respectively the flexible polyol comonomer and aromatic compounds derived from the isocyanate in the foam. The recovered polyol could be used to replace virgin polyol completely. The isocyanate derivatives are treated with steam to produce a precursor for polyol used to make rigid polyurethane foam for insulation. This example illustrates some of the important issues in developing an industrial ecology for polymers, discussed later in this chapter.

---

*It has also been argued that legislation should go further and eliminate the chloralkali industry completely.[9] However, this raises other questions, including what to use in place of chlorine for disinfecting water and how to produce sodium hydroxide for items like soap which are arguably genuine needs.

16.3.3 Service Provision and Dematerialization

It was noted in Section 16.1 of this chapter that Clean Technology includes going beyond improving the environmental performance of a product to include better ways to provide the service which the product provides. A well-known example, shown in Figure 16.12, illustrates how focussing on the service rather than the material product can lead to a simple but very effective Clean Technology.

This example concerns industrial solvents used for cleaning metal surfaces, for example in preparation for surface coating. Solvents have habitually been used once, after which they become waste (Figure 16.12(a)). The simple measure which constitutes a Clean Technology is to collect and reprocess the solvent (Figure 16.12(b)). The reprocessing operation is simple and cheap. It is possible to return more than 90% of the solvent to the original use, so that the demand for fresh solvent is reduced by more than a factor of ten. That this is a Clean Technology, in the sense of Figure 16.8, is confirmed by the financial success of the company which first introduced the practice.[5,15]

This example illustrates many features common to the Clean Technology approach when it goes beyond Cleaner Production or Pollution Prevention. Figures 16.12(a) and 16.12(b) illustrate an instance where, without loss of service provision, it is possible to reduce resource consumption by at least the factor of ten which has been advanced[62] as necessary for human society to be sustainable. The technology involved is simple, in this case merely redistillation. The innovation lies in the commercial practice, which now amounts to leasing the use rather than selling the product. Many similar examples have been documented of step increases in resource efficiency through service provision rather than material sales.[e.g. 63] For manufactured goods, leasing can provide an incentive to extend service life.[64,65] It has commonly been observed that material inefficiencies arise more from custom and habit* than from lack of available technology.[e.g. 5,7,9,15,62–66] Amongst the barriers to this kind of change are the need to shift from an industrial economy where success is measured by throughput and its exchange value[9,64,65] and to find new ways to define ownership and liability.[64]

Even with the recovery and re-use approach shown schematically in Figure 16.12(b) there is still some generation of waste contaminated solvent. Putting this residue to further beneficial use represents a simple example of a further step in Clean Technology, towards an Industrial Ecology. The calorific value of the contaminated organic solvent is high, and it can therefore be used as a fuel, most notably as an auxiliary fuel in cement kilns (where the combustion conditions and rapid quenching are suitable for toxic waste combustion). This example also serves to illustrate some further barriers to this kind of development: the practice became the subject of an enquiry by the House of Commons Environ-

---

*The fact that simple opportunities like this went unexploited for years or decades represents further confirmation that the orthodox economic assumptions are unreliable. Whether, as has been claimed,[66] the shift from material to service represents a new form of capitalism is a different and altogether more debatable proposition.

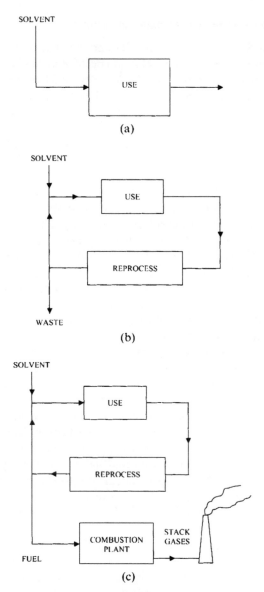

Figure 16.12 *Degreasing solvents as an example of implementing Clean Technology by concentrating on service provision:*[15] *(a)'Once-through' use; (b) recovery and reprocessing; (c) industrial ecology*

ment Committee.[67] The issues raised by the Committee included the need to consider all environmental impacts in defining Best Available Technology (see Sections 16.2.3 and 16.3.1.4 above), to involve all stakeholders in the decision process,[16] and to consider the effect of the practice on the viability of dedicated incinerators for toxic waste. The idea of Industrial Ecology is examined further below.

16.4 INDUSTRIAL ECOLOGY

The Environmental System Analysis approach, represented by Figures 16.1 and 16.2, emphasizes the need to reduce resource use and emissions while supplying the services which sustain human society and to contain materials within the human economy for successive cycles of use. The concept of Industrial Ecology extends beyond Integrated Waste Management and Recycling, but these topics provide a suitable starting point.

16.4.1 Material and Energy Recovery From Waste

16.4.1.1 LCA and Integrated Waste Management. Management of solid waste is discussed in detail in Chapter 15. Attention here is directed at how Life Cycle Assessment can be applied to compare the environmental performance of different waste management strategies. For this purpose, the basic cradle-to-grave approach shown in Figure 16.3 is adapted as in Figure 16.13. As always in LCA, attention to methodology and in particular to system definition is essential. The description here follows that adopted in the UK by the Environment Agency.[68,69] Others have adopted the same general approach but may express it in slightly different form.[e.g. 70-72]

A pragmatic distinction is made between:

Foreground: the set of processes whose selection or mode of operation is affected directly by decisions based on the study;
Background: all other processes which interact with the Foreground, usually by receiving or supplying material or energy. A sufficient (but not necessary) condition for a process or group of processes to be in the Background is that the exchange with the Foreground takes place through a homogeneous market.

This division helps to limit the amount of effort needed to compile the Inventory: Foreground processes are described by primary plant- and process-

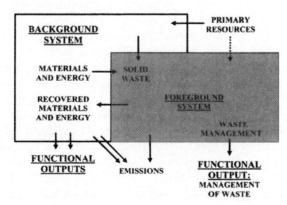

Figure 16.13 *System boundaries for application of Life Cycle Assessment to waste management*[17,19,68,69]

specific data, whereas Background processes are represented by generic industry-average data, usually taken from one of the many commercially available LCA databases. However, the geographical location of the Background processes is undefined, leading to the problems in Impact Assessment outlined in Section 16.2.2 above.

For waste management, the Foreground comprises the waste management activities themselves, from the location where the waste arises, through the beneficial recovery of materials and/or energy (to the point where the material or energy is returned to the Background economy) to final emissions to the environment (as in Figure 16.1). The environmental interventions arising from the Foreground activities are termed *Direct Burdens*. They include emissions from vehicles, from thermal treatment/combustion (gasification or pyrolysis) or from composting, uncontained landfill gas, and long-term leachate emissions from landfill. The resource usages and emissions arising from the Background activities are termed *Indirect Burdens*.

Particular care is needed in defining the benefits arising from recovery of materials or energy from the waste. The explicit assumption[68,69] is that recovery of materials and/or energy in the Foreground does not affect the demand for goods and services in the Background (except for materials and energy supplied to the Foreground). Thus paper produced from recycled paper, for example, is treated as a direct substitution for virgin fibre. The environmental interventions displaced by material or energy recovery are termed *Avoided Burdens*. Like the Indirect Burdens, the location of the Avoided Burdens cannot be defined. The total Life Cycle Inventory for the waste management scheme is thus:

	Direct Burdens	– arising in the Foreground waste management system
plus	*Indirect Burdens*	– arising in the supply chains of materials and energy provided to the Foreground
minus	*Avoided Burdens*	– associated with activities displaced by material and/or energy recovered from the waste.

In this way, LCA can provide a systematic framework for comparing different strategies for Integrated Waste Management. However, it cannot eliminate the uncertainties which remain inherent in assessing the environmental impacts of alternative waste treatment processes. Emissions from thermal treatment – particularly from combustion processes, which can in principle emit chlorinated species such as dioxins – are a notorious concern, with the debate over emission levels and their toxicity not always grounded in the best scientific information. Leachate emissions from landfills, particularly of metals, are also a concern, with the additional uncertainty that release occurs at some time in the future which can usually not be predicted at the time when the landfill is 'active' and receiving solid waste. Emissions from thermal treatment processes occur when the waste is processed, so that there is an argument that thermal processes can be controlled and regulated more effectively. Emissions depend on both the composition of the waste and on the treatment process and its operating conditions. For the specific case of chlorinated compounds such as dioxins, the emission levels depend strongly on the thermal treatment technology and its operation, and hardly at all on its chlorine content unless this is reduced to very low levels.[34] This leads to a distinction between *process-related* burdens which depend on the treatment, and *material-related* burdens which depend on the make-up of the waste.[34,68] It follows in turn that, even if solid waste is burned for energy recovery, there is still an argument for segregating out clean fuel streams such as paper and card for separate treatment.

16.4.1.2 Burn or Recycle? One of the issues which has proved most contentious in waste management is whether combustible waste should be recycled or should be incinerated with energy recovery (*i.e.* used as a fuel, although whether the fuel is correctly termed 'renewable' is questionable). Life Cycle Assessment provides a way to address this question.[73,74] Although, as noted above, LCA alone cannot determine the validity of fears of toxic impacts from incinerators, it can reveal the extent to which energy or material recovery from the waste really offsets the resource use and emissions associated with making fresh material.

The case of paper has been particularly widely studied and discussed, so that it provides an informative example of how LCA can usefully be applied to waste management by positioning the problem within the background economy. A consistent conclusion from the various studies is that the background energy system dominates the comparison between energy recovery and recycling.[74,75] Where energy recovered by burning paper displaces energy from fossil fuels, waste paper should be treated as a fuel. However, if paper recycling 'saves' biomass which can be used for other purposes, then recycling is preferred. This applies even if the biomass is used as an energy source, because recycling of waste paper requires less energy than virgin pulp and paper production.

Translating these conclusions into recommendations for waste management introduces further issues, dependent on the time scale and perspective. It is useful to make a broad distinction between three kinds of scenario:

(1) Immediate decisions, *e.g.* a consumer determining whether to send waste paper for energy recovery or recycling.
(2) Waste policy decisions, *e.g.* whether to invest in waste-to-energy plants (*i.e.* incineration with energy recovery) or to promote increased recycling.
(3) Energy sector decisions, *e.g.* whether to promote waste-to-energy or instead to promote other types of renewable energy.

Given the current dependence on fossil fuels in the UK, the answer to the short term question (no. 1) is that waste paper should be regarded as an energy source.[76] However, moving to the level of waste policy (scenario 2), this conclusion is open to the objection that building waste-to-energy plants leads to long-term dependence on waste as fuel, and therefore acts against waste minimization and recycling. This objection can be overcome if these plants are designed for co-firing of waste and biomass, with the waste to be phased out as biomass fuel becomes available. This conclusion also meets the requirements for development of the energy sector (scenario 3) provided that the waste- and biomass-to-energy plants are built for combined heat and power, with the heat output as the principal economic driver and with the electrical output used to back-up intermittent renewable sources such as wind.[11] This conclusion in turn reinforces the general point that Environmental System Analysis must be applied, so that renewable energy sources are examined in the context of the overall background energy system, not merely considered or advocated in isolation.[11]

16.4.1.3 Transport. Collection of geographically dispersed waste, or *reverse logistics* as it is sometimes called, is an important part of the Foreground of Figure 16.13. The associated fuel use and emissions, plus noise, congestion and traffic accidents, can constitute a large part of the Direct Burdens, and also of the economic costs of waste management. Even for relatively dense high-value products such as telephones, reverse logistics can dominate the environmental and economic assessment of re-use and recycling schemes.[30,77] This has important implications for recycling, especially of low-density low-volume waste, because it can outweigh any benefits from materials recovery or recycling. This point recurs in several of the examples in later sections.

16.4.2 Cascaded and Metabolized Use of Materials

16.4.2.1 Industrial Ecology. The conventional approach to recycling, discussed in Section 16.4.1, essentially assumes that the nature and composition of waste is immutable (although the quantity can be reduced by source waste reduction) and considers how to recycle materials or recover energy from waste. Because the problem is framed in this way, it represents a straightforward extension of Life Cycle Assessment. Industrial Ecology[78,79] looks beyond recycling to systematic management of the material and energy flows through the human economy represented by Figure 16.1,[80] including beneficial use of waste. Allen[52,81] has developed a systematic approach to identifying materials in waste streams which have sufficient value to be recovered economically, and the concentration at which recovery is economic. The analysis shows that metals in many waste streams are significantly under-utilized.

Figure 16.14 shows schematically one approach to Industrial Ecology, based on following one material as it passes through the economy. Following each application, the material may be *re-used*; *i.e.* it may be put through the same application with essentially no mechanical, physical or chemical reprocessing. Refilling a container is an obvious example of re-use. Alternatively, the material may be *recycled*; *i.e.* reprocessed mechanically, physically or chemically for the same application. An obvious example is reprocessing a glass or plastic container into a 'new' container. Because artefacts are damaged and materials inevitably become contaminated or degraded through cycles of use, they cannot be re-used or recycled in the same application indefinitely. They must therefore leave Application 1, to pass to a second Application with lower purity or

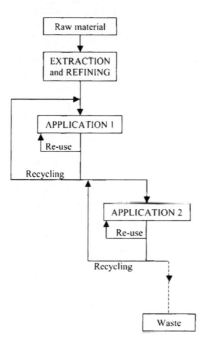

Figure 16.14   *Industrial Ecology: cascaded use of materials*

performance specifications. An example is provided by aluminium, which may be used first for beverage containers, become contaminated (for example with particles which cause 'pin-holes'), and then be used for applications such as engine components or structures where this contamination is acceptable. Thus a material may pass through a 'cascade of uses', as depicted in Figure 16.14. After several cycles of use, re-use and recycling, the material may be too contaminated or degraded for further use. It is then necessary either to reprocess it completely to return to higher specification, or to allow it to become waste. If it has energy value, then the last application might be as a fuel. In this case, the final waste is ash, the residue from combustion. The approach to using industrial solvents introduced in Section 16.3.3 above is a simple example of this kind of cascaded use.

A complementary approach to Industrial Ecology examines the relationship between processing and manufacturing operations which may be located in the same geographical area. Co-operation to improve environmental performance may start from common efforts to reduce waste and to identify and implement process improvements.[9] A more developed form of industrial ecology is provided by the operations around the town of Kalundborg in Denmark, where processes have been deliberately located together so as to use streams which would otherwise be wasted. The principal operations and flows in the Kalundborg 'eco-park' are shown in Figure 16.15. They include the use of low-grade heat, in this case for horticultural production, fish farming, and commercial and domestic heating. A CHP plant burning waste and providing heat, and

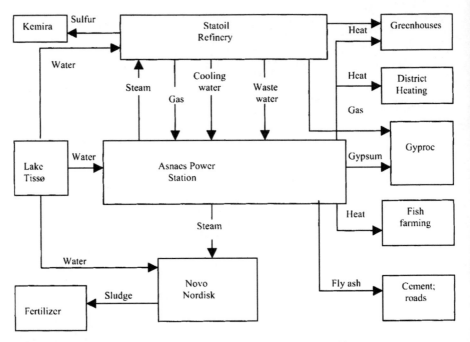

Figure 16.15   *An Industrial Ecosystem: the Kalundborg eco-park*[80]

possibly power, to a number of different industrial, commercial and domestic premises is a simple but common example of an Industrial Ecology.

16.4.2.2   An Example: Plastics.   The application of Industrial Ecology thinking to the use of plastics illustrates the importance of going beyond simplistic thinking about recycling. Figure 16.16 shows schematically a possible Industrial Ecology[8] for plastics. The material inputs are primarily hydrocarbons, which must be extracted and processed to provide the monomer feedstock for polymerization. The raw polymer must then be blended, usually with a range of additives, and formed into a plastic artefact. Following use, the artefact may be simply re-used (*cf.* Figure 16.14). If it has to be recycled, a number of possible routes may be available. If the material can be simply granulated and re-formed, then it may be possible to return it to the same or a similar use. However, if the material has become contaminated or commingled with other plastics, then mechanical recycling is not possible. For some materials it may be possible to depolymerize them to recover monomer which can be repolymerized. This is a possibility for acrylic polymers[57] in particular, even though it is not widely exploited. The example of polyurethane, discussed in Section 16.3.2.3 above, is a further example. If depolymerization is not possible or feasible, then the waste plastic can in principle be subject to chemical recycling (usually by pyrolysis) to recover a hydrocarbon stream for reprocessing, usually for blending into cracker feed. The final option is to use the waste plastic as a fuel, thereby displacing the use of other fuels (see Section 16.4.1.1). All the transfers in Figure

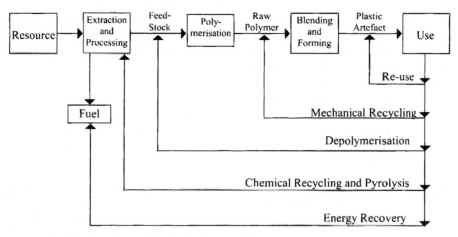

Figure 16.16 *An Industrial Ecology for plastics*[8]

16.16 involve transport, so that the environmental and economic costs of reverse logistics can be important (see Section 16.4.1.3) even for the 'inner' loops.*

Clearly for maximum service to be obtained from plastics through the kind of system shown in Figure 16.16, a well established infrastructure is essential. For metals such as steel, where economic drivers for recovery and re-use are well established, such an Industrial Ecology already exists.[82] For plastics, the driver is at present legislation (notably in Europe) and it is not yet clear how the Industrial Ecology will actually develop with this kind of driver.[83]

In general, the economic costs and environmental impacts increase in passing from the inner loop of re-use outwards to successive recycling loops. There may therefore be an economic and environmental incentive to modify the processing operations to enable simpler re-use and recycling to be used. The example of polyurethane (Section 16.3.2.3) illustrates how appropriate substitution of additives enables the waste material to be depolymerized. However, this approach is relatively new and examples are scarce. Moreover, depolymerization requires waste plastics to be segregated. This is feasible for large objects such as polyurethane mattresses, and is becoming feasible for smaller objects such as bottles. Selecting materials and additives to optimize resource use and minimize environmental impacts over a complete Industrial Ecology requires a decision-support system which incorporates LCA but goes further to examine more than one possible use.[84,85] This is a particular application of the approach known as Life Cycle Product Design.[86–88]

Therefore there remains inevitably a stream of mixed waste plastic, for which the only options are chemical recycling, energy recovery or landfill. Given the low density and low volume of mixed plastic waste, transport over long

---

*Figure 16.16 does not show production waste, which is most commonly recycled within the plant thereby avoiding any transport requirements.

distances would be a clear example of Profligate Environmentalism (see Section 16.3.1.5 above), in this case recycling regardless of the associated economic cost and environmental impact. Apart from relatively densely populated areas where arisings of plastic waste are concentrated, chemical recycling is unlikely to be a real option because it is a capital intensive operation, only viable on a large scale, and it must be located where the hydrocarbon product can be fed to a refinery or petrochemical complex. Referring to Figure 16.16, the next option is energy recovery, preferably in a CHP plant close to the source of the waste,[11] with only landfilling as an alternative. At this point, public acceptance rather than environmental impact determines the option taken.[89]

16.4.2.3   Manufactured Goods.   Life Cycle Product Design is particularly applicable to manufactured products.[86-88] Here the issues concern not just individual materials but potentially also components and sub-assemblies. Not surprisingly, well developed examples are to be found in sectors where leasing rather than selling is the established practice (*cf.* Section 16.3.3 above).

A particularly instructive example is provided by photocopiers.[15] The great majority of photocopiers are leased, and are therefore returned to the supplier after use. The return loop is therefore already closed, and the problems associated with reverse logistics have already been addressed. This also enables multiple-trip packaging to be used. A used machine is not treated just as scrap, but is recovered for re-use and recycling as shown in Figure 16.17. Lightly used machines may be partially dismantled, to remove worn or damaged parts, and returned for 're-engineering'. Machines which have received heavier use are disassembled completely. Components can then be fed back into the assembly line for re-use, after quality checking to ensure that they meet the same specifications as new components. Components which cannot be re-used, for example because they are more heavily worn or because they contain materials which are no longer used or permitted, may be recycled, typically passing back to the manufacturer who can blend the material with fresh material or reprocess it for other 'cascaded' applications.

Management of the kind of system shown in Figure 16.17 requires a decision-support model similar to that needed for plastics.[85] It also emphasizes that the manufacturer needs to see its own operations in the context of their overall life cycle, and to establish relationships not just with the immediate suppliers and customers but also with other organizations in the supply chain. Much as leasing can provide the driver for extending product life (see Section 3.3 above), returning the product to the manufacturer provides a driver to design for re-use and recycling. Well formulated and implemented legislation can promote this change in approach.

16.4.3   Legislative and Economic Drivers

It was noted in Section 16.2.3 that environmental legislation and regulation is progressively shifting towards a life cycle basis, widening the scope of industrial pollution control.[38,40,90,91] A similar trend is seen in extending producer

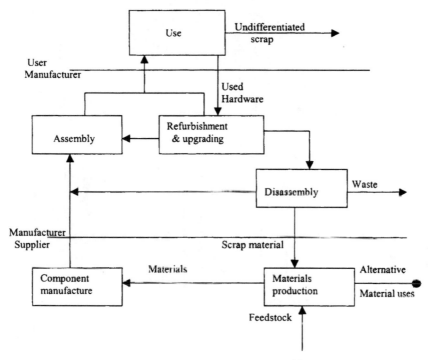

Figure 16.17 *'Metabolized' use of a manufactured product*[15]

responsibility to cover more of the production, use and waste management of a range of industrial products. Figure 16.18 illustrates this trend for the specific case of container glass, showing the scope of legislation as of 1998[92] under Integrated Pollution Prevention and Control (IPPC) which covers the supply chain from cradle to gate, and producer responsibility which covers the life cycle following use of the container. Provisions like these represent a new degree of holism in legislation and regulation. As shown schematically in Figure 16.7, the trend is to extend the scope of legislation and regulation towards covering the whole life cycle.

Potentially the most significant drivers towards development of the Industrial Ecology approach arise from 'take back' legislation which requires material products including packaging, electronic and electrical equipment and vehicles to be returned to the manufacturer or supplier following use.[30,93-97] While the obvious motivation for take-back is to remove products and materials from the waste stream, the greatest environmental benefits will arise if products, components and materials are re-used and recycled, as illustrated by Figure 16.17, so as to avoid environmental impacts and resource usages in the earlier parts of the life cycle. However, the economic costs and environmental impacts of 'reverse logistics' (*i.e.* the operations of collecting and returning used products and materials) counterbalance and can even outweigh the environmental benefits of re-use. This is most obviously the case for low-density low-

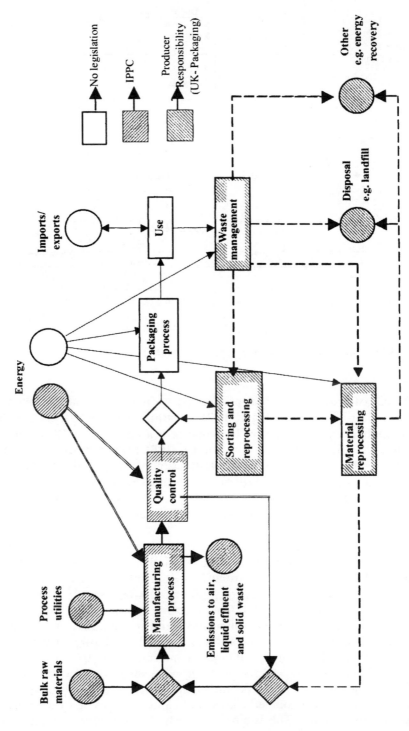

Figure 16.18 *Environmental legislation (1998) covering manufacture of container glass*[92]

value materials such as used packaging (see Section 16.4.2.2 above), but it can also apply for higher value products such as electronic equipment.[e.g. 30,77,95] Take-back legislation therefore needs to drive development of efficient transportation systems, making as much use as possible of return journeys (see Section 16.4.2.3 above), as well as development of the infrastructure for re-using and reprocessing products and materials. Imposition of arbitrary recycling targets which require dispersed waste to be collected and transported over large distances can be an example of the phenomenon of 'profligate environmentalism' illustrated by Figure 16.11. The 'proximity principle', which requires waste to be treated as close as possible to its point of arising, recognizes the environmental impacts of transport and traffic movements.

16.5 CONCLUSIONS: CLEAN TECHNOLOGY AND SUSTAINABLE DEVELOPMENT

Clean Technology is a general approach to providing the services on which human society depends. It goes beyond Pollution Prevention and Cleaner Production in concentrating on services rather than products. Clean Technology recognizes the need to consider the whole supply chains of materials and energy, and is therefore closely linked to Life Cycle approaches and Environmental System Analysis. Clean Technology is also linked to Industrial Ecology – a systematic approach to the use and re-use of materials and energy which looks beyond simple re-use and recycling.

All these approaches imply profound changes in the human economy, to reduce both resource use and emissions and waste. These changes are part of the broad agenda summed up by the idea of Sustainable Development:[98]

> Humanity has the ability to make development sustainable – to ensure that it meets the needs of the present without compromising the ability of future generations to meet their own needs.

Part (but only part) of the Sustainable Development agenda is reduced consumption of material resources.[9,62,63] The contribution of the Clean Technology approach to Sustainable Development lies in its emphasis on providing services with reduced material intensity.

16.6 REFERENCES

1. S. K. Sikdar, J. Drahos and E. Drioli, in 'Tools and Methods for Pollution Prevention', eds. S. K. Sikdar and U. Diweker, Kluwer Academic Publishers, Dordrecht, 1999, p. 1.
2. H. F. Habicht, 'EPA Definition of Pollution Prevention', US Environmental Protection Agency, Washington, DC, 1992.
3. NCE (National Commission on the Environment), 'Choosing a Sustainable Future: The Report of the National Commission on the Environment', Island Press, Washington, D.C., 1993.

4. L. Baas, H. Hoffman, D. Huisingh, J. Huisingh, P. Koppert and F. Newman, 'Protection of the North Sea: Time for Clean Production', Erasmus Centre for Environmental Studies, Erasmus University, Rotterdam, NL, 1990.
5. R. Clift, *J. Chem. Technol. Biotechnol.*, 1995, **62**, 321.
6. R. Clift, *Trans. IChemE, B*, 1998, **76**, 151.
7. R. Clift, *J. Clean Technol. (Korea)*, 1996, **1**, 34.
8. R. Clift, *J. Chem. Technol. Biotechnol.*, 1997, **68**, 347.
9. T. Jackson, 'Material Concerns – Pollution, Profit and Quality of Life', Routledge, London, 1996.
10. S. J. Cowell and R. Clift, 'Life Cycle Assessment for Food Production Systems', The Fertiliser Society, London, 1995.
11. Royal Commission on Environmental Pollution, '22nd Report: Energy – the Changing Climate', The Stationery Office, London, 2000.
12. E. Audsley (ed.), 'Harmonisation of Environmental Life Cycle Assessment for Agriculture', Final Report of EU Concerted Action AIR3-CT94-2028, DGVI, Brussels, 1997.
13. A. Azapagic and R. Clift in 'Proceedings of the European Workshop on Allocation in LCA', eds. G. Huppes and F. Schneider, SETAC-Europe, Brussels, 1994, p. 54.
14. ISO (International Organisation for Standardisation), 'Environmental Management – Life Cycle Assessment – Principles and Framework (ISO 14040) -- Goal and Scope Definition and Life Cycle Analysis (ISO 14041) – Life Cycle Impact Assessment (ISO 14042) – Life Cycle Interpretation (ISO 14043)', ISO, Geneva, 1997, 1998, 1999 and 2000.
15. R. Clift and A. J. Longley in 'Clean Technology and the Environment', eds. R. C. Kirkwood and A. J. Longley, Blackie Academic and Professional, London, 1995, p. 174.
16. Royal Commission on Environmental Pollution, '21st Report: Setting Environmental Standards', The Stationery Office, London, 1998.
17. R. Clift, R. Frischknecht, G. Huppes, A.-M. Tillman and B. Weidema, *SETAC-Eur. News*, 1999, **10**(3), 14.
18. S. J. Cowell, S. Hogan and R. Clift, in 'Life Cycle Assessment: State-of-the-Art and Research Priorities', eds. H. A. Udo de Haes and N. Wrisberg, Eco-Informa Press, Bayreuth, 1997, Vol. 1, p. 33.
19. A. Doig and R. Clift, National Conference on Life Cycle Inventory Analysis for Waste Management, Atomic Energy Research Establishment, Culham, 1995.
20. R. Heijungs (ed.), 'Environmental Life Cycle Assessment of Products: Backgrounds and Guides', Centre of Environmental Science (CML), University of Leiden, NL, 1992.
21. F. Consoli (ed.), 'Guidelines for Life-Cycle Assessment: A 'Code of Practice'', Society of Environmental Toxicology and Chemistry (SETAC), Brussels and Pensacola, 1993.
22. J. B. Guinée, H. A. Udo de Haes and G. Huppes, *J. Cleaner Prod.*, 1993, **1**, 3,
23. J. B. Guinée, R. Heijungs, H. A. Udo de Haes and G. Huppes, *J. Cleaner Prod.*, 1993, **1**, 81.
24. L.-G. Lindfors, K. Christiansen, L. Hoffman, Y. Virtanen, V. Juntilla, O.-J. Hanssen, A. Ronning, T. Ekvall and G. Finnveden, 'Nordic Guidelines on Life-Cycle Assessment', Nordic Council of Ministers Report Nord 1995:20, Copenhagen, 1995.
25. G. Huppes and F. Schneider (eds.), 'Proceedings of the European Workshop on Allocation in LCA', SETAC-Europe, Brussels, 1994.
26. H. A. Udo de Haes (ed.), 'Towards a Methodology for Life-Cycle Impact Assessment', SETAC-Europe, Brussels, 1996.
27. J. Potting and M. Hauschild, *Int. J. LCA*, 1997, **2**, 209.
28. J. Potting, W. Schöpp, K. Blok and M. Hauschild, *J. Ind. Ecol.*, 1998, **2**, 63.
29. M. Wright, D. Allen, R. Clift and H. Sas, *J.Ind. Ecol.*, 1997, **1**, 117.
30. R. Clift and L. Wright, *Technol. Forecasting Social Change*, 2000, **65**, 281.

31. J. Foster (ed.), 'Valuing Nature? – Economics, Ethics and Environment', Routledge, London, 1997.
32. J. O'Neill, *Ambio*, 1997, **26**, 546.
33. R. Clift, *Chem. Technol. Eur.*, July/August 1994, 45.
34. A. Azapagic and R. Clift, *J. Cleaner Prod.*, 1999, **7**, 135.
35. R. Clift and A. Azapagic, in 'Tools and Methods for Pollution Prevention', ed. S. K. Sikdar and U. Diwekar, Kluwer Academic, Dordrecht, 1999, p. 69.
36. A. Azapagic, *Chem. Eng. J.*, 1999, **73**, 1.
37. A. Azapagic and R. Clift, *Comput. Chem. Eng.*, 1999, **23**, 1509.
38. M. J. Nicholas, R. Clift, A. Azapagic, F. Walker and D. E. Porter, *Trans. IChemE, B*, 2000, **78**, 193.
39. Royal Commission on Environmental Pollution, 'Fifth Report: Air Pollution Control – An Integrated Approach', HMSO, London, 1976.
40. M. J. Nicholas, *The Chem. Eng.*, 30 April 1998, 33.
41. I. Christie, H. Rolfe and R. Legard 'Cleaner Production for Industry', Policy Studies Institute, London, 1995.
42. A. M. H. Clayton, G. Spinardi and R. Williams, 'Policies for Cleaner Technology – A New Agenda for Government and Industry', Earthscan, London, 1999.
43. K. A. Golonka and D. J. Brennan, *Trans. IChemE, B*, 1996, **74**, 105.
44. A. Warhurst, *Nat. Resource Forum*, Feb. 1992, 39.
45. M. Stewart and J. G. Petrie, in 'Clean Technology for the Mining Industry', eds. M. A. Sanchez, F. Vergara and S. H. Castro, Andros Ltd., Santiago, Chile, 1996, p. 67.
46. M. Stewart and J. G. Petrie, *Mining Env. Manage.*, 1997, **6**(2), 10.
47. M. Flood, *WARMER Bull.*, May 1993, **37**, 16.
48. R. Clift, *The Chem. Eng.*, 13 Feb. 1992, 3.
49. G. Allen, *Phil. Trans. Roy. Soc.*, 1997, **355**, 1467.
50. M. Overcash, *Phil. Trans. Roy. Soc.*, 1997, **355**, 1299.
51. R. Bahu, B. Crittenden and J. O'Hara, 'Management of Process Industry Waste', Institution of Chemical Engineers, Rugby, 1997.
52. D. T. Allen and K. S. Rosselot, 'Pollution Prevention for Chemical Processes', John Wiley and Sons, New York, 1997.
53. C. Pesso, *J. Cleaner Prod.*, 1993, **1**, 139.
54. G. E. Kniel, K. Delmarco and J. G. Petrie, *Environ. Progress*, 1996, **15**, 221.
55. E. N. Pistikopoulos, S. K. Stefanis and G. Livingston, *AIChE Symp. Ser.*, 1996, **90**(303), 164.
56. P. Barton, B. Ahmad, W. Cheong and J. Tolsma, in 'Tools and Methods for Pollution Prevention', eds. S. K. Sikdar and U. Diwekar, Kluwer Academic, Dordrecht, 1999, p. 205.
57. M. Wright, *Phil. Trans. Roy. Soc.*, 1997, **355**, 1349.
58. B. Linhoff, *Trans. IChemE, A*, 1993, **71**, 503.
59. Y. P. Wang and P. Smith, *Chem. Eng. Sci.*, 1994, **49**, 481.
60. M. M. El-Halwagi and V. Manousiouthakis, *Chem. Eng. Sci.*, 1990, **45**, 2813.
61. V. Markovic and D. A. Hicks, *Phil. Trans. Roy. Soc.*, 1997, **355**, 1415.
62. F. Schmidt-Bleek, 'Wieviel Umwelt braucht der Mensch? MIPS – dass Mass für ökologisches Wirtschaften', Birkhäuser, Basel, 1996.
63. E. von Weizsäcker, A. B. Lovins and L. H. Lovins, 'Factor Four: Doubling Wealth, Halving Resource Use', Earthscan, London, 1997.
64. O. Giarini and W. R. Stahel, 'The Limits to Certainty', 2nd edn., Kluwer Academic Publishers, Dordrecht, 1993.
65. W. R. Stahel, *Phil. Trans. Roy. Soc.*, 1997, **49**, 1309.
66. P. Hawken, A. Lovins and L. Lovins, 'Natural Capitalism', Earthscan, London, 1999.
67. House of Commons Environment Committee, 'Session 1996/97, Third Report: The Environmental Impact of Cement Manufacture', The Stationery Office, London, 1997.

68. P. Nichols and S. Aumônier (eds.), 'Developing Life Cycle Inventories for Waste Management', Report no. CWM 128/97, Environmental Agency, London, 1997.
69. R. Clift, A. Doig and G. Finnveden, *Trans. IChemE, B*, 2000, **78**, 279.
70. P. R. White, M. Franke and P. Hindle, 'Integrated Solid Waste Management – A Life Cycle Inventory', Blackie Academic and Professional, London, 1995.
71. G. Finnveden, *Resources, Conserv. Recyc.*, 1999, **26**, 173.
72. T. Ekvall, *J. Cleaner Prod.*, 1999, **7**, 281.
73. R. A. Denison, *Ann. Rev. Energy Environ.*, 1996, **21**, 191.
74. G. Finnveden and T. Ekvall, *Resources, Conserv. Recyc.*, 1998, **24**, 235.
75. T. Ekvall and G. Finnveden, *Trans. IChemE, B*, 2000, **78**, 288.
76. M. Leach, A. Bauen and N. Lucas, *J. Environ. Planning Manage.*, 1997, **40**, 705.
77. ECTEL, 'End-of-Life Management of Cellular Phones – An Industry Perspective and Response', ECTEL Cellular Phones Takeback Working Group, London, 1997.
78. R. U. Ayres, in 'Clean Production Strategies', ed. T. Jackson, Lewis Publishers, Boca Raton, FL, 1993, p. 165.
79. T. E. Graedel and B. R. Allenby, 'Industrial Ecology', Prentice Hall, Upper Saddle River, NJ, 1999.
80. B. R. Allenby, 'Industrial Ecology: Policy Framework and Implementation', Prentice Hall, Upper Saddle River, NJ, 1999.
81. D. T. Allen and N. Behmanesh, in 'The Greening of Industrial Ecosystems', eds. B. R. Allenby and D. J. Richards, National Academy, Washington, D.C., 1994, p. 69.
82. R. A. Frosch, W. C. Clark, J. Crawford, A. Sagar, F. T. Tschang and A. Webber, *Phil. Trans. Roy. Soc.*, 1997, **49**, 1335.
83. T. Jackson and R. Clift, *J. Ind. Ecol.*, 1998, **2**, 3.
84. K. Geiser, in 'Clean Production Strategies', ed. T. Jackson, Lewis Publishers, Boca Raton, FL, 1993, p. 225.
85. E. Williams, W. Mellor, A. Azapagic, G. Stevens and R. Clift, '7th LCA Case Studies Symposium', SETAC-Europe, Brussels, 1999.
86. G. A. Keoleian and D. Menerey, 'Life Cycle Design Guidance Manual-Environmental Requirements and the Product System', US EPA Report EPA/600/R-92/226, Office of Research and Development, Washington, D.C., 1993.
87 L. Alting, M. Hauschild and H. Wenzel, *Phil. Trans. Roy. Soc.*, 1997, **355**, 1373.
88. H. Wenzel, M. Hauschild and L. Alting, 'Environmental Assessment of Products – Methodology, Tools and Techniques and Case Studies in Product Development', Chapman and Hall, London, 1997.
89. R. Clift, '2nd International Conference on Incineration and Flue Gas Treatment Technologies', Institution of Chemical Engineers, Rugby, 1999.
90. N. Emmot and N. Haigh, *J. Environ. Law*, 1996, **8**(2), 301.
91. J. Geldermann, C. Jahn, T. Spengler and O. Rentz, *Int. J. LCA*, 1999, **4**, 94.
92. M. J. Nicholas, A. Azapagic and R. Clift, *J. Ind. Ecol.*, 1998, **2**(4), 4.
93. R. J. Lifset, *J. Resource Man. Technol.*, 1993, **21**(4), 163.
94. W.-P. Schmidt and H.-M. Beyer, *Int. J. LCA*, 1999, **4**, 121.
95. K. Mayers and C. France, *Greener Manage. Int.*, 1999, **25**, 51.
96. F. Berkhout and D. Smith, *Eur. Environ.*, 1999, **9**, 174.
97. F. Rubik and G. Scholl, *Eur. Environ.*, 1999, **9**, 186.
98. World Commission on Environment and Development, 'Our Common Future', Oxford University Press, Oxford, 1987, p. 89.

CHAPTER 17

The Environmental Behaviour of Persistent Organic Pollutants

S. HARRAD

17.1 INTRODUCTION

The last four decades have seen tremendous growth in public, scientific and governmental interest in the environmental effects of persistent organic pollutants (POPs). This is attributable to a number of factors, principally: the increasing weight of scientific evidence related to their adverse effects, the dramatic development of the analytical technology required to measure such compounds, and enhanced public awareness – the latter promoted by media coverage of incidents such as the 1976 explosion at a chemical plant in Seveso, Italy, alongside books such as Rachel Carson's 'The Silent Spring',[1] that publicized the potentially detrimental effects of pesticide use. In 1992, this concern led to the initiation of an assessment programme under the auspices of the United Nations Economic Commission for Europe (UN/ECE) to determine which chemicals should be targeted for concerted international action. Under the framework of the UN/ECE Convention on Long Range Transboundary Air Pollution (LRTAP), a methodology was developed by the UK Department of the Environment and subsequently refined by Environment Canada, to screen and identify chemicals of potential concern. In June 1998, a POPs Protocol to the LRTAP Convention was signed by 33 countries and the European Community. Sixteen contaminants (or groups of contaminants) were targeted for action. These are listed below:

- Aldrin
- Chlordane
- DDT (+ DDD and DDE)
- Dieldrin
- Polychlorinated dibenzo-p-dioxins (PCDDs)
- Polychlorinated dibenzofurans (PCDFs)
- Endrin
- Heptachlor

445

- Hexachlorobenzene
- Mirex
- Polychlorinated biphenyls (PCBs)
- Toxaphene
- Chlordecone (Kepone)
- Hexabromobiphenyl
- Polycyclic aromatic hydrocarbons (PAHs)
- Lindane (γ-Hexachlorocyclohexane)

In July 1998, a global initiative to take action on POPs was agreed by 92 countries under the auspices of the United Nations Environment Program (UNEP). Under Executive Body Decision 1998/2, a procedure was adopted for selecting POPs for international action. Compounds selected consisted of the first twelve listed above. The second session of the UNEP Intergovernmental Negotiating Committee was held in Nairobi in January 1999, at which it was agreed that "conducting inventories would be an essential step in implementing an international instrument on POPs." The third session of the Committee convened in Geneva in September 1999, at which a draft agreement for implementing international action on the listed POPs was discussed.

Such high-profile international efforts serve to highlight the importance of POPs. This chapter addresses their environmental impact and the factors influencing that impact.

17.1.1 Definition of POPs

Under UNECE Executive Body Decision 1998/2, a procedure was adopted for adding POPs to those targeted for action. For inclusion in any future list, the proposed substance should demonstrate:

(1) *Potential for long range transboundary atmospheric transport*: evidence that the substance has a vapour pressure below 1000 Pa and an atmospheric half-life greater than two days. Alternatively, monitoring data showing that the substance is found in remote regions: *and*

(2) *Toxicity*: potential to adversely affect human health and/or the environment; *and*

(3) *Persistence*: evidence that the substance's half-life in water is greater than two months, or that its half-life in soils is greater than six months, or that its half-life in sediments is greater than six months. Alternatively, evidence that the substance is otherwise sufficiently persistent to be of concern within the scope of the protocol; *and*

(4) *Bioaccumulation*: (a) evidence that the bioconcentration factor (BCF) or bioaccumulation factor (BAF) is greater than 5000 or the log K_{OW} is greater than 5 (see Sections 17.5.1 and 17.6.2); or (b) alternatively, if the bioaccumulative potential is significantly lower than (a) above, other factors, such as the high toxicity of the substance, that make it of concern within the scope of the protocol.

Other factors such as production, uses, socio-economic factors, alternatives, cost and benefits are also to be considered. Summarized, POPs are chemicals that are resistant to environmental degradation, that can accumulate through foodchains, and induce toxic effects in humans and wildlife.

17.1.2 Scope

According to UNEP, over 300 chemicals with properties that would classify them as POPs are subject to bans or other controls in one or more countries worldwide. Clearly, some discrimination of coverage is necessary for the sake of brevity in this chapter. We shall therefore consider the following compounds already classified as POPs by the UNECE and UNEP; *viz.*: PCDDs and PCDFs, (collectively known as PCDD/Fs), PCBs, PAHs, and organochlorine pesticides (OCPs). Also included are compounds which – by virtue of their properties (see Section 17.1.1) – are deemed likely candidates for inclusion in future POPs protocols, *viz.*: polychlorinated naphthalenes (PCNs) and poly-brominated diphenyl ethers (PBDEs).

For each class of compounds, the following key areas pertinent to their environmental impact are addressed: their toxicology – with particular empha-sis on their human effects; the methods used to monitor their presence in the environment; their major sources; their physicochemical properties and the influence of these on their environmental fate and behaviour, and, finally, a brief examination of the use of theoretical models to predict such behaviour.

17.1.3 Chemical Structure and Nomenclature

The environmental fate and behaviour of groups of POPs is strongly struc-turally-dependent, and a brief discussion of their structures and nomenclature is deemed worthwhile.

17.1.3.1 PCDD/Fs. The basic chemical structures of PCDD/Fs are illus-trated in Figure 17.1. The numbers indicate sites of chlorination, for example, 2,3,7,8-tetrachlorodibenzo-*p*-dioxin (2,3,7,8-TCDD) contains four chlorine atoms, one each at the 2, 3, 7 and 8 positions; in total, there are 75 possible PCDDs and 135 possible PCDFs. Each individual PCDD/F is referred to as a congener, whilst those congeners possessing identical empirical formulae are isomers of each other. Each group of isomers (there exists one for each degree of chlorination) constitutes a homologue group.

17.1.3.2 PCBs. Figure 17.2 illustrates the basic chemical structure of PCBs. As with PCDD/Fs, the numbers denote sites of chlorination, *e.g.* 3,3',4,4',5-pentachlorobiphenyl possesses five chlorines, three attached to one biphenyl ring at the 3, 4 and 5 positions, the remainder on the other ring in the 3' and 4' sites. 209 PCB congeners are possible, and the definitions and distinctions between congeners, isomers and homologues are identical to those given above for PCDD/Fs. Unlike PCDD/Fs, a IUPAC system[2] assigning a single unique number to each possible PCB congener is widely – indeed almost universally –

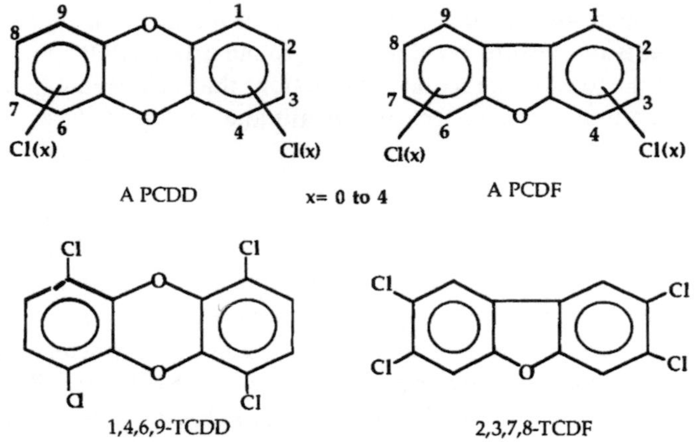

A PCDD   x= 0 to 4   A PCDF

1,4,6,9-TCDD   2,3,7,8-TCDF

Figure 17.1 *PCDD/F nomenclature*

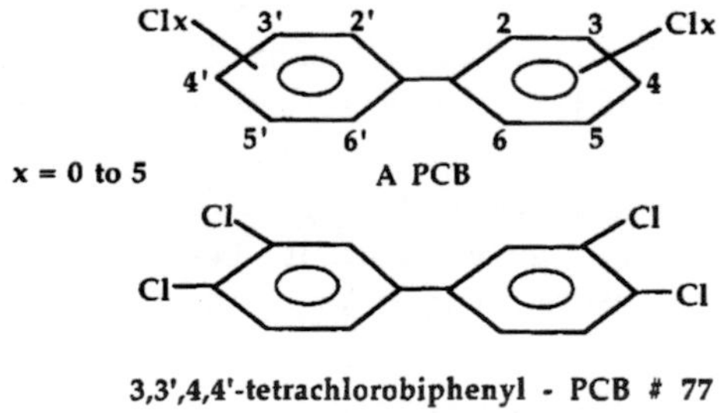

x = 0 to 5   A PCB

3,3',4,4'-tetrachlorobiphenyl - PCB # 77

Figure 17.2 *PCB nomenclature*

utilized and this greatly simplifies reference to individual PCBs. For example, 3,3',4,4'-tetrachlorobiphenyl is referred to as PCB # 77.

17.1.3.3 PAHs. There are three possible PAH isomers consisting of three rings. When this number increases to six rings, 82 isomers are possible. When 12 rings are present, the possible total amounts to 683 101. It is thus unsurprising that only around 45% of these have been identified, and that the complex chemistry of these compounds requires a complex nomenclature system. Figure 17.3 gives the chemical structures of some environmentally significant PAHs, which are an important sub-division of the wider group of compounds known as polyaromatic compounds (PACs). This encompasses a far wider range of compounds, including heterocyclic derivatives such as dibenzothiophenes and carbazoles, along with substituted PAHs such as nitro-PAH.

17.1.3.4 OCPs. As with PAHs, a limit to the number considered in this chapter has been imposed, and Figures 17.4, 17.5, and 17.6 show the chemical structures of aldrin, dieldrin, a toxicologically active toxaphene, DDT and metabolites (DDE and DDD) and the hexachlorocyclohexanes (HCHs).

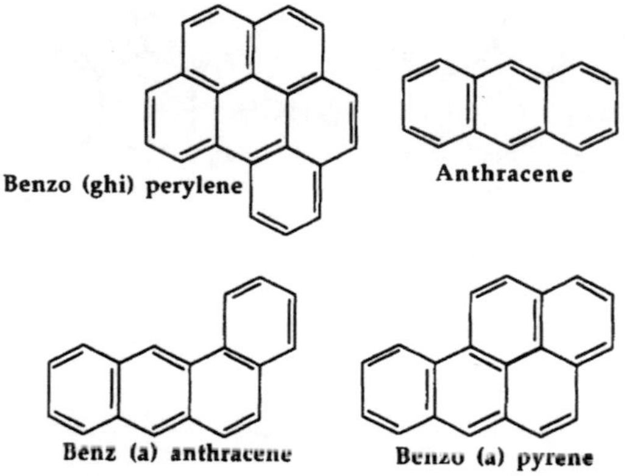
Benzo (ghi) perylene
Anthracene
Benz (a) anthracene
Benzo (a) pyrene

Figure 17.3 *Selected PAH structures*

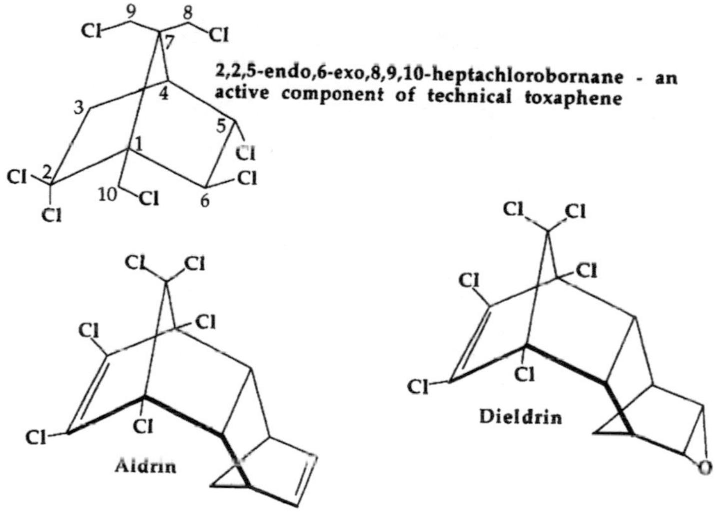
2,2,5-endo,6-exo,8,9,10-heptachlorobornane - an active component of technical toxaphene
Aldrin
Dieldrin

Figure 17.4 *Structures of chlorinated cyclodiene pesticides*

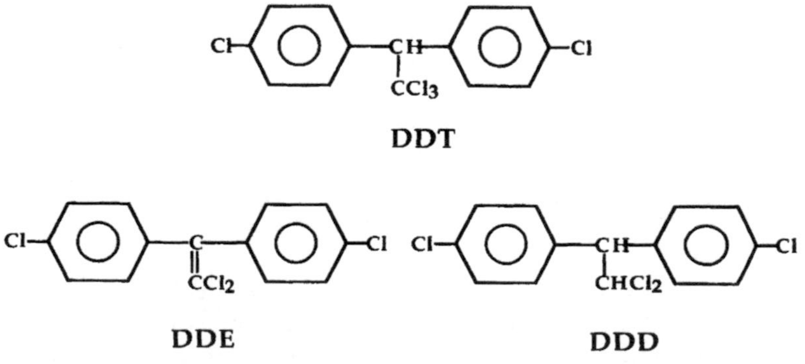
DDT
DDE
DDD

Figure 17.5 *Structures of the DDT 'family'*

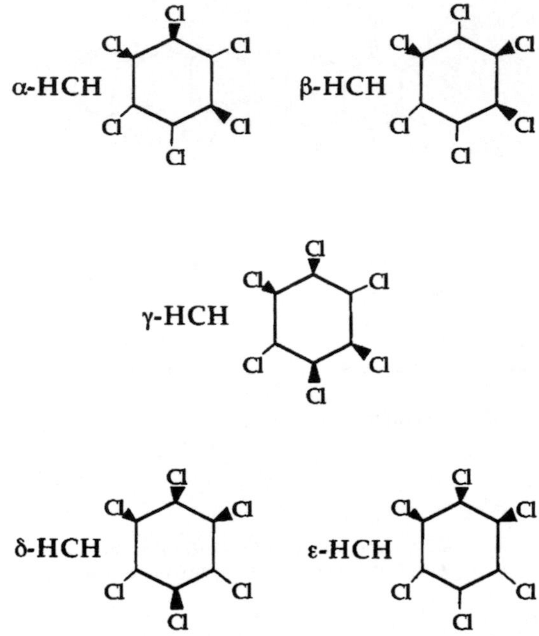

α-HCH   β-HCH

γ-HCH

δ-HCH   ε-HCH

Figure 17.6   *Structures of selected HCHs*

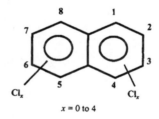

Figure 17.7   *Structures of PCNs*

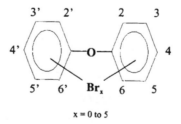

Figure 17.8   *Structures of PBDEs*

17.1.3.5   Other POPs.   The structures of PCNs and PBDEs are illustrated in Figures 17.7 and 17.8 respectively. Note that the numbering system employed for the 209 possible PBDEs is identical to that of PCBs, *i.e.* 3,3′,4,4′-tetra-bromodiphenylether is known as PBDE # 77.

17.2 ADVERSE EFFECTS

An in-depth treatment of the toxicology and ecotoxicology of POPs is beyond the scope of this chapter and only a brief summary of the adverse effects of each compound class is presented. A general consideration in relation to the toxic effects of compounds present at extremely low concentrations in the environment is the difficulty in ascribing health or reproductive abnormalities to contamination by a specific chemical alone, in the presence of other similarly potentially toxic pollutants, often at much higher concentrations. In particular, one cannot discount the possibility of synergistic and/or antagonistic* toxicological effects arising from contamination by a 'cocktail' of pollutants like PAHs, PCBs and PCDD/Fs, all of which are known to evoke similar biochemical responses. This area, which has important health implications, is now receiving much closer scientific scrutiny. In addition to these difficulties, there are considerable problems with assessing the effects in humans, in the understandable absence of clinical data. Instead, evidence for adverse effects in humans is reliant on epidemiological surveys and extrapolations from animal studies, with all the problems inherent in such indirect measurements of human toxicity. A final general point, is the tendency of all of the compounds considered here to partition into fatty tissues. As a result, their adverse effects are compounded by their bioaccumulation in species at the head of food chains, such as marine mammals, birds of prey, and humans.

17.2.1 Polychlorinated Dibenzo-*p*-dioxins and Polychlorinated Dibenzofurans (PCDD/Fs)

Probably no other group of chemicals has been subjected to greater scrutiny with respect to their environmental effects than PCDD/Fs. It is to the Seveso incident in 1976 that present interest may be traced. The episode occurred on July 10th, 1976 at the ICMESA chemical plant in Seveso, near Milan, Italy. The alkaline hydrolysis of 1,2,4,5-tetrachlorobenzene to produce 2,4,5-trichlorophenol (2,4,5-TCP) – an intermediate in the production of the bactericide hexachlorophene – went out of control, and the resultant explosion distributed large quantities of 2,3,7,8-TCDD (formed *via* the self-condensation of 2,4,5-TCP) over the surrounding area. Although the exact environmental impact of the incident has yet to be fully evaluated – whilst widespread animal mortality occurred, the only undisputed adverse effect in humans remains the induction of chloracne (an extremely disfiguring skin complaint) – the psychological effect on the general public was immense. Widespread media coverage both in Italy and beyond – ranging from the responsibly factual to the hysterical (an assertion was made that dioxin could facilitate "the end of Western civilization") – propelled the 'dioxin issue' onto the political agenda of the industrialized world. Since then, a number of incidents have maintained a high public profile for these compounds. In

---

*Synergism is defined as a situation in which the combined effect of two or more substances exceeds the sum of their separate effects. Antagonism is the opposite phenomenon, whereby the combined presence of two or more substances lessens the effects of those substances acting independently.

Table 17.1   *PCDD/F and PCB congeners for which TEFs are defined*

Congener	TEF
2,3,7,8-Tetrachlorodibenzo-*p*-dioxin (2,3,7,8-TCDD)	1
1,2,3,7,8-Pentachlorodibenzo-*p*-dioxin (1,2,3,7,8-PeCDD)	1
1,2,3,4,7,8-Hexachlorodibenzo-*p*-dioxin (1,2,3,4,7,8-HxCDD)	0.1
1,2,3,6,7,8-Hexachlorodibenzo-*p*-dioxin (1,2,3,6,7,8-HxCDD)	0.1
1,2,3,7,8,9-Hexachlorodibenzo-*p*-dioxin (1,2,3,7,8,9-HxCDD)	0.1
1,2,3,4,6,7,8-Heptachlorodibenzo-*p*-dioxin (1,2,3,4,6,7,8-HpCDD)	0.01
Octachlorodibenzo-*p*-dioxin (OCDD)	0.0001
2,3,7,8-Tetrachlorodibenzofuran (2,3,7,8-TCDF)	0.1
1,2,3,7,8-Pentachlorodibenzofuran (1,2,3,7,8-PeCDF)	0.05
2,3,4,7,8-Pentachlorodibenzofuran (2,3,4,7,8-PeCDF)	0.5
1,2,3,4,7,8-Hexachlorodibenzofuran (1,2,3,4,7,8-HxCDF)	0.1
1,2,3,6,7,8-Hexachlorodibenzofuran (1,2,3,6,7,8-HxCDF)	0.1
1,2,3,7,8,9-Hexachlorodibenzofuran (1,2,3,7,8,9-HxCDF)	0.1
2,3,4,6,7,8-Hexachlorodibenzofuran (2,3,4,6,7,8-HxCDF)	0.1
1,2,3,4,6,7,8-Heptachlorodibenzofuran (1,2,3,4,6,7,8-HpCDF)	0.01
1,2,3,4,7,8,9-Heptachlorodibenzofuran (1,2,3,4,7,8,9-HpCDF)	0.01
Octachlorodibenzofuran (OCDF)	0.0001
3,3′,4,4′-Tetrachlorobiphenyl (PCB # 77)	0.0001
3,4,4′,5-Tetrachlorobiphenyl (PCB # 81)	0.0001
3,3′,4,4′,5-Pentachlorobiphenyl (PCB # 126)	0.1
3,3′,4,4′,5,5′-Hexachlorobiphenyl (PCB # 169)	0.01
2,3,3′,4,4′-Pentachlorobiphenyl (PCB # 105)	0.0001
2,3,4,4′,5-Pentachlorobiphenyl (PCB # 114)	0.0005
2,3′,4,4′,5-Pentachlorobiphenyl (PCB # 118)	0.0001
2′,3,4,4′,5-Pentachlorobiphenyl (PCB # 123)	0.0001
2,3,3′,4,4′,5-Hexachlorobiphenyl (PCB # 156)	0.0005
2,3,3′,4,4′5′-Hexachlorobiphenyl (PCB # 157)	0.0005
2,3′,4,4′,5,5′-Hexachlorobiphenyl (PCB # 167)	0.00001
2,3,3′,4,4′,5,5′-Heptachlorobiphenyl (PCB # 189)	0.0001

particular, there has been much controversy in the USA over the exposure of Vietnam War veterans to the defoliant Agent Orange (a mixture of esters of di- and trichlorophenoxyacetic acids and heavily contaminated with PCDD/Fs).

It is important to note that amongst the range of 210 compounds known as PCDD/Fs, there exist wide differences in toxicological potency. In summary, only those PCDD/Fs chlorinated at the 2, 3, 7 and 8 positions are considered of toxicological significance. The seventeen congeners fitting this criterion are listed in Table 17.1, along with their toxic equivalent factors (TEFs*). This

---

*TE (or TEQ) is an acronym for 2,3,7,8-TCDD equivalents. It is a means of expressing the toxicity of a complex mixture of different PCDD/Fs and related compounds (such as coplanar PCBs – see Section 17.2.2) in terms of an equivalent quantity of 2,3,7,8-TCDD. Each compound is assigned a Toxic Equivalency Factor (TEF) based on its toxicity relative to that of 2,3,7,8-TCDD, which is universally assigned a TEF of 1. Multiplication of the concentration of a compound by its assigned TEF gives its concentration in terms of TE and the toxicity of a mixture is the sum of the TEs calculated for all congeners. The weighting scheme referred to in this chapter is that devised by the WHO.[3]

structural-dependence of toxicity is due to the fact that the toxic effects of PCDD/Fs are mediated by initial interaction with a cellular protein – the Ah-receptor. Such binding is subject to the 'lock and key' principle commonly associated with compound–receptor interactions and, hence, only those compounds able to assume a coplanar molecular configuration similar to that of 2,3,7,8-TCDD are capable of binding to the Ah-receptor. This receptor-mediated theory of toxicity also lends a rationale to the considerable species-specific variations in the toxicity of 2,3,7,8-TCDD (guinea-pigs have an LD_{50} of $0.6\ \mu g\ kg^{-1}$ body weight (bw), *cf.* hamsters which have an LD_{50} of $3500\ \mu g\ kg^{-1}$ bw),[4] as the precise nature of the Ah-receptor varies widely between species.

Scientific opinion concerning the exact toxicological potency of PCDD/Fs continues to differ widely. Despite this, there is a significant body of opinion that PCDD/Fs may well present a human cancer hazard, and that they are potent toxins with the potential to elicit a range of non-cancer effects, starting with binding to the Ah-receptor. These effects may be occurring in humans at very low levels comparable with the upper limit of background (*i.e.* non-occupational) exposure, and include reproductive, immunological, and developmental effects. To put this into context, average UK background exposure to PCDD/Fs is estimated at around 1 pg ΣTE kg^{-1} bw per day, compared with the WHO's recently revised guideline exposure limit of 1–4 pg ΣTE kg^{-1} bw per day.[5]

17.2.2 Polychlorinated Biphenyls (PCBs)

PCBs have been linked to a number of toxic responses, including the impairment of immune responses in biota. Other cited effects include hepatotoxicity, carcinogenicity, teratogenicity, neurotoxicity, and reproductive toxicity. Recent concerns have focused on reports of a relationship between perinatal exposure to PCBs and PCDD/Fs and impaired neurological development in infants.[6,7] In evaluating the significance of these effects, it is important to recognize that – as for PCDD/Fs – considerable species- and congener-specific variations exist. In particular, there are indications that their potency in humans is markedly less than in other animals, and it is very important that caution is exercised when extrapolating toxicological data obtained from laboratory animals to humans, and that PCB behaviour is studied on an individual congener basis. Also of relevance is the fact that some PCBs (especially those lacking chlorine substituents at the *ortho* 2, 2′, 6 and 6′ positions) are capable of eliciting similar toxicological effects to the 2,3,7,8-chlorinated PCDD/Fs. This is due to their ability to adopt the coplanar molecular configuration necessary to interact with the Ah-receptor responsible for mediating 'dioxin-like' effects, and has led to the assignation of TEFs to such 'coplanar' PCBs.[3] A significant recent development has been the inclusion of the toxic equivalent contribution of PCBs for the purposes of comparing exposure with the Tolerable Daily Intake (TDI) of 1–4 pg ΣTE kg^{-1} bw per day, with the effect of further narrowing the gap between background human exposure and the TDI. Those PCBs assigned TEFs are included in Table 1. In terms of UK adult human exposure, PCBs and PCDD/Fs make an approximately equal contribution on a ΣTE basis – average UK background exposure to PCDD/Fs *and* PCBs together is estimated at around 2 pg ΣTE kg^{-1} bw per day.

17.2.3 Polycyclic Aromatic Hydrocarbons (PAHs)

The toxicity of PAHs, like PCBs and PCDD/Fs, varies widely on both a compound- and species-specific basis. Their adverse effects have been known for a considerable time, and hindsight allows us to associate them with one of the first recorded cases of occupational cancer, *viz.* the high incidence of scrotal cancer in chimney sweeps reported by Percival Pott in 1775, which was subsequently attributed to the high concentrations of PAHs – in particular, benzo[*a*]pyrene – in chimney soot. Several PAHs have been shown to be acutely toxic. However, health concerns regarding these compounds centre on their metabolic transformation by aquatic and terrestrial organisms into carcinogenic, teratogenic, and mutagenic metabolites such as dihydrodiol epoxides, which bind to and disrupt DNA and RNA, with possible resultant tumour formation. Such metabolic conversion and activation is essential if PAHs are to exhibit their latent carcinogenic properties. The most potent carcinogens of the PAH group – in addition to benzo[*a*]pyrene – include: the benzofluoranthenes, benzo[*a*]anthracene, dibenzo[*ah*]anthracene, and indeno[1,2,3–*cd*]pyrene. Such PAHs are recognized by regulatory agencies such as the EC and the United States Environmental Protection Agency (USEPA) as priority pollutants. In a significant recent development, the UK government's Expert Panel on Air Quality Standards (EPAQS) recommended a UK air quality standard for benzo[*a*]pyrene of 0.25 ng m^{-3}. This standard is of a challenging nature, given that annual mean concentrations in UK cities[8] are typically 0.3–1.8 ng m^{-3}.

17.2.4 Organochlorine Pesticides (OCPs)

A common effect of all three of the compound groups considered, *i.e.* the DDT group, the HCH group, and the chlorinated cyclodiene family (*e.g.* dieldrin and toxaphenes), is an ability to induce reproductive effects including reduced fertility and eggshell thinning in birds (this is as a result of the insecticide's ability to affect calcium metabolism in birds and mammals), as well as direct toxic effects. The combined consequence of such effects, was a marked reduction in populations of birds of prey. Restrictions on the use of the insecticides responsible have since resulted in a welcome recovery in affected populations; however, their environmental persistence, coupled with continued 3rd World usage, means that the danger is considered far from over. Furthermore, as a result of the marked bioaccumulative properties of these compounds, their presence in human fat is of some concern. For example, at the peak of DDT usage, the typical American adult contained 12 mg kg^{-1} DDT in their fat. The International Agency for Research into Cancer (IARC) regards both HCHs and technical toxaphene as "possibly carcinogenic to human beings" – although there is some dissent with this view – whilst the evidence for human carcino-genicity for aldrin, DDT, and dieldrin is considered "inadequate". The general lack of data from human studies is striking however, and the WHO deems further study imperative. In addition, other adverse human effects are known. For example, organochlorine insecticides in general are associated with enhanced induction of drug-metabolizing enzymes in the liver, whilst both DDT and lindane (γ-HCH) are known to elicit skin sensitization. An important fact concerning the relative toxicity of the different HCH isomers (although five are shown in Figure 17.6, eight are possible) is that the γ-isomer is fifty to several thousand times more toxic to insects than the α- and δ-isomers, whilst the β- and ε-isomers are usually almost inert.

17.2.5　Other POPs

Some PCNs elicit similar toxic responses to 2,3,7,8-TCDD and have therefore been assigned TEFs, which are comparable to those assigned to the coplanar PCBs. While there is considerable spatial and matrix-specific variation, it is fair to say that the overall contribution of PCNs to ΣTE concentrations in the environment is of a similar order of magnitude to that made by PCBs. While evidence of adverse effects arising from PBDE exposure is to date limited, the widely reported presence in human tissues, together with their structural similarity to PCDD/Fs and PCBs has raised concerns. Of particular interest are the effects of neonatal exposure, which has been demonstrated to adversely affect learning and memory functions in adult animals.[9] If replicated in humans, such effects are of potential concern in light of the fact that while human exposure and body burdens of organochlorine POPs appears to be declining, PBDE concentrations in human milk from Sweden have increased continuously between 1972 and 1997.[10]

17.3　MEASUREMENT TECHNIQUES

Table 17.2 indicates 'typical' levels of selected POPs in soil, air, human milk and freshwater. From this it is evident that sensitivity is an important prerequisite of any measurement technique for such compounds, particularly for PCDD/Fs, where the quantities involved may justifiably be described as 'ultra-trace'. Given the increasing level of legislation relating to the presence of such compounds in the environment, there is also a clear need for the techniques to be both as accurate and precise as is possible. In achieving these aims, the crucial role of sampling methodology must be recognized. Finally, the measurement techniques employed must be able to distinguish the target compounds from the complex 'soup' of other chemicals present in the sample – in short, they must be selective.

17.3.1　Sampling Methodology

Even using the most sophisticated analytical instrumentation, a measurement will be severely compromised if the sample taken for analysis is unrepresenta-

Table 17.2 *Typical environmental levels of selected POPs*

Compartment	PAHs[a]	PCBs[b]	PCDD/Fs[c]	Toxaphene[d]	PBDEs[e]
		'Typical' concentration of			
Soil	2000 μg kg^{-1}	5 μg kg^{-1}	0.3 μg kg^{-1}	200 μg kg^{-1}	1 μg kg^{-1f}
Outdoor air	100 ng m^{-3}	0.3 ng m^{-3}	0.003 ng m^{-3}	0.03 ng m^{-3}	0.005 ng m^{-3}
Human milk[g]	–[h]	10 μg kg^{-1}	0.015 μg kg^{-1}	5 μg kg^{-1}	0.1 μg kg^{-1}
Freshwater	10 ng dm^{-3}	1 ng dm^{-3}	0.003 ng dm^{-3}	1 ng dm^{-3}	–[i]

[a] Expressed as the sum of the 16 PAH required to be monitored by the United States Environmental Protection Agency (USEPA). These include naphthalene, anthracene, benzo[*a*]pyrene and benzo-[*ghi*]perylene.
[b] Expressed as the sum of the most commonly occurring PCBs.
[c] Reported as the sum of all tetra- through octachlorinated congeners.
[d] Measured as the sum of all congeners present in technical toxaphene.
[e] Expressed as the sum of the most commonly occurring PBDEs.
[f] Data for marine sediments – soils data not available.
[g] Data on a whole milk basis.
[h] Data not available. Human metabolism of PAH considered too fast.
[i] Data not available.

tive. Hence, a good sampling method provides a sample that accurately reflects the levels of the chosen analyte(s) in the matrix studied. It must also provide sufficient sample for the purposes of the study (in this respect, the sensitivity of the analytical instrumentation employed must also be taken into account). Another important consideration for both soil and water sampling is that the glass containers for sample storage must be thoroughly cleaned prior to use, to minimize potential contamination.

In terms of sampling methodology, the major differences occur according to the nature of the matrix under scrutiny, *e.g.* soil, air or water, rather than on a compound-specific basis. As a result, the following sections will discuss the types of sampling strategies used for the determination of trace levels of organic chemicals in air, soil and water, with allusion to compound-specific refinements where necessary.

17.3.1.1 Air Sampling. There are two principal categories of 'air' sampling. One involves sampling ambient air, *i.e.* that to which humans and other animals are exposed in their normal working and recreational lives, whilst the other involves the study of the gaseous and/or dust emissions from a pollutant source such as a waste incinerator. For the purposes of this chapter, we shall limit our focus to ambient air.

A general consideration for all air sampling is the monitoring of sampling efficiency, *i.e.* the proportion of analyte retained by the sampler. This may be achieved by the addition of 'sampling standards' to the sampler prior to commencement of the sampling campaign. The fraction remaining at the end of sampling affords a measure of the sampling efficiency.

Figure 17.9 shows a schematic of the sampling apparatus used to sample POPs in ambient air. The equipment basically comprises a PM$_{10}$-selective

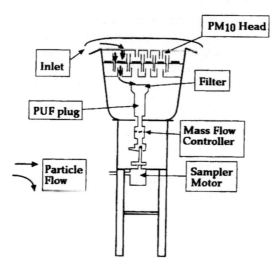

Figure 17.9 *Schematic of hi-vol apparatus for sampling POPs in ambient air*

impactor head (see also Chapter 7) – designed to remove particles above 10 μm diameter – followed by a smaller sampling head housing a filter paper – commonly glass fibre or PTFE – the latter is more expensive, but minimizes analyte reaction on the filter during sampling – followed by one or two cylindrical 'plugs' of polyurethane foam (PUF). POPs present in ambient air will exist in both gaseous and aerosol form (vapour and particulate phase respectively). The aerosol fraction is collected by the filter, while the vapour phase is adsorbed onto the PUF 'plugs'. Owing to the low levels of POPs (particularly PCDD/Fs) in ambient air, very high volumes (100–3000 m³) must be sampled in order to procure measurable quantities of the analyte(s) of interest.

17.3.1.2 Soil Sampling. Procuring soil samples is relatively straightforward. A cylindrical metal corer of *ca.* 10 cm diameter is sunk into the soil, with the material removed being placed into a clean glass jar, which is sealed to minimize adsorption and volatilization of analytes from and to the air.

An important aspect of any soil sampling campaign is the sampling depth. This will depend on the nature of the analyte (most POPs are not prone to movement through the soil column, thus largely restricting contamination to the surface), and on whether the soil has been ploughed or otherwise recently disturbed, as disturbance will enhance analyte distribution throughout the soil column. Typical sampling depths for POPs are 5–10 cm on undisturbed land, as concentrations fall dramatically below such depths. Obtaining a representative sample is also very important, and is commonly achieved by combining sampled soil cores from various points in the vicinity of the sampling site. These points are situated along either an 'X' or a 'W' pattern covering an area typically 10 m by 10 m. The combined samples are homogenized and appropriately sized sub-samples taken for analysis.

17.3.1.3   Water Sampling.   Hydrophobicity is a common property of all POPs dealt with in this chapter. As a result, the levels present in aqueous media are low (see Table 17.2), and much of the burden of such chemicals associated with 'water' samples is in fact bound to particulate matter. Consequently, large sample volumes are necessary (*e.g.* up to 100 dm^3 for the determination of PCBs in relatively uncontaminated river water), and the particulate fraction must be separated from the aqueous fraction if the true 'dissolved phase' concentration is to be measured. Such separation is carried out by pulling the sample through a fine mesh filter (usually of 0.45 μm porosity) to trap particle-bound pollutants. The filtrate is subsequently either solvent extracted or passed through an adsorptive resin-filled tube or PUF plug to concentrate the dissolved phase, which is subsequently solvent extracted.

17.3.2   Analytical Methodology

Once sampling is complete, a vast array of analytical techniques is available to the analyst. That chosen depends on both the nature of the matrix (the determination of POPs in sewage sludge is extremely difficult owing to the high levels of potential interferences that are present), the expected concentrations of the target analyte, and the nature of the analyte itself. In this section therefore, the range of available techniques will be examined on a compound-specific basis.

17.3.2.1   PCDD/Fs.   The universal method of choice for PCDD/F determination is GC/MS. The selectivity afforded by mass spectrometry is essential, given that the levels of potential chemical interferences present in samples will usually be far in excess of those of PCDD/Fs, whilst the requisite sensitivity is partly achieved *via* use of selected ion monitoring (SIM). In general, and particularly for the determination of PCDD/Fs in ambient air (where individual congener concentrations may be as low as a few fg m^{-3} (1 fg = 10^{-15} g)), the use of high resolution mass spectrometers capable of providing 'on-column' detection limits of 25 fg is favoured.

   Quantitation of PCDD/Fs is made *via* reference to ^{13}C$_{12}$-labelled PCDD/F internal standards added to the sample prior to extraction. These standards – which are identical to the corresponding 'native' PCDD/F except for their mass – are used to correct analyte concentrations for losses throughout the extensive extraction and purification procedures employed prior to GC/MS analysis. The quantification method employed is referred to as 'isotope dilution'.

17.3.2.2   PCBs.   There are two principal selectivity requirements for the analysis of PCBs. The first, the separation of individual PCB congeners from each other, is crucial, given the congener-specific variations in toxicity, and is achieved by the use of capillary GC. The second selectivity requirement – *viz.* the ability to differentiate PCBs from co-eluting interferences – is less pivotal than for PCDD/Fs, given the fact that environmental concentrations of individual PCBs are much greater. As a result, a common detection technique for PCBs is the electron capture detector (ECD), which offers excellent sensitivity ($\leqslant 1$ pg 'on-column'), alongside a reasonable degree of selectivity. Mass spectrometric detection – especially when employed alongside the use of $^{13}C_{12}$-labelled PCB internal standards – offers advantages, *e.g.* in the resolution of congeners that coelute on the GC column but possess different numbers of chlorines and hence molecular ions. Mass spectrometric detection also offers enhanced selectivity in terms of differentiating PCBs from co-eluting interferences such as β-HCH and DDE. These advantages have resulted in a marked shift towards GC/MS as the method of choice for the determination of PCBs.

17.3.2.3 PAHs. As environmental concentrations of PAHs are generally around two orders of magnitude greater than those of PCBs, the strict sensitivity and selectivity requirements associated with analytical techniques for the determination of the latter are less important. Perhaps the most commonly used technique for PAH analysis, is reversed-phase HPLC in conjunction with variable-wavelength UV and/or fluorescence detection. Other methods are capillary GC-based, with the flame ionization detector (FID) being the most common method of detection. However, the inherent universality of response of FID means that more specific detectors like ECD and fluorescence are sometimes employed. GC/MS is also finding increasingly widespread use, with isotopically-labelled internal standards (either deuterated or ^{13}C-labelled) and SIM, enhancing accuracy and sensitivity.

17.3.2.4 OCPs. The most common analytical technique for the determination of OCPs is capillary GC/ECD, although detection of low levels in complex matrices such as sewage sludge is greatly facilitated by the use of GC/MS. The analytical chemistry of toxaphenes remains a fertile area for research, with current methods centred on GC/ECD and GC/ECNIMS (electron capture negative ion mass spectrometry).

17.3.2.5 Other POPs. In common with the POPs discussed above, GC/MS and GC/ECD are the most commonly employed analytical techniques for the determination of PCNs and PBDEs.

17.4 SOURCES

Source apportionment is an important research area, the purpose of which is to identify and rank the major sources of pollutants to the environment. Such ranking tables permit the identification of major release pathways, and hence the prioritization of emission control strategies. Indeed, one of the principal aims of the UN/ECE protocol discussed in Section 17.1 is to identify the key sources of the targeted POPs.

The basic strategy of a source inventory is to derive an emission factor for a specific source activity (*e.g.* 10 μg per t of waste burnt), and subsequently to multiply this by an activity factor, *i.e.* the extent to which the activity is practised (*e.g.* 3 million t waste burnt per year). In this way, an estimate of annual pollutant emissions for a specific source can be derived – in the case illustrated, annual emissions would amount to 30 g. The degree of difficulty involved in the construction of a source inventory – and hence its accuracy – varies widely according to the nature of the emission source (*e.g.* quantifying pesticide releases using well-documented industry production figures is more reliable than estimating emission factors for unintentional releases of PAHs from fossil fuel combustion) and, as a result, the accuracy of such source inventories varies widely, with the primary benefit of many proving to be the identification of areas requiring further investigation. The section that follows summarizes present knowledge of the magnitude and sources of current and past releases of POPs.

17.4.1   PCDD/Fs

PCDD/Fs have never been intentionally produced – other than on a laboratory scale. Figure 17.10 illustrates the temporal variation of PCDD/F concentrations in archived soil taken from the Rothamsted experimental crop station in Hertfordshire, England. This study, which covers the period 1844–1986, shows there to have existed a low, essentially constant 'background' of PCDD/Fs throughout the latter part of the 19th century, after which there has been a dramatic rise.[11] The inference of this, is that – whilst some natural sources exist, notably forest fires (which may arguably be of significance in some countries) – the origins of the environmental burden of these compounds are principally as by-products of anthropogenic activities, in particular the manufacture and use of organochlorine chemicals (such as chlorophenols) and combustion processes (such as waste incineration and coal combustion).

   The mechanisms *via* which PCDD/Fs form during combustion activities are complex, and are still not wholly understood. It appears that the mechanism of formation is *via de novo* synthesis, *i.e.* formation from the basic chemical 'building blocks' of carbon, chlorine, oxygen and hydrogen. In summary, this means that, provided the conditions of temperature *etc.* outlined below are met, PCDD/F formation can occur from the combustion of any 'fuel', providing sources of these four elements are present, although formation is enhanced if levels of so-called 'precursor' compounds like chlorophenols and chloro-

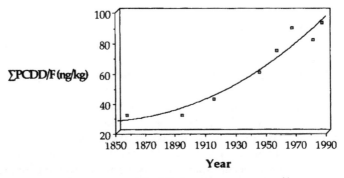

Figure 17.10   *Temporal Trends in ΣPCDD/F soil concentrations*[11]

benzenes are present in the 'fuel'. The evidence of laboratory experiments and studies on working waste incinerators is that PCDD/F formation during combustion occurs in post-combustion zones, *i.e.* oxygen-rich regions where temperatures are in the region 250–350 °C, such as electrostatic precipitators. In these regions, a series of reactions, catalysed by the presence of metal chlorides, occur on the surface of fly ash particles and, as a result, PCDD/Fs are formed.

The presence of PCDDs in chlorophenols and products produced *via* chlorophenol intermediates (such as chlorophenoxyacetic acid derivatives; one of the principal constituents of 'Agent Orange' – a defoliant used in the Vietnam War, and heavily contaminated with 2,3,7,8-TCDD – was the *n*-butyl ester of 2,4,5-trichlorophenoxyacetic acid) is more easily explained. PCDD formation occurs *via* the inadvertent reaction of two chlorophenol molecules (or chlorophenate ions), which yields PCDD(s), the exact identity(ies) of which are dependent on the chlorination pattern of the reactants, and the possibility of Smiles rearrangement products, as illustrated in Figure 17.11.

It should be noted that there exist many other known sources, though a dearth of data prevents estimates of emissions from these. Indeed, the continuing discovery of fresh sources in recent years (including metal smelting, and chlorine bleaching of paper pulp) suggests that much work remains to be done before the origins of the present environmental burden of PCDD/Fs are fully clarified. A further illustration of this is that source inventories of atmospheric emissions of PCDD/Fs conducted in the early 1990s in the UK[12] and Sweden,[13] both failed to account for more than 10–20% of deposition to the surface of these countries. Although more recent and more detailed inventories have accounted for a greater percentage of depositional input,[14] this area remains controversial.

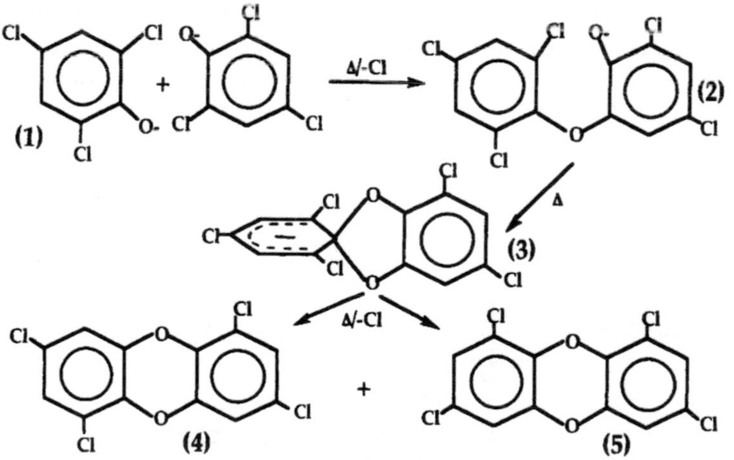

Figure 17.11 *Self-condensation of 2,4,6-trichlorophenate (1) via intermediate species (2) and (3) and Smiles rearrangement of (3) to give 1,3,6,8-TCDD (4) and 1,3,7,9-TCDD (5)*

17.4.2 PCBs

In stark contrast to PCDD/Fs, the origin of the environmental burden of PCBs is almost exclusively their deliberate manufacture – primarily for use as dielectric fluids in electrical transformers and capacitors, but also for use in, *inter alia*, carbonless copy papers and inks. Industrial manufacture commenced in the USA in 1929 – UK production in 1954 – and between then and the late 1970s, when production (although not use of existing stocks) ceased in most western nations, an estimated total of 1.2 million t were produced (67 000 t in the UK). Figure 17.12 shows the temporal variations in PCB levels in soils over the period 1942 to 1992, from the so-called 'Woburn Market Garden' experiment. This study[15] revealed UK PCB levels to have essentially reflected temporal trends in use, with the restrictions on such use producing a precipitous decline in soil concentrations. The principal loss mechanism from soils – which has been estimated to hold > 90% of the UK's burden of PCBs – is considered to be volatilization (degradation in soils is known to be slow) and hence it is unsurprising that re-circulation of the existing environmental burden has been identified as one of the major sources of current PCB input to the UK. Other significant sources include the scrapping of redundant PCB-contaminated electrical goods such as refrigerators (fragmentizing), volatilization of PCBs during the recycling of contaminated scrap metal, and leaks from PCB-filled transformers and capacitors still in use.

Recent indications are that this latter source may be more significant than previously realized. Concentrations of ΣPCB in indoor air in the West Midlands were found to be on average 30 times higher than those in outdoor air.[16] Whilst unequivocal source apportionment of these elevated indoor concentrations remains elusive, leaks from PCB-filled equipment remaining in use – alongside volatilization of PCBs contained in permanently-elastic sealants present within building structures – seem the most likely source. The presence of elevated concentrations in indoor air raises the possibility that ventilation of such air may constitute an appreciable source of PCBs in outdoor air. This has important policy implications, as it implies that the UK government's action plan to phase out PCBs remaining in use may have a significant effect on

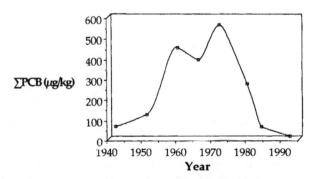

Figure 17.12   *Temporal trends in ΣPCB soil concentrations*[15]

reducing PCB concentrations in outdoor air. Conversely, if volatilization of the existing PCB burden in soils remains the principal source to the atmosphere, then there is little more that can be done, other than to simply wait for concentrations to fall as the existing burden degrades or undergoes atmospheric transport away from the UK. Distinguishing between these two sources: ventilation of indoor air (primary); and volatilization from soil (secondary) is currently a research priority.

A recent development in atmospheric source apportionment of POPs has involved the use of enantiomeric fractions (EFs) of chiral organic contaminants. Nineteen PCBs containing three or four chlorine atoms exist as stable atropisomers – that is, they exist in two forms (known as enantiomers) which are essentially physically and chemically indistinguishable, except that they rotate the plane of polarized light in opposite directions, and may biodegrade at different rates. In commercial PCB formulations, these 19 PCBs are present as racemates, *i.e.* concentrations of the two enantiomers are equal, EF = 0.5 (EF = concentration of the ($+$) enantiomer divided by the sum of the concentrations of both enantiomers). By comparison, due to enantiomeric differences in resistance to biodegradative processes, the EFs of chiral organochlorines found in soils – and which are preserved upon volatilization – deviate from 0.5, in theory varying between close to zero to *ca.* 1. It has therefore been hypothesized that a knowledge of EFs in samples of outdoor air (EF_{OA}), commercial PCB formulations, and soil could provide a useful indicator of the extent to which the contemporary ambient atmospheric burden is due to revolatilized material, and how much arises from fresh emissions from remaining PCB stocks. To illustrate, if the EF for a given chiral congener in soil (EF_{SOIL}) is known, and it is assumed that this EF would be preserved on volatilization, the percentage contribution of primary (*i.e.* racemic) sources to outdoor air levels of that PCB ($PC_{RACEMIC}$) may be calculated *via* the equation below.

$$PC_{RACEMIC} = \frac{(EF_{OA} - EF_{SOIL})}{(EF_{RACEMIC} - EF_{SOIL})} \times 100\% \tag{1}$$

Clearly, this technique has great potential as a source apportionment tool. However, there are appreciable technical difficulties – *e.g.* achieving chromatographic resolution of enantiomers from achiral PCBs – to be overcome before it is likely to find widespread use.

17.4.3 PAHs

Like PCDD/Fs, PAHs have never been intentionally produced. In contrast to PCDD/Fs however, it is widely accepted that natural sources can make a significant contribution to their environmental burden in countries where forest fires and volcanic activity are prevalent. Furthermore, the biogenic formation of perylene in anaerobic sediments is well-established, although the significance of similar mechanisms for other PAHs is unclear. However, the overwhelming bulk of releases to the atmosphere occur *via* incomplete combustion of organic materials, in particular fossil fuels. The major sources to the UK environment on a national basis are domestic coal combustion and motor vehicle fuel combustion, with the volatilization of the existing burden in soils (especially those from contaminated sites, *e.g.* from disused coal gas-works) and from creosote-treated timber considered likely to play important rôles. The relative importance of anthropogenic and natural sources of environmental PAH are illustrated by Figure 17.13, which shows the variation in PAH levels over the period 1846–1986 in soils sampled from the Rothamsted experimental station.[17]

Although other sources may play an important rôle on a national scale, it is now clear that traffic emissions are the single greatest source of PAH in urban areas. An excellent illustration is given in Figure 17.14, which shows the covariation in particulate phase atmospheric concentrations of benzo[*a*]anthra-

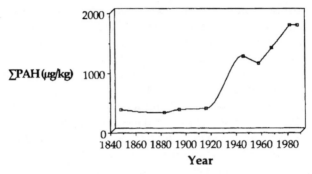

Figure 17.13   *Temporal trends in ΣPAH soil concentrations*[17]

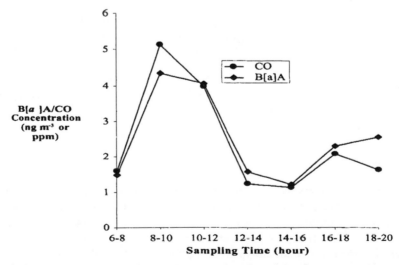

Figure 17.14   *Covariation of particulate phase concentrations of benzo[a]anthracene (B[a]A) and concentrations of CO on 5/12/96 in central Birmingham*[18]

cene and CO – a known traffic pollutant – over the course of a single day in central Birmingham.[18]

Linear regression of the benzo[*a*]anthracene and CO concentrations recorded during this sampling event reveals a statistically highly significant relationship whereby [benzo[*a*]anthracene] = 0.774 × [CO] + 0.657. By assuming that in the absence of traffic [CO] = 0.1 ppmv, and substituting this value into the above relationship, one can estimate the concentration of benzo[*a*]anthracene in the absence of traffic as 0.73 ng m^{-3}. Subtracting this value from the mean benzo[*a*]anthracene concentration (2.53 ng m^{-3}), yields the concentration of benzo[*a*]anthracene due to traffic (1.80 ng m^{-3}), and the percentage of benzo[*a*]anthracene due to traffic at this site on this day = 100 × (1.80/2.53) = 71%. Figures for other PAHs ranged between 59 and 87%, with the mean percentage of all 4–7-ring PAH monitored that arose from traffic calculated to be 71%. Furthermore, in a related study, the same authors showed the contribution of diesel vehicle emissions to overall traffic emissions of PAH to be appreciable (*ca.* 60%).

17.4.4 OCPs

Sources of organochlorine pesticides in the environment are entirely anthropogenic. Production of lindane, 'purified' and 'technical' HCH commenced during World War II and, despite the introduction of use restrictions in the last 25 years, they remain in use throughout the globe. It is difficult to accurately estimate production volumes of the toxicologically important γ-HCH, given the fact that, in addition to its widespread European and N. American usage in its pure γ form (lindane), it has found significant use in the southern hemisphere in the form of 'purified' HCH (60–98.9% γ-HCH) and 'technical' HCH, in which the proportion of γ-HCH ranges between 10–16%. However, cumulative worldwide use of HCH products since their introduction (as insecticides and as a wood preservative) has been put in excess of 10^6 t. The family of toxaphene compounds are also particularly worthy of mention as, in 1971, technical toxaphene accounted for 80% of all chlorinated cyclodiene insecticides still used at that point by US farmers, chiefly to control cotton pests. Despite being banned in the USA in the early 1980s – up to which point, at least 3 × 10^5 t had been used in the USA alone – it is still manufactured and used in developing countries. Estimates of cumulative worldwide toxaphene use amount to a minimum of 5 × 10^5 t, although the true figure may be much higher, given the fact that, between 1955 and 1960, Egypt alone used 54 000 t.

For DDT, figures relating to world production are only very approximate, but an estimate in excess of 3 × 10^6 t would not seem unreasonable. Global production of aldrin and dieldrin will be lower than for the other pesticides considered here, but is still significant, particularly in some countries, where their use may have exceeded that of DDT.

As with PCBs, determination of the relative contributions of 'new' and 'old' (*i.e.* revolatilized) releases of OCPs to the atmosphere is an active research area. Fortunately, some OCPs are chiral, and therefore the source apportionment techniques discussed in Section 17.4.2 for PCBs are applicable. Furthermore, because chromatographic resolution is less difficult, the technique has been successfully applied.[19]

17.4.5 Other POPs

PBDEs are flame retardant chemicals that have found extensive use in manufactured products such as paints, plastics and textiles in many countries. Although production has fallen off in recent years, annual global production of PBDEs in 1992 was 40 000 t.[20] PCNs have been widely employed in similar applications to those of polychlorinated biphenyls (PCBs), *viz.* as plasticisers, as heat-transfer fluids in heat exchangers, as dielectric fluids in electrical equipment, *etc.* Indeed, PCN use actually predates that of PCBs in the UK, with production thought to have occurred between 1919 and the mid-1960s. Although exact production figures are unavailable, total worldwide PCN production has been put at around 10% of that of PCBs.[21]

17.5 IMPORTANT PHYSICOCHEMICAL PROPERTIES AND THEIR INFLUENCE ON ENVIRONMENTAL BEHAVIOUR

The following sections deal with the influence of a number of important physicochemical properties of POPs on their environmental behaviour; in particular, their availability for uptake by plants and animals, routes of human exposure, their relative partitioning between different environmental compartments and their transport throughout the environment. Note that the practical difficulties* of accurately measuring such properties has resulted in considerable uncertainty surrounding values reported for the compounds studied here. As a result, a table of definitive values for physicochemical properties is not included, and selected values are instead quoted to illustrate general trends, and to indicate the order of magnitude of these properties for individual pollutants.

17.5.1 Equilibrium Partitioning Coefficients

17.5.1.1 K_{OW}. Strictly, K_{OW} is a measurement of the equilibrium partition coefficient of a compound between water and octan-1-ol. The significance of this – at first sight, seemingly rather obscure parameter – is greater than might be expected, as it approximates to the lipid–water partition coefficient (K_{LW}). As a

---

*To illustrate, accurate measurement of the K_{OW} for highly hydrophobic POPs is extremely difficult. Measurements of pollutant concentration in the octan-1-ol phase are relatively simple, but accurate determination of the extremely low levels in the water layer is severely hampered by the difficulty in achieving complete separation of the two layers. Given the much higher levels present in the octan-1-ol phase, the presence of even a very small fraction of this phase in the aqueous layer will lead to an exaggeration of the pollutant concentration in the latter. For example, if 50 μg of a substance with a log K_{OW} of 8 is partitioned between 1 dm^3 of both octan-1-ol and water, the equilibrium concentrations in the two layers will be 50 μg dm^{-3} and 0.5 pg dm^{-3} respectively. Inadvertent contamination of the aqueous layer with just 1 μl of the organic phase will raise the apparent aqueous phase concentration to 50.5 pg dm^{-3}, giving rise to an apparent log K_{OW} value of 6.

result, K_{OW} (the log form is most commonly quoted for convenience) is considered an expression of a compound's hydrophobicity, with implications for its bioaccumulative tendencies and movement through the food chain. It is important to note that the 'solubility' of organic chemicals in octan-1-ol only varies within a very narrow range (*i.e.* 200 to 2000 mol m^{-3}). As a consequence, K_{OW} for an organic chemical is predominantly dependent on water solubility, and the common assumption that K_{OW} is an expression of lipophilicity is misleading, as most organic chemicals possess an equal affinity for lipids, whilst having very different affinities for water. As a general rule, K_{OW} increases with increasing molecular weight for a given class of compounds, and hence OCDD (8.2) and benzo[*a*]pyrene (6.3) have higher log K_{OW} values than 2,3,7,8-TCDD (6.8) and naphthalene (3.6) respectively.

17.5.1.2 K_{OC}. The exact definition of K_{OC} is the equilibrium partition coefficient of a chemical between water and natural organic carbon. The wider implications of this parameter (which has units of dm^3 kg^{-1}) is as an indication of the tendency of a chemical to sorb to organic matter. This can have important consequences for – amongst other things – a chemical's ability to leach through soil columns, its availability for uptake by plants and animals from soil and sediments, and its tendency to volatilize from soil. As a general rule for a particular group of chemicals, K_{OC} increases with increasing molecular weight and, hence, OCDD and benzo[*a*]pyrene have higher K_{OC} values than 2,3,7,8-TCDD and naphthalene respectively. As a very general approximation, $K_{OC} = 0.41K_{OW}$, although the dependence of K_{OC} on the chemical form of organic carbon must not be overlooked, and the margin of error associated with this approximation is about a factor of two.

17.5.1.3 K_{OA}. This represents the equilibrium partition coefficient of a chemical between octan-1-ol and air. It has found particular utility in describing the equilibrium partitioning of POPs between plant lipids and the air in which they are immersed (directly analogous to the use of K_{OW} to represent the partitioning of POPs between fish lipid and the water in which they are immersed), and has been incorporated into mathematical modelling frameworks designed to describe POPs transfer through terrestrial food chains (see Section 17.6.2). K_{OA} is the ratio of the octanol–water and air–water partition coefficients, *i.e.*

$$K_{OA} = K_{OW}/K_{AW} \qquad (2)$$

where $K_{AW} = H/RT$, H is the Henry's law constant (see Section 17.5.1.4), R is the ideal gas constant, and T the absolute temperature. Although K_{OA} values are calculable from a knowledge of K_{OW} and H, direct measurements are now possible, with values of log K_{OA} (measured at 25 °C) ranging from *ca.* 6 for the trichlorinated PCBs through to 12.7 for OCDD, with values for PAH falling in the range 6–15. K_{OA} is strongly temperature-dependent, with values inversely related to temperature – *i.e.* low temperatures result in higher K_{OA} values.

17.5.1.4   Henry's Law Constant.   Henry's law constant (H) is a measure of a compound's tendency to partition between a solution and the air above it and can be expressed in different ways. Its dimensionless form H' is directly equivalent to the air–water equilibrium partition coefficient (K_{AW}), and relates the concentration of a chemical in the gas phase C_{sg} to its concentration in the liquid phase C_{sl},

$$H' = C_{sg}/C_{sl} \qquad (3)$$

Henry's law constant can also be written thus,

$$H = P_{vp}/S \qquad (4)$$

where P_{vp} is the vapour pressure of the chemical (in Pa) and S the aqueous solubility or saturation concentration in mol m^{-3}. H therefore has units of Pa m^3 mol^{-1} and can be related to H' by,

$$H' = H/RT \qquad (5)$$

where R is the gas constant (8.31 Pa m^3 mol^{-1} K^{-1}) and T is the temperature (K). The utility of Henry's constant is that it describes the tendency of a chemical to volatilize from water/plant/soil surfaces into the atmosphere, and compounds with a high H value are more prone to volatilization from aqueous solutions. In practical environmental terms, knowledge of H affords an insight into the movement of a chemical into and out of aquatic ecosystems such as ponds, lakes or oceans. Examples of H values (expressed as Pa m^3 mol^{-1}) at 25 °C are: 2,2',3,3'-tetrachlorobiphenyl (22), 2,2',4,4',5,5'-hexachlorobiphenyl (43), 2,3,7,8-TCDD (3.3) and OCDD (0.68).

17.5.2   Aqueous Solubility

As would be expected, an apolar organic compound's aqueous solubility is inversely related to its K_{OW} and K_{OC} and a good approximation is that solubility, at neutral pH and a given temperature, will decrease with increasing molecular weight for a given class of compounds. To illustrate, the water solubilities at 25 °C of phenanthrene, pyrene, and benzo[*a*]pyrene are, respectively: 7.2×10^{-3}, 6.7×10^{-4}, and 1.5×10^{-4} mol m^{-3}. The environmental significance of this property is obvious; not least because more hydrophobic POPs are more prone to food chain transfer.

17.5.3   Environmental Persistence

An exact measure of environmental persistence is impossible to define, as persistence will vary tremendously between different environmental media (*e.g.* atmospheric persistence is appreciably lower than in other environmental compartments – reaction with the hydroxyl radical and dry and wet deposition constituting the principal removal mechanisms) and there are so many confounding variables. Even if environmental persistence is arbitrarily defined as the half-life in soil, measurements of this one parameter vary wildly, depending on the binding of the compound to the soil (which is, in turn, dependent on factors like soil organic matter content, available surface area, K_{OC}, soil moisture content and temperature) and the depth of incorporation below the surface (this influences the intensity of UV radiation received by the chemical and hence its photolysis rate). Even where a definitive half-life can be obtained for a compound there are grave difficulties in reliably assessing the relative importance of each contributory loss mechanism, *e.g.* bacterial degradation, photolysis, leaching, chemical reaction, uptake by biota and volatilization, although recent experimental evidence strongly suggests that, for tri- through pentachlorinated PCBs, volatilization is the major pathway of loss from soils.[22] Despite these caveats, knowledge of the range of a compound's persistence is extremely important in assessing its environmental impact, not least because a longer soil half-life will allow more time for the chemical to transfer into the food chain, and elicit any adverse effects in biota. Illustrative values for soil half-lives range from a few months for lower molecular weight PAHs, PCBs, and OCPs, and up to 10 years or more for PCDD/Fs and higher molecular weight PAHs.

17.5.4 Vapour Pressure

As a general rule, the vapour pressure of a chemical decreases with increasing molecular weight, and hence decachlorobiphenyl possesses a lower vapour pressure (5×10^{-8} Pa) than 2,2',3,3'-tetrachlorobiphenyl (2.3×10^{-3} Pa). Other illustrative measurements of vapour pressure at 25 °C are: 2,3,7,8-TCDD (2×10^{-7} Pa) and OCDD (1.1×10^{-10} Pa). As with all of the other properties dealt with in this section, the influence of vapour pressure on environmental behaviour cannot be considered in isolation, and it is strongly dependent on environmental factors such as temperature. The environmental significance of vapour pressure is tremendous; *inter alia* it influences a compound's partitioning between air and other environmental compartments, thus affecting its rate of atmospheric transport, availability for uptake by biota, and susceptibility to photolytic degradation (compounds spending more time in the vapour phase are generally more susceptible to this). Vapour pressure also exerts a powerful influence on the relevance of inhalation as a human exposure route. This pathway is usually not appreciable for POPs, owing to their low vapour pressures.

17.5.5 General Comments

The combined influence of the above physicochemical properties of individual chemicals, together with environmental factors such as meteorology (temperature, rainfall) and other factors like soil properties *etc.*, governs the environmental fate and behaviour of such chemicals in a predictable fashion. Thus, detailed knowledge of a compound's physicochemical properties, and the range of prevailing environmental conditions which it may experience, affords important insights into how an individual compound may behave once released into the environment.

For example, most POPs display common characteristics in their environmental behaviour. They reside primarily in soil, accumulate through the food chain, and their principal route into humans is *via* food. In addition, they are capable of volatilizing from soils and water and subsequently undergoing long-range atmospheric transport either in the vapour phase or bound to aerosol. It is *via* such atmospheric transport, which occurs in a repeated cycle sometimes referred to as 'the grasshopper effect', that these chemicals distribute in a ubiquitous fashion. This universal distribution is illustrated by the presence of POPs in polar regions, which is a matter of concern, for although inputs to such locations are small, the low temperatures minimize volatilization losses, leading to a steady accumulation of the overall pollutant burden. This accumulation is compounded by the reduced biomass to surface area ratio of polar regions, with the result that the pollutant burden is distributed amongst a much smaller biomass than in industrialized areas. These factors account for the observations of PCB concentrations in the breast milk of Canadian Inuit mothers that exceed those found in women from urban Canada.

17.6 MODELLING ENVIRONMENTAL BEHAVIOUR

The environmental behaviour of a chemical under uniform meteorological and other environmental conditions is largely predictable in terms of a relatively small number of physicochemical properties. Without doubt, the most challenging aspect of predicting the environmental behaviour of a chemical lies in accounting for the uncertainties introduced by fluctuations in meteorological and other environmental parameters. For example, a hot, still and sunny day will result in enhanced volatilization from soil into the atmosphere, where a greater proportion will partition into the vapour as opposed to the aerosol phase, with concomitant susceptibility to reaction with the hydroxyl radical. Similarly, the influence of both the magnitude and chemical composition of the organic content of a soil will exert a profound influence on the environmental fate of a given pollutant. To illustrate, PCBs present in peat-like soils will bind much more strongly to the soil than if they were present in a sandy soil. As a result, movement of pollutants from peat soils – either by volatilization or leaching – is minimal; there is also evidence to suggest that uptake of pollutants from such soils by biota is similarly reduced – in other words, the 'bioavailability' of the chemical is diminished.

17.6.1 The Fugacity Concept

These simple examples serve as illustrations of the complex requirements that a successful environmental modelling technique must meet. Whilst a completely satisfactory model has yet to be developed – there are too many real-world variables for that – the work of Prof. Donald Mackay and co-workers has provided an excellent framework within which the environmental fate of organic contaminants may be predicted with some accuracy. His approach is essentially to construct a 'model world'; for example, a small lake of defined dimensions and with clearly defined values for properties like suspended particulate matter (SPM), the organic content of such matter, and the 'volume' of biota like fish and plants that are present. Assumptions must also be made concerning the depth and physical properties – such as organic content – of lake sediment, and the height of the air layer above the lake surface deemed relevant. Combined with data on the 'model world' dimensions and composition are mass loadings and calculated values of fugacity for the compound(s) of interest; fugacity essentially describing the potential for a substance to move from one environmental compartment to another. Incorporation of all these data into a single computational package allows the construction of a reasonably realistic model, capable of predicting the distribution of a chemical throughout the various environmental compartments of the 'model world'. The model can be further enhanced by inclusion of transport factors relating to processes such as leachability and volatilization, along with reaction properties to provide information on loss processes such as biodegradation and hydrolysis, and this basic approach can be modified to consider smaller or larger 'model worlds' as appropriate, in whatever degree of detail necessary. To illustrate, the concept has been used to assess the global fate and distribution of toxaphenes – dividing the globe into nine separate 'model worlds' such as the arctic, the antarctic, northern hemisphere temperate, tropical, *etc.* and modelling chemical distribution within and between each of the 'model worlds'.[23] One use for this global model has been to assess the relative propensity for chemicals to transfer from temperate zones and 'condense out' at the polar regions. On a smaller scale, the approach can be utilized to predict chemical movement within a soil column following amelioration with sewage sludge, and assess the potential for leaching to groundwater or uptake by crops.

17.6.2 Equilibrium Partitioning Modelling Approaches

In addition to the more sophisticated modelling approaches represented by, *inter alia*, the fugacity approach it is possible to predict POPs behaviour to an acceptable degree of accuracy by simple equilibrium partitioning approaches. The most widely employed example is in predicting POPs transfer from water to fish. Ignoring intake *via* the diet, and assuming that POPs are at equilibrium between fish lipid and the surrounding water, the relationship between POPs concentration in fish lipid (C_f) and that in water (C_w), may be described in terms of K_{OW} thus:

$$\mathrm{Log}(C_f/C_w) = \mathrm{Log}\, K_{OW} \tag{6}$$

where C_f/C_w is referred to as the bioaccumulation factor (BAF). In reality, while experimental observations have demonstrated a linear relationship between the

BAF and K_{OW}, the relationship is of the type shown below, with values of m and c varying on both a location- and species-specific basis:

$$Log \ BAF = m \ Log \ K_{OW} + c \qquad (7)$$

As an illustration, the relationships between BAF and K_{OW} for eels and pike from the river Severn are:[24]

$$Log \ BAF = 0.97 \ Log \ K_{OW} - 0.52 \ (eels) \qquad (8)$$

$$Log \ BAF = 0.94 \ Log \ K_{OW} + 0.04 \ (pike) \qquad (9)$$

A similar approach has been taken to describing POPs partitioning between air and plants, leading to relationships of the form:

$$Log(C_P/C_A) = m \ Log \ K_{OA} + c \qquad (10)$$

where C_P and C_A are the POPs concentrations in plant lipid and air respectively. As with the analogous relationship between fish lipid and water, such relationships assume POPs to exist in an equilibrium between plant lipid and air. This has been shown to be true for the tri- through pentachlorinated PCBs, but it appears that other POPs such as the higher chlorinated PCBs and PCDD/Fs do not reach such an equilibrium. As with the relationship between BAF and K_{OW}, there is evidence of appreciable species- and location-specific variation in both m and c.

In conclusion, it is important to note that although mathematical models are extremely useful in comparing the behaviour of different chemicals under identical environmental conditions, they are not yet sufficiently reliable to provide accurate predictions of actual concentrations.

17.7 REFERENCES

1. R. Carson, 'The Silent Spring', Hamish Hamilton, London, 1963.
2. D. E. Schulz, G. Petrick and J. C. Duinker, *Environ. Sci. Technol.*, 1989, **23**, 852.
3. M. Van den Berg, L. Birnbaum, A. T. C. Bosveld, B. Brunström, P. Cook, M. Feeley, J. P. Giesy, A. Hanberg, R. Hasegawa, S. W. Kennedy, T. Kubiak, J. C. Larsen, F. X. R. van Leeuwen, A. K. D. Liem, C. Nolt, R. E. Peterson, L. Poellinger, S. Safe, D. Schrenk, D. Tillitt, M. Tysklind, M. Younes, F. Waern and T. Zacharewski, *Environ. Health Perspect.*, 1998, **106**, 775.
4. F. W. Karasek and O. Hutzinger, *Anal. Chem.*, 1986, **54**, 309.
5. F. X. R. van Leeuwen and M. Younes, *Organohalogen Compd.*, 1998, **38**, 295.
6. M. Huisman, C. Koopman-Esseboom, V. Fidler, M. Hadders-Algra, C. G. van der Pauw, L. G. M. T. Tuinstra, N. Weisglas-Kuperus, P. J. J. Sauer, B. C. L. Touwen and E. R. Boersma, *Early Hum. Dev.*, 1995, **41**, 111.
7. J. L. Jacobson and S. W. Jacobson, *New Engl. J. Med.*, 1996, **335**, 783.
8. P. J. Coleman, R. G. M. Lee, R. E. Alcock and K. C. Jones, *Environ. Sci. Technol.*, 1997, **31**, 2120.
9. P. Eriksson, H. Viberg, E. Jakobsson, U. Örn and A. Fredriksson, *Organohalogen Compd.*, 1999, **40**, 333.

10. D. Meironyte, Å. Bergman and K. Norén, *Organohalogen Compd.*, 1998, **35**, 387.
11. L.-O. Kjeller, K. C. Jones, A. E. Johnston and C. Rappe, *Environ. Sci. Technol.*, 1991, **25**, 1619.
12. S. J. Harrad and K. C. Jones, *Sci. Total Environ.*, 1992, **26**, 89.
13. C. Rappe, *Organohalogen Compd.*, 1990, **4**, 33.
14. G. H. Eduljee and P. Dyke, *Sci. Total Environ.*, 1996, **177**, 303.
15. R. E. Alcock, A. E. Johnston, S. P. McGrath, M. L. Berrow and K. C. Jones, *Environ. Sci. Technol.*, 1993, **27**, 1918.
16. G. M. Currado and S. Harrad, *Environ. Sci. Technol.*, 1998, **32**, 3043.
17. K. C. Jones, J. A. Stratford, K. S. Waterhouse, E. T. Furlong, W. Giger, R. A. Hites, C. Schaffner and A. E. Johnston, *Environ. Sci. Technol.*, 1989, **23**, 95.
18. L. H. Lim, R. M. Harrison and S. Harrad, *Environ. Sci. Technol.*, 1999, **33**, 3538.
19. T. F. Bidleman and R. L. Falconer, *Environ. Sci. Technol.*, 1999, **33**, 206A.
20. R. Renner, *Environ. Sci. Technol.*, 2000, **34**, 223A.
21. F. A. Beland and R. D. Greer, *J. Chromatgr.*, 1973, **84**, 59.
22. S. Ayris and S. Harrad, *J. Environ. Monit.*, 1999, **1**, 395.
23. F. Wania and D. Mackay, *Chemosphere*, 1993, **27**, 2079.
24. S. Harrad and D. Smith, *Environ. Sci. Pollut. Res.*, 1997, **4**, 189.

CHAPTER 18

Radioactivity in the Environment

C. N. HEWITT

18.1 INTRODUCTION

Pollution of the natural environment by radioactive substances is of concern because of the considerable potential that ionising radiation has for damaging biological material and because of the very long half-lives of some radio-nuclides. Although both the benefits of controlled exposure for medical purposes and the catastrophic effects of large doses of radiation (for example, those received by the inhabitants of Hiroshima and Nagasaki in 1945 and in the vicinity of Chernobyl in 1986) are well understood, what is less clear are the effects of small doses on the general population. For this reason it is necessary to evaluate in detail the exposure received by man from each of the multiplicity of natural and man-made sources of radioactivity in the environment. In order to do this, an understanding is required of the actual and potential source strengths, the pathways and cycling of radioactivity through the environment and their flux rates, and the possible routes of exposure for man.

18.2 RADIATION AND RADIOACTIVITY

18.2.1 Types of Radiation

Radiation arises from a spontaneous rearrangement of the nucleus of an atom. Whilst some nuclei are stable many are not and these can undergo a change, losing mass or energy in the form of radiation. Some unstable nuclei are naturally occurring while others are produced synthetically. The most common forms of these radiations are alpha particles, beta particles, and gamma rays, the physical properties of which are shown in Table 18.1 and are described below.

(a) Alpha (α) particles consist of two protons and two neutrons bound together and so are identical to helium nuclei. They have a mass number of 4 and a charge of $+2$. Because they are so large and heavy, alpha particles travel slowly compared with other types of radiation (at maximum, about 10% the speed of light) and can be stopped relatively easily. Their ability to penetrate into living tissue is therefore limited and damage occurs to animals only when

Table 18.1 *Types of radioactive particles and rays*

Radiation	Symbol	Composition	Charge	Mass number	Approximate tissue penetration (cm)
Alpha	α	Particle containing two protons and two neutrons	$2+$	4	0.01
Beta	β	Particle of one electron	$1-$	0	1
Gamma	γ	Very short wavelength electromagnetic radiation	0	0	100

alpha-emitting isotopes are ingested or inhaled. However, their considerable kinetic energy and the double positive charge which attracts and pulls away electrons from atoms belonging to tissue means that alpha particles can cause the formation of ions and free radicals and hence severe chemical change along their path. The emission of alpha particles is common only for nuclides of mass number greater than 209 and atomic number 82, nuclides of this size having too many protons for stability. An example of α-decay is:

$$^{210}_{84}\text{Po} \longrightarrow {}^{206}_{82}\text{Pb} + {}^{4}_{2}\text{He}$$

Note that the sum of the superscripts (mass numbers or sum of neutrons and protons) and the sum of the subscripts (atomic numbers or sum of protons) remain unchanged during the decay.

(b) Beta (β) particles are simply electrons emitted by the nucleus during the change of a neutron into a proton. They have minimal mass (1.36×10^{-4} times that of an alpha particle) but high velocity (typically 40% the speed of light) and a charge of -1. They may penetrate through skin or surface cells into tissue and may then pass close to the orbital electrons of tissue atoms where the repulsion of the two negative particles may force the orbital electron out of the atom, ionizing the tissue, and forming radicals. Because beta decay involves the change of a neutron into a proton the atomic number of the nuclide increases by one, but the mass number does not change. Beta decay is a common mode of radioactive disintegration and is observed for both natural and synthetic nuclides. An example is:

$$^{90}_{38}\text{Sr} \longrightarrow {}^{90}_{39}\text{Y} + {}_{-1}^{0}\text{e}$$

(c) Gamma (γ) radiation is very short wavelength electromagnetic radiation. It travels at the speed of light, is uncharged, but is highly energetic and so has considerable penetration power. As it passes through biological tissue the electric field surrounding a gamma ray may eject orbital electrons from atoms and so can cause ionization of the tissue and formation of radicals along its path. The emission of gamma radiation does not lead to changes in mass or atomic numbers, and may occur either on its own from an electronically excited nucleus or may accompany other types of radioactive decay. Examples of these two processes are:

$$^{125}_{52}Te^* \longrightarrow {}^{125}_{52}Te + \gamma$$

the asterisk signifying the excited state of the nucleus and:

$$^{137}_{55}Cs \longrightarrow {}^{137}_{56}Ba + {}^{0}_{-1}e \longrightarrow {}^{137}_{56}Ba + \gamma$$

18.2.2 The Energy Changes of Nuclear Reactions

The energy changes associated with nuclear reactions are considerably greater than those associated with ordinary chemical reactions. The sum of the mass of the products of the nuclear reaction is invariably less than the sum of the mass of the reactants and the amount of energy released (ΔE) is equivalent to this difference in mass (Δm). Most of this energy is released as kinetic energy, although some may be used to promote the nucleus to an excited state from where it will lose energy in the form of irradiation and return to the ground state.

The energy equivalent of a given mass can be calculated by means of Einstein's equation:

$$\Delta E = (\Delta m)c^2$$

The energy equivalent of one atomic mass unit (1 u = 1.660566×10^{-27} kg) is:

$$\Delta E = (1.660566 \times 10^{-27} \text{ kg})(2.99792 \times 10^8 \text{ m s}^{-1})^2$$
$$= 1.49244 \times 10^{-10} \text{ J}$$

This is usually expressed in units of electron volts (eV) where:

$$1 \text{ eV} = 1.60219 \times 10^{-19} \text{ J}$$
and $$1 \text{ MeV} = 1.60219 \times 10^{-13} \text{ J}$$

The energy equivalent of 1 u is therefore 931.5 MeV.

The amount of energy released by a decay process can now be calculated. For example, in the alpha decay of polonium-210:

$$^{210}_{84}Po \longrightarrow {}^{206}_{82}Pb + {}^{4}_{2}He$$

$$\Delta m = (\text{mass } {}^{210}_{84}\text{Po}) - (\text{mass } {}^{206}_{82}\text{Pb} + \text{mass } {}^{4}_{2}\text{He})$$
$$= 209.9829 \text{ u} - (205.9745 \text{ u} + 4.0026 \text{ u})$$
$$= 0.0058 \text{ u}$$

The energy released by the decay is therefore:

$$\Delta E = 0.0058 \text{ u} \times 931.5 \text{ MeV/u} = 5.4 \text{ MeV}$$

18.2.3 Rates of Radioactive Decay

The rates of decay of radioactive nuclides are first-order and independent of temperature. This implies that the activation energy of radioactive decay is zero and that the rate of decay depends only on the amount of radioactive substance present. If N is the number of atoms present at time t the rate of change of N is given by

$$dN/dt = -\lambda N$$

where λ is the characteristic (or disintegration or decay) constant for that radionuclide. Integrating between times t_1 and t_2 gives:

$$N_2 = N_1 \exp[-\lambda(t_2 - t_1)]$$

where N_1 and N_2 are the number of atoms of the radionuclide present at times t_1 and t_2 respectively. If t_1 is set to zero then:

$$N = N_0 \exp(-\lambda t) \tag{1}$$

where t is the elapsed time and N_0 is the number of atoms of the radionuclide present when $t_1 = 0$.

When N/N_0 is equal to 0.5 (*i.e.*, half the atoms have decayed) t is defined as being the half-life, $t_{1/2}$. Then:

$$N/N_0 = 0.5 = \exp(-\lambda t_{1/2})$$

$$t_{1/2} = 0.693/\lambda$$

Equation (1) can now be written in terms of the more readily available t_1, rather than λ, to give

$$N = N_0 \exp(-0.693 t_1/t_{1/2}).$$

The half-lives of some selected radionuclides are given in Table 18.2.

Table 18.2   *Half-lives of some environ-*
                mentally important radio-
                nuclides

Radionuclide	Half-life
^{131}I	8.1 d
^{85}Kr	10.8 y
^{3}H	12.3 y
^{90}Sr	28 y
^{137}Cs	30 y
^{239}Pu	2.4×10^4 y
^{238}U	4.5×10^9 y

18.2.4   Activity

The amount of radiation emitted by a source per unit time is known as the activity of the source, expressed in terms of the number of disintegrations per second. The unit of activity, the becquerel (Bq), is defined as being one disintegration per second. The activity of a source is proportional to the number of radioactive atoms present and so diminishes with time according to first order kinetics.

18.2.5   Radioactive Decay Series

Some radioactive decay processes lead in one step to a stable product but frequently a disintegration leads to the formation of another unstable nucleus. This can be repeated several times, producing a radioactive decay series which only terminates on the formation of a stable nuclide. There are three naturally occurring decay series, headed by ^{232}Th, ^{238}U, and ^{235}U. Each of these nuclides has a half-life which is long in relation to the age of the earth and each finally produces stable isotopes of lead, ^{208}Pb, ^{206}Pb and ^{207}Pb respectively. The 14 steps of the $^{238}_{92}$U series are shown in Figure 18.1 and it will be seen that at several points branching occurs as the series proceeds by two different routes which rejoin at a later point.

18.2.6   Production of Artificial Radionuclides

The first artificial transmutation of one element into another was achieved in 1919 when Ernest Rutherford passed α-particles (produced by the radioactive decay of ^{214}Po) through nitrogen:

$$^{14}_{7}N + {}^{4}_{2}He \longrightarrow {}^{17}_{8}O + {}^{1}_{1}H$$

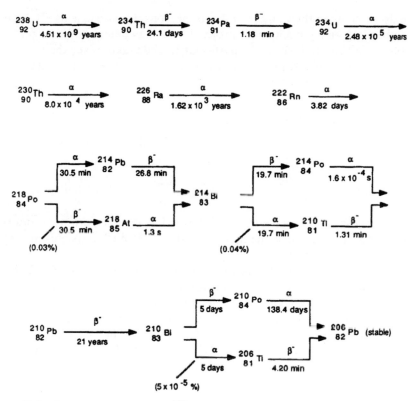

Figure 18.1 *Disintegration series of* $^{238}_{92}U$ *(half-lives of isotopes are indicated)*

Subsequently the first artificial radioactive nuclide, ^{30}P, was produced:

$$^{27}_{13}N + ^4_2He \longrightarrow ^{30}_{15}P + ^1_0n \qquad \text{(neutron)}$$

$$^{30}_{15}P \longrightarrow ^{30}_{14}Si + ^0_1e \qquad \text{(positron)}$$

Bombardment reactions of this type have been used to produce isotopes of elements which do not exist in nature and which are particularly important as pollutants of the environment, for example:

$$^{238}_{92}U + ^1_0n \longrightarrow ^{239}_{92}U + \gamma \qquad \text{uranium}$$

$$^{239}_{92}U \longrightarrow ^{239}_{93}Np + ^0_{-1}e \qquad \text{neptunium}$$

$$^{239}_{93}Np \longrightarrow ^{239}_{94}Pu + ^0_{-1}e \qquad \text{plutonium}$$

$$^{239}_{94}Pu + ^2_1H \longrightarrow ^{240}_{95}Am + ^1_0n \qquad \text{americium}$$

18.2.7 Nuclear Fission

Many heavy nuclei with mass numbers greater than 230 are susceptible to spontaneous fission, or splitting into lighter fragments, as a result of the forces of repulsion between their large number of protons. Fission can also be induced by bombarding heavy nuclei with projectiles such as neutrons, alpha particles, or protons. When the fission of a particular nuclide takes place the nuclei may split in a variety of ways, producing a number of products. For example, some of the possible fission reactions of ^{235}U are:

$$^{235}_{92}U + ^{1}_{0}n \longrightarrow ^{95}_{39}Y + ^{138}_{53}I + 3^{1}_{0}n$$

$$^{235}_{92}U + ^{1}_{0}n \longrightarrow ^{97}_{39}Y + ^{137}_{53}I + 2^{1}_{0}n$$

$$^{235}_{92}U + ^{1}_{0}n \longrightarrow ^{90}_{36}Kr + ^{144}_{56}Ba + 2^{1}_{0}n$$

$$^{235}_{92}U + ^{1}_{0}n \longrightarrow ^{90}_{35}Br + ^{143}_{57}La + 3^{1}_{0}n$$

In each of these reactions neutrons are formed as primary products and these neutrons can, in turn, cause the fission of other ^{235}U nuclei, so causing a chain reaction of the fissile uranium.

The number of nuclei of a particular daughter product formed by the fission of 100 parent nuclei is defined as the yield of the process. Fission yields vary, fission induced by slow neutrons (as in a nuclear power plant) have a different set of yields to that induced by fast neutrons (as in many nuclear weapons). Table 18.3 shows examples of the percentage yields of the fast-neutron induced fission of ^{239}Pu. It also shows some of the yields of ^{235}U fission, which are similar for both slow and fast neutron induced processes.

18.2.8 Beta Decay of Fission Products

The primary daughter products of nuclear fission are almost always β-radioactive and often these also quickly produce β-radioactive products. Only after several such decays are products with long half-lives formed. For example, the fission of ^{235}U produces ^{90}Br:

$$^{235}_{92}U + ^{1}_{0}n \longrightarrow ^{90}_{35}Br + ^{143}_{57}La + 3^{1}_{0}n$$

Table 18.3 *Yields of some long-lived radionuclides following uranium and plutonium fission*

Radioisotope	Half-life	Yield (%)	
		^{235}U	^{239}Pu
Strontium-90	28 y	5.8	2.2
Iodine-131	8 d	3.1	3.8
Caesium-137	30 y	6.1	5.2
Krypton-85	10.3 y	0	–
Cerium-141	33 d	6.0	5.2

which quickly decays to ^{90}Kr with a half-life of 1.4 s. This in turn has a half-life of 33 s and decays to ^{90}Rb ($t_{1/2} = 2.7$ min), and this to ^{90}Sr:

$$^{90}_{35}\text{Br} \longrightarrow {}^{90}_{36}\text{Kr} + {}^{0}_{-1}e \qquad (t_{1/2} = 1.4 \text{ s})$$
$$^{90}_{36}\text{Kr} \longrightarrow {}^{90}_{37}\text{Rb} + {}^{0}_{-1}e \qquad (t_{1/2} = 33 \text{ s})$$
$$^{90}_{37}\text{Br} \longrightarrow {}^{90}_{38}\text{Sr} + {}^{0}_{-1}e \qquad (t_{1/2} = 2.7 \text{ min})$$

Strontium-90 decays to yttrium-90:

$$^{90}_{38}\text{Sr} \longrightarrow {}^{90}_{38}\text{Y} + {}^{0}_{-1}e \qquad (t_{1/2} = 28 \text{ y})$$

and has a half-life of 28 y, which is sufficiently long for it to be widely circulated through the environment, in contrast to its short-lived precursors. After one more beta decay a stable product, ^{90}Zr, is formed. It is therefore radioactive strontium rather than its precursors that is the important environmental pollutant in this series.

Examples of other environmentally-important nuclides formed by beta decay chains are iodine-131 and caesium-137. ^{131}I is formed from ^{131}Sn, itself produced by the fission of ^{235}U, and forms the stable nuclide ^{131}Xe:

$$^{131}_{50}\text{Sn} \longrightarrow {}^{131}_{51}\text{Sb} + {}^{0}_{-1}e \qquad (t_{1/2} = 3.4 \text{ min})$$
$$^{131}_{51}\text{Sb} \longrightarrow {}^{131}_{52}\text{Te} + {}^{0}_{-1}e \qquad (t_{1/2} = 23 \text{ min})$$
$$^{131}_{52}\text{Te} \longrightarrow {}^{131}_{53}\text{I} + {}^{0}_{-1}e \qquad (t_{1/2} = 24 \text{ min})$$
$$^{131}_{53}\text{I} \longrightarrow {}^{131}_{54}\text{Xe} + {}^{0}_{-1}e \qquad (t_{1/2} = 8 \text{ days})$$

^{137}Cs begins as the uranium fission daughter ^{137}I and ends as the stable barium isotope, ^{137}Ba:

$$^{137}_{53}\text{I} \longrightarrow {}^{137}_{54}\text{Xe} + {}^{0}_{-1}e \qquad (t_{1/2} = 24 \text{ s})$$
$$^{137}_{54}\text{Xe} \longrightarrow {}^{137}_{55}\text{Cs} + {}^{0}_{-1}e \qquad (t_{1/2} = 3.9 \text{ min})$$
$$^{137}_{55}\text{Cs} \longrightarrow {}^{137}_{56}\text{Ba} + {}^{0}_{-1}e \qquad (t_{1/2} = 30 \text{ years})$$

18.2.9 Units of Radiation Dose

The amount of biological damage caused by radiation and the probability of the occurrence of damage are directly related to dosage and so before discussing the effects of radiation it is necessary to consider the units of dose.

18.2.9.1 Absorbed Dose. The amount of energy actually absorbed by tissue or other material from radiation is known as the *absorbed dose* and is expressed in a unit called the *gray* (Gy). One gray is equal to the transfer of one joule of energy to one kilogram of material. Table 18.4 shows the relationship of this and other SI units with the older units they have replaced.

Table 18.4 *SI and old radiation units*

Quantity	SI Unit	Old unit	Relationship
Activity	becquerel	curie	$1\ Ci = 3.7 \times 10^{10}\ Bq$
Absorbed dose	gray	rad	$1\ rad = 0.01\ Gy$
Dose equivalent	sievert	rem	$1\ rem = 0.01\ Sv$

18.2.9.2 Dose Equivalent. Because the physical and chemical properties of α, β, and γ radiations vary, equal absorbed doses of radiation do not necessarily have the same biological effects. In order to equate the biological effects of one type of radiation with another it is therefore necessary to multiply the absorbed dose by a quality factor that accounts for the differences in biological damage caused by radioactive particles having the same energy. This is known as the *dose equivalent* and is expressed in units of *sieverts* (Sv). For γ-rays, X-rays, and β-particles the quality factor equals 1, whereas for α-particles it is 20. The dose equivalent resulting from an absorbed dose of 1 Gy of alpha radiation therefore equals 20 Sv.

18.2.9.3 Effective Dose Equivalents. Having established the dose equivalent for each tissue (H_T) it is now necessary to weight this to take account of the differing susceptibilities of different organs and types of tissue to damage. For example the testes and ovaries are more easily damaged than are the lungs or bones. The risk weighting factors (W_T) currently recommended by the International Commission for Radiological Protection (ICRP) are shown in Table 18.5. The effective dose equivalent (H_E) for the body is then expressed as the sum of the weighted dose equivalents:

$$H_E = \sum_T H_T W_T$$

18.2.9.4 Collective Effective Dose Equivalents. As well as quantifying the effective dose equivalent (or *dose*) received by individuals it is also important to have a measure of the total radiation dose by a group of people or population. This is the *collective effective dose equivalent* (or *collective dose*) and is obtained

Table 18.5 *ICRP risk weighting factors*

Tissue or organ	Weighting factor
Testes and ovaries	0.25
Breast	0.15
Red bone marrow	0.12
Lung	0.12
Thyroid	0.03
Bone surfaces	0.03
Remainder	0.30
Whole body total	1.00

by multiplying the mean effective dose equivalent to the group from a particular source by the number of people in that group, to give units of man Sieverts (man Sv).

18.3 BIOLOGICAL EFFECTS OF RADIATION

18.3.1 General Effects

The formation of ions and free radicals in tissue by radiation and the subsequent chemical reactions between these reactive species and the tissue molecules causes a range of short-term and long-term biological effects. As a frame of reference, acute exposure can cause death, an instantaneously absorbed dose of 5 Gy probably being lethal. The range of effects include cataracts, gastrointestinal disorders, blood disorders including leukaemia, damage to the central nervous system, impaired fertility, cancer, genetic damage and changes to chromosomes producing mutations in later generations. Of the long-term effects of radiation to the exposed generation, cancer and especially leukaemia is probably the most important. The number of excess (or extra) cancers observed in an exposed group compared with a non-exposed control group divided by the product of the exposed group size and the mean individual dose gives the *risk factor*. This is the risk of the effect occurring per unit dose equivalent.

Risk factors have been estimated by the United Nations Scientific Committee on the Effects of Atomic Radiation (UNSCEAR) by studying various groups of exposed people. The risk factor found for fatal leukaemia, for example, is about 20×10^{-3} Sv^{-1} or about a 1 in 500 chance of dying of leukaemia following an exposure of 1 Sv. When irradiation of the testes or ovaries occurs before the conception of children there is a risk that damage to the DNA may cause hereditary defects in future generations. The current UNSCEAR estimate of the risk factor for serious hereditary damage to humans is about 2×10^{-2} Sv^{-1}, or about 1 in 50 per Sv, with about half of this damage manifesting itself in the first two generations. When fatal cancers and serious hereditary defects in the first two generations are considered together the current ICRP estimate of the risk factor is 1.65×10^{-2} Sv^{-1}. This is based on the assumption that there is no lower threshold of dose below which the probability of effect is zero. In other words it is assumed that any exposure to radiation carries some risk and that the probability of the effect occurring is proportional to the dose. It is worth noting that the historical trend has been for estimates of risk to increase with time, and hence the estimates of acceptable dose have decreased with time.

18.3.2 Biological Availability and Residence Times

The rate of uptake of a radionuclide into the body depends upon its concentrations in the various environmental media (air, water, food, dust, *etc.*), the rates of intake of these media and the efficiency with which the body absorbs the nuclide from the media. This latter parameter largely depends upon the physical and chemical properties of the radioactive substance in question. For example, the chemical properties of strontium are similar to those of another Group II element, calcium, which is of vital biological significance. Calcium enters the body to form bones and to carry on other physiological functions and ^{90}Sr can readily follow calcium into the bones where it will remain. In the same way the chemical similarity of caesium to potassium, which is present in all cells in the body, allows caesium to be readily transported throughout the entire body. On the other hand the noble gases krypton, argon, xenon and radon, which are all radioactive air pollutants, are not readily absorbed by the body (although they can still contribute to the exposure of an individual by external irradiation and by forming radioactive daughters with different physico-chemical properties).

In any given organism there exists a balance between the intake of an element and its excretion and this controls the concentration of the element in the organism. For radioactive isotopes this balance will include on the debit side the radioactive decay of the isotope. The effective half-life, t_{eff}, of a nuclide in an organism, *i.e.*, the time required to reduce the activity in the body by half, is a function of both the biological half-life, t_{biol}, and the radioactive half-life, t_{rad}:

$$t_{eff} = t_{rad}t_{biol}/(t_{rad} + t_{biol})$$

Whilst t_{rad} is fixed for any given nuclide, t_{biol} will vary from species to species and between individuals with age, sex, physical condition, and metabolic rate. For ^{90}Sr which has $t_{rad} = 28$ y and $t_{biol} =$ about 35 y the effective half-life for man is about 15.5 y. Obviously those nuclides of most concern as pollutants are those present in the environment in the highest concentrations, with the most energetic emissions and with the longest effective biological half-lives.

18.4 NATURAL RADIOACTIVITY

18.4.1 Cosmic Rays

The earth's atmosphere is continuously bombarded by highly energetic protons and alpha particles emitted by the sun and of galactic origin. Their energies range from about 1 MeV to about 10^4 MeV, with a flux rate at the outer edge of the atmosphere of about 2×10^7 MeV m^{-2} s^{-1}. These primary particles have two effects. They cause some radiation exposure directly to man, the magnitude of which varies with altitude and latitude, and they interact with stable components of the atmosphere causing the formation of radionuclides. Of these, ^{3}H and especially ^{14}C are important to the biosphere, while several of the shorter-lived nuclides (*e.g.* ^{7}Be, $t_{1/2} = 53$ days; ^{39}Cl, $t_{1/2} = 55$ min) have applications as tracers in the study of atmospheric dispersion and deposition processes.

Carbon-14 is formed from atmospheric nitrogen by:

$$^{14}_{7}\text{N} + ^{1}_{0}\text{n} \longrightarrow ^{15}_{7}\text{N}$$

$$^{15}_{7}\text{N} \longrightarrow ^{14}_{6}\text{C} + ^{1}_{1}\text{p}$$

The ^{14}C is then oxidized to CO_2 and enters the global biogeochemical cycle of carbon, being incorporated into plants by photosynthesis and into the oceans by absorption. Most natural production of tritium is also by the reaction of cosmic ray neutrons (with energy > 4.4 MeV) with nitrogen:

$$^{14}_{7}N + ^{1}_{0}n \longrightarrow ^{12}_{6}C + ^{3}_{1}H$$

Tritium is then oxidized or exchanges with ordinary hydrogen to form tritiated water and enters the global hydrological cycle. The National Radiological Protection Board (NRPB) estimate that the annual effective dose equivalent from cosmic rays in the UK is about 300 μSv y^{-1} on average. However, people that live at high altitudes and high latitudes can receive rather greater doses.

18.4.2 Terrestrial Gamma Radiation

The earth's crust contains three elements with radioactive isotopes which contribute significantly to man's exposure to radiation. These are ^{40}K, with an average concentration in the upper crust of ~ 3 ppm, ^{232}Th, which is present in granitic rocks at 10–15 ppm, and the three isotopes of uranium which total an average 3–4 ppm in granite. These latter nuclides have relative abundances of 99.274% ^{238}U, 0.7205% ^{235}U, and 0.0056% ^{234}U in the crust.

The NRPB estimate that the gamma rays emitted by these radionuclides and their daughters in soil, sediments, and rocks and in building materials give an average annual effective dose equivalent in the UK of about 400 μSv y^{-1}, with the whole body being irradiated more or less equally. The local geology and types of building material used lead to considerable variation about this figure.

18.4.3 Radon and its Decay Products

The decay series of ^{238}U and ^{232}Th both contain radioactive isotopes of the element radon, ^{222}Rn in the former case and ^{220}Rn (sometimes known as thoron) in the latter. Radon, being a noble gas, readily escapes from soil and porous rock and diffuses into the lower atmosphere. There the nuclides decay with half-lives of 3.8 days (^{222}Rn) and 55 s (^{220}Rn) producing a series of short-lived daughter products. The full ^{238}U decay series is shown in Figure 18.1. These daughters attach themselves to aerosol particles in the atmosphere. These particles are efficiently deposited in the lungs if inhaled. Their subsequent α and β emissions can then irradiate and damage the lung tissue. The mean flux rate of radon from soil to the atmosphere is about 1200 Bq m^{-2} d^{-1} but there are large variations about this value at very small spatial scales.

Ambient outdoor concentrations of radon are low, typically about 3 Bq m^{-3} in the UK. However, when these gases enter a building, either through the floor following soil emissions or from the building's construction materials or by desorption from the water supply, the low air exchange rates in modern buildings causes a substantial increase in concentrations. Considerable variations in indoor concentrations have been observed, a positively-skewed distribution with a mean of ~ 21 Bq m^{-3} being typical in the UK. The concentrations in the outside air, in the soil, and in soil pore water, the rate of emanation from building materials and the building ventilation rate will all affect the indoor concentration, and in a large scale measurement programme a few individual dwellings with concentrations two or three orders of magnitude above the mean might be observed. Because of the large variations in concentration and differences in people's activity patterns the calculation of an individual's exposure to radiation from radon is rather uncertain. The average annual effective dose equivalent in the UK is estimated by NRPB to be ~ 1300 μSv y^{-1}, or 50% of the total annual average individual dose of 2600 μSv, although some individuals may receive very much higher doses from radon than this. In any case radon and its daughters generally give a greater dose to man than any other sources of radiation. In the UK an 'Action Level' has been proposed by NRPB (200 Bq m^{-3}) above which remedial action to lower radon concentrations is recommended. Areas with $>1\%$ of dwellings with radon levels above the Action Level are designated as 'Affected Areas'.

In the USA in particular, attempts are made to reduce indoor radon exposure in the worst affected areas and buildings. This can be done by preventing radon entering the building, by increasing the building ventilation rate or by removing the radon decay products from the air by, for example, using an electrostatic precipitator. Since most radon in buildings comes from the ground below the best method is to reduce the ingress of radon into the building by sealing the floor or by creating or increasing under floor ventilation.

18.4.4 Radioactivity in Food and Water

The most important naturally occurring radionuclides in food and water are ^{226}Ra, formed from ^{238}U, and its α-emitting daughter products ^{222}Rn, ^{218}Po and ^{214}Bi. There is a wide variation in the ^{226}Ra content of public water supplies, it usually being very low and representing a minor source of intake, but some well and spring water may contain 0.04–0.4 Bq L^{-1}. As previously mentioned, the domestic water supply, particularly water used in showers, may act as a source of ^{222}Rn into the indoor environment. For the population in general the main source of radium intake is from food. Wide variations in concentration are found but a typical daily intake from the average diet is about 0.05 Bq per day. Some foods, for example Brazil nuts and Pacific salmon, accumulate radium in preference to calcium and can have very much higher concentrations than the average for food. Unbalanced diets based on these foods could lead to enhanced intakes of radium. Both ^{210}Pb and ^{210}Po can enter food, both from the soil but also by wet and dry deposition from the atmosphere where they are present as daughters of ^{222}Rn. For the average individual in the UK, the NRPB estimate of the total effective dose equivalent of alpha activity from the diet is about 200 μSv y^{-1}.

An additional source of α-activity for some individuals which should be mentioned is cigarette smoke. The decay products of radon emitted from the ground under tobacco plants can adsorb onto the growing plant leaves and hence be incorporated into cigarettes. Dose rates have been estimated to be as high as 6–7 mSv per year from this source for some individuals.

Most of the naturally occurring β activity in food is due to ^{40}K. The availability of potassium to plants is subject to wide variations and hence the ^{40}K activity of foods also varies. However, potassium is an essential element for plants and animals, constituting about 0.2% of the soft tissue of the body. This leads to a total ^{40}K content of ~ 4 kBq for the average person although the individual value depends upon age, weight, sex, and proportion of fat. The NRPB estimate that this gives an average effective dose equivalent of about 170 μSv y^{-1} to individuals in the UK.

18.5 MEDICAL APPLICATIONS OF RADIOACTIVITY

Although not generally regarded as being pollutants, radiations used for medical purposes make a significant contribution to man's exposure. Indeed they provide the major artificial route of exposure to man and so warrant mention here. X-rays are used for a wide range of diagnostic purposes, a typical chest X-ray giving an effective dose equivalent of about 20 μSv. Short-lived radionuclides are also used diagnostically. For example, ^{99}Tc is used for bone and brain scans. The use of radiation for diagnostic purposes gives an average effective dose equivalent in the UK of about 250 μSv per year. Of this about 120 μSv y^{-1} is considered to be genetically significant, compared with about a 1000 μSv y^{-1} genetically significant dose from natural sources. Restrictions on the use of X-rays during pregnancy, limiting the X-ray beam area to the minimum needed, reducing radiation leakage from the source and improving the sensitivity and reliability of the measuring methods used are all aimed at reducing the genetically significant dose to reduce the probability of future genetic effects.

Externally administered beams of X-rays, gamma rays (from ^{60}Co sources) and neutrons and radiations from internally administered radionuclides (*e.g.*, ^{131}I in the thyroid) are all used for therapeutic purposes. Some of the doses involved are very high but the potential adverse effects have to be weighed against the benefits accrued to the individual patient. Regular and stringent calibration and testing of medical radiation sources is vital if patients are to receive the correct dose in the correct place.

18.6 POLLUTION FROM NUCLEAR WEAPONS EXPLOSIONS

Since 1945 the products of nuclear weapons explosions, both of fission and fusion devices, have caused considerable pollution of the globe. The major bomb testing programmes were held in 1954–8 and 1961–2, with the number of documented test explosions now approaching 1000 (21 by the United Kingdom) and totalling about 1000 megatons of explosive force. Both underground and atmospheric tests have been used, with most of the explosive yield being from explosions in the atmosphere. Following an atmospheric explosion there will be some local deposition (or fallout) of activity but the majority of the products are injected into the upper troposphere and stratosphere, allowing their dispersion and subsequent deposition to the earth's surface on a global scale. About 10^{29} Bq of fission products have been injected into the atmosphere by these tests.

Fission bombs, such as those used against Japan, depend upon the rapid formation of a critical mass of ^{235}U or ^{239}Pu from several components of subcritical size. The critical mass for ^{235}U is about 10 kg which on fission produces a very large amount of radioactivity: the total activity released by the 14 kiloton device at Hiroshima was about 8×10^{24} Bq. Most of the released activity is in the form of short-lived nuclides, but considerable amounts of environmentally important fission products, including ^{85}Kr, ^{89}Sr, ^{90}Sr, ^{99}Tc ^{106}Ru, ^{137}Cs, ^{140}Ba, and ^{144}Ce, are also produced. Of particular biological significance are ^{89,90}Sr, ^{131}I, and ^{137}Cs.

As mentioned above, strontium is of importance because it has similar chemical properties to those of calcium. It enters the body by inhalation and ingestion, particularly through milk and vegetables. Iodine differs from strontium in that it is present in the atmosphere in both the gas and aerosol phases. Exposure may be by inhalation and ingestion, but of particular significance is the concentration of iodine in milk caused by cows grazing on contaminated grass and its subsequent consumption by man. Estimates of the amount of ^{137}Cs released into the atmosphere by bomb tests vary but it is probably of the order of 10^{18} Bq, giving rise to human exposure mainly through grain, meat, and milk.

Fusion bombs (also known as thermonuclear or hydrogen bombs) use the reaction of lithium hydride with slow neutrons to generate tritium:

$$^6_3\text{Li} + ^1_0\text{n} \longrightarrow ^3_1\text{H} + ^4_2\text{He}$$

which then reacts with deuterium releasing energy:

$$^2_1\text{H} + ^3_1\text{H} \longrightarrow ^4_2\text{He} + ^1_0\text{n}$$

The slow neutrons are initially supplied by the fission of ^{235}U or ^{239}Pu and the deuterium by the use of lithium-6 deuteride. Fusion bombs therefore release fission products as well as large amounts of tritium.

Tritium is a pure beta emitter with a half-life of 12.3 years. It is readily oxidized in the environment, forming tritiated water and hence enters the hydrological cycle. The amount of tritium injected into the atmosphere by weapons tests probably exceeds 10^{20} Bq, giving rise to estimated dose commitments over an average lifetime of 2×10^{-5} Gy in the Northern Hemisphere and 2×10^{-6} Gy in the Southern Hemisphere.

Various other radionuclides are produced by weapons explosions by the neutron activation of other elements in the soil or surface rocks, the air and the bomb casings. Of these ^{14}C is the most significant pollutant, being formed from stable nitrogen in the air. It is a pure beta emitter of mean energy 49.5 keV and a half-life of 5730 years. The rate of natural production of ^{14}C by cosmic ray interactions is about 1×10^{15} Bq y^{-1} compared with a mean of $\sim 5 \times 10^{15}$ Bq y^{-1} produced by weapons testing since 1945.

The other important group of pollutant nuclides produced by weapons testing are the transuranics, including plutonium. Of these ^{239}Pu is the most significant as it has been produced in large quantities ($\sim 1.5 \times 10^{16}$ Bq) and has a long half-life (24 360 y). It is formed by capture of a neutron by ^{238}U:

$$^{238}_{92}U + ^{1}_{0}n \longrightarrow {}^{239}_{92}U^{\beta^-} \longrightarrow {}^{239}_{93}Np^{\beta^-} \longrightarrow {}^{239}_{94}Pu$$

The most important route to man for plutonium from weapons tests is by inhalation. The UNSCEAR estimates for the population–weighted dose, up to the year 2000, from tests carried out to 1977, are 1×10^{-5} Gy in the Northern Hemisphere and 3×10^{-6} Gy in the Southern Hemisphere.

The current NRPB estimate for the average effective dose equivalent in the UK from weapon test activity is about 10 pSv per year at present, compared with about 80 pSv per year in the early 1960s.

18.7 POLLUTION FROM ELECTRIC POWER GENERATION PLANT AND OTHER NUCLEAR REACTORS

18.7.1 Emissions Resulting from Normal Reactor Operation

There are a large number of nuclear reactors in operation world-wide of many different sizes, designs, and applications. As well as 434 large reactors operational, and 36 under construction (total for 1998), there are many other smaller reactors used for research, isotope production, education, materials testing, and military purposes. Included in this latter category are reactors used to power submarines and other naval vessels. However, the greatest use remains the generation of electricity. The share of electricity generated by nuclear power varies from country to country, with a maxima in Lithuania and France where over 75% of total production was from this source in 1998. In the UK the fraction is 27% (1998) and in the USA 19% (1998).

The production of electric power by using thermal energy derived from nuclear fission involves a chain of steps from the mining and preparation of fissionable fuel through to the disposal of radioactive wastes. There are actual and potential emissions of radioactivity into the environment at each of these steps.

18.7.1.1 Uranium Mining and Concentration. Uranium ores contain, typically, about 0.15% U_2O_3 and require milling, extraction and concentration before shipping to the consumer. At several stages in this process dusts bearing radioactivity are produced which may lead to contamination of the environment in the vicinity of the plant. Large volumes of mill tailings are also produced and as these contain virtually all the radium and thorium isotopes present in the ore they give rise to emissions of radon. Also of environmental significance are liquid releases from the mills which, if uncontrolled, may lead to contamination of surface and groundwaters. It should also be mentioned that the mining and milling of uranium ore inevitably leads to the occupational exposure of workers to uranium and its daughters, especially *via* inhaled ^{222}Rn.

18.7.1.2 Purification, Enrichment, and Fuel Fabrication. The concentrated and purified ore extract, known as yellowcake, contains about 75% uranium and is converted into usable forms of uranium metal and uranium oxide by several processes. First uranium tetrafluoride, UF_4, is produced. The yellow-cake is digested in nitric acid, insoluble impurities removed by filtration and soluble impurities extracted with an organic solvent. The uranyl nitrate solution is concentrated, oxidized to UO_2 and reacted with hydrogen fluoride to form uranium tetrafluoride, UF_4. The UF_4 so produced contains the relative natural abundances of the various uranium isotopes, including about 0.7% of ^{235}U. This is adequate for some reactors, such as the first generation British Magnox reactors, and unenriched uranium metal is produced for them by reaction of the UF_4 with magnesium. The metal is cast and machined into fuel rods, heat treated, and then inserted into magnesium aluminium alloy ('Magnox') cans.

Other designs of reactors, including the Advanced Gas-cooled and Pressurized Water Reactors, require fuel in the form of enriched uranium oxide containing 2–3% of ^{235}U. This is produced by first reacting UF_4 with fluorine gas to form uranium hexafluoride, UF_6. The UF_6 gas is then repeatedly centrifuged or diffused through porous membranes, the small mass differences between the isotopes resulting in their eventual fractionation. Following enrichment the UF_6 is hydrolysed and reacted with hydrogen to form uranium oxide, UO_2, in powder form. This is pressed into pellets and packed in helium-filled stainless steel or zirconium–tin alloy cans.

Radioactive releases into the environment during all these steps should be minimal, although as with all chemical and mechanical processes some emissions are inevitable. The current NRPB estimates of the maximum annual effective dose equivalents arising from fuel preparation for members of the public living near to the relevant plants in the UK are less than 5 μSv from discharges to the air and about 50 μSv from discharges to water. These each give estimated *collective* effective dose equivalents for the UK of about 0.1 man Sv per year.

18.7.1.3 Reactor Operation. An operational nuclear reactor contains, and potentially may emit, radioactivity from three sources, from the fuel, from the products of fission reactions, and from the products of activation reactions. The amount of activity present as fuel itself is of course variable but a Pressurized Water Reactor with 100 tonnes of 3.5% enriched uranium would contain $\sim 0.25 \times 10^{12}$ Bq of ^{235}U and 1.1×10^{12} Bq of ^{238}U. Unless a catastrophic accident occurs, releases of fuel into the environment should not occur.

Fission of the fuel produces a large number of primary products which in turn decay, producing a large number of secondary decay products. Because of their very wide range of half-lives the relative composition and amounts of fission products present in a reactor varies with time in a complex manner. As each nuclide has a different rate of decay the rate of accumulation will vary. However, there reaches a point when the rate of production of a nuclide equals its rate of removal by decay and an equilibrium arises between parent and daughter. If the rate of formation is constant then for all practical purposes equilibrium may be assumed to have been reached after about seven half-lives. For a short-lived nuclide such as ^{85}Kr, equilibrium will be reached about 10 weeks after the start of the reactor, whereas for ^{137}Cs with a half-life of 30.1 y it will take more than 200 years of continuous operation to reach equilibrium. As the reactor continues to operate with a given set of fuel rods the long-lived fission products become relatively more important and eventually the reactor might contain 10^{20} Bq of activity in total.

Most of the fission products are contained within the fuel cans themselves but total or partial failure of the fuel cladding will allow contamination of the coolant. In addition, contamination of the outer surfaces of the fuel rods by uranium fuel will allow fission products to form in the coolant. Most of the fission product activity is in the form of gaseous elements and although various scrubbing systems are used to remove them from the coolant, leakage into the environment can and does occur. The amount of gaseous fission products released will depend upon the number of fuel cladding failures, the design of the ventilation, cooling, and coolant purification systems and the length of operation of the reactor. Of the fission products, ^{85}Kr, which has a half life of 10.8 y, makes the greatest contribution to the global dose commitment from reactor operation.

The third source of activity in the reactor results from the neutron activation of elements present in the fuel and its casing, the moderator, the coolant, and other components of the reactor itself. Those areas most subjected to neutron activation are those where the incident neutron flux is highest, and include the fuel casings and the coolant system. From an environmental point of view the most important activation product is probably tritium. This is formed by various routes, especially by the activation of deuterium present in the water of Heavy Water Reactors but also by the activation of deuterium present in the water of light water-cooled reactors, the activation of boron added as a regulator to the primary coolant of Pressurized Water Reactors and by activation of boron present in the control rods of Boiling Water Reactors. The small size of the tritium nucleus allows it to diffuse through the cladding materials and so be released into the environment where it will be inhaled and absorbed through the skin. A smaller dose will also be received from drinking and cooking water, following the incorporation of released tritium into the water cycle. Estimates have been made of the annual dose equivalent for individuals living 1, 3, and 10 km from a Canadian CANDU Heavy Water Reactor and are of the order of 8, 3, and 2 μSv respectively.

Other important activation products are ^{58,60}Co, ^{65}Zn, ^{59}Fe, ^{14}C and the actinides. This latter group, which includes the three long-lived isotopes of plutonium ^{238}Pu, ^{239}Pu, and ^{240}Pu (with half-lives of 86.4, 24360, and 6580 y respectively), are produced by the neutron activation of fuel uranium. Plutonium has a very low volatility and is not significantly released during normal reactor operation. However, it can potentially enter the environment following a reactor accident and as a result of fuel reprocessing.

Various estimates are available for the total dose arising from these various discharges of radioactivity to the environment during the normal operation of power generation reactors. The current NRPB estimate of the maximum effective dose equivalents resulting from discharges to the atmosphere for the most exposed individuals living near to power stations in the UK is about 100 μSv per year, giving a collective effective dose equivalent in the UK of about 4 man Sv per year. In addition to discharges to the atmosphere there are also discharges of radioactivity from power stations to natural waters. These result from the temporary storage of used fuel under water in specially constructed ponds. The NRPB estimate that these discharges to natural waters give a maximum effective dose equivalent to local individuals of less than 350 μSv per year. The collective effective dose equivalent in the UK, mainly through the eating of contaminated seafood, is about 0.1 man Sv per year.

18.7.2 Pollution Following Reactor Accidents

With the large number of reactors operational world-wide it is inevitable that accidents with environmental consequences should occur. To date five major incidents have taken place in the UK, USA, and the former USSR.

18.7.2.1 Windscale, 1957.   In October, 1957, a graphite moderated reactor at Windscale, UK, overheated, rupturing at least one fuel can and causing the release of fission products into the environment. An estimated 740×10^{12} Bq of ^{131}I, 22×10^{12} Bq of ^{137}Cs, 3×10^{12} Bq of ^{89}Sr, and 3×10^{12} Bq of ^{90}Sr were released. The cause of the accident was the sudden release of a huge amount of Wigner energy from the graphite in the core. When graphite is bombarded by fast neutrons, as in a reactor, an increase in volume and a decrease in thermal and electrical conductivity results. The graphite can then efficiently store energy and, if heated above $300\,^{\circ}$C, this is released in the form of further heat. Normally this Wigner energy is released by slowly heating the graphite, allowing the stored energy to be released under control and reversing the radiation damage. During 7–8 October, 1957, however, an unsuccessful attempt was made to release the Wigner energy at Windscale, following which the reactor rapidly overheated releasing radioactive material into the atmosphere.

Following the accident the inhalation of aerosol and gaseous nuclides probably gave the largest source of exposure to the general population, although the drinking of milk contaminated by ^{131}I was also significant, despite a restriction on the sale of milk from farms in the worst affected area. The total committed effective dose equivalent in the UK has been estimated at about 2000 man Sv, giving about 30 excess deaths from cancer in the UK during the first 40 y after the accident.

18.7.2.2 Kyshtym, 1957.   In 1957 a chemical explosion in a high-level radio-active waste tank caused the release of about 7×10^{17} Bq into the environment.

18.7.2.3 Idaho Falls, 1961.   In January, 1961, during maintenance work on an enriched uranium-fuelled boiling water reactor at the National Reactor

Testing Station, Idaho Springs, USA, the central rod was accidentally removed causing a large explosion, killing three men. Most of the coolant water was ejected together with an estimated 5–10% of the total fission products in the core. About 4×10^{12} Bq of activity escaped from the reactor building.

18.7.2.4 Three Mile Island, 1979. Whilst operating at full power on 28 March, 1979, the cooling water system failed on the pressurized water reactor at Three Mile Island reactor 2 at Harrisburg, Pennsylvania. The reactor automatically shut down and three auxiliary pumps started to provide cooling water. However, valves in this circuit had inadvertently been left closed and the resultant increase in temperature and pressure in the primary circuit caused an automatic relief valve to open, allowing about 30% of the primary coolant to escape. Fresh cooling water was injected into the primary circuit by the emergency cooling system but, following an error in their analysis of the accident, this and the main coolant pumps were deactivated by the operators. It was only when cooling water was later added that the overheating of the core ceased.

Despite the reactor core sustaining serious damage, the amount of activity released into the environment was limited to about 10^{17} Bq, almost entirely as short-lived noble gases, especially ^{133}Xe (which has a half-life of 5.2 days). The resultant total committed effective dose equivalent to the population was estimated to be about 20 man Sv with the maximum individual dose equivalent being about 1 mSv.

18.7.2.5 Chernobyl, 1986. The most serious reactor accident to date began on 26 April, 1986, when an explosion and fire occurred at the Chernobyl number 4 reactor near Kiev in the Ukraine. The reactor came into service in 1984 and was one of 14 RBMK boiling water pressure tube reactors operational in the USSR. These are graphite-moderated reactors of 950–1450 MW electrical power generation capacity fuelled with 2% enriched uranium dioxide encased in zirconium alloy tubing. The normal maximum temperature of the graphite is about 700 °C and in order to prevent this being oxidized the core is surrounded by a thin-walled steel jacket containing an inert helium/nitrogen mixture.

Prior to the accident the reactor had completed a period of full power operation and was being progressively shut down for maintenance when an experiment was begun to see whether the mechanical inertia of one of the turbogenerators could be used to generate electricity for a short period in the event of a power failure. The reactor core contained water at just below the boiling point but when the experiment began some of the main coolant pumps slowed down causing the core water to boil vigorously. The bubbles of steam so formed displaced the water in the core and, because steam absorbs neutrons much less efficiently than water, the number of neutrons in the core began to rise. This increased the power output of the reactor, so increasing the heat output and the amount of steam in the core, which in turn led to a further rise in the neutron density. This positive feedback mechanism led to a rapid surge in power causing the fuel to melt and disintegrate. As the fuel came into contact with the surrounding water, steam explosions occurred destroying the structure of the core and the pile cap, causing radioactive material to be ejected into the atmosphere. The core fires allowed a continuing release of activity which was slowly reduced by the dumping of clay and other materials onto the core debris. However, the core temperature again began to rise and a second peak in activity release occurred on 5 May. After this the core was progressively buried and finally sealed in a concrete sarcophagus.

Estimates of the amount of activity released from the reactor vary but Ukranian measurements suggest that all of the noble gases, 10–20% of the volatile fission products (mainly iodine and caesium), and 3–4% of the fuel activity, giving a total of 1.85×10^{18} Bq, escaped into the environment. On the basis of air concentration and deposition measurements the UK AEA's Harwell Laboratory initially estimated that about 7×10^{16} Bq of ^{137}Cs was released.

The immediate casualties of the accident were 31 killed and about two hundred diagnosed as suffering from acute radiation effects. About 135 000 people and a large number of animals were evacuated from a 30 km radius area surrounding the plant. However, in 1995 the Ukranian Health Ministry announced that a total of 120 000 fatalities had occurred as a result of the accident but the basis of this claim is not yet clear.

The meteorological conditions prevailing over Europe at the time of the accident were rather complex, leading to the dispersion of activity over a very wide area. In the UK peak air concentrations occurred on 2 May when ~ 0.5 Bq m^{-3} of ^{137}Cs were recorded at Harwell. No Chernobyl radioactivity was immediately detected in the Southern Hemisphere. In the year following the accident the annual mean ^{137}Cs concentration in the Northern Hemisphere was about the same as that of 1963 when weapons testing activity was at its highest. The amount of ^{137}Cs deposited on the ground surface obviously varied with air concentration, rainfall, and other parameters but close to Chernobyl (within 30 km) as much as 10^4 kBq m^{-2} of ^{137}Cs was deposited. The average for Austria was 23 kBq m^{-2}, for the UK 1.4 kBq m^{-2}, and for the USA 0.04 kBq m^{-2}. The very heavy but localized rainfall which occurred in parts of Europe and the UK during the time when the plume was overhead led to a very patchy distribution of deposited activity on the ground. In the UK for example it varied from > 10 kBq m^{-2} of ^{137}Cs in parts of Cumbria to < 0.3 kBq m^{-2} in parts of Suffolk.

Following the wet and dry deposition of activity from the atmosphere some contamination of foodstuffs was inevitable and outside of the immediate area around Chernobyl this was the major consequence of the accident. In general the most vulnerable foods in the UK were lamb and milk products. In the UK a ban on the movement and slaughter of lambs was imposed within specified areas until the meat consistently contained less than 1000 Bq kg^{-1} of radiocaesium. No restrictions were placed on the sale of milk.

The main exposure pathways to man following the accident were direct gamma irradiation from the cloud and from activity deposited on the ground (groundshine), inhalation of gaseous and particulate activity from the air and, most importantly, the ingestion of contaminated foods. The Organization for Economic Co-operation and Development Nuclear Energy Agency estimate that the average individual effective dose equivalents received in the first year after the accident range from a few microsieverts or less for Spain, Portugal, and most countries outside of Europe to about 0.7 mSv for Austria. However, hidden within these averages are the higher peak doses received by the most exposed individuals, or critical group, in each country. These vary from a few microsieverts outside Europe to an upper extreme of 2–3 mSv for the Nordic countries and Italy.

The total collective dose in Eastern and Western Europe has been estimated to be about 1.8×10^5 man Sv. The NRPB estimate that the average effective dose equivalent received in the UK during the first year was about 40 μSv with a total of about 20 μSv being received over subsequent years. The total committed effective dose equivalent for the UK population is estimated to be about 2100 man Sv. When compared with the other doses of radioactivity normally received by individuals, for example the 2 mSv per year individual effective dose equivalent received on average in Europe from natural background radiation, these doses are small and probably insignificant. Using the ICRP estimate of risk it is possible to calculate the additional mortality likely to result from Chernobyl. In the 40 years following the accident about 50 excess deaths are likely in the UK (compared with about 145 000 non-radiogenic cancers per year in the UK), indicating that the impact of the accident on future mortality statistics will be small.

The Chernobyl accident presented a unique opportunity for experiments and studies in a wide range of environmental sciences. It allowed the validation and refinement of atmospheric dispersion models, the calculation of washout ratios and deposition velocities, and the study of the behaviour of caesium and iodine in food chains, natural waters, sediments, and in the urban environment. It demonstrated the necessity of better international harmonization of scientific databases and public health protection policies and it provided a vivid example of the long-range transboundary transport of pollutants. In 1994 and 1995, a very large European Union-funded tracer release experiment (ETEX) was carried out to simulate a nuclear reactor accident on the scale of Chernobyl with the aim of testing the predictive capabilities of long range transport models and forecasts. Inert tracers were released from a site in Brittany, western France, and the resultant plume detected across Europe by a network of monitoring sites and by aircraft.

18.7.3 Radioactive Waste Treatments and Disposal

An inevitable consequence of man's use of radioactivity is that radioactive waste material is produced which must then be disposed of. Although there is no universal scheme for the classification of such waste it is usual for it to be categorized in terms of its activity content as low, intermediate and high level.

18.7.3.1 Low Level Waste. Low level wastes are produced in large volumes by all the various medical, industrial, scientific, and military applications of radioactivity. They include contaminated solutions and solids, protective, cleaning and decontamination materials, laboratory ware, and other equipment. They also include gases and liquids operationally discharged from power stations and other facilities. It has been estimated that during the period 1980–2000 about $3.6 \times 10^6 \text{ m}^3$ of such wastes will be produced, some of which is sufficiently low in activity to be directly discharged into the environment, either with or without prior dilution or chemical treatment. Typical maximum activity concentrations in low level waste are $4 \times 10^9 \text{ Bq t}^{-1}$ (alpha) and $12 \times 10^9 \text{ Bq t}^{-1}$ (beta and gamma). Much low level waste is currently disposed of by shallow burial in landfill sites, often with the co-disposal of other, non-radioactive, controlled wastes, or by discharge into surface waters in rivers, lakes, estuaries, or coastal seas or by discharge into the atmosphere. If the environmental biophysicochemical behaviour of the radionuclides in question and their possible pathways back to man are well understood then it is possible to make reliable estimates of the likely resultant doses of radioactivity to those most exposed in the population. If these doses are suitably low then the disposal methods may be deemed to be acceptable. However, at the present time there are many uncertainties in the understanding of such behaviour, pathways and doses, but nevertheless the very large volumes of low level waste being generated will, through lack of economically and environmentally viable alternatives, continue to be disposed of in these relatively uncontrolled ways. In the UK there are 7900 m^3 of low level waste stocks (1994).

18.7.3.2   Intermediate Level Waste.   Intermediate level wastes are sufficiently active to prevent their direct discharge into the environment, with maximum specific activities of typically $2 \times 10^{12} \text{ Bq m}^{-3}$ (α) and $2 \times 10^{14} \text{ Bq m}^{-3}$ (β and γ). They comprise much of the solid and liquid wastes generated during fuel reprocessing, residues from power station effluent plants, and wastes produced by the decommissioning of nuclear facilities. Very large quantities ($\sim 2000 \text{ t y}^{-1}$) of intermediate and low level wastes have been disposed of by dumping in deep ocean waters in the NE Atlantic. Although some authorities still consider this method of disposal to be the best practicable environmental option for these categories of waste it is no longer practised and intermediate level waste produced in the UK is now stored on land, mainly at Sellafield awaiting further policy decisions. The total stock of intermediate level waste in the UK is around $61\,000 \text{ m}^3$ (1994).

18.7.3.3   High Level Waste.   High level wastes mainly consist of spent fuel and its residues and very active liquids generated during fuel reprocessing. Typical maximum activities are $4 \times 10^{14} \text{ Bq m}^{-3}$ (α) and $8 \times 10^{16} \text{ Bq m}^{-3}$ (β and γ). At present such wastes generated in the UK are stored at Sellafield in storage ponds where it is proposed they will be vitrified prior to further storage (to allow the decay of shorter lived nuclides) and finally disposal in deep repositories. No such repository yet exists but deep mines and boreholes on land and sea as well as other more exotic solutions, including extraterrestrial disposal, have all been proposed. Current stocks of high level waste in the UK total 1600 m^3 (1994).

18.7.4 Fuel Reprocessing

In most nuclear reactors the economic lifetime of the fuel in the core is determined not by the depletion of fissile material but by the production and accumulation of fission products which progressively reduce the efficiency of the reactor. In a typical reactor the fuel is changed on a three-year cycle, generating large amounts of partially spent fuel contaminated with fission products. Apart from the economic considerations, which may or may not be in favour of recovering the unreacted fissile material from the spent fuel, depending upon the relative costs of reprocessing and importing further uranium ore, there are several reasons why such reprocessing is carried out: it allows the production of plutonium for military purposes, it reduces dependence on imported ores and it reduces the volume of high-level waste produced by a reactor. It is also a very large scale international commercial operation.

In the reprocessing method currently used in the UK the short-lived activity is allowed to decay by storage for several months, after which the spent fuel is dissolved in nitric acid and a sequential extraction procedure used successively to remove the uranium, plutonium and fission products. The uranium is then re-enriched and fabricated into fuel rods and the plutonium used in the mixed oxide fuel or for military purposes. About 25 000 t of Magnox fuel has been reprocessed at Sellafield, yielding about 10 000 t of uranium for re-enrichment. The THORP thermal oxide reprocessing plant at Sellafield in NW England processes fuel from Advanced Gas-cooled and Pressurised Water Reactors at a projected rate of about 600 t y^{-1} and will eventually handle the 2500 t of waste currently being stored at Sellafield and that being produced now and in the future in the UK and overseas. The current NRPB estimates of the annual effective dose equivalents arising from fuel reprocessing to the UK population are 1 mSv y^{-1} to the most exposed individuals and a collective dose of 80 man Sv y^{-1}. The reprocessing of fuel requires the transport of large quantities of spent and reprocessed fuel around the globe, giving rise to the possibility of further accidental releases into the environment.

18.8 POLLUTION FROM NON-NUCLEAR PROCESSES

Two non-nuclear industrial processes, the burning of fossil fuels and the smelting of non-ferrous metals, release non-trivial quantities of radioactivity into the environment and require brief consideration here. Coal contains uranium and thorium in varying concentrations, typically 1–2 ppm of both ^{238}U and ^{232}Th but at much higher concentrations (100–300 ppm) in some areas, *e.g.*, the western USA. It also contains significant quantities of ^{14}C and ^{40}K. Similarly oil and natural gas both contain members of the naturally-occurring U and Th decay series. When the fuel is burned, as in a conventional power station, these nuclides and any daughters present are either released into the atmosphere in the flue gas and fly ash or retained in the bottom ash. One current estimate of the amount of ^{226}Ra emitted by a typical 1000 MW coal-fired station is 10^9–10^{10} Bq y^{-1} with a larger amount being retained in the ash.

These emissions undoubtedly give rise to elevated environmental concentrations with the whole body dose equivalents for those most exposed living in the vicinity of large coalfired plants being possibly as high as 1 mSv per year.

The second non-nuclear industrial source of radioactivity, the smelting of nonferrous metals, arises because of the natural occurrence of radioactive isotopes of lead. The geochemistry of lead is intimately associated with that of uranium and thorium, there being four radioactive isotopes of lead: ^{210}Pb and ^{214}Pb in the ^{238}U decay chain, ^{211}Pb in the ^{235}U decay chain, and ^{212}Pb in the ^{232}Th decay chain. All except ^{210}Pb have half-lives of less than 12 h, but the 22 y half-life of ^{210}Pb and its subsequent decay to form ^{210}Po, itself an alpha-emitter of half-life 138 days, makes it of environmental significance. During the primary and secondary smelting of lead and other non-ferrous metals and their ores some lead is released into the atmosphere. A fraction of this will be ^{210}Pb together with a similar quantity of ^{210}Po. This aerosol may then be inhaled, giving rise to an exposure of the lung to alpha particles. Investigation and understanding of such 'non-nuclear' pathways of radioactivity to man is at an early stage but it is possible that those living close to large smelters and other sources of ^{210}Po may receive measurable and non-trivial doses of radioactivity.

Apart from the obvious public concerns over possible nuclear weapons proliferation, testing and use, three areas appear to be of growing interest at the present time. The first is the possibility of exposure of the public to radiation resulting from illegal trafficking of radioactive sources. The World Customs Organisation reported 223 cases of the seizure of radioactive sources in the five year period to 1998. Secondly, given the increasing concern over the emissions of greenhouse gases to the atmosphere, particularly carbon dioxide, extending the lifetime of existing nuclear power plants may be the most viable option for the continued use of nuclear power, for a variety of economic and sociopolitical reasons, and this is raising public concerns about nuclear plant safety. Finally, the transport and processing of scrap metal contaminated with radioactive material is also attracting interest.

Acknowledgements. I thank Dr M. Kelly for his constructive and encouraging comments.

18.9   BIBLIOGRAPHY

R. S. Cambray, *et al.*, 'Observations on Radioactivity from the Chernobyl Accident', *Nucl. Energy*, 1987, **26**, 77–101.

K. D. Cliff. J. C. H. Miles and K. Brown, 'Decay Products in Buildings', NRPB-R159, HMSO, London, 1984.

J. H. Gittus, *et al.*, 'The Chernobyl Accident and its Consequences', United Kingdom Atomic Energy Authority, HMSO, London, 1988.

J. S. Hughes and G. C. Roberts, 'The Radiation Exposure of the UK Population – 1984 Review', NRPB-R173, HMSO, London, 1984.

R. Kathren, 'Radioactivity in the Environment: Sources, Distribution, and Surveillance', Harwood, Amsterdam, 1984.

National Radiological Protection Board, 'Living with Radiation', HMSO, London, 1986.

Organization for Economic Co-operation and Development Nuclear Energy Agency, 'The Radiological Impact of the Chernobyl Accident in OECD Countries', OECD, Paris, 1987.

F. Warner and R. M. Harrison, 'Radioecology after Chernobyl', SCOPE 50, John Wiley and Sons, Chichester, 1993.

CHAPTER 19

Health Effects of Environmental Chemicals

P. T. C. HARRISON

19.1 INTRODUCTION

With the established expectation of a long and healthy life, and growing public concern about the impact of man's activities upon the environment, questions on the relationship between environmental quality and human health have come increasingly to the fore. The increasing incidence of childhood asthma, for example, has spawned much popular debate and new research initiatives on possible environmental causes. Similarly, the issue of a possible decline in human fertility has resulted in growing speculation about the role of endocrine disrupting chemicals. The health impact of exposure to small particles in the air is also attracting much attention, and there continues to be concern about environmental exposure to pesticides, toxic metals and asbestos fibres. Memories of severe accidents, such as those at Bhopal and Seveso, serve to highlight the potential consequences to health of chemicals inadvertently released into the environment.

The public's perception of risk is highly coloured by the origin of the hazard. Chemicals seen as natural (or where exposure is by deliberate choice) are not perceived as having the same danger as those emanating, for example, from an industrial source. Thus natural toxins present in plants, for example, are generally not regarded in the same light as residues remaining after pesticide treatment. This issue has recently become somewhat clouded, however, by the development of genetic modification technology which has the capacity to introduce 'natural' substances into 'unnatural' locations.

Industrialization has resulted in increased concentrations of chemicals in air, water and soil. Of course, the presence of a chemical or chemicals in the environment cannot be taken as an indication *per se* that it is harmful to human health; it may have no relevant exposure pathway or be present in too low a concentration to pose any real threat to health. However, for the sake of human health and that of the natural environment, particular attention must be paid to those substances which persist in the environment and in the tissues of plants and animals and which bioaccumulate in the food chain.

The health outcome resulting from exposure to a toxic substance (either directly or indirectly through the food chain) can range from effects on perceived well-being, through to chronic ill-health, cancer, and death. There may be reversible physiological changes or more severe pathological effects on the cardiovascular and respiratory systems, for example. Or there may be adverse reproductive, immunological, neurological and developmental consequences.

The factors which will determine the outcome of exposure to a particular chemical include the inherent toxicity of the substance and its mode of action, the route of exposure, the concentration, timing and duration of exposure, and the inherent susceptibility of the individual exposed. Individuals can be at particular risk because of their genetic makeup, their age, exposure history, current health status or lifestyle – including diet. Indeed, nutritional status is increasingly being recognized as an important modifying factor that can influence health and affect responses to environmental stressors. For example, it is likely that intake of dietary antioxidants can afford some resistance to the effects of air pollutants.

Different types of exposure to environmental chemicals can be distinguished. There is catastrophic exposure, which results from the massive accidental release of material into the environment such as occurred at Seveso or Bhopal. Localized incidents can occur as a result of heavy contamination of the local environment or of the adulteration of food and water; the outbreaks of mercury poisoning in Iraq and elsewhere as the result of the use for flour making of seed grain treated with organic mercurials are examples of this kind of exposure. Then there is the lower-level but more chronic exposure which can occur with air pollutants and food and water contaminants, for example. Sometimes these exposures are inescapable in a society which depends upon the use of chemicals to maintain its way of life, but better understanding of the risks of such exposures can enable decisions to be made about more controlled use of the substances or justifiable replacement by proven safer alternative chemicals or processes. Availability of accurate scientific information on such issues is crucial for a proper assessment of risk and the appropriate application of the 'precautionary principle'.

19.2 CATASTROPHIC EXPOSURE

The two best known examples of catastrophic exposure to chemicals occurred at Seveso and Bhopal.

19.2.1 Seveso

On 10 July, 1976 there was a massive release of 2,3,7,8-tetrachlorodibenzo-*p*-dioxin from a chemical plant in Seveso, near Milan in northern Italy, which was manufacturing 2,4,5-trichlorophenol (2,4,5-TCP). A safety disc in a reaction vessel ruptured and a plume of chemicals containing 2,4,5-TCP blew 30 to 50 m above the factory. As it cooled the material in it was deposited over a cone shaped area, down wind from the factory, about 2 km long and 700 m wide. In all, an area of 3–4 km was contaminated and an estimated 3 to 16 kg of dioxin was released. There were almost 28 000 people living in the vicinity of the factory. Those who lived in the immediate area downwind were evacuated 14 days after the explosion and the area was closed off. About 5000 people in the most heavily contaminated area were allowed to stay in their homes but they were not allowed to cultivate or consume local vegetables or fruit nor to raise or keep poultry or other animals.[1]

Dioxin is both extremely toxic and extremely stable and is known, at sufficient dose levels, to affect fetal development and to have porphyrinogenic effects (often manifested as digestive system and skin disorders), all of which are well documented. It is not used commercially but is found as a contaminant when 2,4,5-TCP is synthesized by the hydrolysis of tetrachlorobenzene at high temperatures. 2,4,5-TCP itself is used to make 2,4,5-trichlorophenoxyacetic acid (2,4,5-T) and 2,4-dichlorophenoxyacetic acid (2,4-D) which are used as herbicides; dioxin is often present in trace amounts in these compounds. Dioxin is also produced as a by-product of waste incineration, and continues to be of major concern as a general environmental pollutant.

The population at Seveso was screened shortly after the accident happened and a number of positive findings were noted. A few months after the accident, 176 individuals, mostly children, were found to have the skin condition chloracne, which is known to be associated with high level exposure to chlorinated hydrocarbons. Chloracne is more severe than the type of acne that occurs in adolescents and it has a rather different distribution on the body. Fifty of the individuals with chloracne (about 7% of the population estimated to be at risk) came from the most contaminated area. A further round of medical screening in February 1977 revealed an additional 137 cases of chloracne, but subsequent follow-ups showed that the incidence had decreased and that the individuals already affected had improved.

In addition to the chloracne, some neurological abnormalities were noted. These included polyneuropathy with some symptoms that were due to effects on the central nervous system. Such observations were more common amongst the people who lived in the most heavily contaminated zone, and the incidence of abnormal nerve conduction tests was significantly increased in those who had chloracne.[2]

Finally, there was evidence of liver enlargement in about 8% of the population, which again was most noticeable amongst the most heavily exposed individuals. Liver enzyme activity showed some abnormalities, but had returned to normal about a year after the explosion. It is interesting and noteworthy that there was no evidence that the immune system had been affected, and there has been no evidence of chromosomal abnormalities or of any damage to the fetus. No deaths were recorded as being caused by the accident;[3,4] however, a recent analysis of cancer incidence in the exposed population has reported increased hepatobiliary (liver) cancer, elevated incidence of leukaemias and other haematological neoplasms in men, increased multiple myeloma and myeloid leukaemia (bone marrow cancers) in women, and evidence for higher incidences of soft tissue tumours and non-Hodgkin's lymphoma. Interestingly (because dioxin is an anti-oestrogen – see Section 19.4.5) breast cancer and endometrial cancer in women were reduced[5] and there have been recent suggestions of changed sex ratio in the offspring of exposed men.

The explosion at Seveso excited a great deal of public alarm, particularly because dioxin was involved. However, major harmful effects were directed towards the environment; many farm animals died and the site became a wasteland of dying plants and deserted homes. Huge amounts of top-soil were removed from the site. The incident prompted the European Community to adopt, in 1982, a Directive aimed at preventing such major chemical accidents (the so-called 'Seveso' Directive, now constituted as the Control of Major Accident Hazards – 'COMAH' – Directive 96/82/EC).

19.2.2 Bhopal

The effects of the incident at Bhopal, India, were directed almost entirely onto the population living around the factory involved in a catastrophic release of methyl isocyanate (MIC).

The accident occurred on 3 December, 1984 at the Union Carbide factory, which had been producing the insecticide carbaryl for about eighteen years; MIC was one of the main ingredients. The MIC itself was produced from monomethylamine (MMA) and phosgene, the latter being produced on site by reacting chlorine and carbon monoxide. The MMA and chlorine were brought by tanker from other plants in India, stored and used when required; chloroform was used as a solvent throughout the process. Thus there was, in this plant, a variety of extremely toxic materials in use. On the night of the accident, it seems that some water inadvertently got into a tank where 41 tonnes of MIC were being stored, causing a runaway chemical reaction. The heat of the reaction, possibly augmented by reactions with other materials present in the tank as contaminants, produced vaporization of such momentum that it could not be contained by the safety systems. The safety valve on the tank blew open and remained open for about two hours, allowing MIC in liquid and vapour form to escape into the surroundings. The prevailing wind carried the cloud towards the north of the plant, and then towards the west, affecting approximately 100 000 people living in the vicinity. There were at least 2000 deaths, although none of the workers on night duty at the plant were harmed. The most frequent symptoms in those who survived were burning of the eyes, coughing, watering of the eyes, and vomiting.[6] Very many individuals have continued to suffer physical and mental trauma as a consequence of this tragedy.

19.3 LOCALIZED CONTAMINATION INCIDENTS

Most incidents in which local communities have suffered overt signs of toxicity have involved exposure to food contaminants. Well-known examples covered in this section include Toxic Oil Syndrome in Spain and methylmercury poisoning in Iraq. In the UK, incidents have occurred in Epping (Greater London) and in North Cornwall, where the water supply was contaminated. There have been other examples where contamination has occurred indirectly as a result of an environmental pollutant entering the food chain, for example at Minamata and Niigata, Japan. The best known example of health effects being associated with environmental contamination around waste disposal sites is Love Canal in the USA, although there have been more recent concerns sparked by findings of increased incidences of congenital abnormalities around landfill sites in Europe.

19.3.1 Toxic Oil Syndrome

Cooking oil was at the heart of a severe poisoning episode in Spain,[7] although the toxic agent responsible has not been identified satisfactorily to this day. The syndrome manifested itself in May 1981 in an eight year old boy who died with acute respiratory insufficiency. He was one of a family of eight living in Madrid, of whom six eventually became ill. By June, 2000 patients with the condition had been admitted to hospitals in Madrid, and a further 600 to hospitals in the provinces. By the end of August, 13 000 people had been treated in hospital and 100 had died. The final toll was about 20 000 persons affected and nearly 400 deaths – a case fatality rate of about 2%.

The illness began with a fever, followed by severe respiratory symptoms and a variety of skin rashes, which led some of the victims to be diagnosed as having measles or German measles (rubella). Many of the patients developed signs of cerebral oedema (swelling of the brain) and many had cardiological abnormalities.

The cause of the disease was traced to adulterated cooking oil which was fraudulently sold to the public as pure olive oil. The oil was sold by door-to-door salesmen in five-litre plastic bottles with no labels. Because olive oil is an expensive commodity in Spain, it was the poorer families in the working class suburbs of Madrid who were visited by the salesmen and it was they who almost entirely bore the brunt of the disease episode. The composition of the oil varied but rapeseed oil accounted for up to 90%; there were varying amounts of soya oil, castor-oil, olive oil, and animal fats. The oil also contained between 1 and 50 ppm of aniline and between 1500 and 2000 ppm of acetanilide.

It seems that those who perpetrated the fraud tried to refine out the aniline and in doing so produced a number of other chemical species. One of these was acetanilide, which then reacted with fatty acids in the oil to produce oleoanilide, which was originally presumed to be the toxic agent.[8] Later work, however, showed the presence of a number of other anilides, including a fatty acid diester of propane-1,2-diol-3-aminophenyl. The other aniline compounds were considered to be hydrolysis products of diacyl propane-1,2-diol-3-aminophenyl or positional isomers. It has been suggested that these compounds may have been responsible for the symptoms produced by the toxic oil,[9] but despite much effort the precise toxicant has not definitively been identified.

19.3.2 Rice Oil Contamination by Polychlorinated Biphenyls (PCBs)

Episodes of human poisoning with PCBs have occurred in Japan (which seems to have had rather more than its share of environmental disasters) and in Taiwan. The disease first made its appearance in 1968 in the western part of Japan, when a number of families were noted to have developed chloracne, the skin condition which affected the victims at Seveso. Epidemiological studies brought other cases to light and it was found that the factor which the cases had in common was exposure to a particular batch of one brand of rice oil. Chemical analysis showed that the oil was contaminated with PCBs.

The PCBs were shown to have leaked into the oil from equipment which had been used to process the oil, the PCBs having been used as an industrial heat-transfer fluid, as was commonly the case at that time. By the end of 1977, 1665 individuals were considered to have met the diagnostic criteria for what has come to be known as Yusho disease.[10]

In addition to chloracne, the patients with Yusho disease had a number of systemic complaints, including loss of appetite, lassitude, nausea and vomiting, weakness, and loss of sensation in the extremities. Some also had hyperpigmentation of the face and nails.

The patients were followed up from 1969 to 1975, and in 64% of cases the skin lesions improved. A number of non-specific symptoms persisted, however, including a feeling of fatigue, headache, abdominal pain, cough with sputum, numbness and pain in the extremities and, in women, changes in menstruation. Objective findings included a sensory neuropathy, retarded growth in children, and abnormal development of the teeth. Children who had been exposed *in utero* had lower birth weight and were hyperpigmented.

Some patients were found to be anaemic and some had other abnormalities, but the most striking observation was a marked increase in serum triglyceride levels. The mean value in the patients was 134 ± 60 mg per 100 ml compared with a mean of 74 ± 29 mg per 100 ml in normal controls. When the PCB concentrations in the serum were measured by gas chromatography it was found that the patients with Yusho disease had an isomeric pattern which was different from that seen in controls whose only exposure had been from the general environment.[11]

Yusho disease appeared in Taiwan in the spring and summer of 1979 in two prefectures in the middle part of the country. The signs and symptoms were indistinguishable from those seen in the Japanese outbreak and by the end of 1980 more than 1800 people had been affected. The source of the PCBs was again contaminated rice oil. The blood levels of those with the disease ranged from 54 to 135 ppb. Later studies showed that polychlorinated dibenzofurans (PCDFs) and polychlorinated quaterphenyls (PCQs) were also present in the blood.[12]

To what extent the symptoms seen in the patients with Yusho disease were *entirely* due to PCBs is difficult to say since they were also exposed to PCDFs and to PCQs (formed when PCBs are heated). In animal models, PCDFs and PCQs are more toxic than PCBs and it seems that there may have been some synergistic effects between the different compounds in the oil. The fact that the PCB isomers in the oil were different from those in the general environment may also be of importance since they may have been more toxic than those to which the population at large is ubiquitously exposed.

19.3.3 Polybrominated Biphenyls (PBBs) in Cattle Feed

PBBs are used mainly in plastics as a fire retardant. In May and June of 1973, some ten to twenty bags of PBB were sent in error (instead of a livestock food additive) to a grain elevator in the state of Michigan, USA. The chemical company which made the PBB normally supplied magnesium oxide to go into the cattle feed but both products were packed in the same colour bag and although the PBB was labelled 'Firemaster' rather than 'Nutrimaster', and although this difference was actually noted by the staff at the grain elevator, it was nevertheless incorporated into the feed and distributed throughout the state to be fed to the unsuspecting cows.

Reports of sick cows began to surface in August, 1973 and towards the end of the year it was realized that the feed was to blame. Despite this the contamination continued, both because there was cross-contamination of otherwise normal feed from the grain elevator and because the tainted feed was resold at a discount after it had been returned to the suppliers. Not until PBB was formally identified in the feed in May 1974 was any attempt made to limit the contamination.

From the time that the feed had become contaminated, dairy products containing PBB had been sold throughout Michigan and cows and other livestock which had been given it had been slaughtered for meat. A representative sample of 2000 people was surveyed and more than half had a concentration of PBB in their fatty tissues exceeding 10 ppb.[13] Farmers and others who consumed produce directly from contaminated farms had the highest levels of PBB.

An initial study of 217 farmers concluded that the exposure had caused no deleterious effects on their health, but the study was criticized on the grounds that the control group had also been exposed to PBB.[14] A second study was then carried out in which over 1000 farmers were compared with unexposed farmers from Wisconsin. This second study did find some adverse effects: acne, dry skin, hyperpigmentation, and discoloration of the nails were all more common in the exposed group, and they also complained more of headaches, nausea, depression, and a number of other non-specific symptoms. Serum levels of hepatic enzymes were higher in the Michigan farmers than their neighbouring controls. Individuals with symptoms were also shown to be more likely to have elevated enzyme levels. Changes in the immune system were also found[15,16] and some individuals had enlargement of the liver and a sensory neuropathy.[17] In follow-up studies, the PBB levels in the serum were found to have decreased, but it was interesting that elevated PCB levels were seen and that these were actually higher than the PBB levels. While there was no relationship between abnormal liver function tests and serum PBB concentrations, there was a slight (statistically insignificant) negative correlation with serum PBB levels and some tests of thyroid function.[18]

It is noteworthy that in none of these studies did the subjective or the objective health effects findings correlate with serum or fat PBB concentrations. This may have been because there was another toxic contaminant present which was acting independently of the PBB, or that the levels of PBB had fallen in the interval between ingestion and the beginning of the studies. It may also be the case that levels of PBB in blood and fat are not good indicators of levels in target organs.

Two other scares involving contaminated animal feed – and the subsequent transfer of pollutants into the food chain – have occurred more recently. Both these incidents occurred in Belgium, first in May 1999, when dioxins (see Section 19.2.1) from contaminated fats were involved, and then again in May 2000 after the discovery of high levels of PCBs in animal feed. The dioxin scare resulted in the widespread recall of products, with countries around the world banning the import of Belgian food products.

19.3.4 Mercury Poisoning in Minamata and Niigata

Mercury in its organic form has accounted for many of the episodes of endemic disease resulting from environmental exposure to this metal. Probably the best known of these is Minamata Bay disease.[19] This disease was first noted at the end of 1953 when an unusual neurological disorder began to affect the villagers who lived on Minamata Bay on the south west coast of Kyushu, the most southerly of the main islands of Japan. It was commonly referred to as *kibyo*, that is, the 'mystery illness'. Both sexes and all ages were affected and presented with a mixture of signs relating to the peripheral and central nervous systems. The prognosis of the condition was poor; many patients became disabled and bedridden and about 40% died. The disorder was associated with the consumption of fish and shellfish caught in the bay, but mercury poisoning was not considered in the early investigations of the condition because many of what were then considered to be the classic symptoms of mercury poisoning were not present. It was only when it was realized that the signs of the disease were similar to those described in a man who had died after being poisoned during the manufacture of alkylmercury fungicides that the possibility was raised.

The source of the mercury was effluent released into the bay from a factory which was manufacturing vinyl chloride using mercuric chloride as a catalyst. It is claimed that *inorganic* mercury was released by the factory and that this was methylated by microorganisms living the in sediments in the bay. However, the rate of conversion is extremely slow, much too slow to have accounted for the large amounts of methylmercury which are calculated to have accumulated in the waters of the bay. It is therefore much more likely that the mercury was actually released in the organic form; at the time there were no regulations forbidding this in Japan.

About 700 people were affected at Minamata Bay, and there was a second outbreak of methylmercury poisoning in Japan in 1965 in Niigata, affecting a further 500 or so individuals. This latter episode followed pollution of the Agano River by industrial effluent and, again, the consumption by the local population of fish in which mercury had bioconcentrated.[20]

19.3.5 Methylmercury Poisoning in Iraq

A major poisoning episode occurred in 1971–72 when the Iraqi government imported a large consignment of seed grain treated with an alkylmercury fungicide, which was then distributed to the largely illiterate rural population. The distribution was accompanied by warnings that the seed was for sowing not for eating, and the sacks were marked with warning labels (in English and Spanish!). The seed had been treated with a red dye to distinguish it from edible grain, but the farmers found that they could remove the dye by washing and they equated this with the removal of the poison. The grain began to be used to make bread in November 1971 and the first cases of poisoning appeared in December. By the end of March 1972 there had been 6530 admissions to hospital and 459 (7%) of these had died.[21] These represented only the most severe cases and the true extent of this outbreak will probably never be known, although it has been suggested that the incidence of the disease may have been as high as 73 per 10 000.[22]

19.3.6 Aluminium Contamination of Drinking Water in North Cornwall

In July 1988, about 200 tonnes of aluminium sulfate were accidentally deposited into the treated water reservoir at Lowermoor Treatment Works, resulting in major contamination of the drinking water supply to Camelford in Cornwall and surrounding district.[20] Aluminium levels increased to over 10 mg l^{-1}, well above the 0.2 mg l^{-1} limit set by the EC on palatability grounds, and the pH of the water dropped below 5.0.

Despite reassuring messages from local sources, residents and holiday makers reported a large number of acute symptoms, and speculation about longer term effects was rife. The expert assessment of the situation, made by the Lowermoor Incident Health Advisory Group,[23] was that the early reported symptoms of gastrointestinal disturbances, rashes and mouth ulcers were indeed probably due to the incident, but that these effects were short-lived. Persistent toxic effects were thought unlikely because of the transient nature of the exposure and because all the known toxic effects of aluminium in man are associated with prolonged exposure.

However, a significant time after the incident, hundreds of people resident in the Lowermoor area continued to attribute health complaints to the contamination event. Such complaints included joint and muscle pains, malaise, fatigue and memory problems. Also, various scientific studies reported unexpected symptoms and clinical findings. The Advisory Group reconvened and concluded that, based on all the evidence available to it, some of the continuing symptoms experienced by the residents were probably induced by the sustained anxiety naturally felt by many people, and that others were wrongly attributed to the incident as a result of heightened awareness provoked by the incident and subsequent events. It was, however, recommended that a developmental follow-up of children exposed *in utero* during the incident should be undertaken, together with formal testing of the possibility of particular individual 'sensitivity' to aluminium.[23]

A complication of this incident was that the low pH of the drinking water resulted in other metals – copper, zinc, and lead – being dissolved from domestic plumbing, so that various additive or synergistic effects may have occurred.

A subsequent revisit of the Camelford incident [24] confirmed that a number of individuals showed consistent evidence of impaired information processing and memory, with no obvious relationship with measurement of anxiety and depression. Although the abnormal neuropsychological findings indicated cognitive impairment, it was uncertain whether this was caused by an acute episode of brain damage or other causes of stress resulting from the accident. A more recent study, completed some 10 years after the event,[25] investigated 55 people affected by the incident and found objective evidence of damage to cerebral function which was not related to anxiety. The authors suggested that the aluminium exposure was indeed the cause of the observed changes. However, the validity of this study has not gone unchallenged, with major criticisms relating to the bias that is inherent in the self-selection of cases.

19.3.7 'Epping Jaundice' – Chemical Contamination of Food During Storage

The ingestion of contaminated wholemeal bread resulted, in February 1965, in an unusual outbreak of jaundice in Epping, UK.[20,26] The outbreak, which affected at least 84 people, was traced to the contamination of flour by 4,4'-diaminophenylmethane. This had spilled from a container onto the floor of a van transporting both the flour and chemicals. The chemicals were absorbed by the flour through the sacking, and the flour was subsequently used to make the bread.

Jaundice and liver enlargement were preceded, in most cases, by severe pains in the upper abdomen and chest. The pains were mostly of acute onset, but in some patients the onset was insidious. Raised levels of serum bilirubin (a blood breakdown product) and the enzymes alkaline phosphatase and aspartate aminotransferase were recorded in most of the 57 patients further investigated. Needle biopsies of the liver showed considerable evidence of inflammation and cholestasis and damage to liver cells. Experimental studies subsequently showed the liver lesions to be reproducible in mice following administrations of 4,4'-diaminophenylmethane.[27] All the patients eventually recovered.

19.3.8 Love Canal

The reporting of chemical odours in the basements of homes in the Love Canal district, USA, led to a toxicological investigation which made this area famous in the history of waste disposal and resulted in much regulatory activity in the USA (and subsequently in other countries of the world).

Love Canal was a waste disposal site containing municipal and chemical waste disposed of over a 30 year period up to 1953.[20] Homes were then built on the site during the 1960s, and leachates began to be detected in the late 1960s. Dibenzofurans and dioxins were among the chemicals detected in the organic phase of the leachates. Animal studies indicated possible risks of immunotoxic, carcinogenic and teratogenic effects.[28,29] The episode resulted in significant fears of ill-health and much psychological stress. Limited follow-up of residents identified low birthweights in the offspring of Love Canal residents,[30] but no causal link has been established for cancer incidence in the area.

Love Canal provides an instance of considerable public anxiety and stress resulting from the identification of a potential environmental toxic hazard. There is evidence that the psychological and other consequences of such incidents might easily outweigh the actual toxic effects of chemical exposure.[31,32] Indeed, having reviewed the literature on hazardous waste disposal sites, Grishan[33] concluded that "there are few published scientific reports of health effects clearly attributable to chemicals from uncontrolled disposal sites".

Concern about possible health impacts of landfill sites has resurfaced in recent years with the publication of studies conducted in the UK and elsewhere in Europe[34,35] that suggest an association between congenital malformations and residence close to landfill sites. However, these studies have produced various anomalous findings and provided no evidence for a causal relationship. It is clear that further work is needed to investigate the question of causality.

19.4 GENERALIZED ENVIRONMENTAL POLLUTION

While local disasters and incidents such as those described above are often significant and can tell us much about the consequences of high level, acute or sub-chronic exposure to environmental contaminants, of greater significance to public health are the lower level chronic exposures to air pollutants and other substances that are more commonly encountered or more widely dispersed in the environment. Human exposure to these pollutants can occur through inhalation, ingestion (from contaminated food and drink) or absorption through the skin and mucous membranes (from soil or dust, for example). Dioxins, PCBs and DDT are well known examples of chemicals that are now ubiquitous in the environment but whose impact on human health remains uncertain. This section details the potential impact on health of indoor air pollutants, toxic metals, asbestos and other fibrous materials in the environment, pesticides, and finally an ill-defined group of substances commonly referred to as 'endocrine disrupters'.

19.4.1   Indoor Air Pollution

The presence of noxious substances in the outside air, coming from factories, domestic fuel combustion and vehicle exhausts, continues to be a major cause of concern. In terms of overall morbidity and mortality, probably nothing can surpass the damage which has been caused historically by exposure to the by-products of coal burning; certainly there been remarkable episodes in which sudden increases in the number of recorded deaths have followed exceptional air pollution events, for example, in the Meuse Valley in 1930, in Donora in 1948, and in London in 1952, when the overall excess in daily deaths was estimated to be 3500 to 4000.[36] Even though the levels of sulfur dioxide and smoke seen during these episodes are no longer experienced (at least in the developed world), there remains considerable concern that present day air pollution, most notably by airborne particulate matter – commonly measured as PM_{10}* or $PM_{2.5}$ – continues to have a real and measurable impact on human health. These concerns about outdoor air quality are reflected in continuing discussions of the health effects of vehicle emissions, calls for tighter emission controls and the setting of standards for the common air pollutants. In the UK, there has been also an extensive programme of government funded research committed to understanding better the impacts and mechanisms of effect of air pollution.[37]

Chapter 11 of this book is dedicated to the health effects of outdoor air pollution, but there is a need here to concentrate on a compartment of the environment where people may spend up to 90% of their time – the indoor environment. Although outdoor air quality has a considerable influence on indoor air, for many pollutants the greatest proportion of total exposure is determined by exposures indoors. Indoor pollution sources, such as gas cookers, can result in pollutant levels much higher indoors than out, and some pollutants are found only indoors. The impetus over recent years to conserve energy has resulted in 'tighter' buildings with much reduced air exchange and therefore a greater propensity for indoor pollutants to build up.

People are exposed to a wide variety of indoor air pollutants both in their workplace and at home. Concern is directed especially to domestic air quality since the home is where many non-healthy individuals, and the very young and old, spend much of their time. Pollutants indoors arise from a variety of sources, most notably from the combustion of fuel for cooking and heating, emissions from building products and furnishings, and from the use of DIY and consumer products.[38 40]

This section briefly reviews the main chemical indoor air pollutants found in homes. Radon (a natural radioactive gas) and house dust mites, bacteria and fungi (biological agents), whilst important, are not included here.

19.4.1.1 Carbon Monoxide. Carbon monoxide is one of the most important indoor air pollutants[41] and continues to kill as many as 100 people a year in the UK through accidental poisoning. It is especially dangerous since it has no colour, smell or taste. Carbon monoxide is produced when fuel burns with an inadequate supply of oxygen. High concentrations can be produced by badly installed or inadequately maintained gas and solid fuel appliances, paraffin (kerosene) heaters, *etc.*, or by fumes leaking from a flue into a poorly ventilated

*Particulate matter of aerodynamic diameter less than 10 μm.

room. Its toxic action is primarily through the displacement of oxygen in haemoglobin in the blood to form carboxyhaemoglobin, thus depriving the tissues of the body of their oxygen supply. Early symptoms of exposure include tiredness, drowsiness, headaches, dizziness, pains in the chest and stomach pains. Excessive exposure can lead to loss of consciousness, coma and death. Exposure of pregnant women can also result in adverse effects on the fetus, and longer term effects of chronic exposure in adults are suspected. People thought to be most at risk are those with existing cardiovascular disease or individuals with compromized blood oxygen levels, including pregnant women and the elderly. Surprisingly, many details of carbon monoxide toxicity remain unresolved, especially those relating to chronic low level exposure. Signs and symptoms of acute carbon monoxide poisoning are often confused with those of food poisoning and this can result in misdiagnosis, with sometimes tragic consequences. Many fatal cases of carbon monoxide poisoning result from the accidental blockage of flues, but leakage of combustion products into the room air and misuse of fuel burning appliances are also important causes.

19.4.1.2 Environmental Tobacco Smoke (ETS). Where it occurs, this is another very significant indoor air pollutant. It contains tar droplets and a cocktail of various other toxic chemicals including carbon monoxide, nitric oxide, ammonia, hydrogen cyanide and acrolein, together with proven animal carcinogens such as *N*-nitrosamines, polycyclic aromatic hydrocarbons and benzene. Environmental tobacco smoke can irritate the eyes, nose and throat, and exposed babies and children are more prone to chest, ear, nose and throat infections. Women exposed during pregnancy tend to have lower birthweight babies, and asthmatics may be adversely affected by acute exposures. The Californian EPA concluded that causal links have been established between ETS exposure and heart disease, lung cancer and nasal sinus cancers in adults, and sudden infant death syndrome (SIDS), asthma and middle ear disease in children.[42]

19.4.1.3 Nitrogen Dioxide (NO_2). NO_2 and other oxides of nitrogen are formed when fuel is burned in air. Thus NO_2 is generated indoors by gas, oil and solid fuel appliances. The main sources are unflued appliances such as gas cookers, gas wall heaters and kerosene (paraffin) heaters. Indoor levels are significantly influenced by outdoor levels, but where an indoor source is present this tends to dominate. Exposure to high levels typically occurs in the kitchen during gas cooking. Nitrogen dioxide can irritate the lungs, but the mechanisms of toxic action at lower levels remain to be fully elucidated, and there is continued uncertainty and debate about the actual impact on the health of occupants of NO_2 levels as typically found indoors. Overall, the weight of published evidence points to a possible hazard of respiratory illness in children, perhaps resulting from increased susceptibility to infections.[40] Interactive effects of NO_2, for example with house dust mite allergen, have also been postulated.[43] The possible impact of NO_2 on potentially susceptible groups such as asthmatics and people with chronic bronchitis is largely unknown.

19.4.1.4 Formaldehyde. Formaldehyde[40] is a colourless gas with a pungent odour that is given off from various furnishings and fittings found in the home. One of the most important sources is pressed wood ('chipboard'), made using bonding materials containing urea–formaldehyde resin, which has become increasingly used in furniture items over the last few decades. Another major source is urea–formaldehyde foam insulation (UFFI) installed in wall cavities. Formaldehyde is also generated during the combustion of fuel and is a component of cigarette smoke. Formaldehyde gas can irritate the mucous membranes of those exposed. The odour threshold is in the region of 0.05– 1.00 ppm and levels found in some homes may reach the threshold, for some individuals, for transient eye, nose and throat irritation. Although formaldehyde is a sensitizing agent, no studies to date have demonstrated the induction of asthma by domestic exposure. Certain individuals do appear peculiarly sensitive to and intolerant of formaldehyde exposure.

19.4.1.5 Volatile Organic Compounds (VOCs). VOCs* in the indoor environment originate from a number of sources including furnishings, furniture and carpet adhesives, building materials, cosmetics, cleaning agents and DIY materials. VOCs also originate from fungi, tobacco smoke and fuel combustion. By far the greatest peak exposure to VOCs occurs during home decorating using solvent-based paints. Glues are another important source of high peak levels. It has been estimated that between 50 and 300 different compounds may occur in a typical non-industrial indoor environment,[44] including aliphatic and aromatic hydrocarbons, halogenated compounds and aldehydes. Because of the diverse range of chemical substances defined as VOCs, determination of health effects is problematic. However, it is known that at levels typically found indoors the major effects are likely to be sensory. Short exposures to high levels of solvent vapours can cause temporary dizziness; lengthy or repeated exposure can irritate the eyes and lungs and may affect the nervous system. Some VOCs, such as benzene, are cancer causing agents, but the consequences of exposure to the levels typically found in homes are uncertain.[40] VOCs have been associated by some with the syndromes known as 'sick building syndrome' and 'multiple chemical sensitivity'[45] and it has been postulated that they may react with ozone and other substances to produce more toxic compounds.[46]

19.4.2 Metals

Episodes or incidents involving exposure to mercury and aluminium, mostly through ingestion, have already been described. Added to this is the evidence for more widespread environmental contamination by metals such as mercury, cadmium and lead. For example, lead concentration in polar ice has increased over 20-fold since 1800, and tissue levels of lead and mercury in Greenland Inuit Eskimos are 4–8 times higher than in preserved ancestors from five centuries

---

*A wide range of compounds with boiling point between approximately 50 °C and 250 °C and which, at room temperature, produce vapours.

ago.[47] Industrial sources of metals include refineries, chemical plants, cement manufacturing, power plants, smelters and incinerators.[48] Until recently, vehicles were a major source of environmental lead through the use of tetra-ethyllead as an anti-knock additive in petrol.

Metals have a variety of effects upon the human body, mostly at the cellular level. Some metals disrupt biochemical reactions while others block essential biological processes, including the absorption of nutrients. Some accumulate in the body giving rise to toxic concentrations after many years of exposure, and yet others (including arsenic, beryllium, cadmium and chromium) are carcinogens. Exposure to methylmercury and high levels of lead can cause gross developmental deformities.[49]

The health effects of lead have been studied extensively and are well documented.[50,51] Environmental lead exposure has been linked to reduced IQ in children and to elevated blood pressure in adults, although the cause–effect relationship for the latter is not especially robust.[52] Airborne lead exposure has been much reduced in recent years by the introduction in many countries of unleaded gasoline. However, other routes of exposure (most notably ingestion) remain significant. The outbreaks of endemic lead poisoning in classical and historical times are well known: suffice it to say that the adulteration of food and drink with lead has been a significant contributor to morbidity and perhaps to mortality in the past. The most serious form of endemic lead poisoning arose from the habit of adulterating wine with lead to improve a poor vintage and make it more saleable. During the eighteenth century there were a number of famous outbreaks of lead poisoning, the 'Devonshire colic' being perhaps the most well known. In this case, the adulteration of cider with lead arose accidentally due to the presence of lead in the pounds and presses used to make the cider. The widespread use of lead in cooking utensils, in glazes, and in pewter added to the burden of lead exposure during the eighteenth and nineteenth centuries.[53] Most human exposure to lead in recent times has arisen from the use of lead in domestic water pipes.

Environmental exposure to cadmium was considered to be the cause of a disease which was first reported in 1955 as occurring in a localized area downstream from a mine on the Juntsu River in the Toyama Prefecture in Japan. It was called *itai-itai* disease meaning 'it hurts'. The condition was almost entirely confined to elderly women who had borne several children; it was characterized by severe bone pain, waddling gait, severe osteomalacia (bone softening), pathological fractures, and some signs of renal impairment. The water which was used to irrigate crops was frequently contaminated by outpourings from the mine which contained zinc, lead, and cadmium. Levels of cadmium in rice samples were shown to be about ten times the amount normally present and the view was gradually formed that it was the cadmium which was responsible for the disease.[54] However, it seems likely that deficiencies of calcium and vitamin D were also at least partially to blame and that cadmium may have been acting only as one factor in what was a multi-factorial aetiology for this disease.[55] Concern is occasionally raised about high levels of cadmium in soils (due either to naturally high levels or historical industrial or mining activity) and the effects this might have on the health of the local population. An example of this has occurred in Shipham, UK.[56]

People who depend upon fish as the staple part of their diet may still be at risk of excessive exposure to mercury, even if not on the scale experienced at Minamata and Niigata. For example, blood methylmercury levels have been found to be almost ten times higher in a Peruvian population who ate on average 10.1 kg of fish per family of (on average) 6.2 persons, compared with a control population whose fish intake was considerably more modest. (The mean in the high fish eating population was 82 ng ml^{-1} and in the control population, 9.9 ng ml^{-1}.) Moreover, 29.5% of the heavily exposed population had signs of a sensory neuropathy.[57] In a fish-eating population in New Guinea, hair mercury concentrations were between two and three times that of a control group (6.4 μg g^{-1} compared with 2.4 μg g^{-1}); there were, however, no demonstrable ill effects in this group.[58]

19.4.3 Asbestos and Man-made Mineral Fibres (MMMF)

The industrial exploitation of asbestos began about a hundred years ago, its use probably peaking in the 1960s. Because the health effects of asbestos were unsuspected, workers in the industry were often exposed to massive levels of airborne fibres, resulting in serious, and often fatal, pulmonary interstitial fibrosis (termed 'asbestosis'), lung cancer and mesothelioma (cancer of the epithelium lining the chest cavity). First reports of these diseases started to appear in the 1900s, 1930s and 1960, respectively.[59,60] The work of Wagner in South Africa in the 1950s and 1960s was especially significant in assessing and understanding the health impacts of asbestos fibres. These health concerns resulted in a reduction, at least in the developed world, of asbestos use. The commercial materials marketed as 'asbestos' were from both the amphibole and serpentine groups of silicate minerals. Those from the amphibole group were identified as being particularly hazardous, and crocidolite (blue asbestos) use was discontinued in the UK following the voluntary ban on its import in the 1970s. Import of amosite ceased in 1980.[61] Use of chrysotile (white asbestos; serpentine group) has also significantly declined but until recently was still allowed in many products, most notably asbestos cement and sheeting and in friction products such as brake linings. In Europe, at least, there are now formal moves to remove chrysotile from the market because of its potential for causing harm and the availability of less hazardous substitutes.[62]

Because asbestos is a cheap and effective insulator, fire-retardant and reinforcing material, its use became widespread and it was incorporated into a wide range of products. The demonstrated health risks of asbestos have resulted in considerable concern about environmental exposure to asbestos fibres as well as worries about the legacy of historical occupational exposures. Exposure to asbestos fibres in smokers is especially relevant because of the synergistic interaction that has been demonstrated between asbestos and smoking in the induction of lung cancer.[63] Of particular interest is the evidence that, contrary to expectations, the age-specific incidence of mesothelioma in men in the UK, instead of falling in parallel with the decrease in asbestos use, is actually increasing.[64] If the projection by Peto *et al.*[64] is accurate, British mesothelioma deaths will approach 3000 per year by 2020. This is in addition to the several hundred deaths per year ascribed to asbestosis and a similar number from lung cancer. Although most of the reported deaths are a result of previous occupational exposure, the level of mortality and morbidity has given rise to concern about environmental exposures, not only to asbestos but to other fibrous materials with similar characteristics.

As the consumption of asbestos has fallen, so the production and use of man-made mineral fibres has steadily increased.[65] While some of these materials have specifically been developed as asbestos substitutes, many of the uses of these fibres are new and unrelated to the historic use of asbestos. The man-made fibres being used in the largest amounts are the insulation wools (glass or fused rock products), which are used for both thermal and acoustic insulation. Energy conservation drives in the last decade or so have resulted in very widespread use of these materials in homes as loft and cavity wall insulation. They also have important applications in horticultural products, and smaller amounts of slag wool and ceramic fibre are used in more specialized applications.

In considering the health hazards posed by asbestos and man-made mineral fibres, the prime features of importance are the physical diameter and shape of the liberated fibres, and their persistence in the lung; the chemical composition of fibres, except insofar as it influences these properties, is generally considered less important, although the crystalline habit seems to be relevant, and some scientists believe that the propensity to give rise to oxygen free radicals is a major factor. Size and shape are important because, to have effect, fibres must be fine enough to penetrate the deep lung; the fibrous shape affects deposition and inhibits removal by the lung's natural defence systems. Many man-made fibres (notably those used domestically) pose little or no risk because they are too coarse to gain access to the lower respiratory tract, or do not liberate fibrous dust during normal handling and use, or are highly soluble and therefore do not persist in the lung. All these factors have to be taken into account when considering the risk to health of fibrous materials, especially of the newer materials, encountered in the environment.[65]

19.4.4 Pesticides

Notwithstanding the obvious beneficial effects of their use, pesticides do present recognized hazards to human health,[66] mostly through high level occupational exposure and accidental poisoning incidents, although low level exposure remains a cause of public concern.

There are a number of ways in which humans can be exposed to pesticides through the environmental route. Pesticides used domestically in wood preservation or as household insecticides may be a particularly important source of exposure for the general public. Possible effects of pesticide residues in food and water probably cause the greatest public anxiety, although reports of clinical poisoning by residues seem to be extremely rare. Analysis of reported consumer poisonings by pesticides show that most arise from spillage of pesticides onto food during storage or transport, eating a food article not intended for human consumption (*e.g.* treated grain or seed potatoes; see Section 19.3.5) and improper application of pesticides.[67] Much of the concern about pesticide use, however, revolves around long-term accumulation in the environment and the low level but chronic human exposure to these compounds which occurs either directly or indirectly through the food chain. This has almost certainly been one of the main reasons for the increased popularity of 'organic' foods.

Pesticides include insecticides, fungicides and herbicides. The main classes of insecticide are organochlorines (*e.g.* DDT, lindane, dieldrin), which are very persistent in the environment, anticholesterinases (the organophosphates, *e.g.* parathion, and the carbamates) pyrethroids and other botanicals, and fumigants (*e.g.* ethylene dibromide). Fungicides include organometals, phenols and carbamates. Most herbicides are bipyridinium compounds (*e.g.* Paraquat), phenoxy compounds, organophosphates and substituted anilines. Various other compounds (not detailed here) are used to control rodents, mites and ticks, molluscs, bacteria, birds and algae.

The best known pesticides are probably the organochlorines and organophosphates. The organochlorines act as neurotoxins to the target organisms. Although, as a class, the organochlorine pesticides are less acutely toxic to humans than some other insecticide classes, they have greater potential for chronic toxicity. Many are now banned or restricted because of their persistence in the environment and their propensity to accumulate in the tissues of living organisms.[48] They are probably the compounds of most concern with respect to chronic environmental exposure, and recently have also been implicated as possible environmental endocrine disrupters (see below). Organophosphates, on the other hand, which function by blocking the activity of acetylcholinesterase, break down rapidly and do not accumulate in tissues. However, they are often extremely toxic and non-selective, and more instances of acute poisoning have occurred with organophosphates than with any other insecticide class.[68] Information on the effects of occupational exposure through their use in sheep dips[69,70] has given rise to concern in some quarters about possible low level environmental exposure to this group of pesticides. However, as with many other environmental contaminants, the lack of exposure data and the uncertainties regarding low dose extrapolation preclude any firm conclusions being made regarding risks to the general public of environmental exposure to pesticides.

19.4.5 Endocrine Disrupters (see also Chapters 1 and 3)

There is continuing public interest in the possible adverse consequences arising from the release into the environment of chemicals with the propensity to disrupt hormone function in humans and in wildlife. Originally described as 'environmental oestrogens', such agents are now known by the broader term 'endocrine disrupters' or 'endocrine disrupting chemicals'.[71] In addition to the original understanding that such substances mimicked, inhibited or otherwise interfered with the action of the sex hormones oestradiol and the androgens, the new term incorporates possible effects on other endocrine systems and organs, including the thyroid, pituitary and adrenals. In this way, growth, development, behaviour, immune response, *etc.* could all conceivably be affected by endocrine disrupters.

In humans there is an increasing body of evidence for changing trends in reproductive health.[72] In particular, elevated incidences of testicular cancer in men and breast cancer in women have been demonstrated, and there is some evidence, albeit incomplete and uncertain, for reduced sperm counts and sperm quality, at least in certain parts of the world. Other effects of concern include cryptorchidism (undescended testes), hypospadias (a congenital malformation of the penis) and prostate cancer. A wide range of environmental pollutants have been implicated as possible endocrine disrupters. Of these, synthetic hormones, organochlorine pesticides, polychlorinated biphenyls, phthalates, alkylphenols and bisphenol-A are perhaps the most widely studied. Considerable attention is also being placed on the role of natural phytoestrogens in the human diet. It is clear that, as more and more chemicals are investigated, the list of those with endocrine disruptive potential will continue to expand.

Tying up with the concern about possible links between changes in human reproductive health and environmental endocrine disrupters is the demonstrated occurrence of adverse reproductive changes in certain wildlife species linked to chemical pollution. Among the best documented of these are the findings in Florida alligators. A decline in the population of alligators in Lake Apopka in the 1980s was linked to poor reproductive success. Investigations revealed abnormalities in the reproductive organs of both males and females. Subsequent studies showed a possible role for organochlorine pesticides which entered the lake from a major spill of dicofol. Induction of vitellogenin (an egg yolk protein) in male fish and masculinization of females are other effects which have been detected and related to environmental pollution. However, because of different physiologies. exposure routes, *etc.*, the relevance of observations in wildlife to those in humans is very uncertain.

The organochlorine pesticides, because of their propensity to build up in the environment and their known effects on wildlife, are a particular focus of attention. The *o,p'*-isomers of DDT have been demonstrated to have oestrogenic activity and to bind to the oestrogen receptor, but these isomers are relatively unstable and are rarely found in the environment. The major and persistent DDT metabolite, *p,p'*-DDE, has no significant interaction with the oestrogen receptor and possesses no inherent oestrogenic activity. However, it does bind strongly to the androgen receptor and inhibits androgen receptor action.

There are many factors to consider when assessing the possible impact of environmental endocrine disrupters on human health. Firstly, the potency of many of the relevant environmental contaminants is low or very low when compared with endogenous oestrogens such as oestradiol, although differential protein binding complicates the picture. As with all toxicological effects, there may be additive or synergistic effects between endocrine disrupting chemicals, though the results of tests for such interactive effects have been variable. Humans have always been exposed to natural phytoestrogens present in food-stuffs and over the last 50 years or so have been exposed to high concentrations of hormonally active substances in the form of medical treatments and oral contraceptives. Against this background, and considering the likely importance of the stage of life at which exposure occurs, it is currently very difficult to assess whether or not environmental endocrine disrupters play a significant part in reproductive disorders in the general population.[73] Further research is clearly needed before a firm assessment can be made.[74]

19.5 CONCLUSIONS

Following the industrial revolution, and especially since the Second World War and the ensuing rapid growth of the chemical industry, an increasing number of foreign substances have been synthesized and either deliberately, accidentally or incidentally released into the environment. In addition, man's activities in utilising the earth's natural resources and modifying the environment for benefit or gain have resulted in the local and global release of pollutants, especially combustion products. The increasing human population and continuing rapid strides in the industrial development of Third World countries, linked with the inherent desire for improved quality of life, have exacerbated the problem. Against this, however, is the developing trend to consider ever more seriously the impacts of man's activities on the environment. This, coupled with increasing concern about how environmental quality can affect human health, has led for example to the development of such axioms as 'sustainable development' and the 'precautionary principle', and of 'cradle-to-grave' environmental assessments, 'eco-labelling' and 'product stewardship'.

The link between environmental chemicals and health is therefore a topic receiving increasing attention. This chapter has looked at some of the ways that sudden (catastrophic) or more incidental exposures can result in adverse effects on human health. In many cases the impacts of low level exposures on human health are suspected rather than proven. This is because of the problems associated with establishing causal links when effects and doses are small, and many other factors that confound the situation in human populations. Exposure to contaminants can occur through three major pathways; ingestion, inhalation and dermal contact. The former is of prime concern, as demonstrated by the cases reviewed here and by the obvious public interest revolving around the presence of pesticides and other chemical residues in food and water. In particular the halogenated hydrocarbons, including dioxins and organochlorine pesticides, continue to attract public, political and scientific attention. We are living now with the legacy of the historic use of chemicals such as DDT and PCBs that persist and circulate within the environment. Understandably, this has led to the current special concern afforded to any persistent organic chemical with the propensity to bioaccumulate.

With respect to exposure by inhalation, pollution from road vehicles (especially fine particles) and the quality of air in the indoor environment are seen to be significant issues. Clearly, total personal exposure (by whatever route) is fundamentally important, as is individual susceptibility to the effects of toxins determined, for example, by genetic makeup, age, current health status and exposure history. Outputs from the human genome mapping project could well have important impacts in the area of individual susceptibility. Developments in toxicology, including, for example, the application of biomarkers and advances in modelling and risk assessment for low dose exposures, are already enabling more accurate estimates to be made of population exposure profiles and health consequences. Increased understanding of total exposure, and the apportionment of exposure, will facilitate the adoption of appropriate control measures.

One of the biggest issues of the moment relates to the ability of some ubiquitous environmental chemicals to act as hormone mimics (or antagonists) or otherwise to interfere with the balance of sex and other hormones and therefore possibly affect human reproductive health and fertility as well as behaviour, growth and development. This question of the endocrine disruptive activity of chemicals in the environment, to which we are all commonly exposed, is likely to continue to be an issue of considerable scientific and public interest. Some generic issues are also certain to continue to provide challenges, for example, the somewhat conflicting expectations of a totally 'safe' environment, good health *and* a high standard of living. The discrepancies between the public perception of risk and scientific risk assessment, the problem of discriminating physical from psychological health effects, and the general issue of assessing exposure to chemicals in the environment and estimating health risks in the face of many confounding factors are all likely to entertain the minds of scientists, regulators and policy makers well into the 21st century.

19.6 REFERENCES

1. A. Giovanardi, in 'Proceedings of the Expert Meeting on the Problems Raised by TCDD Pollution', Milan, 1976, p. 49.
2. G. Fillipini, B. Bordo, P. Crenna, N. Massetto, M. Musicco and R. Boeri, *Scand. J. Work Environ. Health*, 1981, **7**, 257.
3. F. Pocchiari, V. Silano and A. Zampieri, *Ann. N.Y. Acad. Sci.*, 1979, **320**, 311.
4. G. Reggiani, *J. Toxicol. Environ. Health*, 1980, **6**, 27.
5. P. A. Bertazzi, A. C. Pesatori, D. Consomi, A. Tironi, M. J. Landi and C. Zocchetti, *Epidemiology*, 1993, **4**, 398.
6. N. Anderson, M. Kerr, V. Mehra and A. G. Salmon, *Br. J. Ind. Med.*, 1988, **45**, 469.
7. World Health Organisation, 'Toxic Oil Syndrome'. Report of a WHO meeting, Madrid, 21–25 March 1983, WHO, Copenhagen, 1984.
8. J. M. Tabuenca, *Lancet*, 1981, **2**, 567.
9. A. V. Roncero, C. J. del Valle, R. M. Duran and E. G. Constante, *Lancet*, 1983, **2**, 1025.
10. H. Urate, K. Koda and M. Asahi, *Ann. N.Y. Acad. Sci.*, 1979, **320**, 273.
11. P. H. Chen, J. M. Gaw, C. K. Wong and C. J. Chen, *Bull. Environ. Contam. Toxicol.*, 1980, **25**, 325.
12. T. Kashimoto, H. Miyata, S. Kunita, T.-C. Tung, S.-T. Hsu, K.-J. Chang, S.-Y. Tang, G. Ohi, J. Nakagawa and S.-I. Yamamoto, *Arch. Environ. Health*, 1981, **36**, 321.

13. I. J. Selikoff, 'A Survey of the General Population of Michigan for Health Effects of PBB Exposure', Final Report submitted to the Michigan Department of Public Health, Michigan, 1979.
14. H. E. B Humphrey and N. S. Hayner, 'Polybrominated Biphenyls: An Agricultural Incident and its Consequences: An Epidemiological Investigation of Human Exposure', Michigan Department of Health, Michigan, 1975.
15. J. G. Bekesi, J. F. Holland and H. A. Anderson, *Science*, 1978, **199**, 1207.
16. P. J. Landrigan, K. R. Wilcox and J. Silva, *Ann. N.Y. Acad. Sci.*, 1979, **320**, 284.
17. J. E. Stross, R. K. Nixon and M. D. Anderson, *Ann. N.Y. Acad. Sci.*, 1979, **320**, 368.
18. K. Kreiss, C. Roberts and H. H. Humphrey, *Arch. Environ. Health*, 1982, **37**, 141.
19. M. Katsuna, 'Minimata Disease', Kumamoto University, Japan, 1968.
20. H. P. Illing, in 'General and Applied Toxicology', eds. B. Ballantyne, T. Marrs and P. Turner, Macmillan, London, 1995, p. 1287.
21. F. Bakir, S. F. Damluji, L. Amin-Zaki, M. Murtadha, A. Khalidi, N. Y. al-Rawi, S. Tikviti, H. I. Dhahir, T. W. Clarkson, J. C. Smithy and R. A. Doherty, *Science*, 1973, **181**, 230.
22. G. Kazantzis, A. W. al-Mufti, A. al-Jawad, Y. al-Shawani, M. A. Majid, R. M. Mahmoud, M. Soufi, K. Tawfiq, M. A. Ibrahim and H. Dabagh, *Bull. WHO*, 1976, **53**, Suppl. 37.
23. Lowermoor Incident Health Advisory Group, 'Water Pollution at Lowermoor, North Cornwall', 2nd Report, HMSO, London, 1991.
24. T. M. McMillan, A. J. Freemont, A. Herxheimer, J. Denton, A. P. Taylor, M. Pazianis, A. R. C. Cummin and J. B. Eastwood, *Hum. Exp. Toxicol.*, 1993, **12**, 37.
25. P. Altmann, J. Cunningham, U. Dhanesha, M. Ballard, J. Thompson and F. Marsh, *Br. Med. J.*, 1999, **319**, 807.
26. H. Kopelman, M. H. Robertson, P. G. Saunders and I. Ash, *Br. Med. J.*, 1966, **1**, 514.
27. R. Schoental, *Nature*, 1968, **219**, 1162.
28. J. B. Silkworth, D. S. Cutler and G. Sack, *Fundam. Appl. Toxicol.*, 1989, **12**, 302.
29. J. B. Silkworth, D. S. Cutler and L. Antrim, *Fundam. Appl. Toxicol.*, 1989, **13**, 1.
30. N. J. Vianna and A. K. Polan, *Science*, 1984, **226**, 1217.
31. L. H. Roht, S. W. Vernon, F. W. Weir, S. M. Plier, P. Sullivan and L. J. Reed, *Am. J. Epidemiol.*, 1985, **122**, 418.
32. T. F. Jones, A. S. Craig, D. Hoy, E. W. Gunter, D. L. Ashley, D. B. Barr, J. W. Brock and W. Schaffner, *N. Engl. J. Med.*, 2000, **342**, 96.
33. J. W. Grishan (ed.), 'Health Aspects of the Disposal of Waste Chemicals', Pergamon, Oxford, 1986.
34. H. M. P. Fielder, C. M. Poon-King, S. R. Palmer, N. Moss and G. Coleman, *Br. Med. J.*, 2000, **320**, 19.
35. H. Dolk, M. Vrijheid, B. Armstrong, L. Abramski, F. Bianchi, E. Garne, V. Nelen, E. Robert, J. E. Scott, D. Stone and R. Tenconi, *Lancet*, 1988, **352**, 423.
36. W. P. D. Logan, *Lancet*, 1953, **1**, 336.
37. Institute for Environment and Health, 'Research on Air Pollution 1999', IEH, Leicester, 2000.
38. Department of the Environment, 'Good Air Quality in Your Home', Department of the Environment, London, 1992.
39. British Medical Association, 'Living with Risk', John Wiley & Sons, Chichester, 1987.
40. Institute for Environment and Health, 'Indoor Air Quality in the Home: Nitrogen Dioxide, Formaldehyde, Volatile Organic Compounds, House Dust Mites, Fungi and Bacteria', IEH, Leicester, 1995.
41. Institute for Environment and Health, 'Indoor Air Quality in the Home (2): Carbon Monoxide', IEH, Leicester, 1998.
42. P. T. C. Harrison, in 'Air Pollution and Health', eds. R. E. Hester and R. M. Harrison, Royal Society of Chemistry, Cambridge, 1998, p. 101.

43. W. S. Tunnicliffe, P. S. Burge and J. G. Ayres, *Lancet*, 1994, **344**, 1733.
44. L. Mølhave, *Indoor Air*, 1991, **1**, 357.
45. L. Mølhave in 'Environmental Toxicants – Human Exposures and their Health Effects', ed. M. Lippmann, Wiley, New York, 2000, p. 889.
46. P. Wolkoff, P. A. Clausen, B. Jensen, G. D. Nielsen and C. K. Wilkins, *Indoor Air*, 1997, **7**, 92.
47. A. J. McMichael, 'Planetary Overload', Cambridge University Press, Cambridge, 1993.
48. V. T. Covello and M. W. Merkhofer, 'Risk Assessment Methods', Plenum Press, New York, 1993.
49. L. Friberg, G. F. Nordberg and V. B. Vouk, 'Handbook on the Toxicology of Metals', Elsevier, Amsterdam, 1986.
50. D. Krewski, A. Oxman and G. W. Torrance, in 'The Risk Assessment of Environmental Hazards', ed. D. J. Paustenbach, Wiley, New York, 1989, p. 1047.
51. Environmental Protection Agency (EPA), 'Review of the National Ambient Air Quality Standards for Lead: Assessment of Scientific and Technical Information', USEPA, Research Triangle Park, NC, 1986.
52. F. K. Hare (Chairman), 'Lead in the Canadian Environment: Science and Regulation. Final Report of the Commission on Lead in the Environment', Royal Society of Canada, Ottowa, 1986.
53. T. Waldron, in 'Diet and Crafts in Towns', eds. D. Serjeantson and T. Waldron, BAR, Oxford, 1989, p. 55.
54. L. Friberg, T. Kjellstrom, G. Nordberg and M. Piscator, 'Cadmium in the Environment, III', USEPA, Washington, DC, 1975.
55. K. Tsuchiya, *Fed. Proc.*, 1976, **35**, 2412.
56. D. Barltrop and C. D. Strehlow, *Lancet*, 1982, **2**, 1394.
57. M. D. Turner, D. O. Marsh, J. C. Smith, J. B. Inglis, T. W. Clarkson, C. E. Rubio, J. Chiriboga and C. C. Chiriboga, *Arch. Environ. Health*, 1980, **35**, 367.
58. J. H. Kyle and N. Ghani, *Arch. Environ. Health*, 1982, **37**, 266.
59. D. H. K. Lee and I. J. Selikoff, *Environ. Res.*, 1979, **18**, 300–314.
60. J. C. Wagner, C. A. Sleggs and P. Marchand, *Br. J. Indr. Med.*, 1960, **17**, 260.
61. Department of the Environment, Transport and the Regions, 'Asbestos and Manmade Mineral Fibres in Buildings: Practical Guidance', Thomas Telford, London, 1999.
62. P. T. C. Harrison, L. S. Levy, G. Patrick, G. H. Pigott and L. L. Smith, *Environ. Health Perspect.*, 1999, **107**, 607.
63. P. T. C. Harrison and J. C. Heath, in 'Mechanisms in Fibre Carcinogenesis', Plenum Press, New York, 1991, p. 469.
64. J. Peto, J. T. Hodgson, F. E. Matthews and J. R. Jones, *Lancet*, 1995, **345**, 535.
65. Institute for Environment and Health, 'Fibrous Materials in the Environment', IEH, Leicester, 1997.
66. R. Levine, in 'Handbook of Pesticide Toxicology', Academic Press, New York, 1991, Vol. 1, p. 275.
67. T. C. Marrs, in 'General and Applied Toxicology', eds. B. Ballantyne, T. Marrs and P. Turner, Macmillan, London, 1995, p. 1199.
68. W. Stopford, in 'Industrial Toxicology', eds. P. L. Williams and J. L. Burton, van Nostrand Reinhold, New York, 1985, p. 211.
69. R. Stephens, A. Spurgeon, I. A. Calvert, J. Beach, L. S. Levy, H. Berry and J. M. Harrington, *Lancet*, 1995, **345**, 1135.
70. Institute for Environment and Health, 'Organophosphorus Esters', IEH, Leicester, 1998.
71. R. E. Hester and R. M. Harrison (eds.), 'Endocrine Disrupting Chemicals', Royal Society of Chemistry, Cambridge, 1999.
72. Institute for Environment and Health, 'Environmental Oestrogens: Consequences to Human Health and Wildlife', IEH, Leicester, 1995.

73. P. T. C. Harrison, P. Holmes and C. D. N. Humfrey, *Sci. Total Environ.*, 1997, **205**, 97.
74. European Commission, 'European Workshop on the Impact of Endocrine Disrupters on Human Health and Wildlife', Report EUR 17549, EC, Brussels, 1997.

CHAPTER 20

The Legal Control of Pollution

R. MACRORY

20.1 THE FUNCTION OF LAW IN RELATION TO POLLUTION

In a modern industrial society the control and management of pollution is underpinned by law. Pollution law in the United Kingdom has undergone immense structural development in the last twenty years, in part reflecting external institutional changes such as membership of the European Union, constitutional devolution, and privatization of key sectors of industry. At the same time, pollution law has to accommodate political demands for increased transparency and citizen involvement in decision-making. These pressures have led to led to what might be described as an greatly increased formalism in the legal framework – typified by legally binding environmental standards, a reduction in the discretion enjoyed by regulatory authorities, and a more active role by the courts.

The detailed provisions of pollution law are constantly changing, and it is not the purpose of this chapter to summarize the details of current controls.* Instead it analyses key themes which shape the structure and substance of contemporary British pollution law, and will influence its future direction. This first section considers the underlying purposes of law in the control of pollution. Section two outlines the main sources of law that impact on modern pollution regulation with an emphasis on the role of the European Community since Community policies and legal principles now dominate many areas. This is followed by a consideration of the role of civil liability as a means of handling pollution and where principles have been largely developed by the courts. Section 20.4 contains a more detailed examination of the legal controls governing three important areas – integrated licencing of industrial processes, water pollution, and noise and other statutory nuisances – chosen because they each illustrate different legal models for regulating pollution. Law and legisla-

---

*In addition to standard loose-leaf works which are constantly updated, good modern accounts include D. Wooley, J. Pugh-Smith, R. Langham and W. Upton, 'Environmental Law', Oxford University Press, Oxford, 2000; C. Hilson. 'Regulating Pollution', Hart Publishing. Oxford, 2000; L. Kramer, 'EC Environmental Law', Sweet and Maxwell, London, 2000.

tion is constantly reflecting changes in society, and the final section of the chapter highlights a number of key areas which are likely to significantly affect the future direction of pollution control legislation.

Law serves a variety of functions. At one extreme it defines what society has determined to be a criminal activity, and many pollution controls are ultimately backed by criminal sanctions. It also can create remedies available to individuals to protect their own interests which may include a secure and risk-free environment. And in a contemporary society where government and public bodies possess many interventionist powers law acts as a check on the abuse of such power by the state. Government does not always pursue its policy goals by means of law – persuasion, guidance, financial incentives can be equally effective tools. But where it wishes to impinge on or curtail the activities of private individuals or companies the 'rule of law' principle requires that it has explicit legal power to do so. Critical provisions of a typical pollution will therefore be concerned with the administration of the regulatory regime in question. They will identify the particular authority or body responsible for the implementation and enforcement of the controls, and will deal in considerable detail with matters of process – for example, the procedural steps required when making an application for a licence; the extent of public consultation required; the rights if any to appeal against the decision; and the powers available to enter premises and take samples.

The degree of detail required in the legislation reflects the general principle that public authorities have no inherent regulatory power but must act under powers prescribed by law. This factor led to one of the unexpected consequences of the major privatization of the water and electricity industries in the late 1980s in the United Kingdom. Where such bodies were in the public sector, government could ensure that they carried out its environmental goals by non-legalistic means such as informal directions – subject to any restraints in the legislation establishing such bodies Government exercised its influence by ownership rather than legal power. Once these industries passed into the private sector, the only certain way of ensuring they followed government policy was by means of legal rules. In this respect, the process of privatization, far from encouraging a deregulatory climate, led to a higher degree of explicit regulation, especially in the environmental field.

In a country such as the United Kingdom where Government effectively controls the content of legislation by its majority in Parliament, there is a natural tendency to give itself or its agencies widely drafted powers. A further function of the courts is therefore to intervene where government or its agencies have misinterpreted or misused these powers, and in the last decade there has been a rise in the number of legal challenges, often brought by environmental groups, calling on the courts for authoritative rulings on the meaning or use of such powers. Finally, we are increasingly entering an era where individuals expect to have certain rights concerning the environment (not always balanced by obligations). The potential impact of the Human Rights Act 1998 on British pollution law is considered in the final section of this chapter. But the impact of the 'rights' culture has already been felt in the context of European Community legislation. Most pollution law made at Community level has been in the form of 'directives' which impose obligations on Member States to achieve specified goals but, according to the Treaty of Rome at least, give Member States considerable discretion as to how to achieve those goals, whether by national law, government guidance, or other forms of instrument. But over the last decade or so, the European Court of Justice, the ultimate interpreter of Community law, has characterized many of these pollution directives as giving rights to individuals to have specified levels of environmental quality. It follows, according to the reasoning of the European Court, that the obligations in these Directives must be transposed by the Member State into national legislation – 'softer' means such as circulars or guidance are not sufficient where legal rights are concerned. As a result, we have seen the contents of many such directives now transposed into national legislation.

British pollution legislation used to be characterized by the large degree of discretion given to control authorities, with a reluctance to impose legally binding plans or targets and wherever possible the avoidance of national, legally binding standards expressed in specific terms. To take one example, for over one hundred years the term 'best practicable means' was the key criterion in the legislation regulating air emissions from most industrial sources – a phrase clearly dripping with ambiguity, but one which the courts were never formally called upon to interpret. Regulators and those that they regulated were essentially content to exclude the courts from the process of defining the law, in stark contrast to countries such the United States. This is no longer the situation. Pollution law is much more dense and explicit than ever before, for some of the reasons outlined above. Lawyers are increasingly involved in the process of pollution regulation, and the courts are playing a more active role as industries and environmental groups are more prepared to challenge by way of legal action the interpretation and application of modern pollution law. More legalism may mean more work for lawyers but does not necessarily imply less pollution or more effective pollution control. And while we have now built up an impressive legal structure, policy makers are beginning to question the adequacy of conventional legal methods of regulation to cope with contemporary environmental challenges.

20.2 SOURCES OF LAW

20.2.1 National Law

Determining the applicable law to any aspect of pollution is not necessarily straightforward. Unlike some countries, there is no authoritative codification of law in the United Kingdom, nor is there a single comprehensive law on pollution, although the Environmental Protection Act 1990 and the Water Resources Act 1991 contain a considerable bulk of relevant controls. Nearly all relevant legal principles are contained in Acts of Parliament, passed by

Parliament but generally drafted and promoted by Government. Such primary legislation, though, generally contain only a skeleton framework, with powers given to Ministers to make regulations or other forms of subsidiary legislation at a later date. Regulations have obvious administrative attractions since they are much simpler to promote, change and update than Acts of Parliament, and can contain a wealth of detail which would otherwise overload the Parliamentary process. Decisions of the courts interpreting the legislation are equally an essential source of law, and in some areas, such as civil liability discussed in the next section, the courts themselves have largely developed the relevant principles.

Primary and secondary legislation therefore form the main source of pollution law, but they themselves will be supplemented by various forms of government guidance in the form of published Circulars, Codes of Practice or similar instruments. Collectively and rather ungainly known as 'soft law' these types of instruments are proving increasingly common and influential in the field of pollution control, at both a national and international level. They are attractive to the policy maker because they can be expressed in flexible, non-legalistic language which may be useful particularly when one is dealing with a new area of policy or where formalistic rules would be inappropriate. Their exact legal status in part has to be determined from the relevant legislative provisions if they exist. For example, the key environmental regulatory agency in England and Wales, the Environment Agency, has a duty under its constitutive Act of Parliament to carry out its functions with a general objective of attaining the objective of sustainable development, and in doing so "shall have regard to any guidance" given by the Secretary of State.[1] The Guidance thus has some statutory force, and the Agency could find itself legally challenged if it were to wholly disregard the Guidance. Another variation, which is increasingly used, is simply to state that compliance or non-compliance with a specified Code shall be taken into consideration by the relevant authority or court.* Finally, these types of soft law instrument may be given no express legal status in legislation but they may still prove relevant in assisting a court, particularly where it has to consider what are current appropriate standards of behaviour.†

20.2.2 European Community Legislation

Aside from national sources, two forms of supra-national law are highly

---

*Examples include the Code of Good Agricultural Practice in respect of water pollution (s 97 Water Resources Act 1991) and the Code of Practice in relation to the duty of care on waste producers (s 34 Environmental Protection Act 1990).

†In the *Milka* case, for example, (quoted below, p. 534) the judge, in determining whether levels of noise was sufficient to amount to a nuisance in law, was heavily influenced by non-legally binding guidance issued by the British Standards Institute and the World Health Organization: "Without being a slavish follower of such guidance, I have concluded that expert examination in the light of published standards is a very significant factor in this case." (p. 69).

relevant. European Community legislation is now especially significant in the pollution field, both for its substantive content and its distinctive legal status. The Treaty of Rome and its subsequent amendments give the power to make Community legislation to the European Parliament and Member States acting collectively in the Council of Ministers and based on proposals put forward by the European Commission. The original Treaty made no mention of the environment as such, but since 1973 Member States decided that it was appropriate for the Community to develop a Community programme of environmental law. Initial legislation in the 1980s was heavily focussed on traditional forms of pollution, including water and air, and was based on a conventional regulatory approach, specifying environmental standards of various sorts and requiring licence or consent systems to be implemented by Member States in order to achieve those goals. Given the absence of any express environmental Treaty provisions, the legality of some of these early examples of legislation was questionable and the justification given (which generally had to based on the need to avoid economic distortion with the Community) somewhat meretricious But the Treaty was amended in 1987 to given specific reference to the environment, and the environmental legislative programme continues apace; some estimate that over three quarters of current UK environmental policy and law is derived from Community law.

The legal status of Community law is distinctive from both national and international law. Although agreed by Member States acting in an international setting, accession to the Treaty of Rome requires that national courts follow the rulings of the European Court of Justice, and in case law the European Court has held that, where there is a conflict between Community and national law, courts must give precedence to Community law. Some Community pollution laws are in the form of Community Regulations which by definition in the Treaty have immediate legal effect within Member States. Most, though, take the form of Directives which contain binding goals but give more discretion to Member States as to how to implement them within their own country, reflecting the often differing administrative and legal traditions across the Community. In the early days of Community environmental legislation, it was apparent that many governments, and especially the United Kingdom, failed to appreciate the legal significance of such Directives: "The early phase was characterized by a tendency towards regarding EEC Directives as helpful if eccentric recommendations to be gently eased into the United Kingdom scheme of things, ideally by government circular rather than legislation, and ideally without cost."[2] In a series of cases in the 1980s and 1990s the European Court, however, cranked up the legal significance of Directives – most have to be transposed in the form of national legislation for the reasons discussed in the first section of this chapter; provisions which are considered to be unconditional and precise can be pleaded before the national court against the government and other organs of the state even if there are no relevant national laws; and where someone can prove that they have suffered losses due to the complete failure of a Member State to comply with its obligations under a Directive they may now claim damages against the State.*

One result of the development of these doctrines has been a far closer attention paid by Government to ensuring that the provisions of EC environmental directives are faithfully reflected in national legislation, whether Acts of Parliament or regulations, while at the same time giving environmental lawyers greater awareness of the potential for legal claims in respect of EC law. Nevertheless problems of transposition remain, not least because there may be differing interpretations as to the exact meaning of EC legislation. The drafting of EC Directives in particular often gives rise to many ambiguities, and more so than national legislation because it is usually the result of complex intergovernmental negotiation and bargaining. Here again the EC institutional arrangements provide distinctive legal mechanisms. Under the Treaty of Rome the European Court of Justice is the final legal authority on the meaning of Community law, and the Treaty provides than any national court, dealing with a case involving an unclear aspect of Community law, may refer the matter to the European Court of Justice in Luxembourg for an authoritative ruling on the meaning of the law concerned.†

Aside from this unusual relationship between the national courts and the European Court of Justice, the Treaty also gives the European Commission a particular responsibility to ensure that Member States comply with their Community law obligations. In doing so, it has powers to take enforcement action against Member States which can eventually lead to litigation between the Commission and the Member States before the European Court of Justice. Following amendment to the Treaty under the Maastricht agreement in 1992, the European Court now has a sanction to fine Member States which do not comply with its judgments, a unique power among supra-national courts.[3] In the environmental field, the threat of invoking such powers has already achieved considerable success in persuading a number of Member States to remedy breaches of transposition.[4] In the early days of the Community environmental policy, the Commission devoted little attention to the question of enforcement against Member States but since the mid-1980s it has pursued a rather more vigorous and systematic policy, despite understaffing and internal and external political pressures. It has no direct powers of inspection within Member States

---

*The so-called *Frankovitch* doctrine named after the key decision of the European Court in 1991 and subsequently refined in *Brasserie du Pecheur* in 1996. The conditions laid down by the Court and especially the difficulties of proving causation have meant that to date there appears to be no successful state damages claim in Britain concerning an environmental directive. But in 1998 the Court of Appeal held that a fisherman who claimed his livelihood had been damaged as a result of shellfish decline alleged to be caused by failure of public authorities to comply with the EC Shellfish Directive standards could in principle seek a Frankovitch damages claim against the Government : *Bowden v South West Water Services Ltd.* ENDS Report, **288**, 54–55.

†Art. 234 Treaty establishing the European Community. The Court of Justice does not act as a supreme court of appeal as such since its role is purely to advise the national court as to the correct legal principles to apply, with the case then reverted back to the national court to determine the facts in the light of those principles. In practice, both national courts and the parties involved may often be reluctant to see a referral to the Court of Justice since the back-log is such that a decision may not be forthcoming for 18 months or so.

and has encouraged citizens and national environmental groups to send 'complaints' to the Commission where they allege breaches of Community law.[5]

European Community pollution law is now at something of a cross-roads. The early days saw the development of many binding environmental quality and emission standards. These will continue to be of importance where truly tradable products are concerned (vehicle emission standards are a good example) but in general there is a shift to a rather more subtle approach which emphasizes the importance of procedure and process as a means of improving environmental quality. Directives containing provisions concerning environmental assessment of projects and giving general rights of public access to environmental information held by public authorities are good examples. This is in part a recognition of the real limitations of what can be achieved by essentially 'end-of-pipe' solutions, coupled with legal limitations now built into the Treaty reflecting the principle of subsidiarity. Subsidiarity is a general principle inserted by Member States to limit over-intrusiveness of Community policy and requiring that the Community only take action where it is necessary to do so at that level by reason of the scale or nature of the problem at hand. Thus, the regulation of transfrontier movement of wastes within the Community is clearly still appropriate for Community law, whilst the setting, say, of environmental quality standards for inland waters is less so. The proposed new Water Framework Directive reflects these trends, and is more focussed on general obligations on Member States to provide management plans to upgrade the environmental quality of inland waters rather than prescribing precise standards. At the same time, policies in areas such as waste are moving towards goals that attempt to reduce the quantities of waste that arise in a consumer society by imposing responsibilities on producers of products such as cars and electronic equipment to take back used products.

Policy making at Community level is an intensely dynamic and often unpredictable process. Nevertheless there are three important themes that will influence the future outcome of Community environmental legislation. Treaty amendments over the past decade have substantially altered the institutional arrangements for law making. The early years of environmental policy making was conducted under procedures which gave final decision to the Council of Ministers acting unanimously, allowing individual member states to bargain strongly.

Now with a small number of exceptions, environmental Community laws can be agreed by qualified majority making at Council level and, since the Amsterdam Treaty amendments, with the joint approval of the European Parliament. Essentially, if the Council and European Parliament cannot agree to a proposed Directive, it will fall. The increasing influence of the Parliamentary process will be a key factor in the next decade.

Accession of new Member States, particularly those from Eastern Europe, is fully in train. The precise impact on future policies is difficult to predict but clearly it will make decision-making more complex, and, with even more heterogeneity within the Community, will weaken the arguments for uniformly applied pollution standards. Finally, the Community will pay increasing attention to the need to integrate environmental dimensions into its other areas of policy such as agricultural, structural funding, and transport.

20.2.3 International Obligations

The final key source of law derives from international treaties and agreements. Such agreements may be between two states, but in the environmental field tend to be multilateral involving many countries. International environmental agreements have proliferated over the last thirty years both in number and scope, dealing with common resources such as the sea, tradable goods such as waste, and global threats to the environment such as the ozone layer, climate change, and biodiversity. As a matter of law, such treaties strictly are binding on the ratifying states only, and although some modern treaties contain fairly sophisticated internal mechanisms to ensure their review and implementation, there is no systematic institutional means of ensuring enforcement. The International Court of Justice has yet to develop into a body as potentially powerful as the European Court of Justice.

The effect of international legal obligations within a country's legal system very much depends on its own jurisdictional traditions, and the courts in United Kingdom have long operated a 'dualist' approach implying that international obligations cannot be pleaded directly before a national court, in the absence of any transposing national legislation.* To take one example, the United Kingdom has ratified the Climate Change Convention imposing certain general obligations to reduce dangerous emissions of carbon dioxide, but it would not be possible as a matter of national law to raise the Convention directly before a national court as a legal ground of challenging the decision to authorize, say, a new coal-fired powered station. Even if international obligation cannot be relied upon directly, increasingly the national courts are prepared to look at the text of international agreements to assist them in the interpretation of national legislation designed to implement such obligations. To take a recent example in *Merlin v British Nuclear Fuels plc*[6] the court was prepared to examine international conventions concerning nuclear liability to assist it in deciding whether compensation provisions of the Nuclear Installations Act 1965 were designed to cover economic as well as physical losses.

The United Kingdom generally has a good record of implementing international obligations to which it is a party, and a number of national laws such as those concerning the disposal of wastes at sea or regulating marine oil pollution are directly derived from international agreements. The European Community is also a party to many international environmental agreements, usually as a joint party with Member States, though where trade measures are concerned it generally has exclusive power to negotiate and ratify agreements on behalf of all Member States. The web of contemporary international agreements now means that national and Community legislation has to be developed and negotiated in

---

*In contrast to countries such as Germany and the United States where courts apply a 'monist' approach treating international and national legal obligation as more of a single entity.

the light of this international dimension – a good example being the complex Community rules concerning the transfrontier movement of wastes which are heavily influenced by international agreements. And the international agreements may act as a constraint on unilateral initiatives at Community or national level. Obligations concerning the free movement of trade contained in the General Agreement of Trade and Tariffs provide perhaps the clearest contemporary example; the extent to which the rules constrain the United Kingdom and the European Community from developing their own environmental standards remains a source of considerable conflict, and one where as yet the true integration of environmental and economic interests continues to be elusive.*

20.3   THE ROLE OF CIVIL REMEDIES

Before the industrial revolution, there was little comprehensive or systematic governmental legal control of pollution. The role of law in this field was largely focussed on the development of remedies available to private individuals where their interests were harmed or threatened by pollution, and in this country, along with other 'common law' jurisdictions such as the United States and most Commonwealth countries, these principles were almost entirely the creation of judges dealing with individual cases before them. The principal remedy was the nuisance action, later supplemented by actions in negligence. These remedies were potentially powerful and remain so today, but there are inherent limitations. The negligence action requires the person bringing the action to prove some degree of fault on the part of the defendant. The nuisance action essentially may only be brought by those with an interest in land – amenity groups, non-property owners or occupiers have no right of action. Litigation is an expensive, time-consuming, and risky business and legal aid in the civil law field is available only to the very poor.

But there are deeper conceptual problems with these types of civil action. By their very nature they protect human interests, whether persons or property. Such actions cannot assist those parts of the environment which are not the subject of legal ownership as such – wild mammals, fish,† the marine environment, the upper atmosphere, for example. They require the identification of a defendant and proof of causal connection between his activities and the harm suffered. Identification may be no problem for point sources of pollution, but where there is diffuse pollution, or a multitude of sources, action may prove impossible. Negligence or nuisance actions are appropriate where the defendants' actions fall below the normal standard of care or behaviour of other

---

*See for instance the decision of the Appellate Body of the World Trade Organization which held that European Community measures concerning hormones in meat and meat products failed to satisfy the scientific evidential requirements of the GATT Agreement on Sanitary and Phytosanitary Measures: Report of the Appellate Body AB-1997-4, 16 January 1998.

†Fish swimming in inland rivers are not owned until caught, but most inland waters are subject to private fishing rights · it is these rights which a civil law action can protect and compensate where pollution has occurred.

equivalent business, but are less suited as an instrument to raise general environmental standards. In the nuisance action, for example, the courts have long stressed that where non-physical damage is involved (such as noise or smells) they have to consider what would be reasonable, taking into account the nature of the locality: "If one lived next door to a farm, it was no use applying to the volume of noise, the purity of the air and the freedom of insect life, the standards of the most select suburb of Bournemouth".[7] If there has been a gradual decline in the environmental quality of a locality over a number of years, the common law nuisance action is therefore unsuited to reverse the trends. Causation may also be difficult to prove – even on the civil standard of proof, the balance of probabilities – where the case involves novel science or the limits of detection. And even where a causal connection can been proved the House of Lords, the highest court in the United Kingdom, has recently stressed that a defendant should only be liable for the type of damage that was reasonably foreseeable at the time he caused it.[8] The precautionary approach therefore sits uneasily with the common law.

The courts are even reluctant to embrace a policy of prevention. The remedies in a civil action for nuisance or negligence are damages to compensate for the harm done and an injunction to prevent the harm continuing. An injunction is a particularly powerful legal remedy since breach of an injunction is considered an abuse of the court process and can lead to imprisonment or unlimited fines. In the United Kingdom the approach of the courts has generally been to grant an injunction as of right to a plaintiff who has proved his case, even though it may have considerable economic or social consequences on the defendant.* But perhaps because of the very powerful nature of these remedies, and their potential to inhibit private activities, the courts have been very reluctant to entertain a civil action before at least some harm has occurred and can be proved, even if a more anticipatory approach might be environmentally beneficial.†

Civil actions therefore have inherent limitations in meeting the demands of public policy concerning pollution prevention and control. Enter, then, government intervention and regulation by means of legislation – in this country beginning in the mid-nineteenth century with the control of air emissions from certain specified industries. The model of pollution regulation of fixed sources such as industrial emissions has followed a remarkably consistent course since its early days under the Alkali Act 1863. Typically, some form of licence or consent from a specialised public authority to discharge or emit pollution is

---

*See, e.g., V. C. Page-Wood, in *Attorney-General v Birmingham Borough Council* (1858) 4 K&J 528 "So far as this court is concerned it is a matter of almost absolute indifference whether the decision will affect a population of 250,000 or a single individual". In practice, the operation of an injunction may often be delayed to permit, say, a defendant company to install necessary abatement equipment or shift operations, but its power as a remedy nevertheless remains. In contrast to the British approach, US courts have generally been far more prepared to weigh up the economic and social costs of awarding an injunction against the private interests of the defendant.

†The courts do have a power to grant so called '*quia timet*' injunctions before damage has occurred but will only do so where the damage is imminent and potentially substantial (for example, a tree about to fall down on a property).

required before any such discharge may take place, and the law provides that a discharge without such a consent or in breach of consent conditions is a criminal offence. In effect, the state is authorizing some degree of pollution to take place. Increasingly, as discussed above, the public authority concerned will find its own discretion constrained in that it is obliged to ensure that consented discharges meet, at a minimum, certain specified environmental goals, often the result of European Community legislation.

This form of regulation now dominates contemporary pollution law. Yet it would be foolhardy to completely dismiss the relevance of civil remedies. In nuisance actions, at any rate, compliance with a regulatory consent or licence has never been a valid defence to a civil action,* and the open-ended nature of the nuisance action means that it may be the appropriate or even the sole form of legal remedy available where legislation has yet to deal with a particular or novel form of pollutant. Regulatory controls themselves may contain defences not available in civil actions, and regulatory authorities may feel constrained in enforce the full rigour of the law because of other, often legitimate, policy pressures – a private litigant is unlikely to feel any such inhibitions.† The injunction remains an exceptionally strong remedy, more powerful in many cases than the levels of fines imposed for criminal pollution offences. The potential value of civil liability for environmental damage as a public policy tool has now been raised at European Community level with the recent publication by the European Commission of a White Paper discussing the issue.[9] The paper considers various options but essentially shows a preference for liability without fault for environmental damage, and proposes that a wider class of persons, including amenity groups, should be able to initiate legal action. It is likely to be some years before Community legislation is ever agreed on this topic, but the very fact that it has been produced underlines the continuing potential significance of civil liability actions.

20.4   THREE LEGAL MODELS OF POLLUTION REGULATION

20.4.1   Integrated Licencing

The classical legal approach towards the regulation of pollution from fixed sources has been the requirement of a licence or consent granted by a governmental body, with conditions attached to ensure that the level of emissions reflect society's choices as to their acceptability. In common with

---

*In contrast, in a negligence action the fact that a defendant has failed to comply with conditions attached to a statutory licence or consent may well be relevant though not necessarily conclusive evidence that the defendant had fallen below a reasonable standard of care, the key test in such actions

†For a good recent example see *Milka v Chetwynd Animal By-Products (1995) Ltd.* Carmarthen County Court 19 January 2000 (ENDS Report, **301**, p. 56-57), a nuisance case involving an animal rendering plant licenced by the local authority under the Environmental Protection Act. The judge noted that in deciding how to enforce the statutory controls the local authority "was subject to conflicting pressures: on the one hand to allow the plant to operate, and on the other to ensure compliance with the objectives of the Environmental Protection Act (in relation to odour) and the prevent statutory nuisance from noise."

many other countries, British legislation developed on a medium by medium basis with different laws governing the regulation of pollution of waters, air, and waste disposal on land, and usually operated by different authorities. In 1976 the Royal Commission on Environmental Pollution, a standing independent advisory body to government, argued that this approach failed to take account of the interconnectedness of the physical environment and could lead to sub-optimal decisions.[10] The Commission recommended a more integrated system, using the general criterion of the 'Best Practicable Environmental Option' as a basis for decision-making by licencing bodies.

Although the force of the argument was largely accepted by the government of the day, it took some fifteen years for the concept to be reflected in legislation with the introduction of Integrated Pollution Control under Part I of the Environmental Protection Act 1990. This built upon the long established system for regulating air emissions from certain industrialized processes which had been enforced by a specialized central government inspectorate, and requires that a licence be obtained for operating a large number of industrial processes prescribed in regulations. The system was originally operated by the same central government inspectorate which had been responsible for the air emission controls, but the process of institutional integration was greatly strengthened in 1996 with the creation of the Environment Agency in England and Wales which incorporated into a single body the existing government pollution inspectorate, the National Rivers Authority (responsible for water pollution control and many other river-related functions), and the waste management control functions of local authorities.

Integrated Pollution Control is a key regulatory function of the Environment Agency, and encompasses most of the major industrial processes, including power stations, cement works, incinerators, and chemicals works (see also Chapter 21). Central to the system is the requirement of an operating licence, with the Environmental Protection Act providing that the main objective of conditions attached to the licence is to ensure that in carrying on the process the 'best available techniques not entailing excessive costs' are used (a) to prevent the release of prescribed substances, to reduce them to a minimum or render them harmless if they are released and (b) to render harmless any other substances which might be released. Where a process is likely to result in the release of substances to more than one medium, regard must be had to the "best practicable environmental option" as a whole as respect of substances that may be released. The main provisions of the Act therefore are based on a technology based approach towards pollution control aimed at the prevention of harm but tempered with practicable and economic considerations. Such broad objectives, though, have had to be supplemented with a wealth of technical guidance documents providing details of both general approaches towards the approach to be adopted and specific presumptive technical standards for particular sectors of industry.

The powerful arguments for adopting a more integrated approach towards the pollution control from industrial processes have now been accepted at European Community level which in 1996 agreed a Directive on Integrated Pollution

Prevention and Control (IPPC).[11] The Directive came into force at the end of 1999, and national implementing legislation will gradually replace the system of authorizations under integrated pollution control.* The IPPC Directive was heavily influenced by both the British and Dutch approaches towards integrated licencing, and while it does not explicitly adopt a Best Practical Environmental Option approach it does contain a general objective to attain a "high level of protection of the environment as a whole". It is difficult to predict the full nature of the changes that the IPPC regime will bring, though there are some obvious differences. The Directive, for example, uses the term 'Best Available Techniques' as a key criterion of control rather than 'Best Available Techniques Not Entailing Excessive Costs', though the definition of 'Available' makes references to economic viability and the change in terminology may not be as significant as might appear at first sight.[12] In contrast to the Integrated Pollution Control regime, noise and energy efficiency are included with the scope of the Directive's controls, and the Directive requires operators to restore sites to a satisfactory state after activities have ceased. The scope of installations covered are broader, with the notable inclusion of certain waste disposal sites. The Directive makes provision for the possible agreement of legally binding environmental standards in subsequent Directives, but more significantly provides for the exchange of information between Member States on the approach towards the application of best available techniques. Supporting these provisions, the European Commission has now set up a programme to produce 'BAT Reference' (BREF) notes covering various sectors of industry, and these will perform a similar function, on a European wide basis, to the sectoral guidance notes produced in connection with the national Integrated Pollution Control system. These BREF notes are likely to be treated as presumptive standards but the Directive expressly permits emission limits attached to conditions to take into account the geographical location of the process concerned and any local environmental conditions. These local conditions may well in some circumstances justify more stringent conditions, but the discretion available may also permit laxer standards than those contained in BREF notes, and it is an area which will require careful monitoring to minimize abuse.

20.4.2 Water Pollution (see also Chapter 1)

Many discharges of potential pollutants into both surface and groundwater do not derive from processes subject to integrated licences. Direct discharges into inland and coastal surface waters water are also covered by a consent system, but the legal structure is based on a different model. The principal controls, building on previous similarly structured laws, are now contained in the Water Resources Act 1991, and are centred on the creation of a general criminal offence where someone 'causes or knowingly permits' the entry of any

---

*Pollution Prevention Control Act 1999 and accompanying regulations. Under the IPPC Directive new installations are covered by IPPC after 30 October 1999, with existing installations to be phased in over the next eight years.

poisonous, noxious or polluting matter into waters or the discharge of any sewerage or trade effluent into such waters. Various defences are then provided, including the discharge in accordance with a consent, now granted by the Environment Agency. The different structure reflects the more pervasive nature of water pollution, and the greater potential for illegal discharges; criminal prosecutions have been frequently taken by the Agency and its predecessors.* The interpretation given by the courts to the key statutory provisions also underlines a strict approach to the protection of water quality. In the classic case of *Alphacell v Woodward*[13] the House of Lords held that the first limb of the offence ('causes') did not require the prosecutor to prove any guilty intention or recklessness on the part of the discharger, but simply that he had caused the entry in question; any mitigating circumstances, such as lack of intent or recklessness, is reflected in the level of sentence imposed by the court. Case-law had also suggested that 'causation' requires a positive act on the part of the defendant and that where there had been some failure to take precautions (such as mending a leaky pipe) the appropriate offence was one of 'knowingly permitting' which by its terms does require some degree of knowledge if not actual intent. But, more recently, the House of Lords has further strengthened this limb of the offence by rejecting this distinction as being too restrictive. According to Lord Hoffman in *Empress Car Company v National Rivers Authority*,[14] the offence "only requires a finding that something which the defendant did caused the pollution". The activities of a third party such as a trespasser or the occurrence of a natural event might in some cases break the chain of causation, but if these interventions could be considered a normal fact of life then the defendant could still be considered to have 'caused' the entry; judges or magistrates in individual cases are left to determine whether the circumstances can be described as ordinary or extraordinary in this context.

The policy and institutional context in which water pollution controls have operated has also been distinctive from those concerning integrated pollution control and its predecessors. The physical nature of the river and water catchment systems led to the development of a more integrated institutional approach to water management with the creation of regional water authorities in 1972, followed by a National Rivers Authority in 1989 and its eventual incorporation into the Environment Agency in 1995. The arrangements mean that the same body is responsible for regulating water abstraction as well as pollution, and for controlling other activities such as navigation and recreational use which may affect the quality of water systems; it also has jurisdictional control over complete river catchments systems† even though these may cross local government administrative boundaries. The focus on a largely

---

*A rather similar model has been created for waste disposal on land with a basic criminal offence of illegal disposal with a defence of operating a facility in accordance with a waste management licence. The waste management licences, though, are more akin in complexity and scope to IPC licences encompassing complete facilities and including requirements concerning the technical competence and financial probity of the operator: see Part II Environmental Protection Act 1990.
†Until privatization of water and sewerage utilities in 1989, the Regional Water Authorities were also responsible for the delivery of water and the management of sewerage.

contained environmental medium also meant that the controls tended to be based on ambient environmental policy objectives rather than technology-based standards,[15] with discharge consent conditions being established in the light of meeting such quality goals. In line with a traditionally preferred approach, these ambient standards were largely developed outside the formal legal framework, but European Community legislation concerning water has introduced a large number of binding standards for various types of water or in relation to particular chemical substances.

Most of the Community water Directives contain environmental quality standards, but the long debate over the 1976 Directive on the Discharges of Dangerous Substances into waters underlined important different philosophical approaches towards the control of pollution which remain alive today. Under the original proposal, the discharge of certain particularly dangerous chemicals (known as List I substances) were to be subject to minimum Community-wide emission limits essentially on the ground that their entry into waters should be minimized, or eliminated, while the much longer List II substances were to be subject to quality objectives and pollution reduction programmes. But the British Government argued that such an approach failed to take into account the nature of the receiving waters and its physical capacity to disperse and absorb pollutants.* A classic compromise was reached, permitting Member States the option of adopting Community agreed emission limits *or* quality objective standards in relation to such substances. In practice, the agreement to subsequent daughter Directives for black list chemicals has proved much slower than expected, and the Directive provides that, where no such 'daughter' directives have been agreed, Member States must apply their own quality objective approach aimed at improving the state of the waters.†

The European Commission has recognized that neither approach is perfect and that they should be complementary rather than contradictory: "Neither of the two extremes offers an ideal solution. Environmental quality objectives alone are often insufficient to tackle serious pollution problems and can be abused as a 'licence to pollute' up to a defined level. Likewise, a strict emission limit values approach based on Best Available Technology can in some circumstances lead to unnecessary investment without significant benefits to the environment."[16] Technology based standards still form the basis of more recent Community Directives such as that on Urban Wastewater Treatment Plants,[17] but this Directive allows for the possibility of tighter or more relaxed standards in sensitive and less sensitive areas of receiving waters identified as such by Member States. More recently the Community has recognized the need

---

*Similarly, in negotiations over the IPPC Directive the British were able to argue successfully that in deciding the consent conditions the geographical location and the local environment could be taken into account in addition to appropriate technical standards.

†Daughter Directives for 99 out of the original list of 129 'black list' substances have yet to be agreed. In *Commission v Belgium* C-207/97 21 January 1999, the European Court of Justice confirmed this structure to the Directive, and held that pollution reduction programmes for the remaining substances must be comprehensive, coherent, and specific. Under the Water Framework Directive, which will eventually replace the Dangerous Substances Directive, a more integrated approach towards listing of priority substances is planned.

to consider the management of water systems as a more coherent whole with the Water Framework Directive* which will impose general obligations on Member States to draw up management programmes for the improvement of rivers, while at the same time replacing some of the more specific water Directives concerned with water quality.

20.4.3 Noise and Other Forms of Statutory Nuisance

A third and equally distinct model of statutory pollution control is to be found in the provisions concerning noise and other forms of statutory nuisance. In theory the remedies provided by the civil courts should enable individuals to take action to protect their interference from undue pollution caused by noise and other invasive forms of pollution but such civil actions are costly and notoriously lengthy, and really only appropriate for the most serious of cases. As an alternative, legislative procedures have been developed with the aim of providing a rapid solution to particular problem areas, but where licencing procedures exemplified by integrated pollution control or discharge consents for water pollution are ill-suited; these areas are less amenable to the development of environmental standards which can be incorporated into licence conditions and, especially in the case of noise, the sources of pollution are often from domestic premises where consent or licencing regimes would be politically unacceptable. Instead, the approach is to define in law in very general terms activities which are socially unacceptable, and give local authorities powers to ensure remediation, with the use of courts as a back-up deterrent.

The main controls of this type are now to be found in Part IV of the Environmental Protection Act 1990 which defines a number of conditions or activities as a 'statutory nuisance' – these include, amongst other categories "any premises in such a state as to be prejudicial to health or a nuisance" and "noise emitted from premises so as to be prejudicial to health or a nuisance." The definitions are to a degree tautologious in that the term 'nuisance' is not defined, and the courts must look to the type of criteria applied in common law cases, adopting an objective test of what degree of interference should be considered unreasonable in all the circumstances rather than the actual subjective views of the complainant.[18] Local authorities have a general duty to investigate potential statutory nuisances in their area, and once satisfied of the existence or likelihood of a statutory nuisance are then under a duty to serve on the person responsible or the owner or occupier of the relevant premises a notice requiring abatement of the nuisance. At this stage, the process is essentially a quasi-administrative one designed to avoid the need for court involvement and secure a resolution of the problem at hand. But failure to comply with a notice without reasonable excuse is a criminal offence, and local authorities have the power to carry out remedial works themselves, recovering the costs from the person responsible at a later date.

---

*After lengthy negotiations between the European Council and the European Parliament agreement to the Directive was reached on 30 June 2000.

The statutory nuisance provisions also give powers to individuals to make a complaint direct to a magistrates courts concerning an existing nuisance, seeking an order from the court to abate the nuisance. These powers have proved particularly effective where it is the local authority itself which is the cause of the nuisance (poor local authority housing for example), but the solution driven nature of the remedies is emphasized by a requirement on a private complainant to first give advance notice of their intention to go to court, thereby giving the person responsible an opportunity to remedy the situation without the need for court intervention.*

Although the categories of statutory nuisance are limited in number, the broad nature of the definitions allows the remedies to be applied to a wide range of situations which might not otherwise be covered by more specific pollution controls. Their effectiveness, however, can be constrained by elaborate appeals mechanisms and statutory defences built into the legislation, together with inevitable uncertainties as to what amounts to a nuisance or prejudice to health which calls for initial professional judgment by local authority environmental health officers.

Nevertheless, recent case-law has in many ways strengthened the potential of these remedies as a legal means of pollution control. In *R v Carrick DC ex parte Shelley*, for example,[19] the High Court emphasized that once the authority was satisfied of the nuisance they had no discretion as to whether or not to serve a nuisance order. More recently, in *Falmouth and Truro Port Health Authority v South West Water*[20] the Appeal Court held that statutory nuisances provisions could encompass inland waters and seem unconcerned about possible conflicts of approach between the Environment Agency operating specialized consent systems and local authorities applying statutory nuisance provisions to the same bodies of water.

20.5 TRENDS AND ISSUES IN POLLUTION LEGISLATION

20.5.1 Land Use and Planning and Pollution Control

Effective pollution legislation aims to prevent problems occurring before they cause significant harm or damage. Where new land-use developments are involved, it is clear that the system of planning controls under the Town and Country Planning legislation has an important role to play, not so much in the enforcement of detailed operational controls but in ensuring that poor locational choices are not made which would otherwise exacerbate subsequent pollution problems. Potential polluting loads from a proposed development are clearly a relevant issue for planning authorities: as the Court of Appeal recently put it in a case concerning a proposed incineration plant, "... the extent to which discharges from a proposed plant would unnecessarily or probably pollute the atmosphere and/or create an unacceptable risk of harm to human

---

*Three days notice in the case of noise nuisances: twenty-one days notice in the case of other nuisances.

beings, animals or other organisms, was a material consideration to be taken into account when deciding to grant planning permission."[21] Nevertheless it is equally clear that a local authority should not impose detailed planning conditions designed to regulate pollution which were properly the responsibility of a more specialized body such the Environment Agency using specialized controls. Moreover, the goals of the planning controls are not exclusively those of environmental protection, and economic and other factors may play a legitimate role in the decision as to whether or not to grant planning permission.

The need to ensure a successful linkage between the planning system and pollution control regimes has been recognized for over twenty years since the publication of the seminal report by the Royal Commission on Environmental Pollution on the subject in 1976[22] but its achievement has not been easy. Until recently, the interlinking of the two fields of law was secured largely by administrative means rather than positive legal measures but in 1988 formal environmental assessment procedures were introduced within the land-use planning system in order to implement a European Community Directive on the subject.* The requirements do not not apply to all types of development but are confined to two broad lists of project types, the first being subject to mandatory assessment in all cases, and the second and much broader list being subject to assessment where the project in hand is considered likely to have significant environment effects. In the context of the relationship between planning and pollution control, the procedures are significant in that they clearly require potential pollution impacts to be considered and assessed during the planning application stage, and represent a more explicit legal linking between planning and pollution than existed previously. This by no means resolves all the issues of potential overlap and conflict, and the Royal Commission is currently revisiting the whole issue of the relationship between planning and pollution control in its current study on environmental planning whose results are likely to be published at the end of 2001. At the very least there is likely to be pressure for encouraging simultaneous consideration of applications for planning permissions and IPC (and eventually IPPC) licences, with the requirement for closer cooperation between local planning authorities and specialized agencies.

20.5.2 Human Rights

Constitutions in some countries give citizens a general right to a clean or healthy environment, though rights in such general terms are extremely difficult to enforce directly.[23] Highly interventionist courts, though, such as the Indian High Court, have interpreted general constitutional provisions concerning a right to life as encompassing a right to a degree of environmental quality and have felt less inhibited than many courts in encroaching on areas that might be more thought more appropriate for administrators or policy-makers.[24] United

---

*Now contained in Town and Country (Environmental Impact Assessment) (England and Wales) Regulations 1999, S.I. 1999/293.

Kingdom law has never granted equivalent rights, but with the introduction of the Humans Rights Act 1998 in October 2000 the question of general environmental rights and their implications for decision-makers will need to be reconsidered.

The Human Rights Act requires that all national legislation be interpreted as far as possible in accordance with the European Convention on Human Rights, and that all public authorities act in accordance with the Convention. The Convention itself, being a product of the 1950s, does not explicitly use the word 'environment' as such, but the European Court of Human Rights has in recent years increasingly been prepared to recognize that specific provisions of the Convention may have an environmental dimension. For example, in *Lopez Ostra v Spain*[25] the Court considered the case of an individual who had suffered severe disturbance from a neighbouring waste treatment plant and where public authorities had taken no effective regulatory action. The Court ordered the State to pay damages to the applicant, holding that the circumstances amounted to a breach of the general right to respect for private and family life under Article 8 of the Convention with the State failing to strike "a fair balance between the interests of the town's economic well-being -- that of having a waste treatment plant -- and the applicant's effective enjoyment of her right to respect for her home and her private and family life." The same Article has also been held by the Court to imply a positive right on the part of the state to inform residents of the risks of a neighbouring chemical factory.[26]

There is little doubt that in future years the language of human rights will increasingly be of influence in both the design and application of pollution regulation; in reality, a legal challenge based on a claim to substantive rights of environmental quality will succeed in only the most extreme of cases, such as the Lopez case. But in the context of procedural practice, both in relation to applicants for licences or permits and the general public who might be affected, regulatory bodies will need to be fully alive to the new human rights dimension.

20.5.3   A Specialist Environmental Court?

In any system of pollution regulation the role of the courts is critical, both in the interpretation and the enforcement of the law. Much of pollution law is ultimately backed by criminal sanctions, and in England and Wales the majority of criminal cases are heard in magistrates courts. The determination of sentences is left to the discretion of the courts, acting within a broad framework provided by the legislation. Maximum possible sentences have increasingly been raised in the environmental legislation over the last decade, but there is considerable concern at present that in general terms the level of sentences, especially the level of fines, remains at too low a level, and bears little relation to the turn-over or profits of companies involved. Better practice may result from a report of the Government's Sentencing Advisory Panel published in early 2000[27] which provided general criteria for the factors to be taken into account by courts, and advice from the Court of Appeal has encouraged prosecutors to provide more positive advice to the courts on the financial considerations involved.[28]

Nevertheless, even if the courts adopt a tougher and more consistent approach towards sentencing, there are still doubts as to the ability of non-specialized courts to handle some of the complexities of contemporary environmental law. The idea of a specialist environmental court was first raised by Lord Woolf (appointed Lord Chief Justice in 2000) in 1991,[29] The subject was given more impetus in 2000 with the publication of a major research report sponsored by government examining examples of specialist courts and tribunals in other jurisdictions and proposing a range of models.[30] It is unlikely that a single all purpose court will be established; criminal prosecutions, judicial review, and appeals procedures are quite distinct areas of law and probably inappropriate for handling by a single court, even if there is a common environmental theme. But there may well be the development of more specialized environmental tribunals, possibility combined with the current system of land-use planning appeals, with the purpose of dealing with various types of appeals concerning pollution licences and permits.

20.5.4 Public Information and Participation

The extensive controls concerning land-use planning contained in the Town and Country Planning legislation have long contained measures providing for various forms of involvement by the general public in decision-making, ranging from public registers of planning applications, public consultation, and in cases of controversy the holding of public inquiries. Post-war pollution legislation failed to mirror these provisions, mainly on the grounds that pollution control decisions were often considered to involve complex technical issues on which the general public could have little to offer, and that commercial confidentiality might be exposed by greater public access. This tradition of a largely 'closed' system of regulation has substantially changed in the last twenty years, and in many areas of pollution regulation legislation now provides for public registers of applications, consents issued, and some degree of public consultation. Following agreement to a European Directive on the subject, regulations now provide a general right for members of the public to obtain information on environmental matters held by public authorities, though there are many grounds of exception built into the law.[31] And over and above minimum statutory requirements in many areas regulatory authorities are now prepared to act in a far more transparent manner, and to an extent unknown of a generation ago in the United Kingdom.*

These are very real changes to the legislative framework, and are now supported by international principles contained in the 1998 Aarhus Convention.†

---

*For example, both the Environment Agency and the Scottish Environmental Protection Agency hold their board meetings largely in public, and full sets of board papers are available on the Environment Agency's web-site.

†At the time of writing, both the United Kingdom and the European Community have signed the Convention but not enough countries had ratified it for the Convention to come into force.

But they are likely to represent the beginning rather than the end of a process of greater openness, and decision-makers at all levels will find themselves operating in a more challenging climate where professional and technical judgements are treated with considerable public scepticism. In its 1998 study on environmental standards,[32] the Royal Commission on Environmental Pollution recognized a decline in public trust in official decision-makers in the environmental field, and advocated even greater transparency and openness. The Commission emphasized the uncertainties inherent in environmental science, and the extent to which ethical or moral choices underpinned many environmental choices, but felt that existing consultative procedures reflected in the law were often ill-suited to revealing the subtle nature of such values to decision-makers. The next decade will see greater experimentation with new forms of positive public consultation and involvement moving well-beyond the traditional legal models of public advertisement and representations.

20.5.5   Devolution and Environmental Decision-making

Pollution law in the United Kingdom now has to be seen against the background of the constitutional changes brought about by the substantial degree of devolution introduced in 1998 under the Scotland Act, the Government of Wales Act, and the Northern Ireland Act. Scotland in particular had already long developed differences in its legal and administrative framework, but modern pollution legislation was often paralleled by equivalent provisions throughout the United Kingdom. The new constitutional settlement now raises the possibility of quite distinct pathways being followed, at least for those areas falling outside the remit of European Community and international policy.

The Scotland Act essentially devolves environmental policy to the Scottish Parliament and Executive but a complex range of matters listed in the legislation have been reserved to the United Kingdom Parliament, and a number of these potentially overlap with environmental issues. Devolution in Wales is less extensive, and the Welsh Assembly's legislative powers are limited to the making of regulations rather than primary legislation. Devolution will undoubtedly bring new political and legal uncertainties: "A division of power which preserves international and Community affairs for Westminster sits uneasily with the nature of contemporary environmental politics, where there is a dynamic linkage between international, regional, national and local action."[33] We are likely to see a tension between the new political institutions eager on flexing their muscles and showing themselves alive to current environmental concerns and the machinery of government intent on ensuring the smooth development of good administration. It is too early to predict the direction of the changes that will occur. The Scottish Parliament, for example, could decide that it wishes to have a less intensive regime on contaminated land than England and Wales in order to attract inward investment. Alternatively, local political pressures might press for stricter environmental standards. In this context two overarching legal factors will be of increasing importance. First, the interpretation of precisely the extent of powers which have been devolved under the legislation, a matter ultimately for interpretation by the Courts. And, secondly, the scope and content of European Community environmental legislation which acts as an overall constraint on the freedom of devolved administrations to follow their own environmental policies. As already indicated, there is an increasing tendency for such legislation to focus on procedures and process while giving more discretion to Member States to allow local factors to influence the precise choice of standards, and this direction could prove highly significant in the future structure of UK pollution law in the context of devolution.

20.5.6 Towards More Sustainable Pollution Control

The language of sustainable development permeates the current environmental policy agenda, and was given explicit legal recognition in the most recent amendments to the Treaty on the European Community.* Despite the inherent ambiguities in the concept of sustainable development, the underlying ethos which emphasizes a longer term perspective and the need to examine causes as well as effects is one further factor that will increasingly affect the design of pollution legislation. Many of more traditional forms of pollution regulation, examples of which are contained in this chapter, have been concerned with 'end-of-pipe' solutions, which are designed the minimize or control the impact of polluting activities rather than consider their justification and need in the first place.

These types of controls will remain the backbone of pollution regulation for many years. But they will now have to be seen against core trends in society and their impacts on the environment. New emission standards for motor vehicles, for example, may have dramatic short term effects on pollution loads, but these gains will eventually be outmatched should vehicle numbers and usage continue to rise at current rates. New pollution standards for waste disposal facilities have to be viewed against continuing large percentage increases in consumer waste arisings each year. The developments in integrated pollution control marked an important first step towards considering overall impacts of industrial plants on the environment, and require a greater focus on process methods as well as more straightforward emission control. More recent policy initiatives, especially at European Community level, are increasingly attempting to place greater responsibility on producers, manufacturers and retailers of goods for their eventual disposal, in an effort to reduce eventual waste and build in environmentally sounder design of products.[34] These developments are, at present, to be seen as complementary to conventional pollution regulation, but are likely to assume greater significance. At the end of the day it is, however, individuals whose personal and collective choices largely shape the direction of

---

*Article 6 of the Treaty, as amended by the Amsterdam Treaty in 1997, provides that, "Environmental protection requirements must be integrated into the definition and implementation of the Community policies and activities referred to in Article 3 in particular with a view to promoting sustainable development". Article 2 of the overarching Treaty on European Union provides that one of the objectives of the Union is to achieve "balanced and sustainable development."

society and its overall environmental impact; or, as the House of Commons Select Committee on the Environment put it, "the very act of consumption is at the root of many of the problems which have preoccupied this Committee since 1983."[35] Law will continue to be a key instrument for achieving society's changing goals but is ill-suited to determining them.

20.6 REFERENCES

1. s 4(3) and (4) Environment Act 1995.
2. D. Wyatt, *J. Environ. Law*, 1998, **10.1**, 10.
3. This power was first exercised in an environmental case involving a waste disposal site in Crete, and where the European Court of Justice ordered Greece to pay 20 000 Euros a day (about £14 000) as long as it continued to be in breach of Community waste legislation. *Commission v Hellenic Republic* Case C-387/97 July 4 2000, *Times Law Reports*, 7/7/2000.
4. L. Kramer 'EC Environmental Law', Sweet and Maxwell, London, 2000, pp. 291–292.
5. R. B. Macrory, in 'Protecting the European Environment', ed. H. Somsen, Blackstone Press, London, 1996, pp. 9–21.
6. [1990] 3 All England Law Reports 711. See R. B. Macrory, 'Nuclear Installations and the Statutory Duty to Compensate for Loss', *J. Environ. Law*, 1991, **3**, 122–134.
7. J. Fox, *Milner v Spencer* (1976) **239** Estates Gazette 573.
8. *Cambridge Water Company v Eastern Counties Leather plc* [1994] 2 Weekly Law Reports 53.
9. European Commission White Paper on Environmental Liability, adopted by the Commission, 9 February 2000.
10. Royal Commission on Environmental Pollution 5th Report 'Air Pollution – An Integrated Approach', HMSO, London, 1976.
11. Directive 96/61/EC.
12. M. Reed and R. Harris, 'Industrial Pollution Control', in 'Environmental Law', ed. D. Wooley, J. Pugh-Smith, R. Langham and W. Upton, Oxford University Press, Oxford, 2000.
13. [1972] Appeal Court 824.
14. [1998] Environmental Law Reports 396.
15. In contrast to the United States where air pollution controls have been largely based on ambient standards and water pollution controls on technology-based approaches.
16. European Commission, 'Communication on European Community Water Policy', Com (196) 59 final, Brussels, 1996.
17. Directive 91/271/EEC.
18. *Cunningham v Birmingham City Council* [1998] Environmental Law Reports 1.
19. [1996] *J. Planning Law*, 857.
20. ENDS Report **303** 54–56, Appeal Court, 30 March 2000.
21. Lord Justice Glidewell in *Gateshead Metropolitan District Council v Secretary of State for the Environment* [1995] *J. Planning Law*, 432.
22. See ref. 10.
23. For a comprehensive comparative review see E. Brandl and H. Bungert, 'Constitutional Entrenchment of Environmental Protection: A Comparative Analysis of Experiences Abroad', *Harvard Environ. Law Rev.*, 1992, **16.1**, 1–89.
24. K. Vibhute, 'Environment, Development and the Law', *J. Environ. Law*, 1995, **7.2**, 137–148.
25. [1995] 20 European Human Rights Reports 277.
26. *Guerra v Italy* [1998] 26 European Human Rights Reports 357.
27. Sentencing Advisory Panel, 'Sentencing for Environmental Offences', London, 2000.

28. *R v Howe and Sons (Engineers) Ltd* Court of Appeal 6 November 1998 ENDS Report, **288**, 51–52. The case concerned health and safety legislation but the general principles are equally applicable to the field of pollution regulation as confirmed by the Court of Appeal in the *Sea Empress* case, ENDS Report **302**, 46–47.
29. H. Woolf, 'Are the Judiciary Environmentally Myopic?', *J. Environ. Law*, 1992 **4.1**, 1–14.
30. M. Grant, 'Final Report on Environmental Courts', Department of Environment, Transport and Regions, London, 2000.
31. 'Environmental Information Regulations 1992', S.I. 1992/3240 implementing the EC Directive on Freedom of Access to Environmental Information EC/90/313.
32. Royal Commission on Environmental Pollution, 21st Report, 'Setting Environmental Standards', HMSO, London, 1998.
33. R. Macrory, in 'Constitutional Futures – A History of the Next Ten Years', ed. R. Hazell, Oxford University Press, Oxford, 1999, p. 193.
34. See, for example, the proposed Directive on End of Life Vehicles COM (97) 358, amended Com (1999) 330 final; the proposed Directive on Wastes in Electrical and Electronic Equipment Com (2000) 347; and, more fundamentally, the Commission's developing ideas on Integrated Product Policy, based initially on a study carried out by Ernst and Young and the Science Policy Research Unit, Sussex University: European Commission, 'Integrated Product Policy', European Commission, Brussels, 1998.
35. House of Commons Select Committee on the Environment, 'Eco-Labelling', Session 1990–1991. HC 474–1, HMSO, London, 1991 and repeated by the Committee's successor in 1999, House of Commons Select Committee on Environment, Transport, and Regional Affairs, 'Reducing the Environmental Impact of Consumer Products', Session 1998–99, HC 149-1, HMSO, London, 1999.

CHAPTER 21

Integrated Pollution Prevention and Control

D. SLATER and C. JOHN

21.1 INTRODUCTION

When this chapter was written for the third edition, Integrated Pollution Control (IPC) was at the cutting edge of environmental regulation. Much has happened since then and it would not be appropriate merely to provide an updated review of the legislation. It is not just that IPC has now been superseded by IPPC but that we are seeing, in parallel, the introduction of economic instruments, such as the climate change levy and, imminently, carbon trading, which will change, fundamentally, the way in which industry will be regulated.

To appreciate the significance of the changes, we need to step back and look at the history of the development of environmental regulation in the UK. At the end of the 19th century, regulation was a response to essentially local health and nuisance impacts of uncontrolled Victorian entrepreneurial activities. These often involved opencast and open-hearth processes, such as the alkali process. This involved boiling salt in concentrated acids in open vats – to produce, in addition to the product (soda ash), the original acid rain. The Alkali Act of 1874 was thus aimed at addressing local problems by setting and sharing best practice to reduce air emissions, initially by the application of end-of-pipe technologies.

IPC was the next major step forward in that it attempted to address the operation of the process in a way which minimized the impact on the local environment as a whole – the Best Practicable Environmental Option (BPEO), as proposed by the Royal Commission on Environmental Pollution (RCEP), again by promoting best practice. The matter of best practice is explored below in our discussion of the nature of the Best Available Technique Not Entailing Excessive Cost (BATNEEC).

IPC was also a useful vehicle for tackling wider issues such as national SO_2 targets and river Environmental Quality Objectives. But the pressures following Kyoto to address global issues sustainably meant that IPC, essentially a site-specific piece of legislation, needed augmenting to enable it to keep up with the

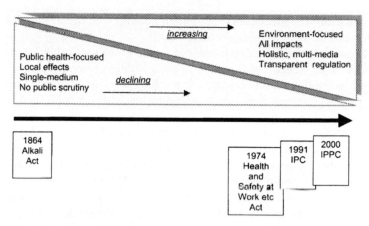

Figure 21.1 *The developing focus of legislation from 1984 to 2000*

growing sophistication of national and international requirements. IPPC, with its ability to consider energy efficiency, is considered to have a contribution to make, but increasingly, governments are turning to supplementary measures, such as carbon trading, to supplement the traditional command and control regulations.

At the beginning of the 20th century, industry was the *problem*. Some commentators think that the aim of environmental regulation was to 'modernize' industry to make it compatible with increasing the quality of life, which the wealth created by that industry made possible.

At the beginning of the 21st century, industry is seen as part of the *solution*, and is being encouraged to develop sustainably and environmentally responsibly. The economic instruments being developed to do this are an attempt to incentivize the market to behave in a more sustainable manner.

Figure 21.1 shows the way that, increasingly, the carrot of partnership, which hopefully will lead to a more cost-effective and lasting solution, is gradually replacing the stick of prosecution. It is, therefore, very appropriate to take another look at IPC and its European extension, IPPC, to see how effective it has been. This seminal legislation could be seen as representing the high water mark of command and control regulation, which has enabled other supplementary or 'smarter' approaches to be introduced to address the global problems which we now face.

21.2   INTEGRATED POLLUTION CONTROL – IPC – IN THEORY

This historical perspective allows us to review over a century and a half of environmental regulation, and we can trace the development of environmental protection policy and legislation as it grows from local in focus to national and international; as it moves from the single medium approach through multi-media to an holistic approach; from tax-payer funded to industry-funded – the 'polluter pays'; and from a straightforward attempt to deal with direct effects on local population to a mechanism capable of dealing with the whole relationship between mankind and his global environment.

The Alkali Acts, and their requirement to use Best Practicable Means (BPM) to prevent and where that was not practicable to minimize, releases of prescribed substances to the atmosphere, had been in operation for over a century by the time our improved understanding of the world, its ecosystems, and our dependence upon them prompted a review of the philosophy and fundamental principles of our regulatory systems. In fairness, during that time, the range of industrial processes subject to regulation had been broadened, as had the number of prescribed substances whose release from those processes was subject to the requirement to use BPM. In 1983 regulations were amended to provide public access to the applications for registration, and to the conditions attached to the registrations issued. Alongside these provisions, legislation to control discharges to the aquatic environment was being put in place, and, last of all, was added a suite of controls on the disposal of waste materials. We had not, therefore, stood still.

Nevertheless, fundamental change was needed. The Royal Commission on Environmental Pollution (RECP) had published a number of seminal reports that provided government with a vision for the future of environmental regulation and its delivery. IPC went a long way to securing the RCEP's vision, including the statutory requirement to achieve the Best Practicable Environmental Option.

What, then, are the main features of the IPC regime?

The Environmental Protection Act 1990 (EPA90) sets out in S 7(2) a number of objectives which must be met in the design and operation of a process. These objectives can be summarized as:

- Use of Best Available Techniques Not Entailing Excessive Cost (BAT-NEEC) to prevent releases to the environment or, where that is not practicable, for minimizing those releases and rendering them harmless.
- Compliance with any directions by the Secretary of State in respect of UK obligations under EU treaties or other international law, and with any plan made by the Secretary of State for the delivery of such obligations.
- Compliance with any relevant statutory environmental quality standards.
- Achieving the Best Practicable Environmental Option (BPEO).

Integral to the IPC system set out in EPA90 was the realization of two other important elements of government policy, namely:

- Recovery by the enforcing authority of the costs of regulation by means of systems of application fees and subsistence charges levied on the process operators.
- Full public access to the information provided by the applicant in support of his application, to the authorizations issued, and to other documents relevant to the ongoing performance of the operator in carrying on his process.

Of the technical objectives, the most contentious was the identification and achievement of BATNEEC.

The Environmental Protection Act 1990 (EPA90) laid a specific duty on the Chief Inspector "to follow developments in technology and techniques for preventing or reducing pollution of the environment due to releases of substances from prescribed processes", and a significant research budget was allocated by the Environment Agency* to commission or conduct the research necessary to underpin its regulatory role. An important area of this research was to find out about available techniques that were in use in the UK and abroad in the process in question, and to evaluate their relative merits. Michael Heseltine, as Secretary of State for the Environment at the time, saw IPC as a driving force for innovation, creating and satisfying demand for new technologies, not only in terms of abatement equipment but in terms of the whole process. What IPC sought to achieve was the right balance between cost to industry of better processes and abatement equipment, and cost to the environment in terms of the damage caused by releases.

Processes regulated under IPC had to employ BATNEEC, which had been a part of European Community legislation for some time, having been first introduced in 1984 in Directive 84/360/EEC on combating air pollution from industrial plant. This states that:

"... and authorisation may be issued only when the competent authority is satisfied (among other things) that all appropriate preventive measures against air pollution have been taken, including the application of best available technology, provided that the application of such measures does not entail excessive costs ....".

For an overall definition of BATNEEC it is helpful to consider the words separately and together. Thus:

" 'Best' must be taken to mean 'most effective in preventing, minimising or rendering polluting releases harmless'. There may be more than one set of techniques that achieves comparable effectiveness – that is, there may be more than one set of 'best' techniques. It implies that the effectiveness has been demonstrated".

'Available' should be taken to mean 'procurable by any operator of the class in question'; that is, the technology should be generally accessible (but not necessarily in use) from sources outside as well as within the UK. It does not imply a multiplicity of sources, but if there is a monopoly supplier, the technique will count as being 'available' provided that all operators can procure it. Furthermore, it includes a technique that has been developed (or proven) at a

---

*IPC and IPPC are enforced by the Environment Agency in England and Wales, by the Scottish Environmental Protection Agency (SEPA) in Scotland, and by the Environment and Heritage Service in Northern Ireland.

scale that allows its implementation in the relevant industrial context and with the necessary business confidence. Industry often believes that for a technology to be recommended as 'available' it would need to have been in commercial use for at least six to twelve months.

'Techniques' is a term that embraces both the plant in which the process is carried on and how that plant is operated. It should be taken to mean the components of which it is made up and the manner in which they are connected together to make the whole. It also includes matters such as numbers and qualifications of staff, working methods, training and supervision and also the design, layout and maintenance of buildings, and will affect the concept and design of the process.

The other North Sea States have also adopted, at least in general, the concept of BATNEEC and are in the process of applying it to industrial sectors discharging hazardous substances, by reviewing discharge consents. Consents specify BATNEEC in terms of treatment efficiency rather than specifying the actual technology to be used, so as not to hamper innovation, and these are continuously revised to include the latest developments.

As examples of the application of BATNEEC, in Sweden, when consents are issued for new or enlarging plants, they also require research and development to further improve treatment technology. The Netherlands and Germany have set two levels of technological requirements. In The Netherlands, Best Technical Means (BTM) applies to all potential EC List I ('Black List') Substances, and Best Practical Means (BPM) for all List II ('Grey List') Substances. BTM, in this case, is defined as the most advanced treatment technology that is practically usable, whereas BPM refers to technologies that are affordable for 'averagely profitable companies'. In Germany, Best Available Technology (BAT), similar to BTM in The Netherlands, must be applied to effluents containing dangerous substances, whereas Generally Acceptable Technology (similar to BPM in the Netherlands) must be applied to effluents containing biodegradable substances.

'Not entailing excessive cost' needs to be taken in two contexts, depending on whether it is applied to new processes or to existing processes. For new processes, the presumption is that the best available techniques will be used. Nevertheless, in all cases, BAT can properly be modified by economic considerations where the costs of applying the best available techniques would be excessive in relation to the nature of the industry and to the environmental protection to be achieved.

In relation to existing processes, the enforcing authority seeks from applicants and operators fully justified proposals for the timescale over which their plants will be upgraded to achieve the same standards as would be expected of new plant. This is why, for example, the very expensive selective catalytic reduction equipment has not been required for a new gas turbine power station: very good control of emissions of nitrogen oxides is achievable by combustor and burner design. This approach also fits the BPEO criteria, as we will see later.

Reasonableness comes into all of the discussions on BATNEEC, in that a situation is achieved where money is spent on abatement only up to the point where the resulting increase in control of pollution justifies the money spent.

BATNEEC judgements refer to what is achievable by way of pollutant release levels by the application of an optimized combination of process and abatement options, which are accepted as representing best practice in the relevant context. The individual officer, when determining on behalf of an enforcing authority what represents BATNEEC for a specific process, must have an awareness of the financial and economic state of, and implications for, the industry sector so that he can gauge how far and how fast the regulated community can commit the expenditure and assimilate process improvements.

For the choice of production techniques, the options are the alternative ways of producing a specified output either for intermediate use or final consumption. The focus of this analysis is on assessing the costs of each technique against the associated environmental effects, so the options may usefully be identified in terms of the ranked environmental performance of each technique. The analysis can then seek to determine either the total or the incremental net cost of each option.

The method of appraisal outlined here aims to identify for a typical firm the incremental costs (unit or total) associated with using different technologies, and thus different levels of environmental impact in order to produce the same final output. The comparison could be made in terms of, say, the difference in net present value of annual output of 1000 units for ten years; or the difference in unit costs in a base year, assuming full accounting costs are covered over project life.

A complete cost benefit analysis would aim to identify and compare the full social costs of two or more different states of the world, so that a decision could be based on the change in Net Social Benefit, *i.e.* the estimated money value of the material gains and losses to all actors, between the different options. Cost effectiveness is the principal economic criterion for identifying the BATNEEC solution. This requires comparison of the relative costs to the operator of alternative environmental quality levels, as determined by the feasible set of production techniques.

The enforcing authority – now the Environment Agency – must determine whether the proposals made by an applicant for an IPC authorization will enable that applicant to achieve these objectives in the operation of his process. It is worth underlining here a particular strength of the IPC framework. With the exception of Environmental Quality Standards and international obligations, IPC does not start from a position of setting blanket numerical limits. This approach has the dual advantage of leaving the applicant free to decide how he is going to meet the objectives, and of enabling the regulator to ratchet the process-specific limits down as the plant and its operation are improved. This site-specific ratcheting should then cascade across the industry sector as BATNEEC is redefined.

The main goal of an authorization is to specify the limits and conditions that are important to achieving the objectives of IPC in a particular circumstance. These are in the main likely to comprise limits and conditions on feed materials, where these would impact upon the releases from the process, such as the sulfur content of coal or oil; the operating parameters, and the nature, quantity and concentration of releases to the various environmental media. The detail of these, such as the period over which they apply, will need to take into account what constitutes BATNEEC, their environmental impact, *e.g.* concentration or load-dependent process characteristics, recognising cyclic variations, fluctuations and practical considerations.

The applicant must provide a strong, detailed justification of his process: for a new or modified process, all the options must be explored and justified. He must list the substances that might cause harm that are used in or that result from the process. He must identify the techniques used to prevent, minimize or render harmless such substances. He must assess the environmental consequences of any proposed releases. He must detail the proposed monitoring of the releases. This comprehensive environmental assessment then becomes the cornerstone of the enforcing authority's determination of the application and paves the way for the setting of conditions in the authorization itself. The ongoing relationship between the application and the authorization represents the 'contract' between the operator and society as a whole – in this case represented by the Environment Agency. The commitment that the operator has made in his application is a real one, and the Environment Agency carries out a continuing programme of inspection and monitoring to ensure that it is honoured. This chapter does not set out to explore the nature of inspection and enforcement: however, in simple terms, the Environment Agency officer will check:

- That the operator is carrying out his process in the manner that he set out in his application, or later alterations.
- That the operator is complying with all the limits and other conditions set out in the authorization.
- That the operator is using BATNEEC for any other aspects of the process which are not specifically mentioned in the authorization.*
- That there are no factors which make it probable that the operator will, in the foreseeable future, fail in any of these three areas.

Integrated Pollution Control will have had a lifetime of just about a decade as it starts to be superseded by Integrated Pollution Prevention and Control. The new regime must be fully in place by October 2007. Before making a brief comparison between IPC and IPPC, it may be interesting, and one hopes instructive for industry, government, regulators and the public, to reflect on the prospectus for IPC and the extent to which that prospectus has been achieved by each of the parties involved. This chapter is written by a former regulator and, while seeking to be objective, no apology is made for illustrating views from that perspective.

*This point is known as the 'residual duty', provided for in S7(6) of the Environmental Protection Act 1990.

21.3 IPC IN PRACTICE

At the outset, then, what were we promised?

- A single regulator
- A 'one-stop shop'
- Cost recovery – the polluter pays for the cost of his regulation
- Public access and transparency
- An holistic system dealing with the potentially most polluting industries
- A better environment

IPC was introduced in the UK as a flagship exercise in environmental protection policy, building on two decades of thinking and debate at the highest levels. We were, demonstrably, a decade ahead of the rest of Europe in adopting the philosophy of integration. What, then has been the outcome? How would we be faring now without IPC? What are the key lessons that we must learn from it and take forward into IPPC? So,

- How good was the law?
- How did industry's approach to environmental management change as a result of IPC?
- How effective were the regulators?
- How has the public, and the pressure groups that represent them, responded to IPC?

21.3.1 How Good Was the Law?

The process of law-making in the UK follows tried and tested methods that, by and large, deliver an adequate result. After a more or less lengthy period of debate, Government publishes its policy and proposals for implementation. Draft legislation emerges and is scrutinized not only by parliament but also by all who have an interest in the outcome. All these parties are able to respond to the Government's consultation, and it is upon these responses that the requisite compromises are built. Simplistically, the legislation that reaches the statute book reflects Government's original intentions, tempered by the restraining influence of those who will be regulated, and bolstered by the pressure groups that want the tightest controls possible. Legislating, therefore, is not an exact science – it is a political fix, and in a democracy who would want it any other way?

It is estimated that 80% of UK environmental legislation is enacted in order to transpose European law. When EPA90 was passed, however, Government was not implementing a Directive, and we had considerably more freedom to manoeuvre, although provision was made to deliver some requirements of EU law through the IPC authorizations.

The structure of the legislation was conventional: an enabling Act – EPA90 – which set out the objectives to be met, the essential mechanisms of the system, the powers and duties of regulators, the obligations on the regulated, and the offences, and gave the Secretary of State powers to make Regulations which put flesh around these bones. The Regulations specify the processes to be regulated, and the precise detail of the application system and the public registers. It was a good package, clear, easy to follow. The legislative process itself had been sound, with the Act, containing the philosophy and essential principles, being fully debated by Parliament.

Nevertheless, EPA90 was novel and groundbreaking. The principal difficulties were encountered in the detail of the descriptions of prescribed processes, and the practical application of the law to real industrial situations, especially in the chemical industry. The Environmental Protection (Prescribed Processes and Substances) Regulations 1991 were amended no fewer than seven times in five years, to ensure that it was possible to regulate the right things in the best possible way. Regulations, of course, are amenable to amendment in this way, and this history illustrates the effectiveness of the active and dynamic relationships between industry, the regulator, and the then Department of Environment.

21.3.2   How Big Is IPC?

Many of the processes regulated under IPC were previously registered works under the Health and Safety at Work *etc.* Act. Some categories were added: a few processes were lost. Government consultation papers on IPC estimated that around 5000 processes would be authorized once IPC was fully implemented. These estimates proved to be wildly inaccurate, and, in absence of other explanations, we can only conclude that it was the move from 'works' to 'process' which was the cause. The number of processes authorized under IPC was around 2000 when implementation was complete, and has hovered around that mark in the four years of full IPC operation. It is difficult, with the benefit of hindsight, to understand how the 5000 figure could ever have been believed.

The industrial processes regulated under IPC fall into six categories:

- Fuel and power
- Minerals
- Ferrous and non-ferrous metals
- Chemicals
- Waste
- Other industries – miscellaneous processes: principally the industries that had not previously been regulated under the Health and Safety at Work *etc.* Act.

Of these six categories, the chemical industry dominates in terms of numbers.

21.3.3   How Effective Were the Regulators?

The separation of state functions into policy and executive arms was an important principle to the Conservative government. The result was a host of 'next steps' agencies, whose activities were more transparent and whose management were more easily held to account in an era that saw great strides in public accountability. HMIP too was destined for 'next steps' status until Government environmental policy determined that a more broadly based Environment Agency was needed to champion the environment and deliver EU and UK environmental policy in the 21st century. The scrutiny to which many of these Agencies are subject is rigorous, often painful, and always beneficial.

The question of resources is well rehearsed. No matter how generously an environmental regulator is resourced, there will always be more work than there is capacity to carry it out. To argue that failure is about under-resourcing, therefore, is dull and old hat. As with any organization, it is for the regulator to take a view about priorities and be hard-nosed about what is possible and what is not, and to seek and achieve ongoing efficiency gains in the way in which the business is run.

There seems, however, to be a major issue here that provides food for thought. All parties to a game must have a shared interest in success. If a team wins, then all its players are winners. You do not win a test match with nine winners and two losers; that concept is clearly nonsensical. And yet there is no obligation upon Departmental policy makers – *i.e.* those who draft the consultation documents and ultimately the Acts and Regulations – to feel any responsibility for the practicability of their products when they hit the streets. The success measure for civil servants, and to a great extent for their Ministers as well, is whether the legislation satisfies the political imperatives of the Minister and his manifesto commitments, and has met the EU deadline and embodies the relevant EU requirements. These measures are of great importance to a democracy, but they are not enough. Industry's understanding of the legislative system has improved by leaps and bounds over the past decade, but in the early days of IPC, even the leading trade associations in the country were ascribing problems to the regulator, which in fact stemmed from the legislation. Arguably the regulator needs to get better at responding to consultation.

21.3.4 How Did Industry's Approach to Environmental Management Change as a Result of IPC?

Environmental regulators are frequently challenged to measure their perform-ance, and it is never easy for a whole variety of reasons, not least that there is no baseline from which to start measuring! But what we have observed over the last decade, and what HMIP and the Environment Agency have been proud to stimulate and promote, has been a huge shift in culture and attitude. Industry does publish information on its environmental performance. Industry did adopt 'joined up thinking' by combining its resources on health, safety and the environment into integrated departments with real clout. IPC was the new product that HMIP took in through the door of industry – very much like the Bettaware Man – and industry, by and large, rose to the challenge. Other initiatives, such as the 3E's project[1] took IPC and extended it to its logical conclusion, and found that there was a huge dividend to be won by integrated and holistic thinking and process management. If industry still has a barrier to overcome, it is to overcome the feeling of disempowerment in the face of regulations and the regulator.

21.3.5 Transparency

The creation and maintenance of the IPC public registers puts an enormous quantity of data, and quite a lot of information, in the public domain. The information that an applicant must supply is set out in Regulations.[2] All this information is placed on the public register, with two exceptions:

- Information that must be withheld in order to safeguard national security: this decision is a matter for the Secretary of State who directs the enforcing authority accordingly.
- Information which is commercially confidential -- *i.e.* its presence in the public register would prejudice to an unreasonable degree the commercial interests of the applicant. This decision is made by the enforcing authority, but with a right of appeal by the applicant to the Secretary of State.

This sea change in public access was, in the early days of IPC, a great worry to industry. The needs to maintain competitive advantage, and to honour contracts containing confidentiality clauses, are real ones, but the fear highlights an important issue. IPC requires information about the process in terms of its actual and potential environmental impact: what is released to the environment. That should not be concealed. Industry had to learn to look at their activities from a different angle – not product, not business, not finance, but environment. Mostly the regulator does not need to know about the things that are genuinely commercially confidential, and so, as both parties learned how to interpret the requirements, and with the exception of one high profile appeal to the Secretary of State, public registers ceased to be a threat and began to be an opportunity.

Transparency comes at a price, however. In this context we would offer the thesis that the public registers might be viewed in two different ways.

First, they might simply provide to the public unrestricted access to the same information as is available to the regulator. With the same training and background, therefore, a member of the public might work through the application and decide whether he would have reached the same conclusions as the regulator. This involves, amongst other things, the disciplined maintenance of files, ensuring that the officer does not take decisions that cannot be supported by the evidence supplied in the application, and provision of facilities for the public. In short, it becomes what is, sadly, disparagingly called a bureaucracy. The system is bureaucratic primarily to ensure that it can even begin to be accessible to the public.

Second, the registers might take the plethora of data and make it genuinely accessible – *i.e.* interpret it in layman's terms – to the man on the Clapham omnibus. Our regulators achieve our first definition of public registers reasonably well. The latter is a major opportunity area for them, and one that could absorb, if they let it, almost infinite resource.

21.3.6 Cost Recovery

The matter of payment of fees and charges was, throughout the five-year IPC implementation period, probably a close second to BATNEEC as the most contentious issue. It will be so again with IPPC, particularly for those industry sectors that come new to rigorous environmental regulation. How do you make a system fair? For what is a company paying? Is it paying for the services that it receives personally from the regulator? Or is it picking up as equitable as possible a chunk of the overall cost of regulating everyone – from the recalcitrant recidivist to the blameless. Should big industry subsidise small and medium sized companies? How transparent should the regulator's accounts be so that his clients can decide whether, as their monopoly supplier, he is providing them with best value? These questions will rumble on.

21.3.7 How Have the General Public, and the Pressure Groups That Represent Them, Responded to IPC?

These days there is always fear and suspicion about 'deals done in smoke-filled rooms'. The closed system that typified the relationship between all large organizations, including government, and the population as a whole, has ceased to work for two main reasons. First, because any decision can only be as good as the information upon which it was based, and on the interpretation of that information in the context of the decision. Second, there is now ample proof that organizations, in pursuit of their own objectives, can and do make decisions which cannot be supported by the evidence and information available at the time, and which therefore necessarily involve the concealment of that evidence and information from the scrutiny of others. From this, it follows that any decision must be reviewed as new information becomes available.

But the pressure groups have certainly matured during the last decade. Their role as angry young people highlighting problems is becoming outdated, and many of them have had the courage to move on to becoming part of the solution. It was not by accident that when Charles Secrett spoke to a conference of the Environment Agency's senior management, there was much common ground to be explored. Progress, however, relies on the creative tension between the interest groups, and it is to be hoped that they will retain their ability to campaign for a better environment.

21.4 INTEGRATED POLLUTION PREVENTION AND CONTROL – WHAT PRICE THE EXTRA P?

The IPPC Directive was piloted through the processes of the European Parliament by the UK, very much with the objective of building on the successes that we had had with IPC. Many of the principles are similar, even where the words differ.

The Directive, and the Act and Regulations which implement it in England*
refer to BAT as opposed to BATNEEC, but the Directive specifically refers to
"taking into consideration the cost and advantages" of a proposed technique.
Subject to any case law, therefore, we may find that this definition is not
significantly different.

The purpose of the IPPC Directive is: "To achieve integrated prevention and
control of pollution arising from the activities listed … It lays down measures
designed to prevent or, where that is not practicable, to reduce emissions in the
air, water and land … in order to achieve a high level of protection of the
environment taken as a whole." So BPEO as an identifiable concept is lost: does
this definition take its place?

IPPC is broader than IPC both in the range of processes that will be brought
under regulation, and in the factors to be considered in relation to each process.

The activities not under IPC that will be covered by IPPC include some
landfill activities, intensive pig and poultry rearing, and food and drink
production. The landfill industry has, of course, been regulated under a
permitting system since 1977 and is accustomed to this approach. Neither
agriculture nor food and drink, however, have been the subject of environmental regulation on the scale of IPPC and will have a very significant learning
curve to deal with. It is to be hoped that they find the new disciplines beneficial
to both the environment and their business success.

The factors to be addressed in an application for an IPPC permit, and
therefore in the enforcing authority's determination, extend IPC by the
addition of:

- The requirement for a site report, to assess the state of the land prior to the
  implementation of the IPPC permit
- Consideration of raw material nature and usage
- Consideration of energy usage and heat losses
- Consideration of odour
- Consideration of noise
- A formal surrender process for the permit – when, *inter alia*, the state of the
  land will be compared with the initial site report.

This extension will enable industry and regulator alike to make a much more
comprehensive and objective assessment of the impact of the process, and to
manage that impact more effectively as time goes on.

IPC and IPPC share another common feature. Neither regime deals with
whether, conceptually, an activity is in the best interests of the environment. For
example, if a company decides to establish its own power plant at a production

---

*Detailed arrangements in Wales, Scotland and Northern Ireland will depend upon their assemblies.

site instead of buying energy in from external providers using remote generation plant – or *vice versa* – there is no regulatory locus for IPC or IPPC.

21.5 CONCLUSIONS

There are unquestionably key lessons to be learned for those who are seeking to improve our environment by means of the law. It is therefore important to learn from the practical experience of IPC, not only from the things that worked but also from the things which in the end did not.

First, the legislation – from the time it was first mooted, in the early 1970s, to its passage through Parliament in 1990, there was ample time to define fully the principles behind the proposals. The fact that the basic ideas were a natural evolution of the regime then current was also helpful. Nevertheless this was a piece of legislation that was relatively straightforward to turn into a practical regulatory tool in a manner satisfactory to all parties.

Second, the institutions – the generous timescale also allowed a special 'integrated' inspectorate to be set up, specifically to implement IPC. HMIP was designed, built and trained to fulfil the precise function required under the Act. The contrast between this approach and the current propensity to allocate duties almost arbitrarily to existing institutions, regardless of their specific competences, is worth noting as an important lesson for the architects of future legislation.

Third, the effectiveness – The Environment Agency, from its pollution inventory, notes that between 1991 and 1996, the total of emissions from IPC processes was reduced by 50%. By any standards this must be seen as a remarkable achievement. It has been ungraciously observed that this reduction was due solely to cleaner process developments, more energy efficient technologies and the 'dash for gas'. A more neutral observer might remind them that the effect was precisely what the legislation was put in place to achieve.

Lastly, the effect on industry – again the neutral observer might note that the main difference between the majority of industry in the 1980s and today was that in the 1980s their priority was to preserve their commercial confidentiality, whereas, today, companies are vying with one another to score highest on one of the many environmental reporting beauty parades, or persuading people about the seriousness of their intent on initiatives such as Responsible Care. Again, an impartial observer might attribute at least some of this to the insistence of the IPC regime on publicly reporting licence applications and enforcement records.

Nevertheless, this public record must be allowed to speak for itself and the fact that the European Commission has given pride of place to IPC as the cornerstone of its IPPC Directive means that other people were impressed also. In the development of environmental awareness and responsibility in industry (modernization), this period, after the implementation of IPC, has probably been the most effective in bringing about a step change in the behaviours observed.

The Royal Commission can take great satisfaction that theirs was the vision that has demonstrably made a significant difference.

21.5.1   What About the Downside?

It is questionable whether any IPC application totally met the expectations of its architects. Did anyone produce a full, by-the-book, BATNEEC case? Did we see a meaningful Best Practicable Environmental Option justification? These are legitimate challenges. But what is undisputed is that all these attempts, however imperfect, are on the public record and have resulted in ever-increasing improvement trends that were the real objective of the Act.

Many academics are still unhappy with the compromises that were necessary to get the authorizations in place and many still suspect that deals in smoke filled rooms continued to occur, of which there was no record. They failed to understand the power of the licence – breaking any condition of which was an automatic criminal offence. With such draconian powers available to guarantee performance, bureaucratic and academic studies in environmental analysis are not crucial to the real needs of the community as a whole. It is this pedantic insistence on ever more sophisticated requirements on operators – almost micro-management by the regulator – which, in the last resort, will limit the further utilization of this approach. As the larger industrial sources of pollution are progressively brought under control and their impact reduced, so the more numerous small and medium-sized industries contribute to the more diffuse sources of pollution, such as transport and farming, which thus become more significant. The practicalities of the regulation of these sources will preclude direct intervention by expert inspectors under regimes such as IPC and IPPC; instead they will rely more on enlightened self-interest in the form of voluntary agreements, tax incentives and scarce resource trading, to allow the markets to deal with the problems they create.

21.6   REFERENCES

1. 'Allied Colloids and HMIP: 3E's Project Concluding Report', Allied Colloids, 1996.
2. The Applications, Appeals and Registers Regulations 1991 SI507 as amended.

Subject Index

Lightning Source UK Ltd.
Milton Keynes UK
07 November 2009

145953UK00001B/2/A

Pollution:
Causes, Effects and Control
Fourth Edition